Studies in Fuzziness and Soft Computing

Volume 439

The series "Studies in Fuzziness and Soft Computing" contains publications on various topics in the area of soft computing, which include fuzzy sets, rough sets, neural networks, evolutionary computation, probabilistic and evidential reasoning, multi-valued logic, and related fields. The publications within "Studies in Fuzziness and Soft Computing" are primarily monographs and edited volumes. They cover significant recent developments in the field, both of a foundational and applicable character. An important feature of the series is its short publication time and world-wide distribution. This permits a rapid and broad dissemination of research results.

Indexed by SCOPUS, DBLP, WTI Frankfurt eG, zbMATH, SCImago.

All books published in the series are submitted for consideration in Web of Science.

Ubaid ur Rehman · Tahir Mahmood

Bipolar Complex Fuzzy Sets

Theory and Applications

Springer

Ubaid ur Rehman
Department of Mathematics
University of Management and Technology
Lahore, Pakistan

Tahir Mahmood
Department of Mathematics and Statistics
International Islamic University Islamabad
Islamabad, Pakistan

ISSN 1434-9922 ISSN 1860-0808 (electronic)
Studies in Fuzziness and Soft Computing
ISBN 978-3-032-07387-7 ISBN 978-3-032-07388-4 (eBook)
https://doi.org/10.1007/978-3-032-07388-4

This Springer imprint is published by the registered company Springer Nature Switzerland AG
The registered company address is: Gewerbestrasse 11, 6330 Cham, Switzerland

To
Prof. Dr. Muhammad Sarwar Kamran

Preface

Lotfi A. Zadeh, a professor of electrical engineering at the University of California, Berkeley, initiated the idea of fuzzy sets in 1965, to handle the uncertainty and ambiguous data. It is intended to deal with circumstances when the distinctions between various elements are not clearly defined and are instead determined by belonging grades. In the traditional theory of sets, an element's binary membership to a set indicates whether it belongs to it or not (0 for not belonging, 1 for belonging). But in many real-world situations, there are shades of grey between fully belonging and not belonging, making membership neither black nor white. Pattern recognition, control systems, linguistics, artificial intelligence, and decision-making, are the areas where fuzzy set theory has been found very useful. It offers a more adaptable and realistic method to model and analyze uncertainty and imprecision in a variety of circumstances, making it an important tool for working with complicated and ambiguous data.

Bipolar or double-sided judgmental thinking on both a positive and a negative side is the foundation of a vast range of human decision-making. As an illustration, consider the effects and side effects, collaboration and competition, likelihood and unlikelihood, friendliness and animosity, similar interests and conflicts of interest, etc. Thus, Wen-Ran Zhang devised the idea of bipolar fuzzy sets in 1994 as an expansion of the theory of fuzzy sets. Bipolar fuzzy sets have been found useful in situations where it is necessary to consider both positive and negative aspects of information when making decisions. They offer a more adaptable and expressive method to depict ambiguity and vagueness in real situations. Bipolar fuzzy sets have numerous uses in areas including pattern recognition, data mining, decision support systems, expert systems, and researchers are still learning more about both their theoretical and practical aspects.

Since the discovery of fuzzy sets, significant focus has been placed on expanding upon the idea of fuzzy sets and creating new theories and on the other hand in physical sciences complex numbers are equally important as the real numbers are. So consequently, Ramot et al. devised the theory of complex fuzzy sets in 2002, by extending the range of traditional fuzzy sets from the unit interval to a unit disc of a complex plane with center at the origin. The interpretation is in the polar form of a complex fuzzy set initiated by Ramot et al. Although the notion of complex fuzzy sets by Ramot et al. looks very natural but Tamir et al. initiated another way, which looks more natural, to define complex fuzzy sets by Cartesian form of a complex number in 2011 by replacing the unit disc with the unit square in first quadrant in the complex plane. The notion of complex fuzzy sets is a significant tool for coping with the information that contains 2nd dimension (i.e., phase terms or unreal parts) and has applications in various fields. The theory of fuzzy sets, bipolar fuzzy sets and complex fuzzy sets may not function properly in a variety of situations. For instance, fuzzy sets, bipolar fuzzy sets, and complex fuzzy sets are impractical when dealing with information that includes both the object's positive and negative views as well as additional fuzzy information (2nd dimension). To handle such type of situation there were a need of the notion which generalize both the notions of complex fuzzy sets and bipolar fuzzy sets. Various researchers tried to cope with this by developing the notion of bipolar complex fuzzy sets but all those theories had some very serious drawbacks and problems. Consequently, Mahmood and Rehman initiated a more accurate concept of bipolar complex fuzzy set in 2022 by removing those drawbacks and problems. The theory of bipolar complex fuzzy set is a perfect mathematical tool for handling the ambiguous and uncertain information that contains the objects or elements of both positive and negative views and extra fuzzy information (2nd dimension). The theory of fuzzy sets, complex fuzzy sets and bipolar fuzzy sets are the special cases of bipolar complex fuzzy sets initiated by Mahmood and Rehman. The notion of the bipolar complex fuzzy set has applications in numerous areas such as medical diagnosis, pattern recognition, decision-making, artificial intelligence, computer science and mathematical sciences etc.

Due to its capacity to successfully manage and represent uncertain, imprecise, and unclear information, bipolar complex fuzzy sets are crucial to Intelligent Data-Driven Systems and Artificial Intelligence. When crisp, binary logic, fuzzy set and its generalizations fail, they offer a more complex representation of the facts and enable improved decision-making. Bipolar complex fuzzy sets make it possible to simulate complex connections and uncertainties in a variety of artificial intelligence applications, including natural language processing, picture recognition, and expert systems, which enhances the resilience and flexibility of intelligent systems. These systems may make more contextually aware, human-like judgments by adding bipolar complex fuzzy logic and degrees of belonging, which improves performance and accuracy in a variety of real-world circumstances. Further, It gives a more adaptable and realistic approach to describe and thinking about uncertainty and imprecision in a variety of scenarios and this makes it an important tool for working with complicated and ambiguous data in the recent complicated and advanced world.

Motivated by the novelty, advantages, and today's world requirement of bipolar complex fuzzy sets, the authors, considered the idea of writing a comprehensive book on *Bipolar Complex Fuzzy Sets* to cover all facts of the *Theory and Applications*. Our initial goal in this direction is to investigate the algebraic structures such as lattices, semigroup, groups, semirings and rings in the structure of bipolar complex fuzzy sets. After that, we devise the distance measures, similarity measures, correlation coefficient and entropy measures and associated properties for bipolar complex fuzzy sets. At last, we inaugurate the aggregation operators such as Hamy mean operators and power generalized Heronian mean operators in the environment of bipolar complex fuzzy sets. The following paragraphs provide a clearer description of the book's structure.

In Chap. 1, we discuss the notion of fuzzy sets, bipolar fuzzy sets, and complex fuzzy sets along with their associated concepts. The aim of Chap. 1 is to review the basic notions before studying the theory of bipolar complex fuzzy sets (BCFSs). Chapter 2 lays the foundation for a detailed investigation of BCFSs. In this chapter, firstly, we investigate prevailing theories of BCFSs and complex bipolar fuzzy sets along with their fundamental operations. We discuss the drawbacks and shortcomings that exists in the prevailing theories and then review the accurate and novel concept of BCFS. Further, we discuss Type-1 partially, and Type-2 partially BCFSs. At last of Chap. 2, we discuss the similarity measures and aggregation operators for BCFSs. In Chap. 3 of this book, we inaugurate the theory of bipolar complex fuzzy (BCF) sublattices, BCF subsemigroup and its ideals. More, we investigate the idea of BCF subgroups, BCF normal subgroups, BCF normalizer, BCF cosets, etc. We also devise the theory of BCF subsemiring and BCF subring and associated results.

Chapter 4 contains a few distance measures and weighted distance measures among BCFSs relying on the well-known Hamming and Hausdorff distances. We also investigate continuous distance measures between BCFSs. This chapter also contains the VIKOR technique relying on the initiated distance measures in the structure of BCF information and an application (automated trading system). At the end of this chapter, we discuss the comparison of initiated work with some of prevailing theories to prove the advantages and supremacy of the initiated work. In Chap. 5, of this book, we inaugurate the similarity measures (SMs) for BCFS based on the initiated distance measures of Chap. 3. We also initiate weighted and continuous SMs in this chapter. More, we devise exponential and weighted exponential SMs for BCFS. At last, this chapter contains the applications of the inaugurated SMs in pattern recognition and medical diagnosis, and the comparative study. Chapter 6 contains the correlation coefficient (CC), weighted correlation coefficient and entropy measures (EMs) under the environment of BCF information. Further, in Chap. 6, we develop the approach of TOPSIS in the setting of BCFS relying on the initiated CC and EMs and the demonstrative examples of the BCF-TOPSIS approach to reveal the applicability of the proposed BCF-TOPSIS approach. we have the comparative analysis in Chap. 6 as well.

In Chap. 7, we investigate BCF Hamy mean, BCF weighted Hamy mean, BCF dual Hamy mean and BCF weighted dual Hamy mean operators along with the associated results. We also develop multi-attribute decision-making (MADM) by employing the initiated aggregation operators (AOs) and then investigate a case study to demonstrate the applicability of the proposed theory. Moreover, we interpret the comparative analysis of the developed work with certain prevailing work in Chap. 7. In Chap. 8, of this book, we investigate the BCF power generalized Heronian mean, BCF weighted power generalized Heronian mean, BCF power generalized geometric Heronian mean BCF weighted power generalized geometric Heronian mean operators, and the associated results. This chapter also contains a weighted sum model and weighted product model (WASPAS) technique in the setting of BCF information and the application (supply chain transportation). The comparative analysis is also part of Chap. 8.

Lahore, Pakistan
Islamabad, Pakistan

Ubaid ur Rehman
Tahir Mahmood

Competing Interests The authors have no competing interests to declare that are relevant to the content of this manuscript.

Contents

About the Authors

Ubaid ur Rehman working as an assistant professor at the Department of Mathematics, University of Management and Technology, Lahore, Pakistan. He received his Ph.D. degree in Mathematics from the International Islamic University Islamabad, Pakistan in 2023. His research interests include fuzzy sets and their generalizations, algebraic structures, aggregation operators, similarity measures, fuzzy decision-making, rough sets, soft sets, decision-making theory, and Machine learning. He has published more than 100 articles in well-reputed peer-reviewed journals with 320+ impact factor. He has been a member of the academic board of Global Open Share Publishing since 2024 and Editorial Board member of *Mathematics and System Science* journal.

Tahir Mahmood is Associate Professor of Mathematics at Department of Mathematics and Statistics, International Islamic University Islamabad, Pakistan. He received his Ph.D. degree in Mathematics from Quaid-i-Azam University Islamabad, Pakistan in 2012 in the field of Fuzzy Algebra. His areas of interest are Algebraic structures, Fuzzy Algebraic structures, Decision making and generalizations of fuzzy sets. Around 390 research publications are on his credit with more than 12500 citations, 900 impact factor, 56 h-index and around 221 i10-index. He has produced 65 MS students and nine Ph.D. students. Currently he is working as editorial board member of three impact factor journals and as editor-in-chief of the journal *Journal of Innovative Research in Mathematical and Computational Sciences.*

Chapter 1
Basic Concepts

This chapter on basic concepts has 3 sections. In Sect. 1.1, we discuss the theory of fuzzy sets and related concepts. In Sect. 1.2, we investigate the notion of a bipolar fuzzy set and its associated basic concepts. Section 1.3, contains both polar and cartesian models of complex fuzzy sets along with other properties of complex fuzzy sets.

1.1 Fuzzy Sets

The notion of fuzzy set (FS) was initiated by Loti A. Zadeh [1] in 1965 to cope with uncertainties and ambiguities. A FS is a mathematical idea that expands on the idea of a traditional set theory. An element can either be a member of a set or not according to classical set theory. However, the elements of a set frequently have varying belonging grades or ambiguity attached to them in real-world circumstances. The notion of FS allows for this flexibility by allocating a belonging grade to every element in the set. The belonging grade in a FS specifies how much an element belongs to the FS. Each element is assigned a value between 0 and 1, where 0 denotes non-belonging grade and 1 denotes full belonging grade, values between 0 and 1 denote signify the belonging grade. For example, consider the collection of tall students in a certain post graduate class. Now it is clear that the term "tall" is not well defined, as its meanings vary from person to person. However, it is a usual sense that a student with height 6 feet and 3 inches or more can be considered as tall and hence surely belongs to the collection and so its membership grade can be assigned 1. Similarly, it is also a usual sense that a student with height 5 feet or less cannot be considered as tall and hence surely does not belong to the collection and so its membership grade can be assigned 0. Now it is clear that the membership grades of the students with heights between 5 and 6 feet 3 inches will be between 0 and 1, for example, a student of height 5 feet 3 inches can be assigned membership grade 0.25, 5 feet 6 inches can

U. ur Rehman and T. Mahmood, *Bipolar Complex Fuzzy Sets*, Studies in Fuzziness and Soft Computing 439, https://doi.org/10.1007/978-3-032-07388-4_1

be assigned membership grade 0.47 and 6 feet 1 inch can be assigned membership grade of 0.90 and so on. Some of the other well-known and ill-defined terms are, intelligent student, wealthy people, big cities, healthy people and beautiful toy etc.

For modeling and coping with ambiguity, vagueness, and incomplete information, FS is helpful. Artificial intelligence, control systems, decision-making, pattern recognition, and expert systems are just a few of the areas in which FSs are used. In circumstances where precise rules or measures are unavailable or inappropriate, fuzzy logic enables more flexible thinking and decision-making.

In this book, $\mathbb{E}$ would be treated as a fixed set and here, in this Section, we basically, revise the definition of FS and related work.

Definition 1.1. [1] The FS over $\mathbb{E}$ is stated as

$$\mathrm{M}_{\mathcal{FS}} = \left\{ \left(e, \mathfrak{F}_{\mathrm{M}_{\mathcal{FS}}}(\mathrm{e})\right) | \mathrm{e} \in \mathbb{E} \right\}$$

where $\mathfrak{F}_{\mathrm{M}_{\mathcal{FS}}}(e)$ is identified as belonging grade and $\mathfrak{F}_{\mathrm{M}_{\mathcal{FS}}}; \mathbb{E} \rightarrow [0, 1]$. The fuzzy number (FN) is signified as $\mathrm{M}_{\mathcal{FS}} = \left(\mathfrak{F}_{\mathrm{M}_{\mathcal{FS}}}\right)$.

Suppose an element $e \in \mathbb{E}$, then e is identified as not contained in FS if $\mathfrak{F}_{\mathrm{M}_{\mathcal{FS}}}(e) = 0$, partially contained if $0 < \mathfrak{F}_{\mathrm{M}_{\mathcal{FS}}}(e) < 1$ and fully contained if $\mathfrak{F}_{\mathrm{M}_{\mathcal{FS}}}(e) = 1$. A FS $\mathrm{M}_{\mathcal{FS}}$ is empty $\Leftrightarrow$ its belonging grade is zero. Two FSs $\mathrm{M}_{\mathcal{FS}-1}$ and $\mathrm{M}_{\mathcal{FS}-2}$ over $\mathbb{E}$, are supposed to be disjoint If there is no element $e \in \mathbb{E}$, such that $\mathfrak{F}_{\mathrm{M}_{\mathcal{FS}-1}}(e) = \mathfrak{F}_{\mathrm{M}_{\mathcal{FS}-2}}(e)$. If $\mathfrak{F}_{\mathrm{M}_{\mathcal{FS}-1}}(\mathrm{e}) = \mathfrak{F}_{\mathrm{M}_{\mathcal{FS}-2}}(\mathrm{e}) \forall \mathrm{e} \in \mathbb{E}$, then $\mathrm{M}_{\mathcal{FS}-1}$ and $\mathrm{M}_{\mathcal{FS}-2}$ are supposed to be equal, written as $\mathrm{M}_{\mathcal{FS}-1} = \mathrm{M}_{\mathcal{FS}-2}$. A FS $\mathrm{M}_{\mathcal{FS}-1}$ over $\mathbb{E}$ is contained in a FS $\mathrm{M}_{\mathcal{FS}-2}$ over $\mathbb{E}$ ($\mathrm{M}_{\mathcal{FS}-1} \subseteq \mathrm{M}_{\mathcal{FS}-2}$) if and only if $\mathfrak{F}_{\mathrm{M}_{\mathcal{FS}-1}}(\mathrm{e}) \leq \mathfrak{F}_{\mathrm{M}_{\mathcal{FS}-2}}(\mathrm{e}) \forall \mathrm{e} \in \mathbb{E}$. Further, $\mathrm{M}_{\mathcal{FS}-1} \subset \mathrm{M}_{\mathcal{FS}-2}$ if and only if $\mathrm{M}_{\mathcal{FS}-1} \subseteq \mathrm{M}_{\mathcal{FS}-2}$ but $\mathrm{M}_{\mathcal{FS}-1} \neq \mathrm{M}_{\mathcal{FS}-2}$. This means that $\mathrm{M}_{\mathcal{FS}-1}$ is properly contained in $\mathrm{M}_{\mathcal{FS}-2}$. $\mathrm{M}^{\delta}_{\mathcal{FS}} = \left(\mathfrak{F}_{\mathrm{M}_{\mathcal{FS}}}\right)$ denotes the fuzzy number FN.

Remark: 1.1.

1. Each fuzzy subset is a subset of itself.
2. The empty fuzzy subset is a subset of each fuzzy subset.

The support of a FS $\mathrm{M}_{\mathcal{FS}}$ is signified and interpreted as $supp(\mathrm{M}_{\mathcal{FS}}) = \left\{e \in \mathbb{E} : \mathfrak{F}_{\mathrm{M}_{\mathcal{FS}}}(\mathrm{e}) > 0\right\}$ and kernal of a FS $\mathrm{M}_{\mathcal{FS}}$ is signified and interpreted as $ker(\mathrm{M}_{\mathcal{FS}}) = \left\{e \in \mathbb{E} : \mathfrak{F}_{\mathrm{M}_{\mathcal{FS}}}(\mathrm{e}) = 1\right\}$.

Let two FSs $\mathrm{M}_{\mathcal{FS}-1}$ and $\mathrm{M}_{\mathcal{FS}-2}$ over $\mathbb{E}$. Then their union is signified as $\mathrm{M}_{\mathcal{FS}-1} \cup \mathrm{M}_{\mathcal{FS}-2}$ and described as

$$\mathrm{M}_{\mathcal{FS}-1} \cup \mathrm{M}_{\mathcal{FS}-2} = \left\{ \left(e, \max\left(\mathfrak{F}_{\mathrm{M}_{\mathcal{FS}-1}}(\mathrm{e}), \mathfrak{F}_{\mathrm{M}_{\mathcal{FS}}}(\mathrm{e})\right)\right) | \mathrm{e} \in \mathbb{E} \right\}$$

Which is again a FS. The union of any collection of$\{\mathrm{M}_{\mathcal{FS}-\sigma} : \sigma \in \mathbb{I}\}$ of FSs $\mathrm{M}_{\mathcal{FS}-\sigma}$ of $\mathbb{E}$ is interpreted as $\left(\bigcup_{\sigma \in \mathbb{I}} \mathrm{M}_{\mathcal{FS}-\sigma}\right) = \left\{\sup\left(\mathfrak{F}_{\mathrm{M}_{\mathcal{FS}-\sigma}}(e)\right) : \sigma \in \mathbb{I}\right\} \forall e \in \mathbb{E}$. It should be emphasized that the smallest FS that includes both $\mathrm{M}_{\mathcal{FS}-1}$ and $\mathrm{M}_{\mathcal{FS}-2}$ is their union. In other words, if $\mathrm{M}_{\mathcal{FS}-3}$ is any FS over $\mathbb{E}$ and $\mathrm{M}_{\mathcal{FS}-3}$ includes both

$\mathrm{M}_{\mathcal{FS}-1}$ and $\mathrm{M}_{\mathcal{FS}-2}$, then $\mathrm{M}_{\mathcal{FS}-3}$ includes their union. Likewise, the intersection of $\mathrm{M}_{\mathcal{FS}-1}$ and $\mathrm{M}_{\mathcal{FS}-2}$ is signified as $\mathrm{M}_{\mathcal{FS}-1} \cap \mathrm{M}_{\mathcal{FS}-2}$ and interpreted as

$$\mathrm{M}_{\mathcal{FS}-1} \cap \mathrm{M}_{\mathcal{FS}-2} = \left\{\left(e, \min\left(\mathfrak{F}_{\mathrm{M}_{\mathcal{FS}-1}}(\mathrm{e}), \mathfrak{F}_{\mathrm{M}_{\mathcal{FS}}}(\mathrm{e})\right)\right) | \mathrm{e} \in \mathbb{E}\right\}$$

Which is again a FS. The intersection of any collection of $\{\mathrm{M}_{\mathcal{FS}-\sigma} : \sigma \in \mathbb{I}\}$ of FSs $\mathrm{M}_{\mathcal{FS}-\sigma}$ of $\mathbb{E}$ is interpreted as $\left(\bigcup_{\sigma \in \mathbb{I}} \mathrm{M}_{\mathcal{FS}-\sigma}\right) = \left\{\inf\left(\mathfrak{F}_{\mathrm{M}_{\mathcal{FS}-\sigma}}(e)\right) : \sigma \in \mathbb{I}\right\} \forall e \in \mathbb{E}$. It should be emphasized that the largest FS that is included in both $\mathrm{M}_{\mathcal{FS}-1}$ and $\mathrm{M}_{\mathcal{FS}-2}$ is their intersection. The complement of a FS $\mathrm{M}_{\mathcal{FS}}$ is signified as $(\mathrm{M}_{\mathcal{FS}})^c$ or $(\mathrm{M}_{\mathcal{FS}})\prime$ and is interpreted as

$$(\mathrm{M}_{\mathcal{FS}})^c = \left\{\left(e, 1 - \mathfrak{F}_{\mathrm{M}_{\mathcal{FS}}}(\mathrm{e})\right) | \mathrm{e} \in \mathbb{E}\right\}$$

The complement, union, and intersection of FSs hold the underneath axioms.

(1) **Commutativity**:

$$\mathrm{M}_{\mathcal{FS}-1} \cup \mathrm{M}_{\mathcal{FS}-2} = \mathrm{M}_{\mathcal{FS}-2} \cup \mathrm{M}_{\mathcal{FS}-1} \quad \text{and} \quad \mathrm{M}_{\mathcal{FS}-1} \cap \mathrm{M}_{\mathcal{FS}-2} = \mathrm{M}_{\mathcal{FS}-2} \cap \mathrm{M}_{\mathcal{FS}-1}$$

(2) **Associativity**:

$$\mathrm{M}_{\mathcal{FS}-1} \cup (\mathrm{M}_{\mathcal{FS}-2} \cup \mathrm{M}_{\mathcal{FS}-3}) = (\mathrm{M}_{\mathcal{FS}-1} \cup \mathrm{M}_{\mathcal{FS}-2}) \cup \mathrm{M}_{\mathcal{FS}-3}$$

and

$$\mathrm{M}_{\mathcal{FS}-1} \cap (\mathrm{M}_{\mathcal{FS}-2} \cap \mathrm{M}_{\mathcal{FS}-3}) = (\mathrm{M}_{\mathcal{FS}-1} \cap \mathrm{M}_{\mathcal{FS}-2}) \cap \mathrm{M}_{\mathcal{FS}-3}$$

(3) **Idempotence**:

$$\mathrm{M}_{\mathcal{FS}} \cup \mathrm{M}_{\mathcal{FS}} = \mathrm{M}_{\mathcal{FS}} \quad \text{and} \quad \mathrm{M}_{\mathcal{FS}} \cap \mathrm{M}_{\mathcal{FS}} = \mathrm{M}_{\mathcal{FS}}$$

(4) **Distributivity**:

$$\mathrm{M}_{\mathcal{FS}-1} \cup (\mathrm{M}_{\mathcal{FS}-2} \cap \mathrm{M}_{\mathcal{FS}-3}) = (\mathrm{M}_{\mathcal{FS}-1} \cup \mathrm{M}_{\mathcal{FS}-2}) \cap (\mathrm{M}_{\mathcal{FS}-1} \cup \mathrm{M}_{\mathcal{FS}-3})$$

and

$$\mathrm{M}_{\mathcal{FS}-1} \cap (\mathrm{M}_{\mathcal{FS}-2} \cup \mathrm{M}_{\mathcal{FS}-3}) = (\mathrm{M}_{\mathcal{FS}-1} \cap \mathrm{M}_{\mathcal{FS}-2}) \cup (\mathrm{M}_{\mathcal{FS}-1} \cap \mathrm{M}_{\mathcal{FS}-3})$$

(5) **Absorption**:

$$\mathrm{M}_{\mathcal{FS}-1} \cup (\mathrm{M}_{\mathcal{FS}-1} \cap \mathrm{M}_{\mathcal{FS}-3}) = \mathrm{M}_{\mathcal{FS}-1} \text{ and } \mathrm{M}_{\mathcal{FS}-1} \cap (\mathrm{M}_{\mathcal{FS}-1} \cup \mathrm{M}_{\mathcal{FS}-3}) = \mathrm{M}_{\mathcal{FS}-1}$$

(6) **De-Morgan's Laws**:

$$(\mathrm{M}_{\mathcal{FS}-1}\cup\mathrm{M}_{\mathcal{FS}-2})^c=(\mathrm{M}_{\mathcal{FS}-1})^c\cap(\mathrm{M}_{\mathcal{FS}-2})^c \text{ and } (\mathrm{M}_{\mathcal{FS}-1}\cap\mathrm{M}_{\mathcal{FS}-2})^c$$
$$=(\mathrm{M}_{\mathcal{FS}-1})^c\cap(\mathrm{M}_{\mathcal{FS}-2})^c$$

(7) **Involution**:

$$\left((\mathrm{M}_{\mathcal{FS}})^c\right)^c=\mathrm{M}_{\mathcal{FS}}$$

Let us assume that $\mathbb{X}_{\mathbb{P}}$ signifies the characteristic function of $\mathbb{P}\subseteq\mathbb{E}$.

Remark 1.2. In the classical set theory, the underneath axioms are true but are not true generally in the notion of FS.

1. $\mathbb{P}\cap\mathbb{P}^c=\emptyset$, but $\mathfrak{F}_{\mathbb{P}}\cap\mathfrak{F}_{\mathbb{P}^c}\neq\mathbb{X}_{\emptyset}$ ($\mathbb{X}_{\emptyset}$ is the empty FS of $\mathbb{E}$)
2. $\mathbb{P}\cup\mathbb{P}^c=\mathbb{E}$, but $\mathfrak{F}_{\mathbb{P}}\cup\mathfrak{F}_{\mathbb{P}^c}\neq\mathbb{X}_{\mathbb{E}}$.

Let $\mathrm{M}_{\mathcal{FS}}$ be a FS and $\delta\in[0,1]$. Then the set

$$\mathrm{M}^{\delta}_{\mathcal{FS}}=\left\{e\in\mathbb{E}:\mathfrak{F}_{\mathrm{M}_{\mathcal{FS}}}(\mathrm{e})\geq\delta\right\}$$

is identified as a level set of FS $\mathrm{M}_{\mathcal{FS}}$. Obviously, $\mathrm{M}^{\delta_1}_{\mathcal{FS}}\subseteq\mathrm{M}^{\delta_2}_{\mathcal{FS}}$ if $\delta_1>\delta_2$. n FS is signified as normalized FS if there exists $e\in\mathbb{E}$ such that $\mathfrak{F}_{\mathrm{M}_{\mathcal{FS}}}(e)=1$. For two FNs $\mathrm{M}_{\mathcal{FS}-1}=\left(\mathfrak{F}_{\mathrm{M}_{\mathcal{FS}-1}}\right)$ and $\mathrm{M}_{\mathcal{FS}-2}=\left(\mathfrak{F}_{\mathrm{M}_{\mathcal{FS}-2}}\right)$, and $\zeta>0$, the operational laws for FNs are

1. $\mathrm{M}_{\mathcal{FS}-1}\oplus\mathrm{M}_{\mathcal{FS}-2}=\left(\mathfrak{F}_{\mathrm{M}_{\mathcal{FS}-1}}+\mathfrak{F}_{\mathrm{M}_{\mathcal{FS}-2}}-\mathfrak{F}_{\mathrm{M}_{\mathcal{FS}-1}}\mathfrak{F}_{\mathrm{M}_{\mathcal{FS}-2}}\right)$
2. $\mathrm{M}_{\mathcal{FS}-1}\otimes\mathrm{M}_{\mathcal{FS}-2}=\left(\mathfrak{F}_{\mathrm{M}_{\mathcal{FS}-1}}\mathfrak{F}_{\mathrm{M}_{\mathcal{FS}-2}}\right)$
3. $\zeta\mathrm{M}_{\mathcal{FS}-1}=1-\left(1-\mathfrak{F}_{\mathrm{M}_{\mathcal{FS}-1}}\right)^{\zeta}$
4. $(\mathrm{M}_{\mathcal{FS}-1})^{\zeta}=\left(\mathfrak{F}_{\mathrm{M}_{\mathcal{FS}-1}}\right)^{\zeta}$

1.2 Bipolar Fuzzy Sets

Bipolar or double-sided judgmental thinking underlies a range of human decision-making (DM) processes, including multiagent decision analysis, and is characterized by both a positive and a negative side (for instance, feedforward and feedback, effect and side effect). It is maintained that if positive and negative causal linkages are not counteractive at the same time, do not originated from the same sources, or do not follow reasonable paths, they should not be included in a summary. Thus, Zhang [2] demonstrated the theory of bipolar fuzzy set (BFS). A BFS is the generalization of FS that allows the interpretation of both positive and negative aspects. On contrary of FS, the elements in BFS demonstrate both positive and negative belonging grades which contained in [0, 1] and $[-1, 0]$ respectively. BFS is a notion that was developed to address circumstances where there is uncertainty regarding both positive and negative aspects of the element. This can be helpful in the fields that take into account both positive and negative facts or opinions.

In many fields, such as DM, pattern recognition, and expert systems, where it is important to model circumstances with conflicting or contradictory information, BFS has found applications. BFS offers a more expressive framework for dealing with ambiguity and uncertainty in these situations. In the following, we discuss the basic concept of BFS and the related results.

The definition of BFS is interpreted as:

Definition 1.2. [2] The BFS over $\mathbb{E}$ is stated as

$$\mathrm{M}_{\mathcal{BFS}} = \left\{\left(e, \left(\mathfrak{F}^{\mathrm{P}}_{\mathrm{M}_{\mathcal{BFS}}}(\mathrm{e}), \mathfrak{F}^{\mathrm{N}}_{\mathrm{M}_{\mathcal{BFS}}}(\mathrm{e})\right)\right) | \mathrm{e} \in \mathbb{E}\right\}$$

where $\mathfrak{F}^{P}_{\mathrm{M}_{\mathcal{BFS}}}(e)$ is identified as a positive belonging grade placed in [0, 1] and $\mathfrak{F}^{N}_{\mathrm{M}_{\mathcal{BFS}}}(e)$ is identified as a negative belonging grade placed in $[-1, 0]$. The set $\mathrm{M}_{\mathcal{BFS}} = \left(\mathfrak{F}^{P}_{\mathrm{M}_{\mathcal{BFS}}}, \mathfrak{F}^{N}_{\mathrm{M}_{\mathcal{BFS}}}\right)$ would be utilized as bipolar fuzzy number (BFN).

A BFS $\mathrm{M}_{\mathcal{BFS}-1}$ over $\mathbb{E}$ is contained (subset) in a BFS $\mathrm{M}_{\mathcal{BFS}-2}$ over $\mathbb{E}$ if and only if $\mathfrak{F}^{P}_{\mathrm{M}_{\mathcal{BFS}-1}}(e) \leq \mathfrak{F}^{P}_{\mathrm{M}_{\mathcal{BFS}-2}}(e)$ and $\mathfrak{F}^{N}_{\mathrm{M}_{\mathcal{BFS}-1}}(e) \leq \mathfrak{F}^{N}_{\mathrm{M}_{\mathcal{BFS}-2}}(e) \forall e \in \mathbb{E}$. Two BFSs $\mathrm{M}_{\mathcal{BFS}-1}$ and $\mathrm{M}_{\mathcal{BFS}-2}$ are equal if and only if $\mathfrak{F}^{P}_{\mathrm{M}_{\mathcal{BFS}-1}}(e) = \mathfrak{F}^{P}_{\mathrm{M}_{\mathcal{BFS}-2}}(e)$ and $\mathfrak{F}^{N}_{\mathrm{M}_{\mathcal{BFS}-1}}(e) = \mathfrak{F}^{N}_{\mathrm{M}_{\mathcal{BFS}-2}}(e)$. More, let $\mathrm{M}_{\mathcal{BFS}-1}$ and $\mathrm{M}_{\mathcal{BFS}-2}$, be two BFSs. Then their union is identified as $\mathrm{M}_{\mathcal{BFS}-1} \cup \mathrm{M}_{\mathcal{BFS}-2}$ and interpreted as

$$\mathrm{M}_{\mathcal{BFS}-1} \cup \mathrm{M}_{\mathcal{BFS}-2} = \left\{\left(e, \left(\max\left(\mathfrak{F}^{\mathrm{P}}_{\mathrm{M}_{\mathcal{BFS}-1}}(\mathrm{e}), \mathfrak{F}^{\mathrm{P}}_{\mathrm{M}_{\mathcal{BFS}-2}}(\mathrm{e})\right), \max\left(\mathfrak{F}^{\mathrm{N}}_{\mathrm{M}_{\mathcal{BFS}-1}}(\mathrm{e}), \mathfrak{F}^{\mathrm{N}}_{\mathrm{M}_{\mathcal{BFS}-2}}(\mathrm{e})\right)\right)\right) | \mathrm{e} \in \mathbb{E}\right\}$$

Similarly, their intersection is identified as $\mathrm{M}_{\mathcal{BFS}-1} \cap \mathrm{M}_{\mathcal{BFS}-2}$ and interpreted as

$$\mathrm{M}_{\mathcal{BFS}-1} \cap \mathrm{M}_{\mathcal{BFS}-2} = \left\{\left(e, \left(\min\left(\mathfrak{F}^{\mathrm{P}}_{\mathrm{M}_{\mathcal{BFS}-1}}(\mathrm{e}), \mathfrak{F}^{\mathrm{P}}_{\mathrm{M}_{\mathcal{BFS}-2}}(\mathrm{e})\right), \min\left(\mathfrak{F}^{\mathrm{N}}_{\mathrm{M}_{\mathcal{BFS}-1}}(\mathrm{e}), \mathfrak{F}^{\mathrm{N}}_{\mathrm{M}_{\mathcal{BFS}-2}}(\mathrm{e})\right)\right)\right) | \mathrm{e} \in \mathbb{E}\right\}$$

Finally, the complement of BFS $\mathrm{M}_{\mathcal{BFS}}$ is signified as $(\mathrm{M}_{\mathcal{BFS}})^{c}$ or $(\mathrm{M}_{\mathcal{BFS}})\prime$ and developed as

After the development of BFS, various researchers started work in this area and developed various theories for BFs. Gul [3], initiated operational laws for BFNs by employing algebraic t-norm and t-conorm which is defined in the underneath definition.

Definition 1.3. [3] For two BFNs $\mathrm{M}_{\mathcal{BFS}-1} = \left(\mathfrak{F}^{P}_{\mathrm{M}_{\mathcal{BFS}-1}}, \mathfrak{F}^{N}_{\mathrm{M}_{\mathcal{BFS}-1}}\right)$ and $\mathrm{M}_{\mathcal{BFS}-2} = \left(\mathfrak{F}^{P}_{\mathrm{M}_{\mathcal{BFS}-2}}, \mathfrak{F}^{N}_{\mathrm{M}_{\mathcal{BFS}-2}}\right)$, and $\zeta > 0$, the fundamental operational laws are inaugurated as.

1. $$\mathrm{M}_{\mathcal{BFS}-1} \oplus \mathrm{M}_{\mathcal{BFS}-2} = \left(\mathfrak{F}^{P}_{\mathrm{M}_{\mathcal{BFS}-1}} + \mathfrak{F}^{P}_{\mathrm{M}_{\mathcal{BFS}-2}} - \mathfrak{F}^{P}_{\mathrm{M}_{\mathcal{BFS}-1}}\mathfrak{F}^{P}_{\mathrm{M}_{\mathcal{BFS}-2}}, -\left|\mathfrak{F}^{N}_{\mathrm{M}_{\mathcal{BFS}-1}}\right|\left|\mathfrak{F}^{N}_{\mathrm{M}_{\mathcal{BFS}-2}}\right|\right)$$
2. $$\mathrm{M}_{\mathcal{BFS}-1} \otimes \mathrm{M}_{\mathcal{BFS}-2} = \left(\mathfrak{F}^{P}_{\mathrm{M}_{\mathcal{BFS}-1}}\mathfrak{F}^{P}_{\mathrm{M}_{\mathcal{BFS}-2}}, \mathfrak{F}^{N}_{\mathrm{M}_{\mathcal{BFS}-1}} + \mathfrak{F}^{N}_{\mathrm{M}_{\mathcal{BFS}-2}} - \mathfrak{F}^{N}_{\mathrm{M}_{\mathcal{BFS}-1}}\mathfrak{F}^{N}_{\mathrm{M}_{\mathcal{BFS}-2}}\right)$$

3. $\zeta \mathrm{M}_{\mathcal{BFS}-1} = \left(1 - \left(1 - \mathfrak{F}^{P}_{\mathrm{M}_{\mathcal{BFS}-1}}\right)^{\zeta}, -\left|\mathfrak{F}^{N}_{\mathrm{M}_{\mathcal{BFS}-1}}\right|^{\zeta}\right)$
4. $(\mathrm{M}_{\mathcal{BFS}-1})^{\zeta} = \left(\left(\mathfrak{F}^{P}_{\mathrm{M}_{\mathcal{BFS}-1}}\right)^{\zeta}, -1 + \left|1 + \mathfrak{F}^{N}_{\mathrm{M}_{\mathcal{BFS}-1}}\right|^{\zeta}\right)$

Example 1.1. Let $\mathrm{M}_{\mathcal{BFS}-1} = (0.4, -1)$ and $\mathrm{M}_{\mathcal{BFS}-2} = (0.6, -1)$, be two BFNs. Then.

$$\mathrm{M}_{\mathcal{BFS}-1} \otimes \mathrm{M}_{\mathcal{BFS}-2} = (0.4 \times 0.6, -1 + (-1) - (-1)(-1)) = (0.24, -3)$$

Obviously, $-3 \notin [-1, 0]$.

Thus, the operation of multiplication $\otimes$ is not giving a BFN again. Thus, we initiate the new operational laws for BFNs as follows.

Definition 1.4. For two BFNs $\mathrm{M}_{\mathcal{BFS}-1} = \left(\mathfrak{F}^{P}_{\mathrm{M}_{\mathcal{BFS}-1}}, \mathfrak{F}^{N}_{\mathrm{M}_{\mathcal{BFS}-1}}\right)$ and $\mathrm{M}_{\mathcal{BFS}-2} = \left(\mathfrak{F}^{P}_{\mathrm{M}_{\mathcal{BFS}-2}}, \mathfrak{F}^{N}_{\mathrm{M}_{\mathcal{BFS}-2}}\right)$, and $\zeta > 0$, the improved fundamental operational laws are inaugurated as.

1. $$\mathrm{M}_{\mathcal{BFS}-1} \oplus \mathrm{M}_{\mathcal{BFS}-2} = \left(\mathfrak{F}^{P}_{\mathrm{M}_{\mathcal{BFS}-1}} + \mathfrak{F}^{P}_{\mathrm{M}_{\mathcal{BFS}-2}} - \mathfrak{F}^{P}_{\mathrm{M}_{\mathcal{BFS}-1}}\mathfrak{F}^{P}_{\mathrm{M}_{\mathcal{BFS}-2}}, -(\mathfrak{F}^{N}_{\mathrm{M}_{\mathcal{BFS}-1}})(\mathfrak{F}^{N}_{\mathrm{M}_{\mathcal{BFS}-2}})\right)$$
2. $$\mathrm{M}_{\mathcal{BFS}-1} \otimes \mathrm{M}_{\mathcal{BFS}-2} = \left(\mathfrak{F}^{P}_{\mathrm{M}_{\mathcal{BFS}-1}}\mathfrak{F}^{P}_{\mathrm{M}_{\mathcal{BFS}-2}}, \mathfrak{F}^{N}_{\mathrm{M}_{\mathcal{BFS}-1}} + \mathfrak{F}^{N}_{\mathrm{M}_{\mathcal{BFS}-2}} + \mathfrak{F}^{N}_{\mathrm{M}_{\mathcal{BFS}-1}}\mathfrak{F}^{N}_{\mathrm{M}_{\mathcal{BFS}-2}}\right)$$
3. $\zeta \mathrm{M}_{\mathcal{BFS}-1} = \left(1 - \left(1 - \mathfrak{F}^{P}_{\mathrm{M}_{\mathcal{BFS}-1}}\right)^{\zeta}, -\left|\mathfrak{F}^{N}_{\mathrm{M}_{\mathcal{BFS}-1}}\right|^{\zeta}\right)$
4. $(\mathrm{M}_{\mathcal{BFS}-1})^{\zeta} = \left(\left(\mathfrak{F}^{P}_{\mathrm{M}_{\mathcal{BFS}-1}}\right)^{\zeta}, -1 + \left(1 + \mathfrak{F}^{N}_{\mathrm{M}_{\mathcal{BFS}-1}}\right)^{\zeta}\right)$

Theorem 1.1. Assume $\mathrm{M}_{\mathcal{BFS}-1} = \left(\mathfrak{F}^{P}_{\mathrm{M}_{\mathcal{BFS}-1}}, \mathfrak{F}^{N}_{\mathrm{M}_{\mathcal{BFS}-1}}\right)$ and $\mathrm{M}_{\mathcal{BFS}-2} = \left(\mathfrak{F}^{P}_{\mathrm{M}_{\mathcal{BFS}-2}}, \mathfrak{F}^{N}_{\mathrm{M}_{\mathcal{BFS}-2}}\right)$ as two BFS, and $\zeta, \zeta_1, \zeta_2 > 0$, then,

1. $\mathrm{M}_{\mathcal{BFS}-1} \oplus \mathrm{M}_{\mathcal{BFS}-2} = \mathrm{M}_{\mathcal{BFS}-2} \oplus \mathrm{M}_{\mathcal{BFS}-1}$
2. $\mathrm{M}_{\mathcal{BFS}-1} \otimes \mathrm{M}_{\mathcal{BFS}-2} = \mathrm{M}_{\mathcal{BFS}-2} \otimes \mathrm{M}_{\mathcal{BFS}-1}$
3. $\zeta(\mathrm{M}_{\mathcal{BFS}-1} \oplus \mathrm{M}_{\mathcal{BFS}-2}) = \zeta \mathrm{M}_{\mathcal{BFS}-1} \oplus \zeta \mathrm{M}_{\mathcal{BFS}-2}$
4. $(\mathrm{M}_{\mathcal{BFS}-1} \otimes \mathrm{M}_{\mathcal{BFS}-2})^{\zeta} = \mathrm{M}^{\zeta}_{\mathcal{BFS}-1} \otimes \mathrm{M}^{\zeta}_{\mathcal{BFS}-2}$
5. $\zeta_1 \mathrm{M}_{\mathcal{BFS}-1} \oplus \zeta_2 \mathrm{M}_{\mathcal{BFS}-1} = (\zeta_1 + \zeta_2)\mathrm{M}_{\mathcal{BFS}-1}$
6. $\mathrm{M}^{\zeta_1}_{\mathcal{BFS}-1} \otimes \mathrm{M}^{\zeta_2}_{\mathcal{BFS}-1} = \mathrm{M}^{\zeta_1+\zeta_2}_{\mathcal{BFS}-1}$
7. $\left(\mathrm{M}^{\zeta_1}_{\mathcal{BFS}-1}\right)^{\zeta_2} = \mathrm{M}^{\zeta_1\zeta_2}_{\mathcal{BFS}-1}$

Proof Trivial

Since in the theory of FS, the two FNs are easily comparable, we can say which FN is greater and finer and we can rank FNs. But in generalized theories of FS, the ranking and comparability of two numbers are not possible in an ordinary way. Thus, for ranking and comparing two numbers in the generalized theories of FS, the notion of score function was introduced. But in some cases, it is noted that the score values of two numbers are the same but the numbers were not the same. Thus, the theory of accuracy function was developed.

For comparing two BFNs, Wei et al. [4] investigated the score and accuracy function which is interpreted underneath.

For a BFN $M_{\mathcal{BFS}} = \left(\mathfrak{F}^{P}_{M_{\mathcal{BFS}}}, \mathfrak{F}^{N}_{M_{\mathcal{BFS}}}\right)$, the formula for determining the score value is signified as

$$\mathbb{S}(M_{\mathcal{BFS}}) = \frac{1}{2}\left(1 + \mathfrak{F}^{P}_{M_{\mathcal{BFS}}} + \mathfrak{F}^{N}_{M_{\mathcal{BFS}}}\right) \tag{1.1}$$

Noticed that $\mathbb{S}(M_{\mathcal{BFS}}) \in [0, 1]$ and the formula for determining the score value is signified as

$$\mathbb{H}(M_{\mathcal{BFS}}) = \frac{\mathfrak{F}^{P}_{M_{\mathcal{BFS}}} - \mathfrak{F}^{N}_{M_{\mathcal{BFS}}}}{2} \tag{1.2}$$

Noticed that $\mathbb{S}(M_{\mathcal{BFS}}) \in [0, 1]$. From Eq. (1.1) and (1.2), the underneath outcomes are achieved.

1. If $\mathbb{S}(M_{\mathcal{BFS}-1}) < \mathbb{S}(M_{\mathcal{BFS}-2})$, then $M_{\mathcal{BFS}-1} < M_{\mathcal{BFS}-2}$;
2. If $\mathbb{S}(M_{\mathcal{BFS}-1}) > \mathbb{S}(M_{\mathcal{BFS}-2})$, then $M_{\mathcal{BFS}-1} > M_{\mathcal{BFS}-2}$;
3. If $\mathbb{S}(M_{\mathcal{BFS}-1}) = \mathbb{S}(M_{\mathcal{BFS}-2})$, then if $\mathbb{H}(M_{\mathcal{BFS}-1}) < \mathbb{H}(M_{\mathcal{BFS}-2})$, then $M_{\mathcal{BFS}-1} < M_{\mathcal{BFS}-2}$; if $\mathbb{H}(M_{\mathcal{BFS}-1}) > \mathbb{H}(M_{\mathcal{BFS}-2})$, then $M_{\mathcal{BFS}-1} > M_{\mathcal{BFS}-2}$; if $\mathbb{H}(M_{\mathcal{BFS}-1}) = \mathbb{H}(M_{\mathcal{BFS}-2})$, then $M_{\mathcal{BFS}-1} = M_{\mathcal{BFS}-2}$.

1.3 Complex Fuzzy Sets

With the phase shift of data, the volume and complexity of the information that is gathered from our daily lives are growing quickly. Additionally, several types of uncertainty are frequently present in knowledge that is signified by challenging dilemmas in a variety of fields, including biology, environmental science, computer science, social science, mathematics, economics, etc. The challenges of DM are getting more and more complex as science and technology advance. Thus, Ramot et al. [5] expanded the theory of FS by adding a phase term and identified it as a complex fuzzy set (CFS). The phase term for CFS is vital in defining the functionality of the structure of CFS. A CFS model is distinguished by this term from all other models in the literature at hand. A CFS excels in handling ambiguous and intuitive information that is common in time-periodic events because it may represent two-dimensional phenomena.

In this section, we review both polar and Cartesian forms of CFS and investigate the related operations and results.

Definition 1.5. [5] The CFS over $\mathbb{E}$ is inaugurated as

$$\mathrm{M}_{\mathcal{CFS}} = \left\{\left(e, \mathfrak{F}_{\mathrm{M}_{\mathcal{CFS}}}(\mathrm{e})\right)|\mathrm{e} \in \mathbb{E}\right\} = \left\{\left(e, \Gamma_{\mathrm{M}_{\mathcal{CFS}}}(e)e^{i\omega_{\mathrm{M}_{\mathcal{CFS}}}(e)}\right)|\mathrm{e} \in \mathbb{E}\right\}$$

where $\mathfrak{F}_{\mathrm{M}_{\mathcal{CFS}}}(e)$ is identified as the amplitude term of belonging grade, that is placed in [0, 1] and $\omega_{\mathrm{M}_{\mathcal{CFS}}}(e)$ is identified as the phase term of belonging grade, that placed in $(0, 2\pi]$. The complex fuzzy number (CFN) would be devised as $\mathrm{M}_{\mathcal{CFS}} = \left(\mathfrak{F}_{\mathrm{M}_{\mathcal{CFS}}}\right) = \left(\Gamma_{\mathrm{M}_{\mathcal{CFS}}} e^{i\,\omega_{\mathrm{M}_{\mathcal{CFS}}}}\right)$.

By ignoring the phase term in the structure of CFS, CFS would be in the shape of FS. Let $\mathrm{M}_{\mathcal{CFS}-1} = \left\{\left(e, \Gamma_{\mathrm{M}_{\mathcal{CFS}-1}}(e)e^{i\,\omega_{\mathrm{M}_{\mathcal{CFS}-1}}(e)}\right)|\mathrm{e} \in \mathbb{E}\right\}$ and $\mathrm{M}_{\mathcal{CFS}-2} = \left\{\left(e, \Gamma_{\mathrm{M}_{\mathcal{CFS}-2}}(e)e^{i\,\omega_{\mathrm{M}_{\mathcal{CFS}-2}}(e)}\right)|\mathrm{e} \in \mathbb{E}\right\}$ be two CFSs. Then $\mathrm{M}_{\mathcal{CFS}-1} = \mathrm{M}_{\mathcal{CFS}-2}$ if and only if $\Gamma_{\mathrm{M}_{\mathcal{CFS}-1}} = \Gamma_{\mathrm{M}_{\mathcal{CFS}-1}}$ and $\omega_{\mathrm{M}_{\mathcal{CFS}-1}}(e) = \omega_{\mathrm{M}_{\mathcal{CFS}-2}}(e)$. The union of $\mathrm{M}_{\mathcal{CFS}-1}$ and $\mathrm{M}_{\mathcal{CFS}-2}$ is identified and demonstrated as

$$\begin{aligned}&\mathrm{M}_{\mathcal{CFS}-1} \cup \mathrm{M}_{\mathcal{CFS}-2}\\&= \left\{\left(e, \max\left(\Gamma_{\mathrm{M}_{\mathcal{CFS}-1}}(e), \Gamma_{\mathrm{M}_{\mathcal{CFS}-2}}(e)\right)e^{\iota\max\left(\omega_{\mathrm{M}_{\mathcal{CFS}-1}}(\mathrm{e}), \omega_{\mathrm{M}_{\mathcal{CFS}-2}}(\mathrm{e})\right)}\right)|\mathrm{e} \in \mathbb{E}\right\}\end{aligned}$$

Similarly, their intersection is signified and interpreted as

$$\begin{aligned}&\mathrm{M}_{\mathcal{CFS}-1} \cap \mathrm{M}_{\mathcal{CFS}-2}\\&= \mathrm{M}_{\mathcal{CFS}-1} \cup \mathrm{M}_{\mathcal{CFS}-2}\\&= \left\{\left(e, \min\left(\Gamma_{\mathrm{M}_{\mathcal{CFS}-1}}(e), \Gamma_{\mathrm{M}_{\mathcal{CFS}-2}}(e)\right)e^{\iota\min\left(\omega_{\mathrm{M}_{\mathcal{CFS}-1}}(e), \omega_{\mathrm{M}_{\mathcal{CFS}-2}}(e)\right)}\right)|\mathrm{e} \in \mathbb{E}\right\}\end{aligned}$$

The complement of CFS $\mathrm{M}_{\mathcal{CFS}}$ is revealed as

$$(\mathrm{M}_{\mathcal{CFS}})^c = \left\{\left(e, 1 - \Gamma_{\mathrm{M}_{\mathcal{CFS}}}(e)e^{\iota\left(2\pi - \omega_{\mathrm{M}_{\mathcal{CFS}}}(\mathrm{e})\right)}\right)|\mathrm{e} \in \mathbb{E}\right\}$$

After that, Tamir et al. [6] inaugurated the definition of CFS in the cartesian form by changing the range from unit disc with center at origin to unit square in first quadrant in the complex plane.

Definition 1.6. [6]. The CFS over $\mathbb{E}$ is stated as

$$\mathrm{M}_{\mathcal{CFS}} = \left\{\left(e, \mathfrak{F}_{\mathrm{M}_{\mathcal{CFS}}}(\mathrm{e})\right)|\mathrm{e} \in \mathbb{E}\right\} = \left\{\left(e, \mathfrak{F}^{\mathrm{P}}_{\mathrm{M}_{\mathcal{CFS}}}(\mathrm{e}) + \iota\mathfrak{F}^{\mathrm{I}}_{\mathrm{M}_{\mathcal{CFS}}}(\mathrm{e})\right)|\mathrm{e} \in \mathbb{E}\right\}$$

where $\mathfrak{F}^{P}_{\mathrm{M}_{\mathcal{CFS}}}(e)$ is identified as a real part of belonging grade, that is placed in [0, 1] and $\mathfrak{F}^{N}_{\mathrm{M}_{\mathcal{CFS}}}(e)$ is identified as an unreal part of belonging grade, that placed in [0, 1]. The belonging grade would be part of the unit square of a complex plane. The CFN would be devised as $\mathrm{M}_{\mathcal{CFS}} = \left(\mathfrak{F}_{\mathrm{M}_{\mathcal{CFS}}}\right) = \left(\mathfrak{F}^{P}_{\mathrm{M}_{\mathcal{CFS}}} + \iota\mathfrak{F}^{I}_{\mathrm{M}_{\mathcal{CFS}}}\right)$.

The complement, union, and intersection of CFNs are inaugurated in the underneath definition.

Definition 1.7. [6]. For two CFSs$\mathrm{M}_{\mathcal{CFS}-1} = \left\{\left(e, \mathfrak{F}_{\mathrm{M}_{\mathcal{CFS}-1}}(\mathrm{e})\right)|\mathrm{e} \in \mathbb{E}\right\} = \left\{\left(e, \mathfrak{F}^{\mathrm{P}}_{\mathrm{M}_{\mathcal{CFS}-1}}(\mathrm{e})+\iota\mathfrak{F}^{\mathrm{I}}_{\mathrm{M}_{\mathcal{CFS}-1}}(\mathrm{e})\right)|\mathrm{e} \in \mathbb{E}\right\}$ and$\mathrm{M}_{\mathcal{CFS}-2} = \left\{\left(e, \mathfrak{F}_{\mathrm{M}_{\mathcal{CFS}-2}}(\mathrm{e})\right)|\mathrm{e} \in \mathbb{E}\right\} = \left\{\left(e, \mathfrak{F}^{\mathrm{P}}_{\mathrm{M}_{\mathcal{CFS}-2}}(\mathrm{e})+\iota\mathfrak{F}^{\mathrm{I}}_{\mathrm{M}_{\mathcal{CFS}-2}}(\mathrm{e})\right)|\mathrm{e} \in \mathbb{E}\right\}$, we have.

1. $(\mathrm{M}_{\mathcal{CFS}-1})^c = \left\{\left(e, 1-\mathfrak{F}^{\mathrm{P}}_{\mathrm{M}_{\mathcal{CFS}-1}}(\mathrm{e})+\iota\left(1-\mathfrak{F}^{\mathrm{I}}_{\mathrm{M}_{\mathcal{CFS}-1}}(\mathrm{e})\right)\right)|\mathrm{e} \in \mathbb{E}\right\}$
2. $\mathrm{M}_{\mathcal{CFS}-1} \cup \mathrm{M}_{\mathcal{CFS}-2} = \left\{\left(e, \max\left(\mathfrak{F}^{\mathrm{P}}_{\mathrm{M}_{\mathcal{CFS}-1}}(\mathrm{e}), \mathfrak{F}^{\mathrm{P}}_{\mathrm{M}_{\mathcal{CFS}-2}}(\mathrm{e})\right)+\iota\max\left(\mathfrak{F}^{\mathrm{I}}_{\mathrm{M}_{\mathcal{CFS}-1}}(\mathrm{e}), \mathfrak{F}^{\mathrm{I}}_{\mathrm{M}_{\mathcal{CFS}-2}}(\mathrm{e})\right)\right)|\mathrm{e} \in \mathbb{E}\right\}$
3. $\mathrm{M}_{\mathcal{CFS}-1} \cap \mathrm{M}_{\mathcal{CFS}-2} = \left\{\left(e, \min\left(\mathfrak{F}^{\mathrm{P}}_{\mathrm{M}_{\mathcal{CFS}-1}}(\mathrm{e}), \mathfrak{F}^{\mathrm{P}}_{\mathrm{M}_{\mathcal{CFS}-2}}(\mathrm{e})\right)+\iota\min\left(\mathfrak{F}^{\mathrm{I}}_{\mathrm{M}_{\mathcal{CFS}-1}}(\mathrm{e}), \mathfrak{F}^{\mathrm{I}}_{\mathrm{M}_{\mathcal{CFS}-2}}(\mathrm{e})\right)\right)|\mathrm{e} \in \mathbb{E}\right\}$

Based on algebraic t-norm and t-conorm, we inaugurated the fundamental operational laws for CFNs.

Definition 1.8. For two CFNs $\mathrm{M}_{\mathcal{CFS}-1} = \left(\mathfrak{F}_{\mathrm{M}_{\mathcal{CFS}-1}}\right) = \left(\mathfrak{F}^{P}_{\mathrm{M}_{\mathcal{CFS}-1}} + \iota\mathfrak{F}^{I}_{\mathrm{M}_{\mathcal{CFS}-1}}\right)$ and $\mathrm{M}_{\mathcal{CFS}-2} = \left(\mathfrak{F}_{\mathrm{M}_{\mathcal{CFS}-2}}\right) = \left(\mathfrak{F}^{P}_{\mathrm{M}_{\mathcal{CFS}-2}} + \iota\mathfrak{F}^{I}_{\mathrm{M}_{\mathcal{CFS}-2}}\right)$, and $\zeta > 0$, the fundamental operational laws are inaugurated as.

1. $\mathrm{M}_{\mathcal{CFS}-1} \oplus \mathrm{M}_{\mathcal{CFS}-2} = \left(\begin{array}{c}\mathfrak{F}^{P}_{\mathrm{M}_{\mathcal{CFS}-1}} + \mathfrak{F}^{P}_{\mathrm{M}_{\mathcal{CFS}-2}} - \mathfrak{F}^{P}_{\mathrm{M}_{\mathcal{CFS}-1}}\mathfrak{F}^{P}_{\mathrm{M}_{\mathcal{CFS}-2}} \\ +\iota\left(\mathfrak{F}^{I}_{\mathrm{M}_{\mathcal{CFS}-1}} + \mathfrak{F}^{I}_{\mathrm{M}_{\mathcal{CFS}-2}} - \mathfrak{F}^{I}_{\mathrm{M}_{\mathcal{CFS}-1}}\mathfrak{F}^{I}_{\mathrm{M}_{\mathcal{CFS}-2}}\right)\end{array}\right)$
2. $\mathrm{M}_{\mathcal{CFS}-1} \otimes \mathrm{M}_{\mathcal{CFS}-2} = \left(\mathfrak{F}^{P}_{\mathrm{M}_{\mathcal{CFS}-1}}\mathfrak{F}^{P}_{\mathrm{M}_{\mathcal{CFS}-2}} + \iota\mathfrak{F}^{I}_{\mathrm{M}_{\mathcal{CFS}-1}}\mathfrak{F}^{I}_{\mathrm{M}_{\mathcal{CFS}-2}}\right)$
3. $\zeta\mathrm{M}_{\mathcal{CFS}-1} = \left(1 - \left(1 - \mathfrak{F}^{P}_{\mathrm{M}_{\mathcal{CFS}-1}}\right)^{\zeta} + \iota\left(1 - \left(1 - \mathfrak{F}^{I}_{\mathrm{M}_{\mathcal{CFS}-1}}\right)^{\zeta}\right)\right)$
4. $(\mathrm{M}_{\mathcal{CFS}-1})^{\zeta} = \left(\left(\mathfrak{F}^{P}_{\mathrm{M}_{\mathcal{CFS}-1}}\right)^{\zeta} + \iota\left(\mathfrak{F}^{I}_{\mathrm{M}_{\mathcal{CFS}-1}}\right)^{\zeta}\right)$

Theorem 1.2 Assume CFNs $\mathrm{M}_{\mathcal{CFS}-1} = \left(\mathfrak{F}_{\mathrm{M}_{\mathcal{CFS}-1}}\right) = \left(\mathfrak{F}^{P}_{\mathrm{M}_{\mathcal{CFS}-1}} + \iota\mathfrak{F}^{I}_{\mathrm{M}_{\mathcal{CFS}-1}}\right)$ and $\mathrm{M}_{\mathcal{CFS}-2} = \left(\mathfrak{F}_{\mathrm{M}_{\mathcal{CFS}-2}}\right) = \left(\mathfrak{F}^{P}_{\mathrm{M}_{\mathcal{CFS}-2}} + \iota\mathfrak{F}^{I}_{\mathrm{M}_{\mathcal{CFS}-2}}\right)$ as two CFS, and $\zeta, \zeta_1, \zeta_2 > 0$, then,

1. $\mathrm{M}_{\mathcal{CFS}-1} \oplus \mathrm{M}_{\mathcal{CFS}-2} = \mathrm{M}_{\mathcal{CFS}-2} \oplus \mathrm{M}_{\mathcal{CFS}-1}$
2. $\mathrm{M}_{\mathcal{CFS}-1} \otimes \mathrm{M}_{\mathcal{CFS}-2} = \mathrm{M}_{\mathcal{CFS}-2} \otimes \mathrm{M}_{\mathcal{CFS}-1}$
3. $\zeta(\mathrm{M}_{\mathcal{CFS}-1} \oplus \mathrm{M}_{\mathcal{CFS}-2}) = \zeta\mathrm{M}_{\mathcal{CFS}-1} \oplus \zeta\mathrm{M}_{\mathcal{CFS}-2}$
4. $(\mathrm{M}_{\mathcal{CFS}-1} \otimes \mathrm{M}_{\mathcal{CFS}-2})^{\zeta} = \mathrm{M}^{\zeta}_{\mathcal{CFS}-1} \otimes \mathrm{M}^{\zeta}_{\mathcal{CFS}-2}$
5. $\zeta_1\mathrm{M}_{\mathcal{CFS}-1} \oplus \zeta_2\mathrm{M}_{\mathcal{CFS}-1} = (\zeta_1 + \zeta_2)\mathrm{M}_{\mathcal{CFS}-1}$
6. $\mathrm{M}^{\zeta_1}_{\mathcal{CFS}-1} \otimes \mathrm{M}^{\zeta_2}_{\mathcal{CFS}-1} = \mathrm{M}^{\zeta_1+\zeta_2}_{\mathcal{CFS}-1}$
7. $\left(\mathrm{M}^{\zeta_1}_{\mathcal{CFS}-1}\right)^{\zeta_2} = \mathrm{M}^{\zeta_1\zeta_2}_{\mathcal{CFS}-1}$

Proof 1. From definition (1.9) we have,

$\mathrm{M}_{\mathcal{CFS}-1} \oplus \mathrm{M}_{\mathcal{CFS}-2}$

$$
\begin{aligned}
&= \left(\mathfrak{F}^P_{M_{\mathcal{CFS}-1}} + \mathfrak{F}^P_{M_{\mathcal{CFS}-2}} - \mathfrak{F}^P_{M_{\mathcal{CFS}-1}}\mathfrak{F}^P_{M_{\mathcal{CFS}-2}} + \iota\left(\mathfrak{F}^I_{M_{\mathcal{CFS}-1}} + \mathfrak{F}^I_{M_{\mathcal{CFS}-2}} - \mathfrak{F}^I_{M_{\mathcal{CFS}-1}}\mathfrak{F}^I_{M_{\mathcal{CFS}-2}}\right)\right)\\
&= \left(\mathfrak{F}^P_{M_{\mathcal{CFS}-2}} + \mathfrak{F}^P_{M_{\mathcal{CFS}-1}} - \mathfrak{F}^P_{M_{\mathcal{CFS}-2}}\mathfrak{F}^P_{M_{\mathcal{CFS}-1}} + \iota\left(\mathfrak{F}^I_{M_{\mathcal{CFS}-2}} + \mathfrak{F}^I_{M_{\mathcal{CFS}-1}} - \mathfrak{F}^I_{M_{\mathcal{CFS}-2}}\mathfrak{F}^I_{M_{\mathcal{CFS}-1}}\right)\right)\\
&= M_{\mathcal{CFS}-2} \oplus M_{\mathcal{CFS}-1}
\end{aligned}
$$

1. Since,

$$
\begin{aligned}
M_{\mathcal{CFS}-1} \otimes M_{\mathcal{CFS}-2} &= \left(\mathfrak{F}^P_{M_{\mathcal{CFS}-1}}\mathfrak{F}^P_{M_{\mathcal{CFS}-2}} + \iota\mathfrak{F}^I_{M_{\mathcal{CFS}-1}}\mathfrak{F}^I_{M_{\mathcal{CFS}-2}}\right)\\
&= \left(\mathfrak{F}^P_{M_{\mathcal{CFS}-2}}\mathfrak{F}^P_{M_{\mathcal{CFS}-1}} + \iota\mathfrak{F}^I_{M_{\mathcal{CFS}-2}}\mathfrak{F}^I_{M_{\mathcal{CFS}-1}}\right) = M_{\mathcal{CFS}-2} \otimes M_{\mathcal{CFS}-1}
\end{aligned}
$$

2. From right-hand side, we have

$$
\zeta M_{\mathcal{CFS}-1} = \left(1 - \left(1 - \mathfrak{F}^P_{M_{\mathcal{CFS}-1}}\right)^{\zeta} + \iota\left(1 - \left(1 - \mathfrak{F}^I_{M_{\mathcal{CFS}-1}}\right)^{\zeta}\right)\right)
$$

and

$$
\zeta M_{\mathcal{CFS}-2} = \left(1 - \left(1 - \mathfrak{F}^P_{M_{\mathcal{CFS}-2}}\right)^{\zeta} + \iota\left(1 - \left(1 - \mathfrak{F}^I_{M_{\mathcal{CFS}-2}}\right)^{\zeta}\right)\right)
$$

then,

$$
\begin{aligned}
&\zeta M_{\mathcal{CFS}-1} \oplus \zeta M_{\mathcal{CFS}-2} = \begin{pmatrix} 1 - \left(1 - \mathfrak{F}^P_{M_{\mathcal{CFS}-1}}\right)^{\zeta} \\ +\iota\left(1 - \left(1 - \mathfrak{F}^I_{M_{\mathcal{CFS}-1}}\right)^{\zeta}\right) \end{pmatrix} \oplus \begin{pmatrix} 1 - \left(1 - \mathfrak{F}^P_{M_{\mathcal{CFS}-2}}\right)^{\zeta} \\ +\iota\left(1 - \left(1 - \mathfrak{F}^I_{M_{\mathcal{CFS}-2}}\right)^{\zeta}\right) \end{pmatrix}\\
&= \begin{pmatrix} 1 - \left(1 - \mathfrak{F}^P_{M_{\mathcal{CFS}-1}}\right)^{\zeta} + 1 - \left(1 - \mathfrak{F}^P_{M_{\mathcal{CFS}-2}}\right)^{\zeta} - \left(1 - \left(1 - \mathfrak{F}^P_{M_{\mathcal{CFS}-1}}\right)^{\zeta}\right)\left(1 - \left(1 - \mathfrak{F}^P_{M_{\mathcal{CFS}-2}}\right)^{\zeta}\right) \\ +\iota\left(1 - \left(1 - \mathfrak{F}^I_{M_{\mathcal{CFS}-1}}\right)^{\zeta} + 1 - \left(1 - \mathfrak{F}^I_{M_{\mathcal{CFS}-2}}\right)^{\zeta} - \left(1 - \left(1 - \mathfrak{F}^I_{M_{\mathcal{CFS}-1}}\right)^{\zeta}\right)\left(1 - \left(1 - \mathfrak{F}^I_{M_{\mathcal{CFS}-2}}\right)^{\zeta}\right)\right) \end{pmatrix}\\
&= \begin{pmatrix} 1 - \left(1 - \mathfrak{F}^P_{M_{\mathcal{CFS}-1}}\right)^{\zeta} + 1 - \left(1 - \mathfrak{F}^P_{M_{\mathcal{CFS}-2}}\right)^{\zeta} - \begin{pmatrix} 1 - \left(1 - \mathfrak{F}^P_{M_{\mathcal{CFS}-2}}\right)^{\zeta} - \left(1 - \mathfrak{F}^P_{M_{\mathcal{CFS}-1}}\right)^{\zeta} \\ +\left(1 - \mathfrak{F}^P_{M_{\mathcal{CFS}-1}}\right)^{\zeta}\left(1 - \mathfrak{F}^P_{M_{\mathcal{CFS}-2}}\right)^{\zeta} \end{pmatrix} \\ +\iota\left(1 - \left(1 - \mathfrak{F}^I_{M_{\mathcal{CFS}-1}}\right)^{\zeta} + 1 - \left(1 - \mathfrak{F}^P_{M_{\mathcal{CFS}-2}}\right)^{\zeta} - \begin{pmatrix} 1 - \left(1 - \mathfrak{F}^I_{M_{\mathcal{CFS}-2}}\right)^{\zeta} - \left(1 - \mathfrak{F}^I_{M_{\mathcal{CFS}-1}}\right)^{\zeta} \\ +\left(1 - \mathfrak{F}^I_{M_{\mathcal{CFS}-1}}\right)^{\zeta}\left(1 - \mathfrak{F}^I_{M_{\mathcal{CFS}-2}}\right)^{\zeta} \end{pmatrix}\right) \end{pmatrix}\\
&= \left(1 - \left(1 - \mathfrak{F}^P_{M_{\mathcal{CFS}-1}}\right)^{\zeta}\left(1 - \mathfrak{F}^P_{M_{\mathcal{CFS}-2}}\right)^{\zeta} + \iota\left(1 - \left(1 - \mathfrak{F}^I_{M_{\mathcal{CFS}-1}}\right)^{\zeta}\left(1 - \mathfrak{F}^I_{M_{\mathcal{CFS}-2}}\right)^{\zeta}\right)\right)\\
&= \begin{pmatrix} 1 - \left(1 - \left(\mathfrak{F}^P_{M_{\mathcal{CFS}-1}} + \mathfrak{F}^P_{M_{\mathcal{CFS}-2}} - \mathfrak{F}^P_{M_{\mathcal{CFS}-1}}\mathfrak{F}^P_{M_{\mathcal{CFS}-2}}\right)\right)^{\zeta} \\ +\iota\left(1 - \left(1 - \left(\mathfrak{F}^I_{M_{\mathcal{CFS}-1}} + \mathfrak{F}^I_{M_{\mathcal{CFS}-2}} - \mathfrak{F}^I_{M_{\mathcal{CFS}-1}}\mathfrak{F}^I_{M_{\mathcal{CFS}-2}}\right)\right)^{\zeta}\right) \end{pmatrix}\\
&= \zeta\left(M_{\mathcal{CFS}-1} \oplus M_{\mathcal{CFS}-2}\right)
\end{aligned}
$$

3. From right-hand side, we have

$$
\left(M_{\mathcal{CFS}-1}\right)^{\zeta} = \left(\left(\mathfrak{F}^P_{M_{\mathcal{CFS}-1}}\right)^{\zeta} + \iota\left(\mathfrak{F}^I_{M_{\mathcal{CFS}-1}}\right)^{\zeta}\right)
$$

and

$$
\left(M_{\mathcal{CFS}-2}\right)^{\zeta} = \left(\left(\mathfrak{F}^P_{M_{\mathcal{CFS}-2}}\right)^{\zeta} + \iota\left(\mathfrak{F}^I_{M_{\mathcal{CFS}-2}}\right)^{\zeta}\right)
$$

then,

$$
\begin{aligned}
&(\mathrm{M}_{\mathcal{CFS}-1})^{\zeta} \otimes (\mathrm{M}_{\mathcal{CFS}-2})^{\zeta} \\
&\quad = \left(\left(\mathfrak{F}^{P}_{\mathrm{M}_{\mathcal{CFS}-1}}\right)^{\zeta} + \iota\left(\mathfrak{F}^{I}_{\mathrm{M}_{\mathcal{CFS}-1}}\right)^{\zeta}\right) \otimes \left(\left(\mathfrak{F}^{P}_{\mathrm{M}_{\mathcal{CFS}-2}}\right)^{\zeta} + \iota\left(\mathfrak{F}^{I}_{\mathrm{M}_{\mathcal{CFS}-2}}\right)^{\zeta}\right) \\
&\quad = \left(\mathfrak{F}^{P}_{\mathrm{M}_{\mathcal{CFS}-1}}\right)^{\zeta}\left(\mathfrak{F}^{P}_{\mathrm{M}_{\mathcal{CFS}-2}}\right)^{\zeta} + \iota\left(\mathfrak{F}^{I}_{\mathrm{M}_{\mathcal{CFS}-1}}\right)^{\zeta}\left(\mathfrak{F}^{I}_{\mathrm{M}_{\mathcal{CFS}-2}}\right)^{\zeta} \\
&\quad = \left(\left(\mathfrak{F}^{P}_{\mathrm{M}_{\mathcal{CFS}-1}}\mathfrak{F}^{P}_{\mathrm{M}_{\mathcal{CFS}-2}}\right)^{\zeta} + \iota\left(\mathfrak{F}^{I}_{\mathrm{M}_{\mathcal{CFS}-1}}\mathfrak{F}^{I}_{\mathrm{M}_{\mathcal{CFS}-2}}\right)^{\zeta}\right) \\
&\quad = (\mathrm{M}_{\mathcal{CFS}-1} \otimes \mathrm{M}_{\mathcal{CFS}-2})^{\zeta}
\end{aligned}
$$

4. Take right-hand side,

$$
\zeta_1 \mathrm{M}_{\mathcal{CFS}-1} = \left(1 - \left(1 - \mathfrak{F}^{P}_{\mathrm{M}_{\mathcal{CFS}-1}}\right)^{\zeta_1} + \iota\left(1 - \left(1 - \mathfrak{F}^{I}_{\mathrm{M}_{\mathcal{CFS}-1}}\right)^{\zeta_1}\right)\right)
$$

and

$$
\zeta_2 \mathrm{M}_{\mathcal{CFS}-1} = \left(1 - \left(1 - \mathfrak{F}^{P}_{\mathrm{M}_{\mathcal{CFS}-1}}\right)^{\zeta_2} + \iota\left(1 - \left(1 - \mathfrak{F}^{I}_{\mathrm{M}_{\mathcal{CFS}-1}}\right)^{\zeta_2}\right)\right)
$$

then,

$$
\begin{aligned}
&\zeta_1 \mathrm{M}_{\mathcal{CFS}-1} \oplus \zeta_2 \mathrm{M}_{\mathcal{CFS}-1} = \begin{pmatrix} 1 - \left(1 - \mathfrak{F}^{P}_{\mathrm{M}_{\mathcal{CFS}-1}}\right)^{\zeta_1} \\ +\iota\left(1 - \left(1 - \mathfrak{F}^{I}_{\mathrm{M}_{\mathcal{CFS}-1}}\right)^{\zeta_1}\right) \end{pmatrix} \oplus \begin{pmatrix} 1 - \left(1 - \mathfrak{F}^{P}_{\mathrm{M}_{\mathcal{CFS}-1}}\right)^{\zeta_1} \\ +\iota\left(1 - \left(1 - \mathfrak{F}^{I}_{\mathrm{M}_{\mathcal{CFS}-1}}\right)^{\zeta_1}\right) \end{pmatrix} \\
&= \begin{pmatrix} 1 - \left(1 - \mathfrak{F}^{P}_{\mathrm{M}_{\mathcal{CFS}-1}}\right)^{\zeta_1} + 1 - \left(1 - \mathfrak{F}^{P}_{\mathrm{M}_{\mathcal{CFS}-1}}\right)^{\zeta_1} - \left(1 - \left(1 - \mathfrak{F}^{P}_{\mathrm{M}_{\mathcal{CFS}-1}}\right)^{\zeta_1}\right)\left(1 - \left(1 - \mathfrak{F}^{P}_{\mathrm{M}_{\mathcal{CFS}-1}}\right)^{\zeta_1}\right) \\ +\iota\left(1 - \left(1 - \mathfrak{F}^{I}_{\mathrm{M}_{\mathcal{CFS}-1}}\right)^{\zeta_1} + 1 - \left(1 - \mathfrak{F}^{I}_{\mathrm{M}_{\mathcal{CFS}-1}}\right)^{\zeta_1} - \left(1 - \left(1 - \mathfrak{F}^{I}_{\mathrm{M}_{\mathcal{CFS}-1}}\right)^{\zeta_1}\right)\left(1 - \left(1 - \mathfrak{F}^{I}_{\mathrm{M}_{\mathcal{CFS}-1}}\right)^{\zeta_1}\right)\right) \end{pmatrix} \\
&= \left(1 - \left(1 - \mathfrak{F}^{P}_{\mathrm{M}_{\mathcal{CFS}-1}}\right)^{\zeta_1}\left(1 - \mathfrak{F}^{P}_{\mathrm{M}_{\mathcal{CFS}-1}}\right)^{\zeta_1} + \iota\left(1 - \left(1 - \mathfrak{F}^{I}_{\mathrm{M}_{\mathcal{CFS}-1}}\right)^{\zeta_1}\left(1 - \mathfrak{F}^{I}_{\mathrm{M}_{\mathcal{CFS}-1}}\right)^{\zeta_1}\right)\right) \\
&= \left(1 - \left(1 - \mathfrak{F}^{P}_{\mathrm{M}_{\mathcal{CFS}-1}}\right)^{\zeta_1+\zeta_2} + \iota\left(1 - \left(1 - \mathfrak{F}^{I}_{\mathrm{M}_{\mathcal{CFS}-1}}\right)^{\zeta_1+\zeta_2}\right)\right) \\
&= (\zeta_1 + \zeta_2)\mathrm{M}_{\mathcal{CFS}-1}
\end{aligned}
$$

5. Take right-hand side,

$$
(\mathrm{M}_{\mathcal{CFS}-1})^{\zeta_1} = \left(\left(\mathfrak{F}^{P}_{\mathrm{M}_{\mathcal{CFS}-1}}\right)^{\zeta_1} + \iota\left(\mathfrak{F}^{I}_{\mathrm{M}_{\mathcal{CFS}-1}}\right)^{\zeta_1}\right)
$$

and

$$
(\mathrm{M}_{\mathcal{CFS}-1})^{\zeta_1} = \left(\left(\mathfrak{F}^{P}_{\mathrm{M}_{\mathcal{CFS}-1}}\right)^{\zeta_2} + \iota\left(\mathfrak{F}^{I}_{\mathrm{M}_{\mathcal{CFS}-1}}\right)^{\zeta_2}\right)
$$

then,

$$
\begin{aligned}
&(\mathrm{M}_{\mathcal{CFS}-1})^{\zeta_1} \otimes (\mathrm{M}_{\mathcal{CFS}-1})^{\zeta_2} \\
&\quad = \left(\left(\mathfrak{F}^{P}_{\mathrm{M}_{\mathcal{CFS}-1}}\right)^{\zeta_1} + \iota\left(\mathfrak{F}^{I}_{\mathrm{M}_{\mathcal{CFS}-1}}\right)^{\zeta_1}\right) \otimes \left(\left(\mathfrak{F}^{P}_{\mathrm{M}_{\mathcal{CFS}-1}}\right)^{\zeta_2} + \iota\left(\mathfrak{F}^{I}_{\mathrm{M}_{\mathcal{CFS}-1}}\right)^{\zeta_2}\right)
\end{aligned}
$$

$$= \left(\mathfrak{F}^{P}_{\mathrm{M}_{\mathcal{CFS}-1}}\right)^{\zeta_1}\left(\mathfrak{F}^{P}_{\mathrm{M}_{\mathcal{CFS}-1}}\right)^{\zeta_2} + \iota\left(\mathfrak{F}^{I}_{\mathrm{M}_{\mathcal{CFS}-1}}\right)^{\zeta_1}\left(\mathfrak{F}^{I}_{\mathrm{M}_{\mathcal{CFS}-1}}\right)^{\zeta_2}$$
$$= \left(\left(\mathfrak{F}^{P}_{\mathrm{M}_{\mathcal{CFS}-1}}\right)^{\zeta_1+\zeta_2} + \iota\left(\mathfrak{F}^{I}_{\mathrm{M}_{\mathcal{CFS}-1}}\right)^{\zeta_1+\zeta_2}\right)$$
$$= (\mathrm{M}_{\mathcal{CFS}-1})^{\zeta_1+\zeta_2}$$

6. We have

$$\mathrm{M}^{\zeta_1}_{\mathcal{CFS}-1} = \left(\left(\mathfrak{F}^{P}_{\mathrm{M}_{\mathcal{CFS}-1}}\right)^{\zeta_1} + \iota\left(\mathfrak{F}^{I}_{\mathrm{M}_{\mathcal{CFS}-1}}\right)^{\zeta_1}\right)$$

$$\left(\mathrm{M}^{\zeta_1}_{\mathcal{CFS}-1}\right)^{\zeta_2} = \left(\left(\left(\mathfrak{F}^{P}_{\mathrm{M}_{\mathcal{CFS}-1}}\right)^{\zeta_1}\right)^{\zeta_2} + \iota\left(\left(\mathfrak{F}^{I}_{\mathrm{M}_{\mathcal{CFS}-1}}\right)^{\zeta_1}\right)^{\zeta_2}\right)$$
$$= \left(\left(\mathfrak{F}^{P}_{\mathrm{M}_{\mathcal{CFS}-1}}\right)^{\zeta_1\zeta_2} + \iota\left(\mathfrak{F}^{I}_{\mathrm{M}_{\mathcal{CFS}-1}}\right)^{\zeta_1\zeta_2}\right) = \mathrm{M}^{\zeta_1\zeta_2}_{\mathcal{CFS}-1}.$$

The formulas for determining the score value and accuracy value for a CFN in the cartesian structure are respectively interpreted below

$$\mathbb{S}(\mathrm{M}_{\mathcal{CFS}}) = \frac{1}{2}\left(1 + \mathfrak{F}^{P}_{\mathrm{M}_{\mathcal{CFS}}} - \mathfrak{F}^{I}_{\mathrm{M}_{\mathcal{CFS}}}\right) \tag{1.3}$$

$$\mathbb{H}(\mathrm{M}_{\mathcal{CFS}}) = \frac{\mathfrak{F}^{P}_{\mathrm{M}_{\mathcal{CFS}}} + \mathfrak{F}^{I}_{\mathrm{M}_{\mathcal{CFS}}}}{2} \tag{1.4}$$

Noted that $\mathbb{S}(\mathrm{M}_{\mathcal{CFS}}), \mathbb{H}(\mathrm{M}_{\mathcal{CFS}}) \in [0, 1]$.

From Eq. (1.3) and (1.4), the underneath outcomes were achieved.

1. If $\mathbb{S}(\mathrm{M}_{\mathcal{CFS}-1}) < \mathbb{S}(\mathrm{M}_{\mathcal{CFS}-2})$, then $\mathrm{M}_{\mathcal{CFS}-1} < \mathrm{M}_{\mathcal{CFS}-2}$;
2. If $\mathbb{S}(\mathrm{M}_{\mathcal{CFS}-1}) > \mathbb{S}(\mathrm{M}_{\mathcal{CFS}-2})$, then $\mathrm{M}_{\mathcal{CFS}-1} > \mathrm{M}_{\mathcal{CFS}-2}$;
3. If $\mathbb{S}(\mathrm{M}_{\mathcal{CFS}-1}) = \mathbb{S}(\mathrm{M}_{\mathcal{CFS}-2})$, then if $\mathbb{H}(\mathrm{M}_{\mathcal{CFS}-1}) < \mathbb{H}(\mathrm{M}_{\mathcal{CFS}-2})$, then $\mathrm{M}_{\mathcal{CFS}-1} < \mathrm{M}_{\mathcal{CFS}-2}$; if $\mathbb{H}(\mathrm{M}_{\mathcal{CFS}-1}) > \mathbb{H}(\mathrm{M}_{\mathcal{CFS}-2})$, then $\mathrm{M}_{\mathcal{CFS}-1} > \mathrm{M}_{\mathcal{CFS}-2}$; if $\mathbb{H}(\mathrm{M}_{\mathcal{CFS}-1}) = \mathbb{H}(\mathrm{M}_{\mathcal{CFS}-2})$, then $\mathrm{M}_{\mathcal{CFS}-1} = \mathrm{M}_{\mathcal{CFS}-2}$.

Chapter 2
Bipolar Complex Fuzzy Sets and Their Fundamental Properties

Due to the inability of FSs, CFSs and BFSs to solve the problems involving information of both positive and negative aspects of the elements or objects as well as the 2nd dimension (extra fuzzy information) of objects, and due to the serious shortcomings of the existing notions of BCFS and complex bipolar fuzzy sets, Mahmood and Rehman [7] initiated one of the most improved and generalized FS models which they named BCFS. In the structure of BCFS, each element or object has positive and negative belonging grades, which are placed in the unit square in first quadrant of the complex plane. Numerous fields, including artificial intelligence, computer science, pattern recognition, decision-making, and medical diagnostics, use the concept of BCFS. It provides a more flexible and realistic method of describing and considering uncertainty and imprecision in a range of settings, making it a crucial tool for dealing with complex and ambiguous data in the modern, complex, and technologically advanced world.

In Sect. 2.1, we discuss the prevailing notions of BCFS and complex bipolar fuzzy sets (CBFS) interpreted by various researchers. In Sect. 2.2, we interpret a novel approach of BCFS initiated by Mahmood and Rehman [7]. Further, we revise the concept of Type-1 partially BCFS, and Type-2 partially BCFS in Sect. 2.2. In Sect. 2.3, we review the fundamental operations and concepts of BCFS such as union, intersection, complement, algebraic structures, SM, AOs, etc.

2.1 Existing Notions of BCFS and CBFS

We discuss the BCFS, its complement, intersection, and union, demonstrated by Alkouri et al. [8] as follows:

Definition 2.1 [8] A BCFS $\Re$. on $\mathbb{E}$ is of the form,

U. ur Rehman and T. Mahmood, *Bipolar Complex Fuzzy Sets*, Studies in Fuzziness and Soft Computing 439, https://doi.org/10.1007/978-3-032-07388-4_2

$$\Re = \{(e, \mathfrak{F}^{P}_{\Re}(e), \mathfrak{F}^{N}_{\Re}(e)) | e \in \mathbb{E}\}$$
$$= \left\{\left(e, \Gamma^{P}_{\Re}(e)e^{\iota\delta\ \omega^{P}_{\Re}(e)}, \Gamma^{N}_{\Re}(e)e^{\iota\delta\ \omega^{N}_{\Re}(e)}\right) | e \in \mathbb{E}\right\}$$

Noticed that $\Gamma^{P}_{\Re}(e)$ and $\omega^{P}_{\Re}(e)$ are recognized as amplitude and phase terms of positive belonging grade $\mathfrak{F}^{P}_{\Re}(e)$ respectively, and $\Gamma^{N}_{\Re}(e)$ and $\omega^{N}_{\Re}(e)$ are recognized as amplitude and phase terms of a negative belonging grade $\mathfrak{F}^{P}_{\Re}(e)$ respectively. Further, $\iota = \sqrt{-1}, \delta \in (0, 2\pi], \Gamma^{P}_{\Re}(e), \omega^{P}_{\Re}(e) \in [0, 1]$, and. Moreover, in this definition, both positive and negative belonging grades are mappings $\mathfrak{F}^{P}_{\Re}(e) : \mathbb{E} \to [0, 1]e^{\iota\delta[0,1]}$, $\mathfrak{F}^{N}_{\Re}(e) : \mathbb{E} \to [-1, 0]e^{\iota\delta[-1,0]}$, which is placed in the unit square of the complex plane.

For two BCFSs $\Re_1 = \{(e, \mathfrak{F}^{P}_{\Re_1}(e), \mathfrak{F}^{N}_{\Re_1}(e)) | e \in \mathbb{E}\} = \left\{\left(e, \Gamma^{P}_{\Re_1}(e)e^{\iota\delta\ \omega^{P}_{\Re_1}(e)}, \Gamma^{N}_{\Re_1}(e)e^{\iota\delta\ \omega^{N}_{\Re_1}(e)}\right) | e \in \mathbb{E}\right\}$ and $\Re_2 = \{(e, \mathfrak{F}^{P}_{\Re_2}(e), \mathfrak{F}^{N}_{\Re_2}(e)) | e \in \mathbb{E}\} = \left\{\left(e, \Gamma^{P}_{\Re_2}(e)e^{\iota\delta\ \omega^{P}_{\Re_2}(e)}, \Gamma^{N}_{\Re_2}(e)e^{\iota\delta\ \omega^{N}_{\Re_2}(e)}\right) | e \in \mathbb{E}\right\}$, we have.

1. $(\Re_1)^c = \left\{\left(e, 1 - \Gamma^{P}_{\Re_1}(e)e^{\iota\delta\left(1-\omega^{P}_{\Re_1}(e)\right)}, -1 - \Gamma^{N}_{\Re_1}(e)e^{\iota\delta\left(-1-\omega^{N}_{\Re_1}(e)\right)}\right) | e \in \mathbb{E}\right\}$
2. $\Re_1 \cup \Re_2 = \left\{\left(\begin{array}{c} e, \max\left(\Gamma^{P}_{\Re_1}(e), \Gamma^{P}_{\Re_2}(e)\right)e^{\iota\delta\max\left(\omega^{P}_{\Re_1}(e), \omega^{P}_{\Re_2}(e)\right)}, \\ \min\left(\Gamma^{P}_{\Re_1}(e), \Gamma^{P}_{\Re_2}(e)\right)e^{\iota\delta\min\left(\omega^{P}_{\Re_1}(e), \omega^{P}_{\Re_2}(e)\right)} \end{array}\right) | e \in \mathbb{E}\right\}$
3. $.\Re_1 \cap \Re_2 = \left\{\left(\begin{array}{c} e, \min\left(\Gamma^{P}_{\Re_1}(e), \Gamma^{P}_{\Re_2}(e)\right)e^{\iota\delta\min\left(\omega^{P}_{\Re_1}(e), \omega^{P}_{\Re_2}(e)\right)}, \\ \max\left(\Gamma^{P}_{\Re_1}(e), \Gamma^{P}_{\Re_2}(e)\right)e^{\iota\delta\max\left(\omega^{P}_{\Re_1}(e), \omega^{P}_{\Re_2}(e)\right)} \end{array}\right) | e \in \mathbb{E}\right\}$

Underneath we demonstrate the notion of BCFS, its complement, union, and intersection deduced by Al-Husban et al. [9].

Definition 2.2 [9] A BCFS $\Re$ on $\mathbb{E}$ is deduced as

$$\Re = \{(e, \mathfrak{F}^{P}_{\Re}(e), \mathfrak{F}^{N}_{\Re}(e)) | e \in \mathbb{E}\}$$
$$= \left\{\left(e, \Gamma^{P}_{\Re}(e)e^{\iota\ \omega^{P}_{\Re}(e)}, \Gamma^{N}_{\Re}(e)e^{\iota\ \omega^{N}_{\Re}(e)}\right) | e \in \mathbb{E}\right\}$$

Noticed that $\Gamma^{P}_{\Re}(e)$ and $\omega^{P}_{\Re}(e)$ are recognized as amplitude and phase terms of positive belonging grade $\mathfrak{F}^{P}_{\Re}(e)$ respectively, and $\Gamma^{N}_{\Re}(e)$ and $\omega^{N}_{\Re}(e)$ are recognized as amplitude and phase terms of a negative belonging grade $\mathfrak{F}^{N}_{\Re}(e)$ respectively. Further, $\iota = \sqrt{-1}, \omega^{P}_{\Re}(e), \omega^{N}_{\Re}(e) \in (0, 2\pi], \Gamma^{P}_{\Re}(e), \in [0, 1]$, and $\Gamma^{N}_{\Re}(e), \in [-1, 0]$. Both positive and negative belonging grades are placed in a unit square of a complex plane.

For two BCFSs $\Re_1 = \{(e, \mathfrak{F}^{P}_{\Re_1}(e), \mathfrak{F}^{N}_{\Re_1}(e)) | e \in \mathbb{E}\} = \left\{\left(e, \Gamma^{P}_{\Re_1}(e)e^{\iota\ \omega^{P}_{\Re_1}(e)}, \Gamma^{N}_{\Re_1}(e)e^{\iota\ \omega^{N}_{\Re_1}(e)}\right) | e \in \mathbb{E}\right\}$ and $\Re_2 = \{(e, \mathfrak{F}^{P}_{\Re_2}(e), \mathfrak{F}^{N}_{\Re_2}(e)) | e \in \mathbb{E}\} = \left\{\left(e, \Gamma^{P}_{\Re_2}(e)e^{\iota\ \omega^{P}_{\Re_2}(e)}, \Gamma^{N}_{\Re_2}(e)e^{\iota\ \omega^{N}_{\Re_2}(e)}\right) | e \in \mathbb{E}\right\}$, we have.

1. $(\Re_1)^c = \left\{\left(e, 1 - \Gamma^{P}_{\Re_1}(e)e^{\iota\left(2\pi-\omega^{P}_{\Re_1}(e)\right)}, -1 - \Gamma^{N}_{\Re_1}(e)e^{\iota\left(2\pi-\omega^{N}_{\Re_1}(e)\right)}\right) | e \in \mathbb{E}\right\}$

2. $\mathfrak{R}_1 \cup \mathfrak{R}_2 = \left\{ \begin{pmatrix} e, \max\left(\Gamma^P_{\mathfrak{R}_1}(e), \Gamma^P_{\mathfrak{R}_2}(e)\right) e^{\iota \max\left(\omega^{\mathrm{P}}_{\mathfrak{R}_1}(\mathrm{e}), \omega^{\mathrm{P}}_{\mathfrak{R}_2}(\mathrm{e})\right)}, \\ \min\left(\Gamma^P_{\mathfrak{R}_1}(e), \Gamma^P_{\mathfrak{R}_2}(e)\right) e^{\iota \min\left(\omega^{\mathrm{P}}_{\mathfrak{R}_1}(\mathrm{e}), \omega^{\mathrm{P}}_{\mathfrak{R}_2}(\mathrm{e})\right)} \end{pmatrix} | \mathrm{e} \in \mathbb{E} \right\}$
3. $\mathfrak{R}_1 \cap \mathfrak{R}_2 = \left\{ \begin{pmatrix} e, \min\left(\Gamma^P_{\mathfrak{R}_1}(e), \Gamma^P_{\mathfrak{R}_2}(e)\right) e^{\iota \min\left(\omega^{\mathrm{P}}_{\mathfrak{R}_1}(\mathrm{e}), \omega^{\mathrm{P}}_{\mathfrak{R}_2}(\mathrm{e})\right)}, \\ \max\left(\Gamma^P_{\mathfrak{R}_1}(e), \Gamma^P_{\mathfrak{R}_2}(e)\right) e^{\iota \max\left(\omega^{\mathrm{P}}_{\mathfrak{R}_1}(\mathrm{e}), \omega^{\mathrm{P}}_{\mathfrak{R}_2}(\mathrm{e})\right)} \end{pmatrix} | \mathrm{e} \in \mathbb{E} \right\}.$

Underneath we demonstrate the notion of CBFS, its complement, union and intersection, and operational laws deduced by Gulistan et al. [10].

Definition 2.3 [10] A CBFS $\mathfrak{R}$ on $\mathbb{E}$ is deduced as

$$\begin{aligned} \mathfrak{R} &= \left\{\left(\mathrm{e}, \mathfrak{F}^{\mathrm{P}}_{\mathfrak{R}}(\mathrm{e}), \mathfrak{F}^{\mathrm{N}}_{\mathfrak{R}}(\mathrm{e})\right) | \mathrm{e} \in \mathbb{E}\right\} \\ &= \left\{\left(e, \Gamma^{\mathrm{P}}_{\mathfrak{R}}(\mathrm{e}) \mathrm{e}^{\iota \omega^{\mathrm{P}}_{\mathfrak{R}}(\mathrm{e})}, \Gamma^{\mathrm{N}}_{\mathfrak{R}}(\mathrm{e}) \mathrm{e}^{\iota \omega^{\mathrm{N}}_{\mathfrak{R}}(\mathrm{e})}\right) | \mathrm{e} \in \mathbb{E}\right\} \end{aligned}$$

Noticed that $\Gamma^P_{\mathfrak{R}}(e)$ and $\omega^{\mathrm{P}}_{\mathfrak{R}}(\mathrm{e})$ are recognized as amplitude and phase terms of positive belonging grade $\mathfrak{F}^{\mathrm{P}}_{\mathfrak{R}}(\mathrm{e})$ respectively, and $\Gamma^N_{\mathfrak{R}}(e)$ and $\omega^{\mathrm{N}}_{\mathfrak{R}}(\mathrm{e})$ are recognized as amplitude and phase terms of a negative belonging grade $\mathfrak{F}^{\mathrm{N}}_{\mathfrak{R}}(\mathrm{e})$ respectively. Further, $\iota = \sqrt{-1}$, $\omega^{\mathrm{P}}_{\mathfrak{R}}(\mathrm{e}), \in [0, 2\pi]$, $\omega^{\mathrm{N}}_{\mathfrak{R}}(\mathrm{e}) \in [-2\pi, 0]$ $\Gamma^P_{\mathfrak{R}}(e), \in [0, 1]$, and $\Gamma^N_{\mathfrak{R}}(e), \in [-1, 0]$ with conditions $-1 \le \Gamma^P_{\mathfrak{R}}(e) + \Gamma^N_{\mathfrak{R}}(e) \le 1$ and $-2\pi \le \omega^{\mathrm{P}}_{\mathfrak{R}}(\mathrm{e}) + \omega^{\mathrm{N}}_{\mathfrak{R}} \le 2\pi$. The complex bipolar fuzzy number (CBFN) would be recognized as $\mathfrak{R} = \left(\mathrm{e}, \mathfrak{F}^{\mathrm{P}}_{\mathfrak{R}}, \mathfrak{F}^{\mathrm{N}}_{\mathfrak{R}}\right) = \left(\Gamma^{\mathrm{P}}_{\mathfrak{R}} \mathrm{e}^{\iota\, \omega^{\mathrm{P}}_{\mathfrak{R}}}, \Gamma^{\mathrm{N}}_{\mathfrak{R}} \mathrm{e}^{\iota\, \omega^{\mathrm{N}}_{\mathfrak{R}}}\right)$.

For two BCFSs $\mathfrak{R}_1 = \left\{\left(\mathrm{e}, \mathfrak{F}^{\mathrm{P}}_{\mathfrak{R}_1}(\mathrm{e}), \mathfrak{F}^{\mathrm{N}}_{\mathfrak{R}_1}(\mathrm{e})\right) | \mathrm{e} \in \mathbb{E}\right\} = \left\{\left(\mathrm{e}, \Gamma^{\mathrm{P}}_{\mathfrak{R}_1}(\mathrm{e}) \mathrm{e}^{\iota\, \omega^{\mathrm{P}}_{\mathfrak{R}_1}(\mathrm{e})}, \Gamma^{\mathrm{N}}_{\mathfrak{R}_1}(\mathrm{e}) \mathrm{e}^{\iota\, \omega^{\mathrm{N}}_{\mathfrak{R}_1}(\mathrm{e})}\right) | \mathrm{e} \in \mathbb{E}\right\}$ and $\mathfrak{R}_2 = \left\{\left(\mathrm{e}, \mathfrak{F}^{\mathrm{P}}_{\mathfrak{R}_2}(\mathrm{e}), \mathfrak{F}^{\mathrm{N}}_{\mathfrak{R}_2}(\mathrm{e})\right) | \mathrm{e} \in \mathbb{E}\right\} = \left\{\left(\mathrm{e}, \Gamma^{\mathrm{P}}_{\mathfrak{R}_2}(\mathrm{e}) \mathrm{e}^{\iota\, \omega^{\mathrm{P}}_{\mathfrak{R}_2}(\mathrm{e})}, \Gamma^{\mathrm{N}}_{\mathfrak{R}_2}(\mathrm{e}) \mathrm{e}^{\iota\, \omega^{\mathrm{N}}_{\mathfrak{R}_2}(\mathrm{e})}\right) | \mathrm{e} \in \mathbb{E}\right\}$, we have.

1. $(\mathfrak{R}_1)^c = \left\{\left(e, 1 - \Gamma^{\mathrm{P}}_{\mathfrak{R}_1}(\mathrm{e}) \mathrm{e}^{\iota\left(2\pi - \omega^{\mathrm{P}}_{\mathfrak{R}_1}(\mathrm{e})\right)}, -1 - \Gamma^{\mathrm{N}}_{\mathfrak{R}_1}(\mathrm{e}) \mathrm{e}^{\iota\left(-2\pi - \omega^{\mathrm{N}}_{\mathfrak{R}_1}(\mathrm{e})\right)}\right) | \mathrm{e} \in \mathbb{E}\right\}$
2. $\mathfrak{R}_1 \cup \mathfrak{R}_2 = \left\{ \begin{pmatrix} e, \max\left(\Gamma^P_{\mathfrak{R}_1}(e), \Gamma^P_{\mathfrak{R}_2}(e)\right) e^{\iota \max\left(\omega^{\mathrm{P}}_{\mathfrak{R}_1}(\mathrm{e}), \omega^{\mathrm{P}}_{\mathfrak{R}_2}(\mathrm{e})\right)}, \\ \min\left(\Gamma^P_{\mathfrak{R}_1}(e), \Gamma^P_{\mathfrak{R}_2}(e)\right) e^{\iota \min\left(\omega^{\mathrm{P}}_{\mathfrak{R}_1}(\mathrm{e}), \omega^{\mathrm{P}}_{\mathfrak{R}_2}(\mathrm{e})\right)} \end{pmatrix} | \mathrm{e} \in \mathbb{E} \right\}$

3 $\mathfrak{R}_1 \cap \mathfrak{R}_2 = \left\{ \begin{pmatrix} e, \min\left(\Gamma^P_{\mathfrak{R}_1}(e), \Gamma^P_{\mathfrak{R}_2}(e)\right) e^{\iota \min\left(\omega^{\mathrm{P}}_{\mathfrak{R}_1}(\mathrm{e}), \omega^{\mathrm{P}}_{\mathfrak{R}_2}(\mathrm{e})\right)}, \\ \max\left(\Gamma^P_{\mathfrak{R}_1}(e), \Gamma^P_{\mathfrak{R}_2}(e)\right) e^{\iota \max\left(\omega^{\mathrm{P}}_{\mathfrak{R}_1}(\mathrm{e}), \omega^{\mathrm{P}}_{\mathfrak{R}_2}(\mathrm{e})\right)} \end{pmatrix} | \mathrm{e} \in \mathbb{E} \right\}$

For two CBFNs $\mathfrak{R}_1 = \left(\mathrm{e}, \mathfrak{F}^{\mathrm{P}}_{\mathfrak{R}_1}, \mathfrak{F}^{\mathrm{N}}_{\mathfrak{R}_1}\right) = \left(\Gamma^{\mathrm{P}}_{\mathfrak{R}_1} \mathrm{e}^{\iota\, \omega^{\mathrm{P}}_{\mathfrak{R}_1}}, \Gamma^{\mathrm{N}}_{\mathfrak{R}_1} \mathrm{e}^{\iota\, \omega^{\mathrm{N}}_{\mathfrak{R}_1}}\right)$ and $\mathfrak{R}_2 = \left(\mathrm{e}, \mathfrak{F}^{\mathrm{P}}_{\mathfrak{R}_2}, \mathfrak{F}^{\mathrm{N}}_{\mathfrak{R}_2}\right) = \left(\Gamma^{\mathrm{P}}_{\mathfrak{R}_2} \mathrm{e}^{\iota\, \omega^{\mathrm{P}}_{\mathfrak{R}_2}}, \Gamma^{\mathrm{N}}_{\mathfrak{R}_2} \mathrm{e}^{\iota\, \omega^{\mathrm{N}}_{\mathfrak{R}_2}}\right)$, and $\zeta > 0$ the operational laws are demonstrated as.

1. $\mathfrak{R}_1 \oplus \mathfrak{R}_2 = \left(\Gamma^{\mathrm{P}}_{\mathfrak{R}_1} \mathrm{e}^{\iota \omega^{\mathrm{P}}_{\mathfrak{R}_1}} + \Gamma^{\mathrm{P}}_{\mathfrak{R}_2} \mathrm{e}^{\iota \omega^{\mathrm{P}}_{\mathfrak{R}_2}} - \left(\Gamma^{\mathrm{P}}_{\mathfrak{R}_1} \Gamma^{\mathrm{P}}_{\mathfrak{R}_2}\right) \mathrm{e}^{\iota\left(\omega^{\mathrm{P}}_{\mathfrak{R}_1} \omega^{\mathrm{P}}_{\mathfrak{R}_2}\right)}, -\left|\Gamma^{\mathrm{N}}_{\mathfrak{R}_1} \mathrm{e}^{\iota \omega^{\mathrm{N}}_{\mathfrak{R}_1}}\right| \left|\Gamma^{\mathrm{N}}_{\mathfrak{R}_2} \mathrm{e}^{\iota\, \omega^{\mathrm{N}}_{\mathfrak{R}_2}}\right|\right)$
2. $\mathfrak{R}_1 \otimes \mathfrak{R}_2 = \left(\left(\Gamma^{\mathrm{P}}_{\mathfrak{R}_1} \Gamma^{\mathrm{P}}_{\mathfrak{R}_2}\right) \mathrm{e}^{\iota\left(\omega^{\mathrm{P}}_{\mathfrak{R}_1} \omega^{\mathrm{P}}_{\mathfrak{R}_2}\right)}, \Gamma^{\mathrm{N}}_{\mathfrak{R}_1} \mathrm{e}^{\iota \omega^{\mathrm{N}}_{\mathfrak{R}_1}} + \Gamma^{\mathrm{N}}_{\mathfrak{R}_2} \mathrm{e}^{\iota \omega^{\mathrm{N}}_{\mathfrak{R}_2}} - \left(\Gamma^{\mathrm{N}}_{\mathfrak{R}_1} \Gamma^{\mathrm{N}}_{\mathfrak{R}_2}\right) \mathrm{e}^{\iota\left(\omega^{\mathrm{N}}_{\mathfrak{R}_1} \omega^{\mathrm{N}}_{\mathfrak{R}_2}\right)}\right)$

3. $\zeta \mathfrak{R}_1 = \left(1 - \left(1 - \Gamma^{P}_{\mathfrak{R}_1} e^{\iota\, \omega^{P}_{\mathfrak{R}_1}}\right)^{\zeta}, -\left|\Gamma^{N}_{\mathfrak{R}_1} e^{\iota\, \omega^{N}_{\mathfrak{R}_1}}\right|^{\zeta}\right)$
4. $(\mathfrak{R}_1)^{\zeta} = \left(\left(\Gamma^{P}_{\mathfrak{R}_1} e^{\iota\, \omega^{P}_{\mathfrak{R}_1}}\right)^{\zeta}, -1 + \left|-1 + \Gamma^{N}_{\mathfrak{R}_1} e^{\iota\, \omega^{N}_{\mathfrak{R}_1}}\right|^{\zeta}\right)$

2.2 A Novel Approach of BCFS

Mahmood and Rehman noted that in the above three definitions of BCFS and CBFS the amplitude term of negative belonging grades $\mathfrak{F}^{N}_{\mathfrak{R}}(e)$ belongs to $[-1, 0]$. This implies that $\mathfrak{F}^{N}_{\mathfrak{R}}(e)$ can take a negative value which is a contradiction to the complex number theory because in the complex number theory a complex number $\mathcal{Z}$ can be represented in two forms i.e., cartesian form and polar form. In cartesian form $\mathcal{Z}$ is in the shape $\mathcal{Z} = \mathfrak{F} + \iota\tau$ and polar form $\mathcal{Z}$ is in the shape $\mathcal{Z} = \gamma e^{\iota\theta}$, where $\gamma = |\mathcal{Z}| = \sqrt{\mathfrak{F}^2 + \tau^2}$ and $\theta = \tan^{-1}\left(\frac{\tau}{\mathfrak{F}}\right)$. it is clear from the polar form of a complex number that the amplitude term γ cannot be negative. To address this shortcoming, Mahmood and Rehman [7] initiated a novel theory of BCFS which is not only accurate but also is closest to the concept of bipolarity.

Definition 2.4. [7] A BCFS $\mathfrak{R}$ is stated as

$$\mathfrak{R} = \left\{\left(e, \mathfrak{F}^{P}_{\mathfrak{R}}(e), \mathfrak{F}^{N}_{\mathfrak{R}}(e)\right) | e \in \mathbb{E}\right\}$$

$$\left\{\left(e, \mathfrak{F}^{RP}_{\mathfrak{R}}(e) + \iota\mathfrak{F}^{IP}_{\mathfrak{R}}(e), \mathfrak{F}^{RN}_{\mathfrak{R}}(e) + \iota\mathfrak{F}^{IN}_{\mathfrak{R}}(e)\right) | e \in \mathbb{E}\right\}$$

where $\mathfrak{F}^{RP}_{\mathfrak{R}}(e)$ is real and $\mathfrak{F}^{IP}_{\mathfrak{R}}(e)$ are unreal parts of positive belonging grades and $\mathfrak{F}^{RN}_{\mathfrak{R}}(e)$ is real and $\mathfrak{F}^{IN}_{\mathfrak{R}}(e)$ are unreal parts of a negative belonging grade, where both positive and negative belonging grades are placed in a unit square in first quadrant of the complex plane, $\mathfrak{F}^{RP}_{\mathfrak{R}}(e), \mathfrak{F}^{IP}_{\mathfrak{R}}(e) \in [0, 1]$, $\mathfrak{F}^{RN}_{\mathfrak{R}}(e), \mathfrak{F}^{IN}_{\mathfrak{R}}(e) \in [-1, 0]$ and $\iota = \sqrt{-1}$.

For easiness, in this book, the BCF number would be interpreted by $\mathfrak{R} = \left(\mathfrak{F}^{P}_{\mathfrak{R}}, \mathfrak{F}^{N}_{\mathfrak{R}}\right) = \left(\mathfrak{F}^{RP}_{\mathfrak{R}} + \iota\mathfrak{F}^{IP}_{\mathfrak{R}}, \mathfrak{F}^{RN}_{\mathfrak{R}} + \iota\mathfrak{F}^{IN}_{\mathfrak{R}}\right)$ until stated otherwise.

In the above definition, Mahmood and Rehman demonstrated the theory of BCFS in the Cartesian structure to overcome the issue of negative absolute value. The geometrical interpretation of the definition of BCFS deduced by Akouri et al. [8] and Mahmood and Rehman [7] is established in Figs. 1.1. and 1.2.(Fig. 2.1) (Fig. 2.2)

While developing the novel concept of BCFS, Mahmood, and Rehman also deduced the theory of T-1PBCFS and T-2PBCFS.

Definition 2.5. [7] A T-1PBCFS $\mathfrak{R}$ is deduced as

$$\mathfrak{R} = \left\{\left(e, \left(\mathfrak{F}^{P}_{\mathfrak{R}}(e), \mathfrak{F}^{N}_{\mathfrak{R}}(e)\right)\right) | e \in \mathbb{E}\right\}$$

$$\left\{\left(e, \left(\mathfrak{F}^{RP}_{\mathfrak{R}}(e) + \iota\mathfrak{F}^{IP}_{\mathfrak{R}}(e), \mathfrak{F}^{RN}_{\mathfrak{R}}(e) + \iota\mathfrak{F}^{IN}_{\mathfrak{R}}(e)\right)\right) | e \in \mathbb{E}\right\}$$

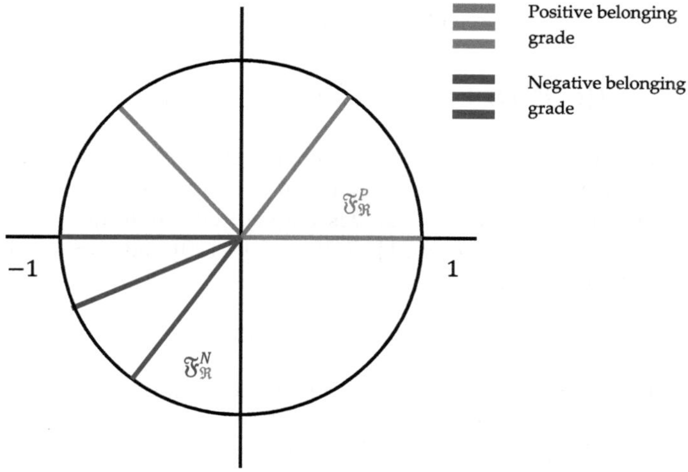

Fig. 2.1 The graphical demonstration of BCFS defined by Alkouri et al. [8]

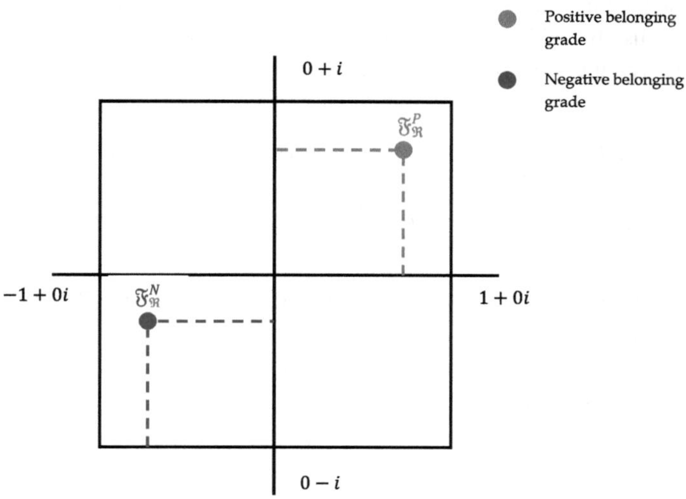

Fig. 2.2 The graphical demonstration of BCFS deduced by Mahmood and Rehman [7]

Noted that $\mathfrak{F}_{\mathfrak{R}}^{RP}(e)$ is real and $\mathfrak{F}_{\mathfrak{R}}^{IP}(e)$ are unreal parts of positive belonging grades and $\mathfrak{F}_{\mathfrak{R}}^{RN}(e)$ is real and $\mathfrak{F}_{\mathfrak{R}}^{IN}(e)$ are unreal parts of a negative belonging grade, where both positive and negative belonging grades are placed in a unit square of a complex plane, $\mathfrak{F}_{\mathfrak{R}}^{RP}(e), \mathfrak{F}_{\mathfrak{R}}^{IP}(e), \mathfrak{F}_{\mathfrak{R}}^{IN}(e) \in [0, 1]$, $\mathfrak{F}_{\mathfrak{R}}^{RN}(e) \in [-1, 0]$ and $\iota = \sqrt{-1}$. Further, $\mathfrak{F}_{\mathfrak{R}}^{P}(e) : \mathbb{E} \to [0, 1] + \iota[0, 1]$ and $\mathfrak{F}_{\mathfrak{R}}^{N}(e) : \mathbb{E} \to [-1, 0] + \iota[1, 0]$

Definition 2.6. [7] A T-2PBCFS $\mathfrak{R}$ is deduced as

$$\mathfrak{R} = \left\{\left(\mathrm{e}, \mathfrak{F}^{\mathrm{P}}_{\mathfrak{R}}(\mathrm{e}), \mathfrak{F}^{\mathrm{N}}_{\mathfrak{R}}(\mathrm{e})\right) | \mathrm{e} \in \mathbb{E}\right\}$$

$$\left\{\left(e, \mathfrak{F}^{\mathrm{RP}}_{\mathfrak{R}}(\mathrm{e}) + \iota\mathfrak{F}^{\mathrm{IP}}_{\mathfrak{R}}(\mathrm{e}), \mathfrak{F}^{\mathrm{RN}}_{\mathfrak{R}}(\mathrm{e}) + \iota\mathfrak{F}^{\mathrm{IN}}_{\mathfrak{R}}(\mathrm{e})\right) | \mathrm{e} \in \mathbb{E}\right\}$$

Noted that $\mathfrak{F}^{\mathrm{RP}}_{\mathfrak{R}}(\mathrm{e})$ is real and $\mathfrak{F}^{\mathrm{IP}}_{\mathfrak{R}}(\mathrm{e})$ are unreal parts of positive belonging grades and $\mathfrak{F}^{\mathrm{RN}}_{\mathfrak{R}}(\mathrm{e})$ is real and $\mathfrak{F}^{\mathrm{IN}}_{\mathfrak{R}}(\mathrm{e})$ are unreal parts of a negative belonging grade, where both positive and negative belonging grades are placed in a unit square of a complex plane, $\mathfrak{F}^{\mathrm{RP}}_{\mathfrak{R}}(\mathrm{e}), \mathfrak{F}^{\mathrm{IP}}_{\mathfrak{R}}(\mathrm{e}), \mathfrak{F}^{\mathrm{RN}}_{\mathfrak{R}}(\mathrm{e}) \in [0, 1]$, $\mathfrak{F}^{\mathrm{IN}}_{\mathfrak{R}}(\mathrm{e}) \in [-1, 0]$ and $\iota = \sqrt{-1}$. Further, $\mathfrak{F}^{\mathrm{P}}_{\mathfrak{R}}(\mathrm{e}) : \mathbb{E} \to [0, 1] + \iota[0, 1]$ and $\mathfrak{F}^{\mathrm{N}}_{\mathfrak{R}}(\mathrm{e}) : \mathbb{E} \to [0, 1] + \iota[-1, 0]$

Theorem 2.1 Every BFS can be demonstrated as BCFS.

Proof: Let $\mathfrak{R}$ be a BFS characterized by a positive belonging grade $\mathfrak{F}^{\mathrm{P}}_{\mathfrak{R}}(\mathrm{e})$ and negative belonging grade $\mathfrak{F}^{\mathrm{N}}_{\mathfrak{R}}(\mathrm{e})$. The change of a BFS $\mathfrak{R}$ to a BCFS is accomplished by placing $\mathfrak{F}^{\mathrm{P}}_{\mathfrak{R}}(\mathrm{e}) = \mathfrak{F}^{\mathrm{RP}}_{\mathfrak{R}}(\mathrm{e})$, $\mathfrak{F}^{\mathrm{N}}_{\mathfrak{R}}(\mathrm{e}) = \mathfrak{F}^{\mathrm{RN}}_{\mathfrak{R}}(\mathrm{e})$, and put the unreal part equal to zero $\forall \mathrm{e} \in \mathbb{E}$.

The proposed BCFS is converted to BFS without the need for an imaginary part. The BCFS thus modifies BFS by asserting that, in several instances, a second dimension must be added to the expression of positive and negative belonging grades However, the fundamental concept of BFS remains unchanged by the second dimension.

Definition 2.7 Let $\mathfrak{R}_1 = \left\{\left(\mathrm{e}, \mathfrak{F}^{\mathrm{P}}_{\mathfrak{R}_1}(\mathrm{e}), \mathfrak{F}^{\mathrm{N}}_{\mathfrak{R}_1}(\mathrm{e})\right) | \mathrm{e} \in \mathbb{E}\right\} = \left\{\left(\mathrm{e}, \mathfrak{F}^{\mathrm{RP}}_{\mathfrak{R}_1}(\mathrm{e}) + \iota\mathfrak{F}^{\mathrm{IP}}_{\mathfrak{R}_1}(\mathrm{e}), \mathfrak{F}^{\mathrm{RN}}_{\mathfrak{R}_1}(\mathrm{e}) + \iota\mathfrak{F}^{\mathrm{IN}}_{\mathfrak{R}_1}(\mathrm{e})\right) | \mathrm{e} \in \mathbb{E}\right\}$.and $\mathfrak{R}_2 = \left\{\left(\mathrm{e}, \mathfrak{F}^{\mathrm{P}}_{\mathfrak{R}_2}(\mathrm{e}), \mathfrak{F}^{\mathrm{N}}_{\mathfrak{R}_2}(\mathrm{e})\right) | \mathrm{e} \in \mathbb{E}\right\} = \left\{\left(\mathrm{e}, \mathfrak{F}^{\mathrm{RP}}_{\mathfrak{R}_2}(\mathrm{e}) + \iota\mathfrak{F}^{\mathrm{IP}}_{\mathfrak{R}_2}(\mathrm{e}), \mathfrak{F}^{\mathrm{RN}}_{\mathfrak{R}_2}(\mathrm{e}) + \iota\mathfrak{F}^{\mathrm{IN}}_{\mathfrak{R}_2}(\mathrm{e})\right) | \mathrm{e} \in \mathbb{E}\right\}$ be two BCFSs. Then.

1. $\mathfrak{R}_1 \subset \mathfrak{R}_2$ if and only if $\mathfrak{F}^{\mathrm{RP}}_{\mathfrak{R}_1}(\mathrm{e}) < \mathfrak{F}^{\mathrm{RP}}_{\mathfrak{R}_2}(\mathrm{e})$, $\mathfrak{F}^{\mathrm{IP}}_{\mathfrak{R}_1}(\mathrm{e}) < \mathfrak{F}^{\mathrm{IP}}_{\mathfrak{R}_2}(\mathrm{e})$, $\mathfrak{F}^{\mathrm{RN}}_{\mathfrak{R}_1}(\mathrm{e}) < \mathfrak{F}^{\mathrm{RN}}_{\mathfrak{R}_2}(\mathrm{e})$, and $\mathfrak{F}^{\mathrm{IN}}_{\mathfrak{R}_1}(\mathrm{e}) < \mathfrak{F}^{\mathrm{IN}}_{\mathfrak{R}_2}(\mathrm{e}) \forall \mathrm{e} \in \mathbb{E}$.
2. $\mathfrak{R}_1 = \mathfrak{R}_2$ if and only if $\mathfrak{F}^{\mathrm{RP}}_{\mathfrak{R}_1}(\mathrm{e}) = \mathfrak{F}^{\mathrm{RP}}_{\mathfrak{R}_2}(\mathrm{e})$, $\mathfrak{F}^{\mathrm{IP}}_{\mathfrak{R}_1}(\mathrm{e}) = \mathfrak{F}^{\mathrm{IP}}_{\mathfrak{R}_2}(\mathrm{e})$, $\mathfrak{F}^{\mathrm{RN}}_{\mathfrak{R}_1}(\mathrm{e}) = \mathfrak{F}^{\mathrm{RN}}_{\mathfrak{R}_2}(\mathrm{e})$, and $\mathfrak{F}^{\mathrm{IN}}_{\mathfrak{R}_1}(\mathrm{e}) = \mathfrak{F}^{\mathrm{IN}}_{\mathfrak{R}_2}(\mathrm{e}) \forall \mathrm{e} \in \mathbb{E}$.

2.3 Elementary Operations and Concepts of BCFSs

After the deduction of the novel definition of BCFS, Mahmood, and Rehman [7] demonstrated various operations in the environment of BCFS.

Definition 2.8 For two BCFSs $\mathfrak{R}_1 = \left\{\left(\mathrm{e}, \mathfrak{F}^{\mathrm{P}}_{\mathfrak{R}_1}(\mathrm{e}), \mathfrak{F}^{\mathrm{N}}_{\mathfrak{R}_1}(\mathrm{e})\right) | \mathrm{e} \in \mathbb{E}\right\} = \left\{\left(\mathrm{e}, \mathfrak{F}^{\mathrm{RP}}_{\mathfrak{R}_1}(\mathrm{e}) + \iota\mathfrak{F}^{\mathrm{IP}}_{\mathfrak{R}_1}(\mathrm{e}), \mathfrak{F}^{\mathrm{RN}}_{\mathfrak{R}_1}(\mathrm{e}) + \iota\mathfrak{F}^{\mathrm{IN}}_{\mathfrak{R}_1}(\mathrm{e})\right) | \mathrm{e} \in \mathbb{E}\right\}$ and $\mathfrak{R}_2 = \left\{\left(\mathrm{e}, \mathfrak{F}^{\mathrm{P}}_{\mathfrak{R}_2}(\mathrm{e}), \mathfrak{F}^{\mathrm{N}}_{\mathfrak{R}_2}(\mathrm{e})\right) | \mathrm{e} \in \mathbb{E}\right\} = \left\{\left(\mathrm{e}, \mathfrak{F}^{\mathrm{RP}}_{\mathfrak{R}_2}(\mathrm{e}) + \iota\mathfrak{F}^{\mathrm{IP}}_{\mathfrak{R}_2}(\mathrm{e}), \mathfrak{F}^{\mathrm{RN}}_{\mathfrak{R}_2}(\mathrm{e}) + \iota\mathfrak{F}^{\mathrm{IN}}_{\mathfrak{R}_2}(\mathrm{e})\right) | \mathrm{e} \in \mathbb{E}\right\}$, we have.

1.
$$(\Re_1)^c = \left\{\left(e, \mathfrak{F}^{+}_{\Re_1^c}(e), \mathfrak{F}^{-}_{\Re_1^c}(e)\right)|e \in \mathbb{E}\right\}$$
$$= \left\{e, \begin{pmatrix} 1 - \mathfrak{F}^{RP}_{\Re_1}(e) + \iota(1 - \mathfrak{F}^{IP}_{\Re_1}(e)), \\ -1 - \mathfrak{F}^{RN}_{\Re_1}(e) + \iota(-1 - \mathfrak{F}^{IN}_{\Re_1}(e)) \end{pmatrix}|e \in \mathbb{E}\right\}$$

2.
$$\Re_1 \cup \Re_2 = \left\{\left(e, \mathfrak{F}^{+}_{\Re_1\cup\Re_2}(e), \mathfrak{F}^{-}_{\Re_1\cup\Re_2}(e)\right)|e \in \mathbb{E}\right\}$$
$$= \left\{e, \begin{pmatrix} \max\left(\mathfrak{F}^{RP}_{\Re_1}(e), \mathfrak{F}^{RP}_{\Re_2}(e)\right) + \iota\max\left(\mathfrak{F}^{IP}_{\Re_1}(e), \mathfrak{F}^{IP}_{\Re_2}(e)\right), \\ \max\left(\mathfrak{F}^{RN}_{\Re_1}(e), \mathfrak{F}^{RN}_{\Re_2}(e)\right) + \iota\max\left(\mathfrak{F}^{RP}_{\Re_1}(e), \mathfrak{F}^{RP}_{\Re_2}(e)\right) \end{pmatrix}|e \in \mathbb{E}\right\}$$

3.
$$\Re_1 \cap \Re_2 = \left\{\left(e, \mathfrak{F}^{+}_{\Re_1\cap\Re_2}(e), \mathfrak{F}^{-}_{\Re_1\cap\Re_2}(e)\right)|e \in \mathbb{E}\right\}$$
$$= \left\{e, \begin{pmatrix} \min\left(\mathfrak{F}^{RP}_{\Re_1}(e), \mathfrak{F}^{RP}_{\Re_2}(e)\right) + \iota\min\left(\mathfrak{F}^{IP}_{\Re_1}(e), \mathfrak{F}^{IP}_{\Re_2}(e)\right), \\ \min\left(\mathfrak{F}^{RN}_{\Re_1}(e), \mathfrak{F}^{RN}_{\Re_2}(e)\right) + \iota\min\left(\mathfrak{F}^{RP}_{\Re_1}(e), \mathfrak{F}^{RP}_{\Re_2}(e)\right) \end{pmatrix}|e \in \mathbb{E}\right\}$$

The next Proposition would show that initiated union, complement, and intersection hold various properties.

Proposition 2.1 Assume that $\Re = \{(e, \mathfrak{F}^{P}_{\Re}(e), \mathfrak{F}^{N}_{\Re}(e))|e \in \mathbb{E}\} = \{(e, \mathfrak{F}^{RP}_{\Re}(e) + \iota\mathfrak{F}^{IP}_{\Re}(e), \mathfrak{F}^{RN}_{\Re}(e) + \iota\mathfrak{F}^{IN}_{\Re}(e))|e \in \mathbb{E}\}$, $\Re_1 = \{(e, \mathfrak{F}^{P}_{\Re_1}(e), \mathfrak{F}^{N}_{\Re_1}(e))|e \in \mathbb{E}\} = \{(e, \mathfrak{F}^{RP}_{\Re_1}(e) + \iota\mathfrak{F}^{IP}_{\Re_1}(e), \mathfrak{F}^{RN}_{\Re_1}(e) + \iota\mathfrak{F}^{IN}_{\Re_1}(e))|e \in \mathbb{E}\}$ and $\Re_2 = \{(e, \mathfrak{F}^{P}_{\Re_2}(e), \mathfrak{F}^{N}_{\Re_2}(e))|e \in \mathbb{E}\} = \{(e, \mathfrak{F}^{RP}_{\Re_2}(e) + \iota\mathfrak{F}^{IP}_{\Re_2}(e), \mathfrak{F}^{RN}_{\Re_2}(e) + \iota\mathfrak{F}^{IN}_{\Re_2}(e))|e \in \mathbb{E}\}$, then the underneath hold.

1. $(\Re^c)^c = \Re$
2. $\Re \cup \Re = \Re$
3. $\Re \cap \Re = \Re$
4. $\Re_1 \cup \Re_2 = \Re_2 \cup \Re_1$
5. $\Re_1 \cap \Re_2 = \Re_2 \cap \Re_1$
6. $\Re_1 \cup (\Re_2 \cap \Re_3) = (\Re_1 \cup \Re_2) \cap (\Re_1 \cup \Re_3)$
7. $\Re_1 \cap (\Re_2 \cup \Re_3) = (\Re_1 \cap \Re_2) \cup (\Re_1 \cap \Re_3)$
8. $\Re_1 \cup (\Re_2 \cup \Re_3) = (\Re_1 \cup \Re_2) \cup \Re_3$
9. $\Re_1 \cap (\Re_2 \cap \Re_3) = (\Re_1 \cap \Re_2) \cap \Re_3$

Proof: 1. By definition (2.8) we have

$$\Re^c = \left\{\left(e, \mathfrak{F}^{P}_{\Re^c}(e), \mathfrak{F}^{N}_{\Re^c}(e)\right)|e \in \mathbb{E}\right\}$$
$$= \left\{\left(e, \left(1 - \mathfrak{F}^{RP}_{\Re}(e) + \iota(1 - \mathfrak{F}^{IP}_{\Re}(e)\right), -1 - \mathfrak{F}^{RN}_{\Re}(e) + \iota(-1 - \mathfrak{F}^{IN}_{\Re}(e)))\right)|e \in \mathbb{E}\right\}$$

Now

$$(\Re^c)^c = \left\{\left(e, \begin{pmatrix} 1 - \left(1 - \mathfrak{F}^{RP}_{\Re}(e)\right) + \iota\left(1 - \left(1 - \mathfrak{F}^{IP}_{\Re}(e)\right)\right), \\ 1 - \left(-1 - \mathfrak{F}^{RN}_{\Re}(e)\right) + \iota\left(1 - \left(-1 - \mathfrak{F}^{IN}_{\Re}(e)\right)\right) \end{pmatrix}\right)|e \in \mathbb{E}\right\}$$
$$= \left\{\left(e, \mathfrak{F}^{RP}_{\Re}(e) + \iota\mathfrak{F}^{IP}_{\Re}(e), \mathfrak{F}^{RN}_{\Re}(e) + \iota\mathfrak{F}^{IN}_{\Re}(e)\right)|e \in \mathbb{E}\right\}$$
$$= \left\{\left(e, \mathfrak{F}^{P}_{\Re}(e), \mathfrak{F}^{N}_{\Re}(e)\right)|e \in \mathbb{E}\right\} = \Re$$

2. By definition (2.8) we have

$$\begin{aligned}\Re \cup \Re &= \left\{\left(\mathrm{e}, \mathfrak{F}^{\mathrm{P}}_{\Re\cup\Re}(\mathrm{e}), \mathfrak{F}^{\mathrm{N}}_{\Re\cup\Re}(\mathrm{e})\right)|\mathrm{e} \in \mathbb{E}\right\}\\ &= \left\{\begin{pmatrix} e, \max\left(\mathfrak{F}^{\mathrm{RP}}_{\Re}(\mathrm{e}), \mathfrak{F}^{\mathrm{RP}}_{\Re}(\mathrm{e})\right) + \iota \max\left(\mathfrak{F}^{\mathrm{IP}}_{\Re}(\mathrm{e}), \mathfrak{F}^{\mathrm{IP}}_{\Re}(\mathrm{e})\right), \\ \max\left(\mathfrak{F}^{\mathrm{RN}}_{\Re}(\mathrm{e}), \mathfrak{F}^{\mathrm{RN}}_{\Re}(\mathrm{e})\right) + \iota \max\left(\mathfrak{F}^{\mathrm{IN}}_{\Re}(\mathrm{e}), \mathfrak{F}^{\mathrm{IN}}_{\Re}(\mathrm{e})\right)\end{pmatrix}|\mathrm{e} \in \mathbb{E}\right\}\\ &= \left\{\left(e, \mathfrak{F}^{\mathrm{RP}}_{\Re}(\mathrm{e}) + \iota\mathfrak{F}^{\mathrm{IP}}_{\Re}(\mathrm{e}), \mathfrak{F}^{\mathrm{RN}}_{\Re}(\mathrm{e}) + \iota\mathfrak{F}^{\mathrm{IN}}_{\Re}(\mathrm{e})\right)|\mathrm{e} \in \mathbb{E}\right\}\\ &= \left\{\left(e, \mathfrak{F}^{\mathrm{P}}_{\Re}(\mathrm{e}), \mathfrak{F}^{\mathrm{N}}_{\Re}(\mathrm{e})\right)|\mathrm{e} \in \mathbb{E}\right\} = \Re\end{aligned}$$

3. By definition (2.8) we have

$$\begin{aligned}\Re \cap \Re &= \left\{\left(\mathrm{e}, \mathfrak{F}^{\mathrm{P}}_{\Re\cap\Re}(\mathrm{e}), \mathfrak{F}^{\mathrm{N}}_{\Re\cap\Re}(\mathrm{e})\right)|\mathrm{e} \in \mathbb{E}\right\}\\ &= \left\{\begin{pmatrix} e, \min\left(\mathfrak{F}^{\mathrm{RP}}_{\Re}(\mathrm{e}), \mathfrak{F}^{\mathrm{RP}}_{\Re}(\mathrm{e})\right) + \iota \min\left(\mathfrak{F}^{\mathrm{IP}}_{\Re}(\mathrm{e}), \mathfrak{F}^{\mathrm{IP}}_{\Re}(\mathrm{e})\right), \\ \min\left(\mathfrak{F}^{\mathrm{RN}}_{\Re}(\mathrm{e}), \mathfrak{F}^{\mathrm{RN}}_{\Re}(\mathrm{e})\right) + \iota \min\left(\mathfrak{F}^{\mathrm{IN}}_{\Re}(\mathrm{e}), \mathfrak{F}^{\mathrm{IN}}_{\Re}(\mathrm{e})\right)\end{pmatrix}|\mathrm{e} \in \mathbb{E}\right\}\\ &= \left\{\left(e, \mathfrak{F}^{\mathrm{RP}}_{\Re}(\mathrm{e}) + \iota\mathfrak{F}^{\mathrm{IP}}_{\Re}(\mathrm{e}), \mathfrak{F}^{\mathrm{RN}}_{\Re}(\mathrm{e}) + \iota\mathfrak{F}^{\mathrm{IN}}_{\Re}(\mathrm{e})\right)|\mathrm{e} \in \mathbb{E}\right\}\\ &= \left\{\left(e, \mathfrak{F}^{\mathrm{P}}_{\Re}(\mathrm{e}), \mathfrak{F}^{\mathrm{N}}_{\Re}(\mathrm{e})\right)|\mathrm{e} \in \mathbb{E}\right\} = \Re\end{aligned}$$

4. By definition (2.8) we have

$$\begin{aligned}\Re_1 \cup \Re_2 &= \left\{\left(\mathrm{e}, \mathfrak{F}^{\mathrm{P}}_{\Re_1\cup\Re_2}(\mathrm{e}), \mathfrak{F}^{\mathrm{N}}_{\Re_1\cup\Re_2}(\mathrm{e})\right)|\mathrm{e} \in \mathbb{E}\right\}\\ &= \left\{\begin{pmatrix} e, \max\left(\mathfrak{F}^{\mathrm{RP}}_{\Re_1}(\mathrm{e}), \mathfrak{F}^{\mathrm{RP}}_{\Re_2}(\mathrm{e})\right) + \iota \max\left(\mathfrak{F}^{\mathrm{IP}}_{\Re_1}(\mathrm{e}), \mathfrak{F}^{\mathrm{IP}}_{\Re_2}(\mathrm{e})\right), \\ \max\left(\mathfrak{F}^{\mathrm{RN}}_{\Re_1}(\mathrm{e}), \mathfrak{F}^{\mathrm{RN}}_{\Re_2}(\mathrm{e})\right) + \iota \max\left(\mathfrak{F}^{\mathrm{IN}}_{\Re_1}(\mathrm{e}), \mathfrak{F}^{\mathrm{IN}}_{\Re_2}(\mathrm{e})\right)\end{pmatrix}|\mathrm{e} \in \mathbb{E}\right\}\\ &= \left\{\begin{pmatrix} e, \max\left(\mathfrak{F}^{\mathrm{RP}}_{\Re_2}(\mathrm{e}), \mathfrak{F}^{\mathrm{RP}}_{\Re_1}(\mathrm{e})\right) + \iota \max\left(\mathfrak{F}^{\mathrm{IP}}_{\Re_2}(\mathrm{e}), \mathfrak{F}^{\mathrm{IP}}_{\Re_1}(\mathrm{e})\right), \\ \max\left(\mathfrak{F}^{\mathrm{RN}}_{\Re_2}(\mathrm{e}), \mathfrak{F}^{\mathrm{RN}}_{\Re_1}(\mathrm{e})\right) + \iota \max\left(\mathfrak{F}^{\mathrm{IN}}_{\Re_2}(\mathrm{e}), \mathfrak{F}^{\mathrm{IN}}_{\Re_1}(\mathrm{e})\right)\end{pmatrix}|\mathrm{e} \in \mathbb{E}\right\}\\ &= \left\{\left(e, \mathfrak{F}^{\mathrm{P}}_{\Re_2\cup\Re_1}(\mathrm{e}), \mathfrak{F}^{\mathrm{N}}_{\Re_2\cup\Re_1}(\mathrm{e})\right)|\mathrm{e} \in \mathbb{E}\right\} = \Re_2 \cup \Re_1\end{aligned}$$

5. By definition (2.8) we have

$$\begin{aligned}\Re_1 \cap \Re_2 &= \left\{\left(\mathrm{e}, \mathfrak{F}^{\mathrm{P}}_{\Re_1\cap\Re_2}(\mathrm{e}), \mathfrak{F}^{\mathrm{N}}_{\Re_1\cap\Re_2}(\mathrm{e})\right)|\mathrm{e} \in \mathbb{E}\right\}\\ &= \left\{\begin{pmatrix} e, \min\left(\mathfrak{F}^{\mathrm{RP}}_{\Re_1}(\mathrm{e}), \mathfrak{F}^{\mathrm{RP}}_{\Re_2}(\mathrm{e})\right) + \iota \min\left(\mathfrak{F}^{\mathrm{IP}}_{\Re_1}(\mathrm{e}), \mathfrak{F}^{\mathrm{IP}}_{\Re_2}(\mathrm{e})\right), \\ \min\left(\mathfrak{F}^{\mathrm{RN}}_{\Re_1}(\mathrm{e}), \mathfrak{F}^{\mathrm{RN}}_{\Re_2}(\mathrm{e})\right) + \iota \min\left(\mathfrak{F}^{\mathrm{IN}}_{\Re_1}(\mathrm{e}), \mathfrak{F}^{\mathrm{IN}}_{\Re_2}(\mathrm{e})\right)\end{pmatrix}|\mathrm{e} \in \mathbb{E}\right\}\\ &= \left\{\begin{pmatrix} e, \min\left(\mathfrak{F}^{\mathrm{RP}}_{\Re_2}(\mathrm{e}), \mathfrak{F}^{\mathrm{RP}}_{\Re_1}(\mathrm{e})\right) + \iota \min\left(\mathfrak{F}^{\mathrm{IP}}_{\Re_2}(\mathrm{e}), \mathfrak{F}^{\mathrm{IP}}_{\Re_1}(\mathrm{e})\right), \\ \min\left(\mathfrak{F}^{\mathrm{RN}}_{\Re_2}(\mathrm{e}), \mathfrak{F}^{\mathrm{RN}}_{\Re_1}(\mathrm{e})\right) + \iota \min\left(\mathfrak{F}^{\mathrm{IN}}_{\Re_2}(\mathrm{e}), \mathfrak{F}^{\mathrm{IN}}_{\Re_1}(\mathrm{e})\right)\end{pmatrix}|\mathrm{e} \in \mathbb{E}\right\}\\ &= \left\{\left(e, \mathfrak{F}^{\mathrm{P}}_{\Re_2\cap\Re_1}(\mathrm{e}), \mathfrak{F}^{\mathrm{N}}_{\Re_2\cap\Re_1}(\mathrm{e})\right)|\mathrm{e} \in \mathbb{E}\right\} = \Re_2 \cap \Re_1\end{aligned}$$

6. From definition (2.8), we have

$$\begin{aligned}\Re_2 \cap \Re_2 &= \left\{\left(\mathrm{e}, \mathfrak{F}^{\mathrm{P}}_{\Re_2\cap\Re_2}(\mathrm{e}), \mathfrak{F}^{\mathrm{N}}_{\Re_2\cap\Re_2}(\mathrm{e})\right)|\mathrm{e} \in \mathbb{E}\right\}\\ &= \left\{\begin{pmatrix} e, \min\left(\mathfrak{F}^{\mathrm{RP}}_{\Re_2}(\mathrm{e}), \mathfrak{F}^{\mathrm{RP}}_{\Re_3}(\mathrm{e})\right) + \iota \min\left(\mathfrak{F}^{\mathrm{IP}}_{\Re_2}(\mathrm{e}), \mathfrak{F}^{\mathrm{IP}}_{\Re_3}(\mathrm{e})\right), \\ \min\left(\mathfrak{F}^{\mathrm{RN}}_{\Re_2}(\mathrm{e}), \mathfrak{F}^{\mathrm{RN}}_{\Re_3}(\mathrm{e})\right) + \iota \min\left(\mathfrak{F}^{\mathrm{IN}}_{\Re_2}(\mathrm{e}), \mathfrak{F}^{\mathrm{IN}}_{\Re_3}(\mathrm{e})\right)\end{pmatrix}|\mathrm{e} \in \mathbb{E}\right\}\end{aligned}$$

Now

$$\begin{aligned}\mathfrak{R}_1 \cup (\mathfrak{R}_2 \cap \mathfrak{R}_3) &= \left\{\left(\mathrm{e}, \mathfrak{F}^{\mathrm{P}}_{\mathfrak{R}_1 \cup (\mathfrak{R}_2 \cap \mathfrak{R}_3)}(\mathrm{e}), \mathfrak{F}^{\mathrm{N}}_{\mathfrak{R}_1 \cup (\mathfrak{R}_2 \cap \mathfrak{R}_3)}(\mathrm{e})\right) | \mathrm{e} \in \mathbb{E}\right\} \\ &= \left\{\begin{pmatrix} e, \max(\mathfrak{F}^{\mathrm{RP}}_{\mathfrak{R}_1}(\mathrm{e}), \min(\mathfrak{F}^{\mathrm{RP}}_{\mathfrak{R}_2}(\mathrm{e}), \mathfrak{F}^{\mathrm{RP}}_{\mathfrak{R}_2}(\mathrm{e}))) + \iota \max(\mathfrak{F}^{\mathrm{IP}}_{\mathfrak{R}_1}(\mathrm{e}), \min(\mathfrak{F}^{\mathrm{IP}}_{\mathfrak{R}_2}(\mathrm{e}), \mathfrak{F}^{\mathrm{IP}}_{\mathfrak{R}_2}(\mathrm{e}))), \\ \max(\mathfrak{F}^{\mathrm{RN}}_{\mathfrak{R}_1}(\mathrm{e}), \min(\mathfrak{F}^{\mathrm{RN}}_{\mathfrak{R}_2}(\mathrm{e}), \mathfrak{F}^{\mathrm{RN}}_{\mathfrak{R}_3}(\mathrm{e}))) + \iota \max(\mathfrak{F}^{\mathrm{IN}}_{\mathfrak{R}_1}(\mathrm{e}), \min(\mathfrak{F}^{\mathrm{IN}}_{\mathfrak{R}_2}(\mathrm{e}), \mathfrak{F}^{\mathrm{IN}}_{\mathfrak{R}_3}(\mathrm{e}))) \end{pmatrix} | \mathrm{e} \in \mathbb{E}\right\}\end{aligned} \tag{2.1}$$

again we have

$$\begin{aligned}\mathfrak{R}_1 \cup \mathfrak{R}_2 &= \left\{\left(\mathrm{e}, \mathfrak{F}^{\mathrm{P}}_{\mathfrak{R}_1 \cup \mathfrak{R}_2}(\mathrm{e}), \mathfrak{F}^{\mathrm{N}}_{\mathfrak{R}_1 \cup \mathfrak{R}_2}(\mathrm{e})\right) | \mathrm{e} \in \mathbb{E}\right\} \\ &= \left\{\begin{pmatrix} e, \max(\mathfrak{F}^{\mathrm{RP}}_{\mathfrak{R}_1}(\mathrm{e}), \mathfrak{F}^{\mathrm{RP}}_{\mathfrak{R}_2}(\mathrm{e})) + \iota \max(\mathfrak{F}^{\mathrm{IP}}_{\mathfrak{R}_1}(\mathrm{e}), \mathfrak{F}^{\mathrm{IP}}_{\mathfrak{R}_2}(\mathrm{e})), \\ \max(\mathfrak{F}^{\mathrm{RN}}_{\mathfrak{R}_1}(\mathrm{e}), \mathfrak{F}^{\mathrm{RN}}_{\mathfrak{R}_2}(\mathrm{e})) + \iota \max(\mathfrak{F}^{\mathrm{IN}}_{\mathfrak{R}_1}(\mathrm{e}), \mathfrak{F}^{\mathrm{IN}}_{\mathfrak{R}_2}(\mathrm{e})) \end{pmatrix} | \mathrm{e} \in \mathbb{E}\right\}\end{aligned}$$

and

$$\begin{aligned}\mathfrak{R}_1 \cup \mathfrak{R}_3 &= \left\{\left(\mathrm{e}, \mathfrak{F}^{\mathrm{P}}_{\mathfrak{R}_1 \cup \mathfrak{R}_3}(\mathrm{e}), \mathfrak{F}^{\mathrm{N}}_{\mathfrak{R}_1 \cup \mathfrak{R}_3}(\mathrm{e})\right) | \mathrm{e} \in \mathbb{E}\right\} \\ &= \left\{\begin{pmatrix} e, \max(\mathfrak{F}^{\mathrm{RP}}_{\mathfrak{R}_1}(\mathrm{e}), \mathfrak{F}^{\mathrm{RP}}_{\mathfrak{R}_3}(\mathrm{e})) + \iota \max(\mathfrak{F}^{\mathrm{IP}}_{\mathfrak{R}_1}(\mathrm{e}), \mathfrak{F}^{\mathrm{IP}}_{\mathfrak{R}_3}(\mathrm{e})), \\ \max(\mathfrak{F}^{\mathrm{RN}}_{\mathfrak{R}_1}(\mathrm{e}), \mathfrak{F}^{\mathrm{RN}}_{\mathfrak{R}_3}(\mathrm{e})) + \iota \max(\mathfrak{F}^{\mathrm{IN}}_{\mathfrak{R}_1}(\mathrm{e}), \mathfrak{F}^{\mathrm{IN}}_{\mathfrak{R}_3}(\mathrm{e})) \end{pmatrix} | \mathrm{e} \in \mathbb{E}\right\}\end{aligned}$$

Now

$$\begin{aligned}(\mathfrak{R}_1 \cup \mathfrak{R}_2) \cap (\mathfrak{R}_1 \cup \mathfrak{R}_3) &= \left\{\left(e, \mathfrak{F}^{\mathrm{P}}_{(\mathfrak{R}_1 \cup \mathfrak{R}_2) \cap (\mathfrak{R}_1 \cup \mathfrak{R}_3)}(\mathrm{e}), \mathfrak{F}^{\mathrm{N}}_{(\mathfrak{R}_1 \cup \mathfrak{R}_2) \cap (\mathfrak{R}_1 \cup \mathfrak{R}_3)}(\mathrm{e})\right) | \mathrm{e} \in \mathbb{E}\right\} \\ &= \left\{\left(\begin{array}{l} e, \min\begin{pmatrix} \max(\mathfrak{F}^{\mathrm{RP}}_{\mathfrak{R}_1}(\mathrm{e}), \mathfrak{F}^{\mathrm{RP}}_{\mathfrak{R}_2}(\mathrm{e})), \\ \max(\mathfrak{F}^{\mathrm{RP}}_{\mathfrak{R}_1}(\mathrm{e}), \mathfrak{F}^{\mathrm{RP}}_{\mathfrak{R}_3}(\mathrm{e})) \end{pmatrix} + \iota \min\begin{pmatrix} \max(\mathfrak{F}^{\mathrm{IP}}_{\mathfrak{R}_1}(\mathrm{e}), \mathfrak{F}^{\mathrm{IP}}_{\mathfrak{R}_2}(\mathrm{e})), \\ \max(\mathfrak{F}^{\mathrm{IP}}_{\mathfrak{R}_1}(\mathrm{e}), \mathfrak{F}^{\mathrm{IP}}_{\mathfrak{R}_3}(\mathrm{e})) \end{pmatrix}, \\ \min\begin{pmatrix} \max(\mathfrak{F}^{\mathrm{RN}}_{\mathfrak{R}_1}(\mathrm{e}), \mathfrak{F}^{\mathrm{RN}}_{\mathfrak{R}_2}(\mathrm{e})), \\ \max(\mathfrak{F}^{\mathrm{RN}}_{\mathfrak{R}_1}(\mathrm{e}), \mathfrak{F}^{\mathrm{RN}}_{\mathfrak{R}_3}(\mathrm{e})) \end{pmatrix} + \iota \min\begin{pmatrix} \max(\mathfrak{F}^{\mathrm{IN}}_{\mathfrak{R}_1}(\mathrm{e}), \mathfrak{F}^{\mathrm{IN}}_{\mathfrak{R}_2}(\mathrm{e})), \\ \max(\mathfrak{F}^{\mathrm{IN}}_{\mathfrak{R}_1}(\mathrm{e}), \mathfrak{F}^{\mathrm{IN}}_{\mathfrak{R}_3}(\mathrm{e})) \end{pmatrix} \end{array}\right) | \mathrm{e} \in \mathbb{E}\right\} \\ &= \left\{\begin{pmatrix} e, \max\left(\mathfrak{F}^{\mathrm{RP}}_{\mathfrak{R}_1}(\mathrm{e}), \min\left(\mathfrak{F}^{\mathrm{RP}}_{\mathfrak{R}_2}(\mathrm{e}), \mathfrak{F}^{\mathrm{RP}}_{\mathfrak{R}_2}(\mathrm{e})\right)\right) + \iota \max\left(\mathfrak{F}^{\mathrm{IP}}_{\mathfrak{R}_1}(\mathrm{e}), \min\left(\mathfrak{F}^{\mathrm{IP}}_{\mathfrak{R}_2}(\mathrm{e}), \mathfrak{F}^{\mathrm{IP}}_{\mathfrak{R}_2}(\mathrm{e})\right)\right), \\ \max\left(\mathfrak{F}^{\mathrm{RN}}_{\mathfrak{R}_1}(\mathrm{e}), \min\left(\mathfrak{F}^{\mathrm{RN}}_{\mathfrak{R}_2}(\mathrm{e}), \mathfrak{F}^{\mathrm{RN}}_{\mathfrak{R}_3}(\mathrm{e})\right)\right) + \iota \max\left(\mathfrak{F}^{\mathrm{IN}}_{\mathfrak{R}_1}(\mathrm{e}), \min\left(\mathfrak{F}^{\mathrm{IN}}_{\mathfrak{R}_2}(\mathrm{e}), \mathfrak{F}^{\mathrm{IN}}_{\mathfrak{R}_3}(\mathrm{e})\right)\right) \end{pmatrix} | \mathrm{e} \in \mathbb{E}\right\}\end{aligned} \tag{2.2}$$

From Eq. (2.2) and (2.2) we have that

$$\mathfrak{R}_1 \cup (\mathfrak{R}_2 \cap \mathfrak{R}_3) = (\mathfrak{R}_1 \cup \mathfrak{R}_2) \cap (\mathfrak{R}_1 \cup \mathfrak{R}_3).$$

7. From definition (2.8), we have

$$\begin{aligned}\mathfrak{R}_2 \cup \mathfrak{R}_3 &= \left\{\left(\mathrm{e}, \mathfrak{F}^{\mathrm{P}}_{\mathfrak{R}_2 \cup \mathfrak{R}_3}(\mathrm{e}), \mathfrak{F}^{\mathrm{N}}_{\mathfrak{R}_2 \cup \mathfrak{R}_3}(\mathrm{e})\right) | \mathrm{e} \in \mathbb{E}\right\} \\ &= \left\{\begin{pmatrix} e, \max(\mathfrak{F}^{\mathrm{RP}}_{\mathfrak{R}_2}(\mathrm{e}), \mathfrak{F}^{\mathrm{RP}}_{\mathfrak{R}_3}(\mathrm{e})) + \iota \max(\mathfrak{F}^{\mathrm{IP}}_{\mathfrak{R}_2}(\mathrm{e}), \mathfrak{F}^{\mathrm{IP}}_{\mathfrak{R}_3}(\mathrm{e})), \\ \max(\mathfrak{F}^{\mathrm{RN}}_{\mathfrak{R}_2}(\mathrm{e}), \mathfrak{F}^{\mathrm{RN}}_{\mathfrak{R}_3}(\mathrm{e})) + \iota \max(\mathfrak{F}^{\mathrm{IN}}_{\mathfrak{R}_2}(\mathrm{e}), \mathfrak{F}^{\mathrm{IN}}_{\mathfrak{R}_3}(\mathrm{e})) \end{pmatrix} | \mathrm{e} \in \mathbb{E}\right\}\end{aligned}$$

Now

$$\begin{aligned}\mathfrak{R}_1 \cap (\mathfrak{R}_2 \cup \mathfrak{R}_3) &= \left\{\left(\mathrm{e}, \mathfrak{F}^{\mathrm{P}}_{\mathfrak{R}_1 \cap (\mathfrak{R}_2 \cup \mathfrak{R}_3)}(\mathrm{e}), \mathfrak{F}^{\mathrm{N}}_{\mathfrak{R}_1 \cap (\mathfrak{R}_2 \cup \mathfrak{R}_3)}(\mathrm{e})\right) | \mathrm{e} \in \mathbb{E}\right\} \\ &= \left\{\begin{pmatrix} e, \min\left(\mathfrak{F}^{\mathrm{RP}}_{\mathfrak{R}_1}(\mathrm{e}), \max\left(\mathfrak{F}^{\mathrm{RP}}_{\mathfrak{R}_2}(\mathrm{e}), \mathfrak{F}^{\mathrm{RP}}_{\mathfrak{R}_2}(\mathrm{e})\right)\right) + \iota \min\left(\mathfrak{F}^{\mathrm{IP}}_{\mathfrak{R}_1}(\mathrm{e}), \max\left(\mathfrak{F}^{\mathrm{IP}}_{\mathfrak{R}_2}(\mathrm{e}), \mathfrak{F}^{\mathrm{IP}}_{\mathfrak{R}_2}(\mathrm{e})\right)\right), \\ \min\left(\mathfrak{F}^{\mathrm{RN}}_{\mathfrak{R}_1}(\mathrm{e}), \max\left(\mathfrak{F}^{\mathrm{RN}}_{\mathfrak{R}_2}(\mathrm{e}), \mathfrak{F}^{\mathrm{RN}}_{\mathfrak{R}_3}(\mathrm{e})\right)\right) + \iota \min\left(\mathfrak{F}^{\mathrm{IN}}_{\mathfrak{R}_1}(\mathrm{e}), \max\left(\mathfrak{F}^{\mathrm{IN}}_{\mathfrak{R}_2}(\mathrm{e}), \mathfrak{F}^{\mathrm{IN}}_{\mathfrak{R}_3}(\mathrm{e})\right)\right) \end{pmatrix} | \mathrm{e} \in \mathbb{E}\right\}\end{aligned} \tag{2.3}$$

Again from definition (2.8), we have

$$\begin{aligned}\Re_1 \cap \Re_2 &= \left\{ \left(\mathrm{e}, \mathfrak{F}^{\mathrm{P}}_{\Re_1 \cap \Re_2}(\mathrm{e}), \mathfrak{F}^{\mathrm{N}}_{\Re_1 \cap \Re_2}(\mathrm{e}) \right) | \mathrm{e} \in \mathbb{E} \right\} \\ &= \left\{ \begin{pmatrix} e, \min\left(\mathfrak{F}^{\mathrm{RP}}_{\Re_1}(\mathrm{e}), \mathfrak{F}^{\mathrm{RP}}_{\Re_2}(\mathrm{e})\right) + \iota \min\left(\mathfrak{F}^{\mathrm{IP}}_{\Re_1}(\mathrm{e}), \mathfrak{F}^{\mathrm{IP}}_{\Re_2}(\mathrm{e})\right), \\ \min\left(\mathfrak{F}^{\mathrm{RN}}_{\Re_1}(\mathrm{e}), \mathfrak{F}^{\mathrm{RN}}_{\Re_2}(\mathrm{e})\right) + \iota \min\left(\mathfrak{F}^{\mathrm{IN}}_{\Re_1}(\mathrm{e}), \mathfrak{F}^{\mathrm{IN}}_{\Re_2}(\mathrm{e})\right) \end{pmatrix} | \mathrm{e} \in \mathbb{E} \right\}\end{aligned}$$

and

$$\begin{aligned}\Re_1 \cap \Re_2 &= \left\{ \left(\mathrm{e}, \mathfrak{F}^{\mathrm{P}}_{\Re_1 \cap \Re_2}(\mathrm{e}), \mathfrak{F}^{\mathrm{N}}_{\Re_1 \cap \Re_2}(\mathrm{e}) \right) | \mathrm{e} \in \mathbb{E} \right\} \\ &= \left\{ \begin{pmatrix} e, \min\left(\mathfrak{F}^{\mathrm{RP}}_{\Re_1}(\mathrm{e}), \mathfrak{F}^{\mathrm{RP}}_{\Re_2}(\mathrm{e})\right) + \iota \min\left(\mathfrak{F}^{\mathrm{IP}}_{\Re_1}(\mathrm{e}), \mathfrak{F}^{\mathrm{IP}}_{\Re_2}(\mathrm{e})\right), \\ \min\left(\mathfrak{F}^{\mathrm{RN}}_{\Re_1}(\mathrm{e}), \mathfrak{F}^{\mathrm{RN}}_{\Re_2}(\mathrm{e})\right) + \iota \min\left(\mathfrak{F}^{\mathrm{IN}}_{\Re_1}(\mathrm{e}), \mathfrak{F}^{\mathrm{IN}}_{\Re_2}(\mathrm{e})\right) \end{pmatrix} | \mathrm{e} \in \mathbb{E} \right\}\end{aligned}$$

Now

$$\begin{aligned}(\Re_1 \cap \Re_2) \cup (\Re_1 \cap \Re_3) &= \left\{ \left(e, \mathfrak{F}^{\mathrm{P}}_{(\Re_1 \cap \Re_2) \cup (\Re_1 \cap \Re_3)}(\mathrm{e}), \mathfrak{F}^{\mathrm{N}}_{(\Re_1 \cap \Re_2) \cup (\Re_1 \cap \Re_3)}(\mathrm{e}) \right) | \mathrm{e} \in \mathbb{E} \right\} \\ &= \left\{ \begin{pmatrix} e, \max\begin{pmatrix} \min\left(\mathfrak{F}^{\mathrm{RP}}_{\Re_1}(\mathrm{e}), \mathfrak{F}^{\mathrm{RP}}_{\Re_2}(\mathrm{e})\right), \\ \min\left(\mathfrak{F}^{\mathrm{RP}}_{\Re_1}(\mathrm{e}), \mathfrak{F}^{\mathrm{RP}}_{\Re_3}(\mathrm{e})\right) \end{pmatrix} + \iota \max\begin{pmatrix} \min\left(\mathfrak{F}^{\mathrm{IP}}_{\Re_1}(\mathrm{e}), \mathfrak{F}^{\mathrm{IP}}_{\Re_2}(\mathrm{e})\right), \\ \min\left(\mathfrak{F}^{\mathrm{IP}}_{\Re_1}(\mathrm{e}), \mathfrak{F}^{\mathrm{IP}}_{\Re_3}(\mathrm{e})\right) \end{pmatrix}, \\ \max\begin{pmatrix} \min\left(\mathfrak{F}^{\mathrm{RN}}_{\Re_1}(\mathrm{e}), \mathfrak{F}^{\mathrm{RN}}_{\Re_2}(\mathrm{e})\right), \\ \min\left(\mathfrak{F}^{\mathrm{RN}}_{\Re_1}(\mathrm{e}), \mathfrak{F}^{\mathrm{RN}}_{\Re_3}(\mathrm{e})\right) \end{pmatrix} + \iota \max\begin{pmatrix} \min\left(\mathfrak{F}^{\mathrm{IN}}_{\Re_1}(\mathrm{e}), \mathfrak{F}^{\mathrm{IN}}_{\Re_2}(\mathrm{e})\right), \\ \min\left(\mathfrak{F}^{\mathrm{IN}}_{\Re_1}(\mathrm{e}), \mathfrak{F}^{\mathrm{IN}}_{\Re_3}(\mathrm{e})\right) \end{pmatrix} \end{pmatrix} | \mathrm{e} \in \mathbb{E} \right\} \\ &= \left\{ \begin{pmatrix} e, \min\left(\mathfrak{F}^{\mathrm{RP}}_{\Re_1}(\mathrm{e}), \max\left(\mathfrak{F}^{\mathrm{RP}}_{\Re_2}(\mathrm{e}), \mathfrak{F}^{\mathrm{RP}}_{\Re_2}(\mathrm{e})\right)\right) + \iota \min\left(\mathfrak{F}^{\mathrm{IP}}_{\Re_1}(\mathrm{e}), \max\left(\mathfrak{F}^{\mathrm{IP}}_{\Re_2}(\mathrm{e}), \mathfrak{F}^{\mathrm{IP}}_{\Re_2}(\mathrm{e})\right)\right), \\ \min\left(\mathfrak{F}^{\mathrm{RN}}_{\Re_1}(\mathrm{e}), \max\left(\mathfrak{F}^{\mathrm{RN}}_{\Re_2}(\mathrm{e}), \mathfrak{F}^{\mathrm{RN}}_{\Re_3}(\mathrm{e})\right)\right) + \iota \min\left(\mathfrak{F}^{\mathrm{IN}}_{\Re_1}(\mathrm{e}), \max\left(\mathfrak{F}^{\mathrm{IN}}_{\Re_2}(\mathrm{e}), \mathfrak{F}^{\mathrm{IN}}_{\Re_3}(\mathrm{e})\right)\right) \end{pmatrix} | \mathrm{e} \in \mathbb{E} \right\}\end{aligned}$$

From Eq. (2.3) and (2.4) we have that

$$\Re_1 \cap (\Re_2 \cup \Re_3) = (\Re_1 \cap \Re_2) \cup (\Re_1 \cap \Re_3)$$

8. We have

$$\begin{aligned}\Re_2 \cup \Re_3 &= \left\{ \left(\mathrm{e}, \mathfrak{F}^{\mathrm{P}}_{\Re_2 \cup \Re_3}(\mathrm{e}), \mathfrak{F}^{\mathrm{N}}_{\Re_2 \cup \Re_3}(\mathrm{e}) \right) | \mathrm{e} \in \mathbb{E} \right\} \\ &= \left\{ \begin{pmatrix} e, \max\left(\mathfrak{F}^{\mathrm{RP}}_{\Re_2}(\mathrm{e}), \mathfrak{F}^{\mathrm{RP}}_{\Re_3}(\mathrm{e})\right) + \iota \max\left(\mathfrak{F}^{\mathrm{IP}}_{\Re_2}(\mathrm{e}), \mathfrak{F}^{\mathrm{IP}}_{\Re_3}(\mathrm{e})\right), \\ \max\left(\mathfrak{F}^{\mathrm{RN}}_{\Re_2}(\mathrm{e}), \mathfrak{F}^{\mathrm{RN}}_{\Re_3}(\mathrm{e})\right) + \iota \max\left(\mathfrak{F}^{\mathrm{IN}}_{\Re_2}(\mathrm{e}), \mathfrak{F}^{\mathrm{IN}}_{\Re_3}(\mathrm{e})\right) \end{pmatrix} | \mathrm{e} \in \mathbb{E} \right\}\end{aligned}$$

Now

$$\begin{aligned}\Re_1 \cup (\Re_2 \cup \Re_3) &= \left\{ \left(\mathrm{e}, \mathfrak{F}^{\mathrm{P}}_{\Re_1 \cup (\Re_2 \cup \Re_3)}(\mathrm{e}), \mathfrak{F}^{\mathrm{N}}_{\Re_1 \cup (\Re_2 \cup \Re_3)}(\mathrm{e}) \right) | \mathrm{e} \in \mathbb{E} \right\} \\ &= \left\{ \begin{pmatrix} e, \max\left(\mathfrak{F}^{\mathrm{RP}}_{\Re_1}(\mathrm{e}), \max\left(\mathfrak{F}^{\mathrm{RP}}_{\Re_2}(\mathrm{e}), \mathfrak{F}^{\mathrm{RP}}_{\Re_2}(\mathrm{e})\right)\right) + \iota \max\left(\mathfrak{F}^{\mathrm{IP}}_{\Re_1}(\mathrm{e}), \max\left(\mathfrak{F}^{\mathrm{IP}}_{\Re_2}(\mathrm{e}), \mathfrak{F}^{\mathrm{IP}}_{\Re_2}(\mathrm{e})\right)\right), \\ \max\left(\mathfrak{F}^{\mathrm{RN}}_{\Re_1}(\mathrm{e}), \max\left(\mathfrak{F}^{\mathrm{RN}}_{\Re_2}(\mathrm{e}), \mathfrak{F}^{\mathrm{RN}}_{\Re_3}(\mathrm{e})\right)\right) + \iota \max\left(\mathfrak{F}^{\mathrm{IN}}_{\Re_1}(\mathrm{e}), \max\left(\mathfrak{F}^{\mathrm{IN}}_{\Re_2}(\mathrm{e}), \mathfrak{F}^{\mathrm{IN}}_{\Re_3}(\mathrm{e})\right)\right) \end{pmatrix} | \mathrm{e} \in \mathbb{E} \right\} \\ &= \left\{ \begin{pmatrix} e, \max\left(\mathfrak{F}^{\mathrm{RP}}_{\Re_1}(\mathrm{e}), \mathfrak{F}^{\mathrm{RP}}_{\Re_2}(\mathrm{e}), \mathfrak{F}^{\mathrm{RP}}_{\Re_3}(\mathrm{e})\right) + \iota \max\left(\mathfrak{F}^{\mathrm{IP}}_{\Re_1}(\mathrm{e}), \mathfrak{F}^{\mathrm{IP}}_{\Re_2}(\mathrm{e}), \mathfrak{F}^{\mathrm{IP}}_{\Re_3}(\mathrm{e})\right), \\ \max\left(\mathfrak{F}^{\mathrm{RN}}_{\Re_1}(\mathrm{e}), \mathfrak{F}^{\mathrm{RN}}_{\Re_2}(\mathrm{e}), \mathfrak{F}^{\mathrm{RN}}_{\Re_3}(\mathrm{e})\right) + \iota \max\left(\mathfrak{F}^{\mathrm{IN}}_{\Re_1}(\mathrm{e}), \mathfrak{F}^{\mathrm{IN}}_{\Re_2}(\mathrm{e}), \mathfrak{F}^{\mathrm{IN}}_{\Re_3}(\mathrm{e})\right) \end{pmatrix} | \mathrm{e} \in \mathbb{E} \right\} \\ &= \left\{ \begin{pmatrix} e, \max\left(\max\left(\mathfrak{F}^{\mathrm{RP}}_{\Re_1}(\mathrm{e}), \mathfrak{F}^{\mathrm{RP}}_{\Re_2}(\mathrm{e})\right), \mathfrak{F}^{\mathrm{RP}}_{\Re_3}(\mathrm{e})\right) + \iota \max\left(\max\left(\mathfrak{F}^{\mathrm{IP}}_{\Re_1}(\mathrm{e}), \mathfrak{F}^{\mathrm{IP}}_{\Re_2}(\mathrm{e})\right), \mathfrak{F}^{\mathrm{IP}}_{\Re_3}(\mathrm{e})\right), \\ \max\left(\max\left(\mathfrak{F}^{\mathrm{RN}}_{\Re_1}(\mathrm{e}), \mathfrak{F}^{\mathrm{RN}}_{\Re_2}(\mathrm{e})\right), \mathfrak{F}^{\mathrm{RN}}_{\Re_3}(\mathrm{e})\right) + \iota \max\left(\max\left(\mathfrak{F}^{\mathrm{IN}}_{\Re_1}(\mathrm{e}), \mathfrak{F}^{\mathrm{IN}}_{\Re_2}(\mathrm{e})\right), \mathfrak{F}^{\mathrm{IN}}_{\Re_3}(\mathrm{e})\right) \end{pmatrix} | \mathrm{e} \in \mathbb{E} \right\} \\ &= \left\{ \left(e, \mathfrak{F}^{\mathrm{P}}_{(\Re_1 \cup \Re_2) \cup \Re_3}(\mathrm{e}), \mathfrak{F}^{\mathrm{N}}_{(\Re_1 \cup \Re_2) \cup \Re_3}(\mathrm{e}) \right) | \mathrm{e} \in \mathbb{E} \right\} = (\Re_1 \cup \Re_2) \cup \Re_3\end{aligned}$$

9. Same as 8, so omitted.

Underneath proposition reveal that BCFSs hold De-Morgan laws.

Proposition 2.2 Let $\mathfrak{R}_1$, and $\mathfrak{R}_1$ be BCFSs. Then.

1. $(\mathfrak{R}_1 \cup \mathfrak{R}_2)^c = \mathfrak{R}_1^c \cap \mathfrak{R}_2^c$
2. $(\mathfrak{R}_1 \cap \mathfrak{R}_2)^c = \mathfrak{R}_1^c \cup \mathfrak{R}_2^c$

Proof: From definition (2.8), we obtain

$$\begin{aligned}\mathfrak{R}_1^c &= \left\{\left(\mathrm{e}, \mathfrak{F}^{\mathrm{P}}_{\mathfrak{R}_1^c}(\mathrm{e}), \mathfrak{F}^{\mathrm{N}}_{\mathfrak{R}_1^c}(\mathrm{e})\right) | \mathrm{e} \in \mathbb{E}\right\} \\ &= \left\{\left(e, \left(1 - \mathfrak{F}^{\mathrm{RP}}_{\mathfrak{R}_1}(\mathrm{e}) + \iota\left(1 - \mathfrak{F}^{\mathrm{IP}}_{\mathfrak{R}_1}(\mathrm{e})\right), -1 - \mathfrak{F}^{\mathrm{RN}}_{\mathfrak{R}_1}(\mathrm{e}) + \iota\left(-1 - \mathfrak{F}^{\mathrm{IN}}_{\mathfrak{R}_1}(\mathrm{e})\right)\right)\right) | \mathrm{e} \in \mathbb{E}\right\}\end{aligned}$$

and

$$\begin{aligned}\mathfrak{R}_2^c &= \left\{\left(\mathrm{e}, \mathfrak{F}^{\mathrm{P}}_{\mathfrak{R}_2^c}(\mathrm{e}), \mathfrak{F}^{\mathrm{N}}_{\mathfrak{R}_2^c}(\mathrm{e})\right) | \mathrm{e} \in \mathbb{E}\right\} \\ &= \left\{\left(e, \left(1 - \mathfrak{F}^{\mathrm{RP}}_{\mathfrak{R}_2}(\mathrm{e}) + \iota\left(1 - \mathfrak{F}^{\mathrm{IP}}_{\mathfrak{R}_2}(\mathrm{e})\right), -1 - \mathfrak{F}^{\mathrm{RN}}_{\mathfrak{R}_2}(\mathrm{e}) + \iota\left(-1 - \mathfrak{F}^{\mathrm{IN}}_{\mathfrak{R}_2}(\mathrm{e})\right)\right)\right) | \mathrm{e} \in \mathbb{E}\right\}\end{aligned}$$

then

$$\begin{aligned}\mathfrak{R}_1^c \cap \mathfrak{R}_2^c &= \left\{\left(\begin{array}{c} e, \min\left(1 - \mathfrak{F}^{\mathrm{RP}}_{\mathfrak{R}_1}(\mathrm{e}), 1 - \mathfrak{F}^{\mathrm{RP}}_{\mathfrak{R}_2}(\mathrm{e})\right) + \iota \min\left(1 - \mathfrak{F}^{\mathrm{IP}}_{\mathfrak{R}_1}(\mathrm{e}), 1 - \mathfrak{F}^{\mathrm{IP}}_{\mathfrak{R}_2}(\mathrm{e})\right) \\ \min\left(-1 - \mathfrak{F}^{\mathrm{RN}}_{\mathfrak{R}_1}(\mathrm{e}), -1 - \mathfrak{F}^{\mathrm{RN}}_{\mathfrak{R}_2}(\mathrm{e})\right) + \iota \min\left(-1 - \mathfrak{F}^{\mathrm{IN}}_{\mathfrak{R}_1}(\mathrm{e}), -1 - \mathfrak{F}^{\mathrm{IN}}_{\mathfrak{R}_2}(\mathrm{e})\right) \end{array}\right)\right\} \\ &= \left\{\left(e, \left(\begin{array}{c} 1 - \max\left(\mathfrak{F}^{\mathrm{RP}}_{\mathfrak{R}_1}(\mathrm{e}), \mathfrak{F}^{\mathrm{RP}}_{\mathfrak{R}_2}(\mathrm{e})\right) + \iota\left(1 - \max\left(\mathfrak{F}^{\mathrm{IP}}_{\mathfrak{R}_1}(\mathrm{e}), \mathfrak{F}^{\mathrm{IP}}_{\mathfrak{R}_2}(\mathrm{e})\right)\right), \\ -1 - \max\left(\mathfrak{F}^{\mathrm{RN}}_{\mathfrak{R}_1}(\mathrm{e}), \mathfrak{F}^{\mathrm{RN}}_{\mathfrak{R}_2}(\mathrm{e})\right) + \iota\left(-1 - \max\left(\mathfrak{F}^{\mathrm{IN}}_{\mathfrak{R}_1}(\mathrm{e}), \mathfrak{F}^{\mathrm{IN}}_{\mathfrak{R}_2}(\mathrm{e})\right)\right) \end{array}\right)\right) | \mathrm{e} \in \mathbb{E}\right\} \\ &= \left\{\left(e, \mathfrak{F}^{\mathrm{P}}_{(\mathfrak{R}_1 \cup \mathfrak{R}_2)^c}(\mathrm{e}), \mathfrak{F}^{\mathrm{N}}_{(\mathfrak{R}_1 \cup \mathfrak{R}_2)^c}(\mathrm{e})\right) | \mathrm{e} \in \mathbb{E}\right\} = (\mathfrak{R}_1 \cup \mathfrak{R}_2)^c \\ &\Rightarrow (\mathfrak{R}_1 \cup \mathfrak{R}_2)^c = \mathfrak{R}_1^c \cap \mathfrak{R}_2^c\end{aligned}$$

2. From definition (2.8), we obtain

$$\begin{aligned}\mathfrak{R}_1^c &= \left\{\left(\mathrm{e}, \mathfrak{F}^{\mathrm{P}}_{\mathfrak{R}_1^c}(\mathrm{e}), \mathfrak{F}^{\mathrm{N}}_{\mathfrak{R}_1^c}(\mathrm{e})\right) | \mathrm{e} \in \mathbb{E}\right\} \\ &= \left\{\left(e, \left(1 - \mathfrak{F}^{\mathrm{RP}}_{\mathfrak{R}_1}(\mathrm{e}) + \iota\left(1 - \mathfrak{F}^{\mathrm{IP}}_{\mathfrak{R}_1}(\mathrm{e})\right), -1 - \mathfrak{F}^{\mathrm{RN}}_{\mathfrak{R}_1}(\mathrm{e}) + \iota\left(-1 - \mathfrak{F}^{\mathrm{IN}}_{\mathfrak{R}_1}(\mathrm{e})\right)\right)\right) | \mathrm{e} \in \mathbb{E}\right\}\end{aligned}$$

and

$$\begin{aligned}\mathfrak{R}_2^c &= \left\{\left(\mathrm{e}, \mathfrak{F}^{\mathrm{P}}_{\mathfrak{R}_2^c}(\mathrm{e}), \mathfrak{F}^{\mathrm{N}}_{\mathfrak{R}_2^c}(\mathrm{e})\right) | \mathrm{e} \in \mathbb{E}\right\} \\ &= \left\{\left(e, \left(1 - \mathfrak{F}^{\mathrm{RP}}_{\mathfrak{R}_2}(\mathrm{e}) + \iota\left(1 - \mathfrak{F}^{\mathrm{IP}}_{\mathfrak{R}_2}(\mathrm{e})\right), -1 - \mathfrak{F}^{\mathrm{RN}}_{\mathfrak{R}_2}(\mathrm{e}) + \iota\left(-1 - \mathfrak{F}^{\mathrm{IN}}_{\mathfrak{R}_2}(\mathrm{e})\right)\right)\right) | \mathrm{e} \in \mathbb{E}\right\}\end{aligned}$$

then

$$\mathfrak{R}_1^c \cup \mathfrak{R}_2^c = \left\{ \left(e, \begin{pmatrix} \max\left(1 - \mathfrak{F}_{\mathfrak{R}_1}^{RP}(e), 1 - \mathfrak{F}_{\mathfrak{R}_2}^{RP}(e)\right) + \iota \max\left(1 - \mathfrak{F}_{\mathfrak{R}_1}^{IP}(e), 1 - \mathfrak{F}_{\mathfrak{R}_2}^{IP}(e)\right) \\ \max\left(-1 - \mathfrak{F}_{\mathfrak{R}_1}^{RN}(e), -1 - \mathfrak{F}_{\mathfrak{R}_2}^{RN}(e)\right) + \iota \max\left(-1 - \mathfrak{F}_{\mathfrak{R}_1}^{IN}(e), -1 - \mathfrak{F}_{\mathfrak{R}_2}^{IN}(e)\right) \end{pmatrix} \right) \right\}$$

$$= \left\{ \left(e, \begin{pmatrix} 1 - \min\left(\mathfrak{F}_{\mathfrak{R}_1}^{RP}(e), \mathfrak{F}_{\mathfrak{R}_2}^{RP}(e)\right) + \iota\left(1 - \min\left(\mathfrak{F}_{\mathfrak{R}_1}^{IP}(e), \mathfrak{F}_{\mathfrak{R}_2}^{IP}(e)\right)\right), \\ -1 - \min\left(\mathfrak{F}_{\mathfrak{R}_1}^{RN}(e), \mathfrak{F}_{\mathfrak{R}_2}^{RN}(e)\right) + \iota\left(-1 - \min\left(\mathfrak{F}_{\mathfrak{R}_1}^{IN}(e), \mathfrak{F}_{\mathfrak{R}_2}^{IN}(e)\right)\right) \end{pmatrix} \right) | e \in \mathbb{E} \right\}$$

$$= \left\{ \left(e, \mathfrak{F}_{(\mathfrak{R}_1 \cap \mathfrak{R}_2)^c}^{P}(e), \mathfrak{F}_{(\mathfrak{R}_1 \cap \mathfrak{R}_2)^c}^{N}(e) \right) | e \in \mathbb{E} \right\} = (\mathfrak{R}_1 \cap \mathfrak{R}_2)^c$$

$$\Rightarrow (\mathfrak{R}_1 \cap \mathfrak{R}_2)^c = \mathfrak{R}_1^c \cup \mathfrak{R}_2^c.$$

Mahmood and Rehman [11] also initiated the basic algebraic operation, score, and accuracy function for BCFNs as follows.

For a BCFN $\mathfrak{R} = \left(\mathfrak{F}_{\mathfrak{R}}^{P}, \mathfrak{F}_{\mathfrak{R}}^{N}\right) = \left(\mathfrak{F}_{\mathfrak{R}}^{RP} + \iota \mathfrak{F}_{\mathfrak{R}}^{IP}, \mathfrak{F}_{\mathfrak{R}}^{RN} + \iota \mathfrak{F}_{\mathfrak{R}}^{IN}\right)$, the formula for determining the score value and accuracy value is respectively signified as

$$\mathbb{S}_{\mathbb{SV}}(\mathfrak{R}) = \frac{1}{4}\left(2 + \mathfrak{F}_{\mathfrak{R}}^{RP} + \mathfrak{F}_{\mathfrak{R}}^{IP} + \mathfrak{F}_{\mathfrak{R}}^{RN} + \mathfrak{F}_{\mathfrak{R}}^{IN}\right) \tag{2.5}$$

$$\mathbb{H}_{\mathbb{SV}}(\mathfrak{R}) = \frac{1}{4}\left(\mathfrak{F}_{\mathfrak{R}}^{RP} + \mathfrak{F}_{\mathfrak{R}}^{IP} - \mathfrak{F}_{\mathfrak{R}}^{RN} - \mathfrak{F}_{\mathfrak{R}}^{IN}\right) \tag{2.6}$$

Noticed that $\mathbb{S}_{\mathbb{SV}}(\mathfrak{R}), \mathbb{H}_{\mathbb{SV}}(\mathfrak{R}) \in [0, 1]$.

From Eq. (2.5) and (2.6), the underneath outcomes were achieved.

1. If $\mathbb{S}_{\mathbb{SV}}(\mathfrak{R}_1) < \mathbb{S}_{\mathbb{SV}}(\mathfrak{R}_2)$, then $\mathfrak{R}_1 < \mathfrak{R}_2$;
2. If $\mathbb{S}_{\mathbb{SV}}(\mathfrak{R}_1) > \mathbb{S}_{\mathbb{SV}}(\mathfrak{R}_2)$, then $\mathfrak{R}_1 > \mathfrak{R}_2$;
3. If $\mathbb{S}_{\mathbb{SV}}(\mathfrak{R}_1) = \mathbb{S}_{\mathbb{SV}}(\mathfrak{R}_2)$, then if $\mathbb{H}_{\mathbb{SV}}(\mathfrak{R}_1) < \mathbb{H}_{\mathbb{SV}}(\mathfrak{R}_2)$, then $\mathfrak{R}_1 < \mathfrak{R}_2$; if $\mathbb{H}_{\mathbb{SV}}(\mathfrak{R}_1) > \mathbb{H}_{\mathbb{SV}}(\mathfrak{R}_2)$, then $\mathfrak{R}_1 > \mathfrak{R}_2$; if $\mathbb{H}_{\mathbb{SV}}(\mathfrak{R}_1) = \mathbb{H}_{\mathbb{SV}}(\mathfrak{R}_2)$, then $\mathfrak{R}_1 = \mathfrak{R}_2$.

Definition 2.9. For two BCFNs $\mathfrak{R}_1 = \left(\mathfrak{F}_{\mathfrak{R}_1}^{P}, \mathfrak{F}_{\mathfrak{R}_1}^{N}\right) = \left(\mathfrak{F}_{\mathfrak{R}_1}^{RP} + \iota \mathfrak{F}_{\mathfrak{R}_1}^{IP}, \mathfrak{F}_{\mathfrak{R}_1}^{RN} + \iota \mathfrak{F}_{\mathfrak{R}_1}^{IN}\right)$, and $\mathfrak{R}_2 = \left(\mathfrak{F}_{\mathfrak{R}_2}^{P}, \mathfrak{F}_{\mathfrak{R}_2}^{N}\right) = \left(\mathfrak{F}_{\mathfrak{R}_2}^{RP} + \iota \mathfrak{F}_{\mathfrak{R}_2}^{IP}, \mathfrak{F}_{\mathfrak{R}_2}^{RN} + \iota \mathfrak{F}_{\mathfrak{R}_2}^{IN}\right)$, $\zeta > 0$, the operational laws for BCFNs are.

1. $\mathfrak{R}_1 \oplus \mathfrak{R}_2 = \left(\mathfrak{F}_{\mathfrak{R}_1}^{RP} + \mathfrak{F}_{\mathfrak{R}_2}^{RP} - \mathfrak{F}_{\mathfrak{R}_1}^{RP}\mathfrak{F}_{\mathfrak{R}_2}^{RP} + \iota\left(\mathfrak{F}_{\mathfrak{R}_1}^{IP} + \mathfrak{F}_{\mathfrak{R}_2}^{IP} - \mathfrak{F}_{\mathfrak{R}_1}^{IP}\mathfrak{F}_{\mathfrak{R}_2}^{IP}\right), -\left(\mathfrak{F}_{\mathfrak{R}_1}^{RN}\mathfrak{F}_{\mathfrak{R}_2}^{RN}\right) + \iota\left(-\left(\mathfrak{F}_{\mathfrak{R}_1}^{IN}\mathfrak{F}_{\mathfrak{R}_2}^{IN}\right)\right)\right)$
2. $\mathfrak{R}_1 \otimes \mathfrak{R}_2 = \left(\mathfrak{F}_{\mathfrak{R}_1}^{RP}\mathfrak{F}_{\mathfrak{R}_2}^{RP} + \iota \mathfrak{F}_{\mathfrak{R}_1}^{IP}\mathfrak{F}_{\mathfrak{R}_2}^{IP}, \mathfrak{F}_{\mathfrak{R}_1}^{RN} + \mathfrak{F}_{\mathfrak{R}_2}^{RN} + \mathfrak{F}_{\mathfrak{R}_1}^{RN}\mathfrak{F}_{\mathfrak{R}_2}^{RN} + \iota\left(\mathfrak{F}_{\mathfrak{R}_1}^{IN} + \mathfrak{F}_{\mathfrak{R}_2}^{IN} + \mathfrak{F}_{\mathfrak{R}_1}^{IN}\mathfrak{F}_{\mathfrak{R}_2}^{IN}\right)\right)$
3. $\zeta \mathfrak{R}_1 = \left(1 - \left(1 - \mathfrak{F}_{\mathfrak{R}_1}^{RP}\right)^{\zeta} + \iota\left(1 - \left(1 - \mathfrak{F}_{\mathfrak{R}_1}^{IP}\right)^{\zeta}\right), -\left|\mathfrak{F}_{\mathfrak{R}_1}^{RN}\right|^{\zeta} + \iota\left(-\left|\mathfrak{F}_{\mathfrak{R}_1}^{RN}\right|^{\zeta}\right)\right)$
4. $\mathfrak{R}_1^{\zeta} = \left(\left(\mathfrak{F}_{\mathfrak{R}_1}^{RP}\right)^{\zeta} + \iota\left(\mathfrak{F}_{\mathfrak{R}_1}^{IP}\right)^{\zeta}, -1 + \left(1 + \mathfrak{F}_{\mathfrak{R}_1}^{RN}\right)^{\zeta} + \iota\left(-1 + \left(1 + \mathfrak{F}_{\mathfrak{R}_1}^{IN}\right)^{\zeta}\right)\right)$

Theorem 2.2 For two BCFNs $\mathfrak{R}_1 = \left(\mathfrak{F}_{\mathfrak{R}_1}^{P}, \mathfrak{F}_{\mathfrak{R}_1}^{N}\right) = \left(\mathfrak{F}_{\mathfrak{R}_1}^{RP} + \iota \mathfrak{F}_{\mathfrak{R}_1}^{IP}, \mathfrak{F}_{\mathfrak{R}_1}^{RN} + \iota \mathfrak{F}_{\mathfrak{R}_1}^{IN}\right)$, and $\mathfrak{R}_2 = \left(\mathfrak{F}_{\mathfrak{R}_2}^{P}, \mathfrak{F}_{\mathfrak{R}_2}^{N}\right) = \left(\mathfrak{F}_{\mathfrak{R}_2}^{RP} + \iota \mathfrak{F}_{\mathfrak{R}_2}^{IP}, \mathfrak{F}_{\mathfrak{R}_2}^{RN} + \iota \mathfrak{F}_{\mathfrak{R}_2}^{IN}\right)$, $\zeta, \zeta_1, \zeta_2 > 0$,

1. $\mathfrak{R}_1 \oplus \mathfrak{R}_2 = \mathfrak{R}_2 \oplus \mathfrak{R}_1$
2. $\mathfrak{R}_1 \otimes \mathfrak{R}_2 = \mathfrak{R}_2 \otimes \mathfrak{R}_1$
3. $\zeta(\mathfrak{R}_1 \oplus \mathfrak{R}_2) = \zeta \mathfrak{R}_1 \oplus \zeta \mathfrak{R}_2$
4. $(\mathfrak{R}_1 \otimes \mathfrak{R}_2)^{\zeta} = \mathfrak{R}_1^{\zeta} \otimes \mathfrak{R}_2^{\zeta}$
5. $\zeta_1 \mathfrak{R}_1 \oplus \zeta_2 \mathfrak{R}_1 = (\zeta_1 + \zeta_2)\mathfrak{R}_1$

6. $\mathfrak{R}_1^{\zeta_1} \otimes \mathfrak{R}_1^{\zeta_2} = \mathfrak{R}_2^{\zeta_1+\zeta_2}$
7. $\left(\mathfrak{R}_1^{\zeta_1}\right)^{\zeta_2} = \mathfrak{R}_1^{\zeta_1\zeta_2}$

Proof: Trivial.

After the development of a novel theory of BCFS, various researchers started research in this direction and developed various AOs, SMs, etc. in the setting of BCFS. Some of the prevailing AOs and SMs are demonstrated as underneath.

Definition 2.10. [12] The BCFWA operator relying on the gathering of BCFNs $\mathfrak{R}_\sigma = \left(\mathfrak{F}^P_{\mathfrak{R}_\sigma}, \mathfrak{F}^N_{\mathfrak{R}_\sigma}\right) = \left(\mathfrak{F}^{RP}_{\mathfrak{R}_\sigma} + \iota\mathfrak{F}^{IP}_{\mathfrak{R}_\sigma}, \mathfrak{F}^{RN}_{\mathfrak{R}_\sigma} + \iota\mathfrak{F}^{IN}_{\mathfrak{R}_\sigma}\right)\sigma = 1, 2, \ldots, y$, is spotted as

$$BCFWA\left(\mathfrak{R}_1, \mathfrak{R}_2, .., \mathfrak{R}_y\right) = \underset{\sigma=1}{\overset{y}{\oplus}} \mathring{w}_\sigma \mathfrak{R}_\sigma = \mathring{w}_1 \mathfrak{R}_1 \oplus \mathring{w}_2 \mathfrak{R}_2 \oplus \ldots \oplus \mathring{w}_y \mathfrak{R}_y$$

$$= \begin{pmatrix} 1 - \prod_{\sigma=1}^{y} \left(1 - \mathfrak{F}^{RP}_{\mathfrak{R}_\sigma}\right)^{\mathring{w}_\sigma} + \iota\left(1 - \prod_{\sigma=1}^{y} \left(1 - \mathfrak{F}^{IP}_{\mathfrak{R}_\sigma}\right)^{\mathring{w}_\sigma}\right), \\ -\prod_{\sigma=1}^{y} \left|\mathfrak{F}^{RRN}_{\mathfrak{R}_\sigma}\right|^{\mathring{w}_\sigma} + \iota\left(-\prod_{\sigma=1}^{y} \left|\mathfrak{F}^{IN}_{\mathfrak{R}_\sigma}\right|^{\mathring{w}_\sigma}\right) \end{pmatrix}$$

Noted that $\mathring{w} = \left(\mathring{w}_{1,} \mathring{w}_2, \ldots, \mathring{w}_y\right)^T$ is a weight vector and $0 \le \mathring{w}_\sigma \le 1$ for all σ with $\sum_{\sigma=1}^{y} \mathring{w}_\sigma = 1$.

Definition 2.11. [12] The BCFWG operator relying on the gathering of BCFNs $\mathfrak{R}_\sigma = \left(\mathfrak{F}^{P}_{\mathfrak{R}_\sigma}, \mathfrak{F}^{N}_{\mathfrak{R}_\sigma}\right) = \left(\mathfrak{F}^{RP}_{\mathfrak{R}_\sigma} + \iota\mathfrak{F}^{IP}_{\mathfrak{R}_\sigma}, \mathfrak{F}^{RN}_{\mathfrak{R}_\sigma} + \iota\mathfrak{F}^{IN}_{\mathfrak{R}_\sigma}\right)\sigma = 1, 2, \ldots, y$, is spotted as

$$BCFWGA\left(\mathfrak{R}_1, \mathfrak{R}_2, .., \mathfrak{R}_y\right) = \underset{\sigma=1}{\overset{y}{\otimes}} \mathfrak{R}_\sigma^{\mathring{w}_\sigma} = \mathfrak{R}_1^{\mathring{w}_1} \otimes \mathfrak{R}_2^{\mathring{w}_2} \otimes \ldots \otimes \mathfrak{R}_y^{\mathring{w}_y}$$

$$= \begin{pmatrix} \prod_{\sigma=1}^{y} \left(\mathfrak{F}^{RP}_{\mathfrak{R}_\sigma}\right)^{\mathring{w}_\sigma} + \iota\left(\prod_{\sigma=1}^{y} \left(\mathfrak{F}^{IP}_{\mathfrak{R}_\sigma}\right)^{\mathring{w}_\sigma}\right), \\ -1 + \prod_{\sigma=1}^{y} \left(1 + \mathfrak{F}^{RN}_{\mathfrak{R}_\sigma}\right)^{\mathring{w}_\sigma} + \iota\left(-1 + \prod_{\sigma=1}^{y} \left(1 + \mathfrak{F}^{IN}_{\mathfrak{R}_\sigma}\right)^{\mathring{w}_\sigma}\right) \end{pmatrix}$$

Noted that $\mathring{w} = \left(\mathring{w}_1, \mathring{w}_2, \ldots, \mathring{w}_y\right)^T$ is a weight vector and $0 \le \mathring{w}_\sigma \le 1$ for all σ. with $\sum_{\sigma=1}^{y} \mathring{w}_\sigma = 1$.

An SM between BCFSs is a real-valued mapping $\mathcal{S} : \mathfrak{R}(\mathbb{E}) \times \mathfrak{R}(\mathbb{E}) \to [0, 1]$, which holds the underneath axioms.

1. $0 \le \mathcal{S}_{\text{BCFS}}(\mathfrak{R}_1, \mathfrak{R}_2) \le 1$
2. $\mathcal{S}_{\text{BCFS}}(\mathfrak{R}_1, \mathfrak{R}_2) = 1 \Leftrightarrow \mathfrak{R}_1 = \mathfrak{R}_2$

3. $\mathcal{S}_{\mathrm{BCFS}}(\mathfrak{R}_1, \mathfrak{R}_2) = \mathcal{S}_{\mathrm{BCFS}}(\mathfrak{R}_2, \mathfrak{R}_1)$
4. If $\mathfrak{R}_1 \subset \mathfrak{R}_2 \subset \mathfrak{R}_3$, then $\mathcal{S}_{\mathrm{BCFS}}(\mathfrak{R}_1, \mathfrak{R}_2) \geq \mathcal{S}^1_{\mathrm{BCFS}}(\mathfrak{R}_1, \mathfrak{R}_3)$ and $\mathcal{S}_{\mathrm{BCFS}}(\mathfrak{R}_2, \mathfrak{R}_3) \geq \mathcal{S}_{\mathrm{BCFS}}(\mathfrak{R}_1, \mathfrak{R}_3)$

The Dice and weighted Dice SMs for BCFS were initiated by Rehman and Mahmood [13].

Definition 2.12. **[13]** Assume two BCFSs

$$\mathfrak{R}_1 = \left\{e_\sigma, \left(\mathfrak{F}^P_{\mathfrak{R}_1}(e_\sigma), \mathfrak{F}^N_{\mathfrak{R}_1}(e_\sigma)\right) \mid e_\sigma \in \mathbb{E}, \sigma = 1, 2, .., y\right\}$$
$$= \left\{e_\sigma, \left(\mathfrak{F}^{RP}_{\mathfrak{R}_1}(e_\sigma) + \iota\mathfrak{F}^{IP}_{\mathfrak{R}_1}(e_\sigma), \mathfrak{F}^{RN}_{\mathfrak{R}_1}(e_\sigma) + \iota\mathfrak{F}^{IN}_{\mathfrak{R}_1}(e_\sigma)\right) \mid e_\sigma \in \mathbb{E}, \sigma = 1, 2, \ldots, y\right\}$$

and

$$\mathfrak{R}_2 = \left\{e, \left(\mathfrak{F}^P_{\mathfrak{R}_2}(e_\sigma), \mathfrak{F}^N_{\mathfrak{R}_2}(e_\sigma)\right) \mid e \in \mathbb{E}, \sigma = 1, 2, \ldots, y\right\}$$
$$= \left\{e_\sigma, \left(\mathfrak{F}^{RP}_{\mathfrak{R}_2}(e_\sigma) + \iota\mathfrak{F}^{IP}_{\mathfrak{R}_2}(e_\sigma), \mathfrak{F}^{RN}_{\mathfrak{R}_2}(e_\sigma) + \iota\mathfrak{F}^{IN}_{\mathfrak{R}_2}(e_\sigma)\right) \mid e_\sigma \in \mathbb{E}, \sigma = 1, 2, \ldots, y\right\}$$

over $\mathbb{E}$, then the Dice SM between $\mathfrak{R}_1$ and $\mathfrak{R}_2$ is demonstrated as

$$\mathfrak{D}_{\mathcal{IC}}(\mathfrak{R}_1, \mathfrak{R}_2) = \frac{1}{y}\sum_{\sigma=1}^{y} \frac{2\left(\mathfrak{F}^{RP}_{\mathfrak{R}_1}(e_\sigma)\mathfrak{F}^{RP}_{\mathfrak{R}_2}(e_\sigma) + \mathfrak{F}^{IP}_{\mathfrak{R}_1}(e_\sigma)\mathfrak{F}^{IP}_{\mathfrak{R}_2}(e_\sigma) + \mathfrak{F}^{RN}_{\mathfrak{R}_1}(e_\sigma)\mathfrak{F}^{RN}_{\mathfrak{R}_2}(e_\sigma) + \mathfrak{F}^{IN}_{\mathfrak{R}_1}(e_\sigma)\mathfrak{F}^{IN}_{\mathfrak{R}_2}(e_\sigma)\right)}{\left(\begin{array}{c}\left((\mathfrak{F}^{RP}_{\mathfrak{R}_1}(e_\sigma))^2 + (\mathfrak{F}^{IP}_{\mathfrak{R}_1}(e_\sigma))^2 + (\mathfrak{F}^{RN}_{\mathfrak{R}_1}(e_\sigma))^2 + (\mathfrak{F}^{IN}_{\mathfrak{R}_1}(e_\sigma))^2\right) \\ +\left((\mathfrak{F}^{RP}_{\mathfrak{R}_2}(e_\sigma))^2 + (\mathfrak{F}^{IP}_{\mathfrak{R}_2}(e_\sigma))^2 + (\mathfrak{F}^{RN}_{\mathfrak{R}_2}(e_\sigma))^2 + (\mathfrak{F}^{IN}_{\mathfrak{R}_2}(e_\sigma))^2\right)\end{array}\right)}$$

Another form of Dice SM is demonstrated as

$$\mathfrak{D}''_{\mathcal{IC}}(\mathfrak{R}_1, \mathfrak{R}_2) = \frac{\sum_{\sigma=1}^{y} 2\left(\mathfrak{F}^{RP}_{\mathfrak{R}_1}(e_\sigma)\mathfrak{F}^{RP}_{\mathfrak{R}_2}(e_\sigma) + \mathfrak{F}^{IP}_{\mathfrak{R}_1}(e_\sigma)\mathfrak{F}^{IP}_{\mathfrak{R}_2}(e_\sigma) + \mathfrak{F}^{RN}_{\mathfrak{R}_1}(e_\sigma)\mathfrak{F}^{RN}_{\mathfrak{R}_2}(e_\sigma) + \mathfrak{F}^{IN}_{\mathfrak{R}_1}(e_\sigma)\mathfrak{F}^{IN}_{\mathfrak{R}_2}(e_\sigma)\right)}{\left(\begin{array}{c}\sum_{\sigma=1}^{y}\left((\mathfrak{F}^{RP}_{\mathfrak{R}_1}(e_\sigma))^2 + (\mathfrak{F}^{IP}_{\mathfrak{R}_1}(e_\sigma))^2 + (\mathfrak{F}^{RN}_{\mathfrak{R}_1}(e_\sigma))^2 + (\mathfrak{F}^{IN}_{\mathfrak{R}_1}(e_\sigma))^2\right) \\ +\sum_{\sigma=1}^{y}\left((\mathfrak{F}^{RP}_{\mathfrak{R}_2}(e_\sigma))^2 + (\mathfrak{F}^{IP}_{\mathfrak{R}_2}(e_\sigma))^2 + (\mathfrak{F}^{RN}_{\mathfrak{R}_2}(e_\sigma))^2 + (\mathfrak{F}^{IN}_{\mathfrak{R}_2}(e_\sigma))^2\right)\end{array}\right)}$$

Definition 2.13. **[13]** Assume two BCFSs

$$\mathfrak{R}_1 = \{e_\sigma, (\mathfrak{F}^{P}_{\mathfrak{R}_1}(e_\sigma), \mathfrak{F}^{N}_{\mathfrak{R}_1}(e_\sigma)) | e_\sigma \in \mathbb{E}, \sigma = 1, 2, \ldots, y\}$$
$$= \{e_\sigma, (\mathfrak{F}^{RP}_{\mathfrak{R}_1}(e_\sigma) + \iota\mathfrak{F}^{IP}_{\mathfrak{R}_1}(e_\sigma), \mathfrak{F}^{RN}_{\mathfrak{R}_1}(e_\sigma) + \iota\mathfrak{F}^{IN}_{\mathfrak{R}_1}(e_\sigma)) | e_\sigma \in \mathbb{E}, \sigma = 1, 2, \ldots, y\}$$

and

$$\mathfrak{R}_2 = \{e, (\mathfrak{F}^{P}_{\mathfrak{R}_2}(e_\sigma), \mathfrak{F}^{N}_{\mathfrak{R}_2}(e_\sigma)) | e_\sigma \in \mathbb{E}, \sigma = 1, 2, \ldots, y\}$$
$$= \{e_\sigma, (\mathfrak{F}^{RP}_{\mathfrak{R}_2}(e_\sigma) + \iota\mathfrak{F}^{IP}_{\mathfrak{R}_2}(e_\sigma), \mathfrak{F}^{RN}_{\mathfrak{R}_2}(e_\sigma) + \iota\mathfrak{F}^{IN}_{\mathfrak{R}_2}(e_\sigma)) | e_\sigma \in \mathbb{E}, \sigma = 1, 2, \ldots, y\}$$

over $\mathbb{E}$, then the weighted Dice SM between $\mathfrak{R}_1$ and $\mathfrak{R}_2$ is demonstrated as

$$\mathfrak{W}\mathfrak{D}_{\mathcal{IC}}(\mathfrak{R}_1, \mathfrak{R}_2) = \sum_{\sigma=1}^{y} \mathring{w}_\sigma \frac{2\left(\mathfrak{F}^{RP}_{\mathfrak{R}_1}(e_\sigma)\mathfrak{F}^{RP}_{\mathfrak{R}_2}(e_\sigma) + \mathfrak{F}^{IP}_{\mathfrak{R}_1}(e_\sigma)\mathfrak{F}^{IP}_{\mathfrak{R}_2}(e_\sigma) + \mathfrak{F}^{RN}_{\mathfrak{R}_1}(e_\sigma)\mathfrak{F}^{RN}_{\mathfrak{R}_2}(e_\sigma) + \mathfrak{F}^{IN}_{\mathfrak{R}_1}(e_\sigma)\mathfrak{F}^{IN}_{\mathfrak{R}_2}(e_\sigma)\right)}{\left(\begin{array}{c}\left(\left(\mathfrak{F}^{RP}_{\mathfrak{R}_1}(e_\sigma)\right)^2 + \left(\mathfrak{F}^{IP}_{\mathfrak{R}_1}(e_\sigma)\right)^2 + \left(\mathfrak{F}^{RN}_{\mathfrak{R}_1}(e_\sigma)\right)^2 + \left(\mathfrak{F}^{IN}_{\mathfrak{R}_1}(e_\sigma)\right)^2\right) \\ +\left(\left(\mathfrak{F}^{RP}_{\mathfrak{R}_2}(e_\sigma)\right)^2 + \left(\mathfrak{F}^{IP}_{\mathfrak{R}_2}(e_\sigma)\right)^2 + \left(\mathfrak{F}^{RN}_{\mathfrak{R}_2}(e_\sigma)\right)^2 + \left(\mathfrak{F}^{IN}_{\mathfrak{R}_2}(e_\sigma)\right)^2\right)\end{array}\right)}$$

Another form of weighted Dice SM is demonstrated as

$$\mathfrak{WD}''_{\mathcal{IC}}(\mathfrak{R}_1,\mathfrak{R}_2)$$
$$=\frac{\sum_{\sigma=1}^{y}2\left(\overset{\circ}{\underset{\sigma}{w}}\right)^2\left(\mathfrak{F}^{RP}_{\mathfrak{R}_1}(e_\sigma)\mathfrak{F}^{RP}_{\mathfrak{R}_2}(e_\sigma)+\mathfrak{F}^{IP}_{\mathfrak{R}_1}(e_\sigma)\mathfrak{F}^{IP}_{\mathfrak{R}_2}(e_\sigma)+\mathfrak{F}^{RN}_{\mathfrak{R}_1}(e_\sigma)\mathfrak{F}^{RN}_{\mathfrak{R}_2}(e_\sigma)+\mathfrak{F}^{IN}_{\mathfrak{R}_1}(e_\sigma)\mathfrak{F}^{IN}_{\mathfrak{R}_2}(e_\sigma)\right)}{\left(\begin{array}{c}\sum_{\sigma=1}^{y}\left(\overset{\circ}{\underset{\sigma}{w}}\right)^2\left(\left(\mathfrak{F}^{RP}_{\mathfrak{R}_1}(e_\sigma)\right)^2+\left(\mathfrak{F}^{IP}_{\mathfrak{R}_1}(e_\sigma)\right)^2+\left(\mathfrak{F}^{RN}_{\mathfrak{R}_1}(e_\sigma)\right)^2+\left(\mathfrak{F}^{IN}_{\mathfrak{R}_1}(e_\sigma)\right)^2\right)\\+\sum_{\sigma=1}^{y}\left(\overset{\circ}{\underset{\sigma}{w}}\right)^2\left(\left(\mathfrak{F}^{RP}_{\mathfrak{R}_2}(e_\sigma)\right)^2+\left(\mathfrak{F}^{IP}_{\mathfrak{R}_2}(e_\sigma)\right)^2+\left(\mathfrak{F}^{RN}_{\mathfrak{R}_2}(e_\sigma)\right)^2+\left(\mathfrak{F}^{IN}_{\mathfrak{R}_2}(e_\sigma)\right)^2\right)\end{array}\right)}$$

Noted that $\overset{\circ}{w}=\left(\overset{\circ}{\underset{1,}{w}}\ \overset{\circ}{\underset{2}{w}},\ldots,\overset{\circ}{\underset{y}{w}}\right)^T$ is a weight vector and $0\le\overset{\circ}{\underset{\sigma}{w}}\le 1$ for all σ with $\sum_{\sigma=1}\overset{\circ}{\underset{\sigma}{w}}=1$.

Remark 2.1 If $\mathcal{S}$ is an SM between two BCFSs that is $\mathfrak{R}_1$ and $\mathfrak{R}_2$, then $d_{BCFS}(\mathfrak{R}_1,\mathfrak{R}_2)=1-\mathcal{S}_{\mathrm{BCFS}}(\mathfrak{R}_1,\mathfrak{R}_2)$ is a distance measure between $\mathfrak{R}_1$ and $\mathfrak{R}_2$.

The distance measure among BCFSs also holds the underneath axioms.

1. $0\le d_{BCFS}(\mathfrak{R}_1,\mathfrak{R}_2)\le 1$
2. $d_{BCFS}(\mathfrak{R}_1,\mathfrak{R}_2)=0\Leftrightarrow\mathfrak{R}_1=\mathfrak{R}_2$
3. $d_{BCFS}(\mathfrak{R}_1,\mathfrak{R}_2)=d_{BCFS}(\mathfrak{R}_2,\mathfrak{R}_1)$
4. If $\mathfrak{R}_1\subset\mathfrak{R}_2\subset\mathfrak{R}_3$, then $d_{BCFS}(\mathfrak{R}_1,\mathfrak{R}_2)\le d_{BCFS}(\mathfrak{R}_1,\mathfrak{R}_3)$ and $d_{BCFS}(\mathfrak{R}_2,\mathfrak{R}_3)\le d_{BCFS}\mathfrak{R}_1 rm\mathfrak{R}_3$

Chapter 3
Algebraic Structures Associated with Bipolar Complex Fuzzy Sets

Mathematical fundamental concepts termed as algebraic structures provide a structured way to comprehend and examine the operations in mathematics and associated sciences. An algebraic structure can be described as a set with one or more binary operations provided that certain axioms or rules are satisfied. Semigroups, lattices, groups, semirings, rings, fields, vector spaces, modules etc. are some classic examples of these structures. Algebraic structures, as they can be used to form and describe mathematical concepts in various fields such as number theory, abstract algebra, geometry, and even in computer science, are important. Fundamental links between ideas that at first appearance seem not connected can be formed by mathematicians by explaining these structures and examining their properties. In addition, Significant insights into the behavior of mathematical objects and systems are provided by the algebraic structures by making it simpler to understand underlying symmetries and transformations in those systems and objects.

In many areas of engineering and sciences, algebraic structures are important. In physics, group theory is used to examine symmetries and conservation rules, however, for understanding the physical quantities and their transformation vector spaces are as important as soul is important for a body or as important as software is important to prove the effectiveness of hardware. Algorithms, coding theory, and cryptography make use of algebraic structures in computer science. The study of algebraic structures also confines advanced topics like abstract algebra and category theory providing useful tools for solving difficult mathematical problems and making significant progress in several fields. In a nutshell, for understanding the mathematical objects and their interaction, algebraic structures offer a crucial foundation. They are crucial because they provide a challenging and structured technique for studying different mathematical concepts, which has practical applications in a range of engineering and scientific disciplines and facilitate further achievements in mathematics as a whole.

In Sect. 3.1 of this chapter, we devise the theory of BCF sublattices and ideals. In Sect. 3.2, we discuss the notion of BCF subsemigroup and its ideals. In Sect. 3.3,

U. ur Rehman and T. Mahmood, *Bipolar Complex Fuzzy Sets*, Studies in Fuzziness and Soft Computing 439, https://doi.org/10.1007/978-3-032-07388-4_3

we investigate the idea of BCF subgroups, BCF normal subgroups, BCF normalizer, BCF cosets, etc. In Sect. 3.4, we devise the theory of BCF subsemiring and related results and in Sect. 3.5, we introduce the notion of BCF subring and associated results. In Sect. 3.6, we discuss other algebraic structures related to the notion of BCFS.

3.1 Bipolar Complex Fuzzy Sublattices and Ideals

When preparing a revised and expanded edition of Dirichlet's Vorlesungen über Zahlentheorie in the early 1890s, Richard Dedekind pondered the following: How many different subgroups can you obtain from the three subgroups A, B, and C of an abelian group G by taking intersections and sums, such as $A+B$, $(A+B)\cap C$, etc. By examining this and similar issues, Dedekind was directed to create the fundamental theory of lattices, which he named Dualgruppen. Inopportunely, Dedekind made this connection before his time, which led to a lack of progress in lattice theory for the following 30 years. Then, Dedekind's work on lattices was rediscovered with the invention of universal algebra in the 1930s by Oystein Ore, Garrett Birkhoff, and others. Since then, lattice theory has been a lively and expanding field, both in terms of its use in algebra and the fundamental issues it raises.

Let $\mathbb{L}$ be a non-empty set and $''\preccurlyeq''$ be a relation on $\mathbb{L}$. Then $\preccurlyeq$ is identified as a partial order on $\mathbb{L}$, if and only if.

1. $\preccurlyeq$ is reflexive
2. $\preccurlyeq$ is antisymmetric
3. $\preccurlyeq$ is transitive.

Then, the pair $(\mathbb{L}, \preccurlyeq)$ is identified as a partially ordered set (poset). A poset $\mathbb{L}$ is identified as lattice if and only if for each $x_1, x_2 \in \mathbb{L}$, $Sup\{x_1, x_2\} = x_1 \vee x_2 \in \mathbb{L}$ and $Inf\{x_1, x_2\} = x_1 \wedge x_2 \in \mathbb{L}$.

Here, we initiate BCF sublattices and ideal and related results. In this section, $\mathbb{L}$ would be identified as a lattice.

Theorem 3.1 Assume that $\mathfrak{R} = \left(\mathrm{F}^{P}_{\mathfrak{R}}, \mathrm{F}^{N}_{\mathfrak{R}}\right) = \left(\mathrm{F}^{RP}_{\mathfrak{R}} + \iota \mathrm{F}^{IP}_{\mathfrak{R}}, \mathrm{F}^{RN}_{\mathfrak{R}} + \iota \mathrm{F}^{IN}_{\mathfrak{R}}\right)$ is a BCFS in $\mathbb{L}$, and $x_1, x_2 \in \mathbb{L}$, then the underneath is equivalent.

1. $x_1 \leq x_2 x \mathrm{F}^{P}_{\mathfrak{R}}(x_1) \geq \mathrm{F}^{P}_{\mathfrak{R}}(x_2)$ means that $\mathrm{F}^{RP}_{\mathfrak{R}}(x_1) \geq \mathrm{F}^{RP}_{\mathfrak{R}}(x_2)$, $\mathrm{F}^{IP}_{\mathfrak{R}}(x_1) \geq \mathrm{F}^{IP}_{\mathfrak{R}}(x_2)$, and $\mathrm{F}^{N}_{\mathfrak{R}}(x_1) \leq \mathrm{F}^{N}_{\mathfrak{R}}(x_2)$ means that $\mathrm{F}^{RN}_{\mathfrak{R}}(x_1) \leq \mathrm{F}^{RN}_{\mathfrak{R}}(x_2)$, $\mathrm{F}^{IN}_{\mathfrak{R}}(x_1) \leq \mathrm{F}^{IN}_{\mathfrak{R}}(x_2)$.
2. $\mathrm{F}^{P}_{\mathfrak{R}}(x \wedge x_2) \geq \max\left(\mathrm{F}^{P}_{\mathfrak{R}}(x_1), \mathrm{F}^{P}_{\mathfrak{R}}(x_2)\right)$ means that $\mathrm{F}^{RP}_{\mathfrak{R}}(x_1 \wedge x_2) \geq \max\left(\mathrm{F}^{RP}_{\mathfrak{R}}(x_1), \mathrm{F}^{RP}_{\mathfrak{R}}(x_2)\right)$, $\mathrm{F}^{IP}_{\mathfrak{R}}(x_1 \wedge x_2) \geq \max\left(\mathrm{F}^{IP}_{\mathfrak{R}}(x_1), \mathrm{F}^{IP}_{\mathfrak{R}}(x_2)\right)$ and $\mathrm{F}^{N}_{\mathfrak{R}}(x_1 \wedge x_2) \leq \min\left(\mathrm{F}^{N}_{\mathfrak{R}}(x_1), \mathrm{F}^{N}_{\mathfrak{R}}(x_2)\right)$ means that $\mathrm{F}^{RN}_{\mathfrak{R}}(x_1 \wedge x_2) \leq \min\left(\mathrm{F}^{RN}_{\mathfrak{R}}(x_1), \mathrm{F}^{RN}_{\mathfrak{R}}(x_2)\right)$, $\mathrm{F}^{IN}_{\mathfrak{R}}(x_1 \wedge x_2) \leq \min\left(\mathrm{F}^{IN}_{\mathfrak{R}}(x_1), \mathrm{F}^{IN}_{\mathfrak{R}}(x_2)\right)$
3. $\mathrm{F}^{P}_{\mathfrak{R}}(x_1 \vee x_2) \leq \min\left(\mathrm{F}^{P}_{\mathfrak{R}}(x_1), \mathrm{F}^{P}_{\mathfrak{R}}(x_2)\right)$ means that $\mathrm{F}^{RP}_{\mathfrak{R}}(x_1 \vee x_2) \leq \min\left(\mathrm{F}^{RP}_{\mathfrak{R}}(x_1), \mathrm{F}^{RP}_{\mathfrak{R}}(x_2)\right)$, $\mathrm{F}^{IP}_{\mathfrak{R}}(x_1 \vee x_2) \leq \min\left(\mathrm{F}^{IP}_{\mathfrak{R}}(x_1), \mathrm{F}^{IP}_{\mathfrak{R}}(x_2)\right)$ and $\mathrm{F}^{N}_{\mathfrak{R}}(x_1 \vee x_2) \geq$

$\max\left(\mathrm{F}_{\Re}^{N}(x_1), \mathrm{F}_{\Re}^{N}(x_2)\right)$ means that $\mathrm{F}_{\Re}^{RN}(x_1 \vee x_2) \geq \max\left(\mathrm{F}_{\Re}^{RN}(x_1), \mathrm{F}_{\Re}^{RN}(x_2)\right)$, $\mathrm{F}_{\Re}^{IN}(x_1 \vee x_2) \geq \max\left(\mathrm{F}_{\Re}^{IN}(x_1), \mathrm{F}_{\Re}^{IN}(x_2)\right)$.

Proof $1 \Rightarrow 2$. $\forall x_1, x_2 \in \mathbb{L}$, we have that $x_1 \wedge x_2 \leq x_1$ and $x_1 \wedge x_2 \leq x_2$, then by employing (1), we have that

$$\mathrm{F}_{\Re}^{P}(x_1 \wedge x_2) \geq \mathrm{F}_{\Re}^{P}(x_1) \Rightarrow \mathrm{F}_{\Re}^{RP}(x_1 \wedge x_2) \geq \mathrm{F}_{\Re}^{RP}(x_1), \mathrm{F}_{\Re}^{IP}(x_1 \wedge x_2) \geq \mathrm{F}_{\Re}^{IP}(x_1)$$

$$\mathrm{F}_{\Re}^{P}(x_1 \wedge x_2) \geq \mathrm{F}_{\Re}^{P}(x_2) \Rightarrow \mathrm{F}_{\Re}^{RP}(x_1 \wedge x_2) \geq \mathrm{F}_{\Re}^{RP}(x_2), \mathrm{F}_{\Re}^{IP}(x_1 \wedge x_2) \geq \mathrm{F}_{\Re}^{IP}(x_2)$$

This means that

$$\mathfrak{F}_{\Re}^{RP}(x_1 \wedge x_2) \geq \max\left(\mathfrak{F}_{\Re}^{RP}(x_1), \mathfrak{F}_{\Re}^{RP}(x_2)\right), \mathfrak{F}_{\Re}^{IP}(x_1 \wedge x_2) \geq \max\left(\mathfrak{F}_{\Re}^{IP}(x_1), \mathfrak{F}_{\Re}^{IP}(x_2)\right)$$
$$\Rightarrow \mathfrak{F}_{\Re}^{P}(x_1 \wedge x_2) \geq \max\left(\mathfrak{F}_{\Re}^{P}(x_1), \mathfrak{F}_{\Re}^{P}(x_2)\right)$$

Similarly,

$$\mathfrak{F}_{\Re}^{N}(x_1 \wedge x_2) \leq \mathfrak{F}_{\Re}^{N}(x_1) \Rightarrow \mathfrak{F}_{\Re}^{RN}(x_1 \wedge x_2) \leq \mathfrak{F}_{\Re}^{RN}(x_1), \mathfrak{F}_{\Re}^{IN}(x_1 \wedge x_2) \leq \mathfrak{F}_{\Re}^{IN}(x_1),$$

$$\mathfrak{F}_{\Re}^{N}(x_1 \wedge x_2) \leq \mathfrak{F}_{\Re}^{N}(x_2) \Rightarrow \mathfrak{F}_{\Re}^{RN}(x_1 \wedge x_2) \leq \mathfrak{F}_{\Re}^{RN}(x_2), \mathfrak{F}_{\Re}^{IN}(x_1 \wedge x_2) \leq \mathfrak{F}_{\Re}^{IN}(x_2),$$

This means that

$$\mathfrak{F}_{\Re}^{RN}(x_1 \wedge x_2) \leq \min\left(\mathfrak{F}_{\Re}^{RN}(x_1), \mathfrak{F}_{\Re}^{RN}(x_2)\right), \mathfrak{F}_{\Re}^{IN}(x_1 \wedge x_2) \leq \min\left(\mathfrak{F}_{\Re}^{IN}(x_1), \mathfrak{F}_{\Re}^{IN}(x_2)\right)$$
$$\Rightarrow \mathfrak{F}_{\Re}^{N}(x_1 \wedge x_2) \leq \min\left(\mathfrak{F}_{\Re}^{N}(x_1), \mathfrak{F}_{\Re}^{N}(x_2)\right).$$

$1 \Rightarrow 3$. $\forall x_1, x_2 \in \mathbb{L}$, we have that $x_1 \vee x_2 \geq x_1$ and $x_1 \vee x_2 \geq x_2$, then by employing (1), we have that

$$\mathfrak{F}_{\Re}^{P}(x_1 \vee x_2) \leq \mathfrak{F}_{\Re}^{P}(x_1) \Rightarrow \mathfrak{F}_{\Re}^{IP}(x_1 \vee x_2) \leq \mathfrak{F}_{\Re}^{IP}(x_1), \mathfrak{F}_{\Re}^{RP}(x_1 \vee x_2) \leq \mathfrak{F}_{\Re}^{RP}(x_1)$$

$$\mathfrak{F}_{\Re}^{P}(x_1 \vee x_2) \leq \mathfrak{F}_{\Re}^{P}(x_2) \Rightarrow \mathfrak{F}_{\Re}^{RP}(x_1 \vee x_2) \leq \mathfrak{F}_{\Re}^{RP}(x_2), \mathfrak{F}_{\Re}^{IP}(x_1 \vee x_2) \leq \mathfrak{F}_{\Re}^{IP}(x_2)$$

This means that

$$\mathfrak{F}_{\Re}^{RP}(x_1 \vee x_2) \leq \min\left(\mathfrak{F}_{\Re}^{RP}(x_1), \mathfrak{F}_{\Re}^{RP}(x_2)\right), \mathfrak{F}_{\Re}^{IP}(x_1 \vee x_2) \leq \min\left(\mathfrak{F}_{\Re}^{IP}(x_1), \mathfrak{F}_{\Re}^{IP}(x_2)\right)$$
$$\Rightarrow \mathfrak{F}_{\Re}^{P}(x_1 \vee x_2) \leq \min\left(\mathfrak{F}_{\Re}^{P}(x_1), \mathfrak{F}_{\Re}^{P}(x_2)\right)$$

Similarly,

$$\mathfrak{F}_{\Re}^{N}(x_1 \vee x_2) \geq \mathfrak{F}_{\Re}^{N}(x_1) \Rightarrow \mathfrak{F}_{\Re}^{RN}(x_1 \vee x_2) \geq \mathfrak{F}_{\Re}^{RN}(x_1), \mathfrak{F}_{\Re}^{IN}(x_1 \vee x_2) \geq \mathfrak{F}_{\Re}^{IN}(x_1)$$

$$\mathfrak{F}^{N}_{\mathfrak{R}}(x_1 \vee x_2) \geq \mathfrak{F}^{N}_{\mathfrak{R}}(x_2) \Rightarrow \mathfrak{F}^{RN}_{\mathfrak{R}}(x_1 \vee x_2) \geq \mathfrak{F}^{RN}_{\mathfrak{R}}(x_2), \mathfrak{F}^{IN}_{\mathfrak{R}}(x_1 \vee x_2) \geq \mathfrak{F}^{IN}_{\mathfrak{R}}(x_2)$$

This means that

$$\mathfrak{F}^{RN}_{\mathfrak{R}}(x_1 \vee x_2) \geq \max\big(\mathfrak{F}^{RN}_{\mathfrak{R}}(x_1), \mathfrak{F}^{RN}_{\mathfrak{R}}(x_2)\big), \mathfrak{F}^{IN}_{\mathfrak{R}}(x_1 \vee x_2) \geq \max\big(\mathfrak{F}^{IN}_{\mathfrak{R}}(x_1), \mathfrak{F}^{IN}_{\mathfrak{R}}(x_2)\big)$$
$$\Rightarrow \mathfrak{F}^{N}_{\mathfrak{R}}(x_1 \vee x_2) \geq \max\big(\mathfrak{F}^{N}_{\mathfrak{R}}(x_1), \mathfrak{F}^{N}_{\mathfrak{R}}(x_2)\big).$$

$2 \Rightarrow 1$. Consider $x_1, x_2 \in \mathbb{L}$ such that $x_1 \leq x_2$, then $x_1 \wedge x_2 = x_1$,

$$\begin{aligned}
&\Rightarrow \mathfrak{F}^{RP}_{\mathfrak{R}}(x_1) = \mathfrak{F}^{RP}_{\mathfrak{R}}(x_1 \wedge x_2)\\
&\geq \max\big(\mathfrak{F}^{RP}_{\mathfrak{R}}(x_1), \mathfrak{F}^{RP}_{\mathfrak{R}}(x_2)\big), \mathfrak{F}^{IP}_{\mathfrak{R}}(x_1)\\
&= \mathfrak{F}^{IP}_{\mathfrak{R}}(x_1 \wedge x_2) \geq \max\big(\mathfrak{F}^{IP}_{\mathfrak{R}}(x_1), \mathfrak{F}^{IP}_{\mathfrak{R}}(x_2)\big)\\
&\Rightarrow \mathfrak{F}^{P}_{\mathfrak{R}}(x_1) \geq \max\big(\mathfrak{F}^{P}_{\mathfrak{R}}(x_1), \mathfrak{F}^{P}_{\mathfrak{R}}(x_2)\big)
\end{aligned}$$

and

$$\begin{aligned}
\Rightarrow \mathfrak{F}^{RN}_{\mathfrak{R}}(x_1) &= \mathfrak{F}^{RN}_{\mathfrak{R}}(x_1 \wedge x_2) \leq \min\big(\mathfrak{F}^{RN}_{\mathfrak{R}}(x_1), \mathfrak{F}^{RN}_{\mathfrak{R}}(x_2)\big), \mathfrak{F}^{IN}_{\mathfrak{R}}(x_1)\\
&= \mathfrak{F}^{IN}_{\mathfrak{R}}(x_1 \wedge x_2) \leq \min\big(\mathfrak{F}^{IN}_{\mathfrak{R}}(x_1), \mathfrak{F}^{IN}_{\mathfrak{R}}(x_2)\big)\\
&\Rightarrow \mathfrak{F}^{N}_{\mathfrak{R}}(x_1) \leq \min\big(\mathfrak{F}^{N}_{\mathfrak{R}}(x_1), \mathfrak{F}^{N}_{\mathfrak{R}}(x_2)\big)
\end{aligned}$$

$3 \Rightarrow 1$. Consider $x_1, x_2 \in \mathbb{L}$ such that $x_1 \leq x_2$, then $x_1 \vee x_2 = x_1$,

$$\begin{aligned}
\Rightarrow \mathfrak{F}^{RP}_{\mathfrak{R}}(x_1) &= \mathfrak{F}^{RP}_{\mathfrak{R}}(x_1 \vee x_2) \leq \min\big(\mathfrak{F}^{RP}_{\mathfrak{R}}(x_1), \mathfrak{F}^{RP}_{\mathfrak{R}}(x_2)\big), \mathfrak{F}^{IP}_{\mathfrak{R}}(x_1)\\
&= \mathfrak{F}^{IP}_{\mathfrak{R}}(x_1 \vee x_2) \leq \min\big(\mathfrak{F}^{IP}_{\mathfrak{R}}(x_1), \mathfrak{F}^{IP}_{\mathfrak{R}}(x_2)\big)\\
&\Rightarrow \mathfrak{F}^{P}_{\mathfrak{R}}(x_1) \leq \min\big(\mathfrak{F}^{P}_{\mathfrak{R}}(x_1), \mathfrak{F}^{P}_{\mathfrak{R}}(x_2)\big)
\end{aligned}$$

and

$$\begin{aligned}
\Rightarrow \mathfrak{F}^{RN}_{\mathfrak{R}}(x_1) &= \mathfrak{F}^{RN}_{\mathfrak{R}}(x_1 \vee x_2) \geq \max\big(\mathfrak{F}^{RN}_{\mathfrak{R}}(x_1), \mathfrak{F}^{RN}_{\mathfrak{R}}(x_2)\big), \mathfrak{F}^{IN}_{\mathfrak{R}}(x_1)\\
&= \mathfrak{F}^{IN}_{\mathfrak{R}}(x_1 \vee x_2) \geq \max\big(\mathfrak{F}^{IN}_{\mathfrak{R}}(x_1), \mathfrak{F}^{IN}_{\mathfrak{R}}(x_2)\big)\\
&\Rightarrow \mathfrak{F}^{N}_{\mathfrak{R}}(x_1) \geq \max\big(\mathfrak{F}^{N}_{\mathfrak{R}}(x_1), \mathfrak{F}^{N}_{\mathfrak{R}}(x_2)\big)
\end{aligned}$$

In a similar pattern, we have the next Theorem.

Theorem 3.2 Assume that $\mathfrak{R} = \big(\mathfrak{F}^{P}_{\mathfrak{R}}, \mathfrak{F}^{N}_{\mathfrak{R}}\big) = \big(\mathfrak{F}^{RP}_{\mathfrak{R}} + \iota\mathfrak{F}^{IP}_{\mathfrak{R}}, \mathfrak{F}^{RN}_{\mathfrak{R}} + \iota\mathfrak{F}^{IN}_{\mathfrak{R}}\big)$ is a BCFS in $\mathbb{L}$, $x_1, x_2 \in \mathbb{L}$, then the underneath are equivalent.

1. $x_1 < x_2 \Rightarrow \mathfrak{F}^{P}_{\mathfrak{R}}(x_1) < \mathfrak{F}^{P}_{\mathfrak{R}}(x_2) \Leftrightarrow \mathfrak{F}^{RP}_{\mathfrak{R}}(x_1) < \mathfrak{F}^{RP}_{\mathfrak{R}}(x_2)$, $\mathfrak{F}^{IP}_{\mathfrak{R}}(x_1) < \mathfrak{F}^{IP}_{\mathfrak{R}}(x_2)$, and $\mathfrak{F}^{N}_{\mathfrak{R}}(x_1) > \mathfrak{F}^{N}_{\mathfrak{R}}(x_2) \Leftrightarrow \mathfrak{F}^{RN}_{\mathfrak{R}}(x_1) > \mathfrak{F}^{RN}_{\mathfrak{R}}(x_2)$, $\mathfrak{F}^{IN}_{\mathfrak{R}}(x_1) > \mathfrak{F}^{IN}_{\mathfrak{R}}(x_2)$.
2. $\mathfrak{F}^{P}_{\mathfrak{R}}(x_1 \wedge x_2) \leq \min\big(\mathfrak{F}^{P}_{\mathfrak{R}}(x_1), \mathfrak{F}^{P}_{\mathfrak{R}}(x_2)\big) \Leftrightarrow \mathfrak{F}^{RP}_{\mathfrak{R}}(x_1 \wedge x_2) \leq \min\big(\mathfrak{F}^{RP}_{\mathfrak{R}}(x_1), \mathfrak{F}^{RP}_{\mathfrak{R}}(x_2)\big)$, $\mathfrak{F}^{IP}_{\mathfrak{R}}(x_1 \wedge x_2) \leq \min\big(\mathfrak{F}^{IP}_{\mathfrak{R}}(x_1), \mathfrak{F}^{IP}_{\mathfrak{R}}(x_2)\big)$

and $\mathfrak{F}^N_{\mathfrak{R}}(x_1 \wedge x_2) \geq \max(\mathfrak{F}^N_{\mathfrak{R}}(x_1), \mathfrak{F}^N_{\mathfrak{R}}(x_2)) \Leftrightarrow \mathfrak{F}^{RN}_{\mathfrak{R}}(x_1 \wedge x_2) \geq \max(\mathfrak{F}^{RN}_{\mathfrak{R}}(x_1), \mathfrak{F}^{RN}_{\mathfrak{R}}(x_2)), \mathfrak{F}^{IN}_{\mathfrak{R}}(x_1 \wedge x_2) \geq \max(\mathfrak{F}^{IN}_{\mathfrak{R}}(x_1), \mathfrak{F}^{IN}_{\mathfrak{R}}(x_2))$

3. $\mathfrak{F}^P_{\mathfrak{R}}(x_1 \vee x_2) \geq \max(\mathfrak{F}^P_{\mathfrak{R}}(x_1), \mathfrak{F}^P_{\mathfrak{R}}(x_2)) \Leftrightarrow \mathfrak{F}^{RP}_{\mathfrak{R}}(x_1 \vee x_2) \geq \max(\mathfrak{F}^{RP}_{\mathfrak{R}}(x_1), \mathfrak{F}^{RP}_{\mathfrak{R}}(x_2)), \mathfrak{F}^{IP}_{\mathfrak{R}}(x_1 \vee x_2) \geq \max(\mathfrak{F}^{IP}_{\mathfrak{R}}(x_1), \mathfrak{F}^{IP}_{\mathfrak{R}}(x_2))$ and $\mathfrak{F}^N_{\mathfrak{R}}(x_1 \vee x_2) \leq \min(\mathfrak{F}^N_{\mathfrak{R}}(x_1), \mathfrak{F}^N_{\mathfrak{R}}(x_2)) \Leftrightarrow \mathfrak{F}^{RN}_{\mathfrak{R}}(x_1 \vee x_2) \leq \min(\mathfrak{F}^{RN}_{\mathfrak{R}}(x_1), \mathfrak{F}^{RN}_{\mathfrak{R}}(x_2)), \mathfrak{F}^{IN}_{\mathfrak{R}}(x_1 \vee x_2) \leq \min(\mathfrak{F}^{IN}_{\mathfrak{R}}(x_1), \mathfrak{F}^{IN}_{\mathfrak{R}}(x_2))$.

Proof Omitted.

In the next definition, we introduce the theory of BCF sublattice (BCFSL) of $\mathbb{L}$.

Definition 3.1 Assume that $\mathfrak{R} = (\mathfrak{F}^P_{\mathfrak{R}}, \mathfrak{F}^N_{\mathfrak{R}}) = (\mathfrak{F}^{RP}_{\mathfrak{R}} + \iota\mathfrak{F}^{IP}_{\mathfrak{R}}, \mathfrak{F}^{RN}_{\mathfrak{R}} + \iota\mathfrak{F}^{IN}_{\mathfrak{R}})$ is a BCFS in $\mathbb{L}$, then $\mathfrak{R}$ is identified as BCFSL of $\mathbb{L}$, if $\forall x_1, x_2 \in \mathbb{L}$,, the underneath properties hold.

1. $\mathfrak{F}^P_{\mathfrak{R}}(x_1 \vee x_2) \geq \min(\mathfrak{F}^P_{\mathfrak{R}}(x_1), \mathfrak{F}^P_{\mathfrak{R}}(x_2))$ this means that $\mathfrak{F}^{RP}_{\mathfrak{R}}(x_1 \vee x_2) \geq \min(\mathfrak{F}^{RP}_{\mathfrak{R}}(x_1), \mathfrak{F}^{RP}_{\mathfrak{R}}(x_2)), \mathfrak{F}^{IP}_{\mathfrak{R}}(x_1 \vee x_2) \geq \min(\mathfrak{F}^{IP}_{\mathfrak{R}}(x_1), \mathfrak{F}^{IP}_{\mathfrak{R}}(x_2))$.
2. $\mathfrak{F}^P_{\mathfrak{R}}(x_1 \wedge x_2) \geq \min(\mathfrak{F}^P_{\mathfrak{R}}(x_1), \mathfrak{F}^P_{\mathfrak{R}}(x_2))$ this means that $\mathfrak{F}^{RP}_{\mathfrak{R}}(x_1 \wedge x_2) \geq \min(\mathfrak{F}^{RP}_{\mathfrak{R}}(x_1), \mathfrak{F}^{RP}_{\mathfrak{R}}(x_2)), \mathfrak{F}^{IP}_{\mathfrak{R}}(x_1 \wedge x_2) \geq \min(\mathfrak{F}^{IP}_{\mathfrak{R}}(x_1), \mathfrak{F}^{IP}_{\mathfrak{R}}(x_2))$
3. $\mathfrak{F}^N_{\mathfrak{R}}(x_1 \vee x_2) \leq \max(\mathfrak{F}^N_{\mathfrak{R}}(x_1), \mathfrak{F}^N_{\mathfrak{R}}(x_2))$ this means that $\mathfrak{F}^{RN}_{\mathfrak{R}}(x_1 \vee x_2) \leq \max(\mathfrak{F}^{RN}_{\mathfrak{R}}(x_1), \mathfrak{F}^{RN}_{\mathfrak{R}}(x_2)), \mathfrak{F}^{IN}_{\mathfrak{R}}(x_1 \vee x_2) \leq \max(\mathfrak{F}^{IN}_{\mathfrak{R}}(x_1), \mathfrak{F}^{IN}_{\mathfrak{R}}(x_2))$
4. $\mathfrak{F}^N_{\mathfrak{R}}(x_1 \wedge x_2) \leq \max(\mathfrak{F}^N_{\mathfrak{R}}(x_1), \mathfrak{F}^N_{\mathfrak{R}}(x_2))$ this means that $\mathfrak{F}^{RN}_{\mathfrak{R}}(x_1 \wedge x_2) \leq \max(\mathfrak{F}^{RN}_{\mathfrak{R}}(x_1), \mathfrak{F}^{RN}_{\mathfrak{R}}(x_2)), \mathfrak{F}^{IN}_{\mathfrak{R}}(x_1 \wedge x_2) \leq \max(\mathfrak{F}^{IN}_{\mathfrak{R}}(x_1), \mathfrak{F}^{IN}_{\mathfrak{R}}(x_2))$.

Example 3.1 Assume that $\mathbb{L} = \{1, 2, 3, 6\}$ is a lattice, where one element is related to another element by divisibility i.e., x_1 divides x_2. Define a BCFS $\mathfrak{R} = (\mathfrak{F}^P_{\mathfrak{R}}, \mathfrak{F}^N_{\mathfrak{R}}) = (\mathfrak{F}^{RP}_{\mathfrak{R}} + \iota\mathfrak{F}^{IP}_{\mathfrak{R}}, \mathfrak{F}^{RN}_{\mathfrak{R}} + \iota\mathfrak{F}^{IN}_{\mathfrak{R}})$ as.

$$\mathfrak{R} = \left\{ \begin{array}{l} (1, 0.9 + \iota 0.82, -0.25 - \iota 0.31), (2, 0.56 + \iota 0.67, -0.43 - \iota 0.51), \\ (3, 0.56 + \iota 0.67, -0.43 - \iota 0.51), (6, 0.45 + \iota 0.39, -0.6 - \iota 0.55) \end{array} \right\}$$

Then $\mathfrak{R}$ is a BCFSL.

Definition 3.2 Assume that $\mathfrak{R} = (\mathfrak{F}^P_{\mathfrak{R}}, \mathfrak{F}^N_{\mathfrak{R}}) = (\mathfrak{F}^{RP}_{\mathfrak{R}} + \iota\mathfrak{F}^{IP}_{\mathfrak{R}}, \mathfrak{F}^{RN}_{\mathfrak{R}} + \iota\mathfrak{F}^{IN}_{\mathfrak{R}})$ is a BCFS in $\mathbb{L}$, then $\mathfrak{R}$ is identified as BCF ideal (BCF-I) of $\mathbb{L}$, if $\forall x_1, x_2 \in \mathbb{L}$, the underneath properties hold.

1. $\mathfrak{F}^P_{\mathfrak{R}}(x_1 \vee x_2) \geq \min(\mathfrak{F}^P_{\mathfrak{R}}(x_1), \mathfrak{F}^P_{\mathfrak{R}}(x_2))$ this means that $\mathfrak{F}^{RP}_{\mathfrak{R}}(x_1 \vee x_2) \geq \min(\mathfrak{F}^{RP}_{\mathfrak{R}}(x_1), \mathfrak{F}^{RP}_{\mathfrak{R}}(x_2)), \mathfrak{F}^{IP}_{\mathfrak{R}}(x_1 \wedge x_2) \geq \min(\mathfrak{F}^{IP}_{\mathfrak{R}}(x_1), \mathfrak{F}^{IP}_{\mathfrak{R}}(x_2))$.
2. $\mathfrak{F}^P_{\mathfrak{R}}(x_1 \wedge x_2) \geq \max(\mathfrak{F}^P_{\mathfrak{R}}(x_1), \mathfrak{F}^P_{\mathfrak{R}}(x_2))$ this means that $\mathfrak{F}^{RP}_{\mathfrak{R}}(x_1 \wedge x_2) \geq \max(\mathfrak{F}^{RP}_{\mathfrak{R}}(x_1), \mathfrak{F}^{RP}_{\mathfrak{R}}(x_2)), \mathfrak{F}^{IP}_{\mathfrak{R}}(x_1 \wedge x_2) \geq \max(\mathfrak{F}^{IP}_{\mathfrak{R}}(x_1), \mathfrak{F}^{IP}_{\mathfrak{R}}(x_2))$
3. $\mathfrak{F}^N_{\mathfrak{R}}(x_1 \vee x_2) \leq \max(\mathfrak{F}^N_{\mathfrak{R}}(x_1), \mathfrak{F}^N_{\mathfrak{R}}(x_2))$ this means that $\mathfrak{F}^{RN}_{\mathfrak{R}}(x_1 \vee x_2) \leq \max(\mathfrak{F}^{RN}_{\mathfrak{R}}(x_1), \mathfrak{F}^{RN}_{\mathfrak{R}}(x_2)), \mathfrak{F}^{IN}_{\mathfrak{R}}(x_1 \vee x_2) \leq \max(\mathfrak{F}^{IN}_{\mathfrak{R}}(x_1), \mathfrak{F}^{IN}_{\mathfrak{R}}(x_2))$
4. $\mathfrak{F}^N_{\mathfrak{R}}(x_1 \wedge x_2) \leq \min(\mathfrak{F}^N_{\mathfrak{R}}(x_1), \mathfrak{F}^N_{\mathfrak{R}}(x_2))$ this means that $\mathfrak{F}^{RN}_{\mathfrak{R}}(x_1 \wedge x_2) \leq \min(\mathfrak{F}^{RN}_{\mathfrak{R}}(x_1), \mathfrak{F}^{RN}_{\mathfrak{R}}(x_2)), \mathfrak{F}^{IN}_{\mathfrak{R}}(x_1 \wedge x_2) \leq \min(\mathfrak{F}^{IN}_{\mathfrak{R}}(x_1), \mathfrak{F}^{IN}_{\mathfrak{R}}(x_2))$.

Example 3.2 Reconsider the lattice $\mathbb{L}$ of example (3.1) and define a BCFS $\mathfrak{R} = \left(\mathfrak{F}_{\mathfrak{R}}^{P}, \mathfrak{F}_{\mathfrak{R}}^{N}\right) = \left(\mathfrak{F}_{\mathfrak{R}}^{RP} + \iota\mathfrak{F}_{\mathfrak{R}}^{IP}, \mathfrak{F}_{\mathfrak{R}}^{RN} + \iota\mathfrak{F}_{\mathfrak{R}}^{IN}\right)$ as.

$$\mathfrak{R} = \left\{ \begin{array}{l} (1, 0.85 + \iota 0.72, -0.63 - \iota 0.65), (2, 0.66 + \iota 0.57, -0.53 - \iota 0.61), \\ (3, 0.66 + \iota 0.57, -0.53 - \iota 0.61), (6, 0.47 + \iota 0.41, -0.21 - \iota 0.25) \end{array} \right\}$$

Then $\mathfrak{R}$ is a BCF-I.

Theorem 3.3 Assume that $\mathfrak{R}_1 = \left(\mathfrak{F}_{\mathfrak{R}_1}^{P}, \mathfrak{F}_{\mathfrak{R}_1}^{N}\right) = \left(\mathfrak{F}_{\mathfrak{R}_1}^{RP} + \iota\mathfrak{F}_{\mathfrak{R}_1}^{IP}, \mathfrak{F}_{\mathfrak{R}_1}^{RN} + \iota\mathfrak{F}_{\mathfrak{R}_1}^{IN}\right)$ and $\mathfrak{R}_2 = \left(\mathfrak{F}_{\mathfrak{R}_2}^{P}, \mathfrak{F}_{\mathfrak{R}_2}^{N}\right) = \left(\mathfrak{F}_{\mathfrak{R}_2}^{RP} + \iota\mathfrak{F}_{\mathfrak{R}_2}^{IP}, \mathfrak{F}_{\mathfrak{R}_2}^{RN} + \iota\mathfrak{F}_{\mathfrak{R}_2}^{IN}\right)$ are two BCFSLs(BCF-Is) of a lattice $\mathbb{L}$, then $\mathfrak{R}_1 \cap \mathfrak{R}_2$ is a BCFSL(BCF-I) of $\mathbb{L}$.

Proof As $\mathfrak{R}_1 = \left(\mathfrak{F}_{\mathfrak{R}_1}^{P}, \mathfrak{F}_{\mathfrak{R}_1}^{N}\right) = \left(\mathfrak{F}_{\mathfrak{R}_1}^{RP} + \iota\mathfrak{F}_{\mathfrak{R}_1}^{IP}, \mathfrak{F}_{\mathfrak{R}_1}^{RN} + \iota\mathfrak{F}_{\mathfrak{R}_1}^{IN}\right)$ and $\mathfrak{R}_2 = \left(\mathfrak{F}_{\mathfrak{R}_2}^{P}, \mathfrak{F}_{\mathfrak{R}_2}^{N}\right) = \left(\mathfrak{F}_{\mathfrak{R}_2}^{RP} + \iota\mathfrak{F}_{\mathfrak{R}_2}^{IP}, \mathfrak{F}_{\mathfrak{R}_2}^{RN} + \iota\mathfrak{F}_{\mathfrak{R}_2}^{IN}\right)$ are two BCFSLs of $\mathbb{L}$, then.

1.
$$\begin{aligned} &\mathrm{F}_{\mathfrak{R}_1 \cap \mathfrak{R}_2}^{RP}(x_1 \vee x_2) \\ &\quad = \min\left(\mathrm{F}_{\mathfrak{R}_1}^{RP}(x_1 \vee x_2), \mathrm{F}_{\mathfrak{R}_2}^{RP}(x_1 \vee x_2)\right) \\ &\quad \geq \min\left(\min\left(\mathrm{F}_{\mathfrak{R}_1}^{RP}(x_1), \mathrm{F}_{\mathfrak{R}_1}^{RP}(x_2)\right), \min\left(\mathrm{F}_{\mathfrak{R}_2}^{RP}(x_1), \mathrm{F}_{\mathfrak{R}_2}^{RP}(x_2)\right)\right) \\ &\quad = \min\left(\min\left(\mathrm{F}_{\mathfrak{R}_1}^{RP}(x_1), \mathrm{F}_{\mathfrak{R}_2}^{RP}(x_1)\right), \min\left(\mathrm{F}_{\mathfrak{R}_1}^{RP}(x_2), \mathrm{F}_{\mathfrak{R}_2}^{RP}(x_2)\right)\right) \\ &\quad = \min\left(\mathrm{F}_{\mathfrak{R}_1 \cap \mathfrak{R}_2}^{RP}(x_1), \mathrm{F}_{\mathfrak{R}_1 \cap \mathfrak{R}_2}^{RP}(x_2)\right) \\ &\quad \Rightarrow \mathrm{F}_{\mathfrak{R}_1 \cap \mathfrak{R}_2}^{RP}(x_1 \vee x_2) \\ &\quad \geq \min\left(\mathrm{F}_{\mathfrak{R}_1 \cap \mathfrak{R}_2}^{RP}(x_1), \mathrm{F}_{\mathfrak{R}_1 \cap \mathfrak{R}_2}^{RP}(x_2)\right), \forall x_1, x_2 \in \mathrm{L} \end{aligned}$$

Likewise,

$$\Rightarrow \mathfrak{F}_{\mathfrak{R}_1 \cap \mathfrak{R}_2}^{IP}(x_1 \vee x_2) \geq \min\left(\mathfrak{F}_{\mathfrak{R}_1 \cap \mathfrak{R}_2}^{IP}(x_1), \mathfrak{F}_{\mathfrak{R}_1 \cap \mathfrak{R}_2}^{IP}(x_2)\right), \forall x_1, x_2 \in \mathbb{L},$$

2.
$$\begin{aligned} \mathfrak{F}_{\mathfrak{R}_1 \cap \mathfrak{R}_2}^{RP}(x_1 \wedge x_2) &= \min\left(\mathfrak{F}_{\mathfrak{R}_1}^{RP}(x_1 \wedge x_2), \mathfrak{F}_{\mathfrak{R}_2}^{RP}(x_1 \wedge x_2)\right) \\ &\geq \min\left(\min\left(\mathfrak{F}_{\mathfrak{R}_1}^{RP}(x_1), \mathfrak{F}_{\mathfrak{R}_1}^{RP}(x_2)\right), \min\left(\mathfrak{F}_{\mathfrak{R}_2}^{RP}(x_1), \mathfrak{F}_{\mathfrak{R}_2}^{RP}(x_2)\right)\right) \\ &= \min\left(\min\left(\mathfrak{F}_{\mathfrak{R}_1}^{RP}(x_1), \mathfrak{F}_{\mathfrak{R}_2}^{RP}(x_1)\right), \min\left(\mathfrak{F}_{\mathfrak{R}_1}^{RP}(x_2), \mathfrak{F}_{\mathfrak{R}_2}^{RP}(x_2)\right)\right) \\ &= \min\left(\mathfrak{F}_{\mathfrak{R}_1 \cap \mathfrak{R}_2}^{RP}(x_1), \mathfrak{F}_{\mathfrak{R}_1 \cap \mathfrak{R}_2}^{RP}(x_2)\right) \\ &\Rightarrow \mathfrak{F}_{\mathfrak{R}_1 \cap \mathfrak{R}_2}^{RP}(x_1 \wedge x_2) \geq \min\left(\mathfrak{F}_{\mathfrak{R}_1 \cap \mathfrak{R}_2}^{RP}(x_1), \mathfrak{F}_{\mathfrak{R}_1 \cap \mathfrak{R}_2}^{RP}(x_2)\right), \forall x_1, x_2 \in \mathbb{L}, \end{aligned}$$

Similarly,

$$\Rightarrow \mathfrak{F}_{\mathfrak{R}_1 \cap \mathfrak{R}_2}^{IP}(x_1 \wedge x_2) \geq \min\left(\mathfrak{F}_{\mathfrak{R}_1 \cap \mathfrak{R}_2}^{IP}(x_1), \mathfrak{F}_{\mathfrak{R}_1 \cap \mathfrak{R}_2}^{IP}(x_2)\right), \forall x_1, x_2 \in \mathbb{L},$$

3.
$$\begin{aligned}\mathfrak{F}^{RN}_{\mathfrak{R}_1\cap\mathfrak{R}_2}(x_1\vee x_2) &= \max\Big(\mathfrak{F}^{RN}_{\mathfrak{R}_1}(x_1\vee x_2), \mathfrak{F}^{RN}_{\mathfrak{R}_2}(x_1\vee x_2)\Big)\\ &\le \max\Big(\max\Big(\mathfrak{F}^{RN}_{\mathfrak{R}_1}(x_1), \mathfrak{F}^{RN}_{\mathfrak{R}_1}(x_2)\Big), \max\Big(\mathfrak{F}^{RN}_{\mathfrak{R}_2}(x_1), \mathfrak{F}^{RN}_{\mathfrak{R}_2}(x_2)\Big)\Big)\\ &= \max\Big(\max\Big(\mathfrak{F}^{RN}_{\mathfrak{R}_1}(x_1), \mathfrak{F}^{RN}_{\mathfrak{R}_2}(x_1)\Big), \max\Big(\mathfrak{F}^{RN}_{\mathfrak{R}_1}(x_2), \mathfrak{F}^{RN}_{\mathfrak{R}_2}(x_2)\Big)\Big)\\ &= \max\Big(\mathfrak{F}^{RN}_{\mathfrak{R}_1\cap\mathfrak{R}_2}(x_1), \mathfrak{F}^{RN}_{\mathfrak{R}_1\cap\mathfrak{R}_2}(x_2)\Big)\\ &\Rightarrow \mathfrak{F}^{RN}_{\mathfrak{R}_1\cap\mathfrak{R}_2}(x_1\vee x_2) \le \max\Big(\mathfrak{F}^{RN}_{\mathfrak{R}_1\cap\mathfrak{R}_2}(x_1), \mathfrak{F}^{RN}_{\mathfrak{R}_1\cap\mathfrak{R}_2}(x_2)\Big), \forall x_1, x_2 \in \mathbb{L}\end{aligned}$$

Similarly,

$$\Rightarrow \mathfrak{F}^{IN}_{\mathfrak{R}_1\cap\mathfrak{R}_2}(x_1\vee x_2) \le \max\big(\mathfrak{F}^{IN}_{\mathfrak{R}_1\cap\mathfrak{R}_2}(x_1), \mathfrak{F}^{IN}_{\mathfrak{R}_1\cap\mathfrak{R}_2}(x_2)\big), \forall x_1, x_2 \in \mathbb{L}$$

4.
$$\begin{aligned}\mathfrak{F}^{RN}_{\mathfrak{R}_1\cap\mathfrak{R}_2}(x_1\wedge x_2) &= \max\Big(\mathfrak{F}^{RN}_{\mathfrak{R}_1}(x_1\wedge x_2), \mathfrak{F}^{RN}_{\mathfrak{R}_2}(x_1\wedge x_2)\Big)\\ &\le \max\Big(\max\Big(\mathfrak{F}^{RN}_{\mathfrak{R}_1}(x_1), \mathfrak{F}^{RN}_{\mathfrak{R}_1}(x_2)\Big), \max\Big(\mathfrak{F}^{RN}_{\mathfrak{R}_2}(x_1), \mathfrak{F}^{RN}_{\mathfrak{R}_2}(x_2)\Big)\Big)\\ &= \max\Big(\max\Big(\mathfrak{F}^{RN}_{\mathfrak{R}_1}(x_1), \mathfrak{F}^{RN}_{\mathfrak{R}_2}(x_1)\Big), \max\Big(\mathfrak{F}^{RN}_{\mathfrak{R}_1}(x_2), \mathfrak{F}^{RN}_{\mathfrak{R}_2}(x_2)\Big)\Big)\\ &= \max\Big(\mathfrak{F}^{RN}_{\mathfrak{R}_1\cap\mathfrak{R}_2}(x_1), \mathfrak{F}^{RN}_{\mathfrak{R}_1\cap\mathfrak{R}_2}(x_2)\Big)\\ &\Rightarrow \mathfrak{F}^{RN}_{\mathfrak{R}_1\cap\mathfrak{R}_2}(x_1\wedge x_2) \le \max\Big(\mathfrak{F}^{RN}_{\mathfrak{R}_1\cap\mathfrak{R}_2}(x_1), \mathfrak{F}^{RN}_{\mathfrak{R}_1\cap\mathfrak{R}_2}(x_2)\Big), \forall x_1, x_2 \in \mathbb{L}\end{aligned}$$

Similarly,

$$\Rightarrow \mathfrak{F}^{IN}_{\mathfrak{R}_1\cap\mathfrak{R}_2}(x_1\wedge x_2) \le \max\big(\mathfrak{F}^{IN}_{\mathfrak{R}_1\cap\mathfrak{R}_2}(x_1), \mathfrak{F}^{IN}_{\mathfrak{R}_1\cap\mathfrak{R}_2}(x_2)\big), \forall x_1, x_2 \in \mathbb{L}$$

Thus, $\mathfrak{R}_1 \cap \mathfrak{R}_2$ is a BCFSL of $\mathbb{L}$.

Remark 3.1 The union of any two BCFSLs of $\mathbb{L}$ is not necessarily a BCFSL.

Remark 3.2 Every BCF-I of a lattice $\mathbb{L}$ is a BCFSL but the converse is not true in general. For example, the BCF-I given in example (3.2) is a BCFSL and the BCFSL interpreted in example (3.1) is not a BCF-I.

Remark 3.3 The union of any two BCF-Is of $\mathbb{L}$ is not necessarily a BCF-I.

Theorem 3.4 Assume that $\mathfrak{R} = \big(\mathfrak{F}^P_{\mathfrak{R}}, \mathfrak{F}^N_{\mathfrak{R}}\big) = \big(\mathfrak{F}^{RP}_{\mathfrak{R}} + \iota\mathfrak{F}^{IP}_{\mathfrak{R}}, \mathfrak{F}^{RN}_{\mathfrak{R}} + \iota\mathfrak{F}^{IN}_{\mathfrak{R}}\big)$ is a BCFSL of $\mathbb{L}$, then $\forall x_1, x_2 \in \mathbb{L}$, the underneath statements hold.

1. $\mathfrak{F}^P_{\mathfrak{R}}(x_1\wedge x_2) \ge \max\big(\mathfrak{F}^P_{\mathfrak{R}}(x_1), \mathfrak{F}^P_{\mathfrak{R}}(x_2)\big)$ this means that $\mathfrak{F}^{RP}_{\mathfrak{R}}(x_1\wedge x_2) \ge \max\big(\mathfrak{F}^{RP}_{\mathfrak{R}}(x_1), \mathfrak{F}^{RP}_{\mathfrak{R}}(x_2)\big)$, $\mathfrak{F}^{IP}_{\mathfrak{R}}(x_1\wedge x_2) \ge \max\big(\mathfrak{F}^{IP}_{\mathfrak{R}}(x_1), \mathfrak{F}^{IP}_{\mathfrak{R}}(x_2)\big)$ iff $x_1 \le x_2 \Rightarrow \mathfrak{F}^P_{\mathfrak{R}}(x_1) \ge \mathfrak{F}^P_{\mathfrak{R}}(x_2)$, this means that $\mathfrak{F}^{RP}_{\mathfrak{R}}(x_1) \ge \mathfrak{F}^{RP}_{\mathfrak{R}}(x_2)$, $\mathfrak{F}^{IP}_{\mathfrak{R}}(x_1) \ge \mathfrak{F}^{IP}_{\mathfrak{R}}(x_2)$.
2. $\mathfrak{F}^P_{\mathfrak{R}}(x_1\vee x_2) \ge \max\big(\mathfrak{F}^P_{\mathfrak{R}}(x_1), \mathfrak{F}^P_{\mathfrak{R}}(x_2)\big)$ this means that $\mathfrak{F}^{RP}_{\mathfrak{R}}(x_1\vee x_2) \ge \max\big(\mathfrak{F}^{RP}_{\mathfrak{R}}(x_1), \mathfrak{F}^{RP}_{\mathfrak{R}}(x_2)\big)$, $\mathfrak{F}^{IP}_{\mathfrak{R}}(x_1\vee x_2) \ge \max\big(\mathfrak{F}^{IP}_{\mathfrak{R}}(x_1), \mathfrak{F}^{IP}_{\mathfrak{R}}(x_2)\big)$ iff $x_1 \le x_2 \Rightarrow \mathfrak{F}^P_{\mathfrak{R}}(x_1) \le \mathfrak{F}^P_{\mathfrak{R}}(x_2)$, this means that $\mathfrak{F}^{RP}_{\mathfrak{R}}(x_1) \le \mathfrak{F}^{RP}_{\mathfrak{R}}(x_2)$, $\mathfrak{F}^{IP}_{\mathfrak{R}}(x_1) \le \mathfrak{F}^{IP}_{\mathfrak{R}}(x_2)$.
3. $\mathfrak{F}^N_{\mathfrak{R}}(x_1\wedge x_2) \le \min\big(\mathfrak{F}^N_{\mathfrak{R}}(x_1), \mathfrak{F}^N_{\mathfrak{R}}(x_2)\big)$ this means that $\mathfrak{F}^{RN}_{\mathfrak{R}}(x_1\wedge x_2) \le \min\big(\mathfrak{F}^{RN}_{\mathfrak{R}}(x_1), \mathfrak{F}^{RN}_{\mathfrak{R}}(x_2)\big)$, $\mathfrak{F}^{IN}_{\mathfrak{R}}(x_1\wedge x_2) \le \min\big(\mathfrak{F}^{IN}_{\mathfrak{R}}(x_1), \mathfrak{F}^{IN}_{\mathfrak{R}}(x_2)\big)$ iff $x_1 \le x_2 \Rightarrow \mathfrak{F}^N_{\mathfrak{R}}(x_1) \le \mathfrak{F}^N_{\mathfrak{R}}(x_2)$, this means that $\mathfrak{F}^{RN}_{\mathfrak{R}}(x_1) \le \mathfrak{F}^{RN}_{\mathfrak{R}}(x_2)$, $\mathfrak{F}^{IN}_{\mathfrak{R}}(x_1) \le \mathfrak{F}^{IN}_{\mathfrak{R}}(x_2)$.

4. $\mathfrak{F}^{N}_{\mathfrak{R}}(x_1 \vee x_2) \leq \min\big(\mathfrak{F}^{N}_{\mathfrak{R}}(x_1), \mathfrak{F}^{N}_{\mathfrak{R}}(x_2)\big)$ this means that $\mathfrak{F}^{RN}_{\mathfrak{R}}(x_1 \vee x_2) \leq \min\big(\mathfrak{F}^{RN}_{\mathfrak{R}}(x_1), \mathfrak{F}^{RN}_{\mathfrak{R}}(x_2)\big)$, $\mathfrak{F}^{IN}_{\mathfrak{R}}(x_1 \vee x_2) \leq \min\big(\mathfrak{F}^{IN}_{\mathfrak{R}}(x_1), \mathfrak{F}^{IN}_{\mathfrak{R}}(x_2)\big)$ iff $x_1 \leq x_2 \Rightarrow \mathfrak{F}^{N}_{\mathfrak{R}}(x_1) \geq \mathfrak{F}^{N}_{\mathfrak{R}}(x_2)$, this means that $\mathfrak{F}^{RN}_{\mathfrak{R}}(x_1) \geq \mathfrak{F}^{RN}_{\mathfrak{R}}(x_2)$, $\mathfrak{F}^{IN}_{\mathfrak{R}}(x_1) \geq \mathfrak{F}^{IN}_{\mathfrak{R}}(x_2)$.

Proof Assume $x_1, x_2 \in \mathbb{L}$,

1. Let us consider that $\mathfrak{F}^{P}_{\mathfrak{R}}(x_1 \wedge x_2) \geq \max\big(\mathfrak{F}^{P}_{\mathfrak{R}}(x_1), \mathfrak{F}^{P}_{\mathfrak{R}}(x_2)\big)$ that is $\mathfrak{F}^{RP}_{\mathfrak{R}}(x_1 \wedge x_2) \geq \max\big(\mathfrak{F}^{RP}_{\mathfrak{R}}(x_1), \mathfrak{F}^{RP}_{\mathfrak{R}}(x_2)\big)$, $\mathfrak{F}^{IP}_{\mathfrak{R}}(x_1 \wedge x_2) \geq \max\big(\mathfrak{F}^{IP}_{\mathfrak{R}}(x_1), \mathfrak{F}^{IP}_{\mathfrak{R}}(x_2)\big)$. If $x_1 \leq x_2$, then $x_1 \wedge x_2 = x_1$. As $\mathfrak{F}^{RP}_{\mathfrak{R}}(x_1 \wedge x_2) \geq \max\big(\mathfrak{F}^{RP}_{\mathfrak{R}}(x_1), \mathfrak{F}^{RP}_{\mathfrak{R}}(x_2)\big)$, $\mathfrak{F}^{IP}_{\mathfrak{R}}(x_1 \wedge x_2) \geq \max\big(\mathfrak{F}^{IP}_{\mathfrak{R}}(x_1), \mathfrak{F}^{IP}_{\mathfrak{R}}(x_2)\big)$, thus, $\mathfrak{F}^{RP}_{\mathfrak{R}}(x_1) = \mathfrak{F}^{RP}_{\mathfrak{R}}(x_1 \wedge x_2) \geq \max\big(\mathfrak{F}^{RP}_{\mathfrak{R}}(x_1), \mathfrak{F}^{RP}_{\mathfrak{R}}(x_2)\big)$, $\mathfrak{F}^{IP}_{\mathfrak{R}}(x_1) = \mathfrak{F}^{IP}_{\mathfrak{R}}(x_1 \wedge x_2) \geq \max\big(\mathfrak{F}^{IP}_{\mathfrak{R}}(x_1), \mathfrak{F}^{IP}_{\mathfrak{R}}(x_2)\big)$ this implies that $\mathfrak{F}^{RP}_{\mathfrak{R}}(x_1) \geq \mathfrak{F}^{RP}_{\mathfrak{R}}(x_2)$, $\mathfrak{F}^{IP}_{\mathfrak{R}}(x_1) \geq \mathfrak{F}^{IP}_{\mathfrak{R}}(x_2)$ and hence $\mathfrak{F}^{P}_{\mathfrak{R}}(x_1) \geq \mathfrak{F}^{P}_{\mathfrak{R}}(x_2)$.

Conversely, assume that $x_1 \leq x_2 \Rightarrow \mathfrak{F}^{P}_{\mathfrak{R}}(x_1) \geq \mathfrak{F}^{P}_{\mathfrak{R}}(x_2)$, that is $\mathfrak{F}^{RP}_{\mathfrak{R}}(x_1) \geq \mathfrak{F}^{RP}_{\mathfrak{R}}(x_2)$, $\mathfrak{F}^{IP}_{\mathfrak{R}}(x_1) \geq \mathfrak{F}^{IP}_{\mathfrak{R}}(x_2)$ then $\mathfrak{F}^{RP}_{\mathfrak{R}}(x_1 \wedge x_2) \geq \mathfrak{F}^{RP}_{\mathfrak{R}}(x_1)$, $\mathfrak{F}^{IP}_{\mathfrak{R}}(x_1 \wedge x_2) \geq \mathfrak{F}^{IP}_{\mathfrak{R}}(x_1)$ and $\mathfrak{F}^{RP}_{\mathfrak{R}}(x_1 \wedge x_2) \geq \mathfrak{F}^{RP}_{\mathfrak{R}}(x_2)$, $\mathfrak{F}^{IP}_{\mathfrak{R}}(x_1 \wedge x_2) \geq \mathfrak{F}^{IP}_{\mathfrak{R}}(x_2)$. Thus, $\mathfrak{F}^{RP}_{\mathfrak{R}}(x_1 \wedge x_2) \geq \max\big(\mathfrak{F}^{RP}_{\mathfrak{R}}(x_1), \mathfrak{F}^{RP}_{\mathfrak{R}}(x_2)\big)$ and $\mathfrak{F}^{IP}_{\mathfrak{R}}(x_1 \wedge x_2) \geq \max\big(\mathfrak{F}^{IP}_{\mathfrak{R}}(x_1), \mathfrak{F}^{IP}_{\mathfrak{R}}(x_2)\big)$. This implies that $\mathfrak{F}^{P}_{\mathfrak{R}}(x_1 \wedge x_2) \geq \max\big(\mathfrak{F}^{P}_{\mathfrak{R}}(x_1), \mathfrak{F}^{P}_{\mathfrak{R}}(x_2)\big)$.

2. Let us consider that $\mathfrak{F}^{P}_{\mathfrak{R}}(x_1 \vee x_2) \geq \max\big(\mathfrak{F}^{P}_{\mathfrak{R}}(x_1), \mathfrak{F}^{P}_{\mathfrak{R}}(x_2)\big)$ that is $\mathfrak{F}^{RP}_{\mathfrak{R}}(x_1 \vee x_2) \geq \max\big(\mathfrak{F}^{RP}_{\mathfrak{R}}(x_1), \mathfrak{F}^{RP}_{\mathfrak{R}}(x_2)\big)$, $\mathfrak{F}^{IP}_{\mathfrak{R}}(x_1 \vee x_2) \geq \max\big(\mathfrak{F}^{IP}_{\mathfrak{R}}(x_1), \mathfrak{F}^{IP}_{\mathfrak{R}}(x_2)\big)$. If $x_1 \leq x_2$, then $x_1 \vee x_2 = x_2$. As $\mathfrak{F}^{RP}_{\mathfrak{R}}(x_1 \vee x_2) \geq \max\big(\mathfrak{F}^{RP}_{\mathfrak{R}}(x_1), \mathfrak{F}^{RP}_{\mathfrak{R}}(x_2)\big)$, $\mathfrak{F}^{IP}_{\mathfrak{R}}(x_1 \vee x_2) \geq \max\big(\mathfrak{F}^{IP}_{\mathfrak{R}}(x_1), \mathfrak{F}^{IP}_{\mathfrak{R}}(x_2)\big)$, thus, $\mathfrak{F}^{RP}_{\mathfrak{R}}(x_2) = \mathfrak{F}^{RP}_{\mathfrak{R}}(x_1 \vee x_2) \geq \max\big(\mathfrak{F}^{RP}_{\mathfrak{R}}(x_1), \mathfrak{F}^{RP}_{\mathfrak{R}}(x_2)\big)$, $\mathfrak{F}^{IP}_{\mathfrak{R}}(x_2) = \mathfrak{F}^{IP}_{\mathfrak{R}}(x_1 \vee x_2) \geq \max\big(\mathfrak{F}^{IP}_{\mathfrak{R}}(x_1), \mathfrak{F}^{IP}_{\mathfrak{R}}(x_2)\big)$ this implies that $\mathfrak{F}^{RP}_{\mathfrak{R}}(x_1) \leq \mathfrak{F}^{RP}_{\mathfrak{R}}(x_2)$, $\mathfrak{F}^{IP}_{\mathfrak{R}}(x_1) \leq \mathfrak{F}^{IP}_{\mathfrak{R}}(x_2)$ and hence $\mathfrak{F}^{P}_{\mathfrak{R}}(x_1) \leq \mathfrak{F}^{P}_{\mathfrak{R}}(x_2)$.

Conversely, assume that $x_1 \leq x_2 \Rightarrow \mathfrak{F}^{P}_{\mathfrak{R}}(x_1) \leq \mathfrak{F}^{P}_{\mathfrak{R}}(x_2)$, that is $\mathfrak{F}^{RP}_{\mathfrak{R}}(x_1) \leq \mathfrak{F}^{RP}_{\mathfrak{R}}(x_2)$, $\mathfrak{F}^{IP}_{\mathfrak{R}}(x_1) \leq \mathfrak{F}^{IP}_{\mathfrak{R}}(x_2)$ then $\mathfrak{F}^{RP}_{\mathfrak{R}}(x_1 \vee x_2) \geq \mathfrak{F}^{RP}_{\mathfrak{R}}(x_1)$, $\mathfrak{F}^{IP}_{\mathfrak{R}}(x_1 \vee x_2) \geq \mathfrak{F}^{IP}_{\mathfrak{R}}(x_1)$ and $\mathfrak{F}^{RP}_{\mathfrak{R}}(x_1 \vee x_2) \geq \mathfrak{F}^{RP}_{\mathfrak{R}}(x_2)$, $\mathfrak{F}^{IP}_{\mathfrak{R}}(x_1 \vee x_2) \geq \mathfrak{F}^{IP}_{\mathfrak{R}}(x_2)$. Thus, $\mathfrak{F}^{RP}_{\mathfrak{R}}(x_1 \vee x_2) \geq \max\big(\mathfrak{F}^{RP}_{\mathfrak{R}}(x_1), \mathfrak{F}^{RP}_{\mathfrak{R}}(x_2)\big)$ and $\mathfrak{F}^{IP}_{\mathfrak{R}}(x_1 \vee x_2) \geq \max\big(\mathfrak{F}^{IP}_{\mathfrak{R}}(x_1), \mathfrak{F}^{IP}_{\mathfrak{R}}(x_2)\big)$. This implies that $\mathfrak{F}^{P}_{\mathfrak{R}}(x_1 \vee x_2) \geq \max\big(\mathfrak{F}^{P}_{\mathfrak{R}}(x_1), \mathfrak{F}^{P}_{\mathfrak{R}}(x_2)\big)$.

3. Let us consider that $\mathfrak{F}^{N}_{\mathfrak{R}}(x_1 \wedge x_2) \leq \min\big(\mathfrak{F}^{N}_{\mathfrak{R}}(x_1), \mathfrak{F}^{N}_{\mathfrak{R}}(x_2)\big)$ that is $\mathfrak{F}^{RN}_{\mathfrak{R}}(x_1 \wedge x_2) \leq \min\big(\mathfrak{F}^{RN}_{\mathfrak{R}}(x_1), \mathfrak{F}^{RN}_{\mathfrak{R}}(x_2)\big)$, $\mathfrak{F}^{IN}_{\mathfrak{R}}(x_1 \wedge x_2) \leq \min\big(\mathfrak{F}^{IN}_{\mathfrak{R}}(x_1), \mathfrak{F}^{IN}_{\mathfrak{R}}(x_2)\big)$. If $x_1 \leq x_2$, then $x_1 \wedge x_2 = x_1$. As $\mathfrak{F}^{RN}_{\mathfrak{R}}(x_1 \wedge x_2) \leq \min\big(\mathfrak{F}^{RN}_{\mathfrak{R}}(x_1), \mathfrak{F}^{RN}_{\mathfrak{R}}(x_2)\big)$, $\mathfrak{F}^{IN}_{\mathfrak{R}}(x_1 \wedge x_2) \leq \min\big(\mathfrak{F}^{IN}_{\mathfrak{R}}(x_1), \mathfrak{F}^{IN}_{\mathfrak{R}}(x_2)\big)$, thus, $\mathfrak{F}^{RN}_{\mathfrak{R}}(x_1) = \mathfrak{F}^{RN}_{\mathfrak{R}}(x_1 \wedge x_2) \leq \min\big(\mathfrak{F}^{RN}_{\mathfrak{R}}(x_1), \mathfrak{F}^{RN}_{\mathfrak{R}}(x_2)\big)$, $\mathfrak{F}^{IN}_{\mathfrak{R}}(x_1) = \mathfrak{F}^{IN}_{\mathfrak{R}}(x_1 \wedge x_2) \leq \min\big(\mathfrak{F}^{IN}_{\mathfrak{R}}(x_1), \mathfrak{F}^{IN}_{\mathfrak{R}}(x_2)\big)$ this implies that $\mathfrak{F}^{RN}_{\mathfrak{R}}(x_1) \leq \mathfrak{F}^{RN}_{\mathfrak{R}}(x_2)$, $\mathfrak{F}^{IN}_{\mathfrak{R}}(x_1) \leq \mathfrak{F}^{IN}_{\mathfrak{R}}(x_2)$ and hence $\mathfrak{F}^{N}_{\mathfrak{R}}(x_1) \leq \mathfrak{F}^{N}_{\mathfrak{R}}(x_2)$.

Conversely, assume that $x_1 \leq x_2 \Rightarrow \mathfrak{F}^{N}_{\mathfrak{R}}(x_1) \leq \mathfrak{F}^{N}_{\mathfrak{R}}(x_2)$, that is $\mathfrak{F}^{RN}_{\mathfrak{R}}(x_1) \leq \mathfrak{F}^{RN}_{\mathfrak{R}}(x_2)$, $\mathfrak{F}^{IN}_{\mathfrak{R}}(x_1) \leq \mathfrak{F}^{IN}_{\mathfrak{R}}(x_2)$ then $\mathfrak{F}^{RN}_{\mathfrak{R}}(x_1 \wedge x_2) \leq \mathfrak{F}^{RN}_{\mathfrak{R}}(x_1)$, $\mathfrak{F}^{IN}_{\mathfrak{R}}(x_1 \wedge x_2) \leq \mathfrak{F}^{IN}_{\mathfrak{R}}(x_1)$ and $\mathfrak{F}^{RN}_{\mathfrak{R}}(x_1 \wedge x_2) \leq \mathfrak{F}^{RN}_{\mathfrak{R}}(x_2)$, $\mathfrak{F}^{IN}_{\mathfrak{R}}(x_1 \wedge x_2) \leq \mathfrak{F}^{IN}_{\mathfrak{R}}(x_2)$. Thus, $\mathfrak{F}^{RN}_{\mathfrak{R}}(x_1 \wedge x_2) \leq \min\big(\mathfrak{F}^{RN}_{\mathfrak{R}}(x_1), \mathfrak{F}^{RN}_{\mathfrak{R}}(x_2)\big)$ and $\mathfrak{F}^{IN}_{\mathfrak{R}}(x_1 \wedge x_2) \leq \min\big(\mathfrak{F}^{IN}_{\mathfrak{R}}(x_1), \mathfrak{F}^{IN}_{\mathfrak{R}}(x_2)\big)$. This implies that $\mathfrak{F}^{N}_{\mathfrak{R}}(x_1 \wedge x_2) \leq \min\big(\mathfrak{F}^{N}_{\mathfrak{R}}(x_1), \mathfrak{F}^{N}_{\mathfrak{R}}(x_2)\big)$.

4. Let us consider that $\mathfrak{F}^{N}_{\mathfrak{R}}(x_1 \vee x_2) \leq \min\big(\mathfrak{F}^{N}_{\mathfrak{R}}(x_1), \mathfrak{F}^{N}_{\mathfrak{R}}(x_2)\big)$ that is $\mathfrak{F}^{RN}_{\mathfrak{R}}(x_1 \vee x_2) \leq \min\big(\mathfrak{F}^{RN}_{\mathfrak{R}}(x_1), \mathfrak{F}^{RN}_{\mathfrak{R}}(x_2)\big)$, $\mathfrak{F}^{IN}_{\mathfrak{R}}(x_1 \vee x_2) \leq \min\big(\mathfrak{F}^{IN}_{\mathfrak{R}}(x_1), \mathfrak{F}^{IN}_{\mathfrak{R}}(x_2)\big)$. If $x_1 \leq x_2$, then $x_1 \vee x_2 = x_2$. As $\mathfrak{F}^{RN}_{\mathfrak{R}}(x_1 \vee x_2) \leq \min\big(\mathfrak{F}^{RN}_{\mathfrak{R}}(x_1), \mathfrak{F}^{RN}_{\mathfrak{R}}(x_2)\big)$,

$\mathfrak{F}^{IN}_{\Re}(x_1 \vee x_2) \leq \min\big(\mathfrak{F}^{IN}_{\Re}(x_1), \mathfrak{F}^{IN}_{\Re}(x_2)\big)$, thus, $\mathfrak{F}^{RN}_{\Re}(x_2) = \mathfrak{F}^{RN}_{\Re}(x_1 \vee x_2) \leq \min\big(\mathfrak{F}^{RN}_{\Re}(x_1), \mathfrak{F}^{RN}_{\Re}(x_2)\big)$, $\mathfrak{F}^{IN}_{\Re}(x_2) = \mathfrak{F}^{IN}_{\Re}(x_1 \vee x_2) \leq \min\big(\mathfrak{F}^{IN}_{\Re}(x_1), \mathfrak{F}^{IN}_{\Re}(x_2)\big)$ this implies that $\mathfrak{F}^{RN}_{\Re}(x_1) \geq \mathfrak{F}^{RN}_{\Re}(x_2)$, $\mathfrak{F}^{IN}_{\Re}(x_1) \geq \mathfrak{F}^{IN}_{\Re}(x_2)$ and hence $\mathfrak{F}^{N}_{\Re}(x_1) \geq \mathfrak{F}^{N}_{\Re}(x_2)$.

Conversely, assume that $x_1 \leq x_2 \Rightarrow \mathfrak{F}^{N}_{\Re}(x_1) \geq \mathfrak{F}^{N}_{\Re}(x_2)$, that is $\mathfrak{F}^{RN}_{\Re}(x_1) \geq \mathfrak{F}^{RN}_{\Re}(x_2)$, $\mathfrak{F}^{IN}_{\Re}(x_1) \geq \mathfrak{F}^{IN}_{\Re}(x_2)$ then $\mathfrak{F}^{RN}_{\Re}(x_1 \vee x_2) \leq \mathfrak{F}^{RN}_{\Re}(x_1)$, $\mathfrak{F}^{IN}_{\Re}(x_1 \vee x_2) \leq \mathfrak{F}^{IN}_{\Re}(x_1)$ and $\mathfrak{F}^{RN}_{\Re}(x_1 \vee x_2) \leq \mathfrak{F}^{RN}_{\Re}(x_2)$, $\mathfrak{F}^{IN}_{\Re}(x_1 \vee x_2) \leq \mathfrak{F}^{IN}_{\Re}(x_2)$. Thus, $\mathfrak{F}^{RN}_{\Re}(x_1 \vee x_2) \geq \min\big(\mathfrak{F}^{RN}_{\Re}(x_1), \mathfrak{F}^{RN}_{\Re}(x_2)\big)$ and $\mathfrak{F}^{IN}_{\Re}(x_1 \vee x_2) \geq \min\big(\mathfrak{F}^{IN}_{\Re}(x_1), \mathfrak{F}^{IN}_{\Re}(x_2)\big)$. This implies that $\mathfrak{F}^{N}_{\Re}(x_1 \vee x_2) \geq \min\big(\mathfrak{F}^{N}_{\Re}(x_1), \mathfrak{F}^{N}_{\Re}(x_2)\big)$.

Theorem 3.5 Assume that $\Re = \big(\mathfrak{F}^{P}_{\Re}, \mathfrak{F}^{N}_{\Re}\big) = \big(\mathfrak{F}^{RP}_{\Re} + \iota\mathfrak{F}^{IP}_{\Re}, \mathfrak{F}^{RN}_{\Re} + \iota\mathfrak{F}^{IN}_{\Re}\big)$ is a BCFSL of $\mathbb{L}$, then $\Re$ would be identified as BCFI if and only if the underneath two conditions hold $\forall x_1, x_2 \in \mathbb{L}$,

1. $\mathfrak{F}^{P}_{\Re}(x_1 \vee x_2) = \min\big(\mathfrak{F}^{P}_{\Re}(x_1), \mathfrak{F}^{P}_{\Re}(x_2)\big)$ this means that $\mathfrak{F}^{RP}_{\Re}(x_1 \vee x_2) = \min\big(\mathfrak{F}^{RP}_{\Re}(x_1), \mathfrak{F}^{RP}_{\Re}(x_2)\big)$, $\mathfrak{F}^{IP}_{\Re}(x_1 \vee x_2) = \min\big(\mathfrak{F}^{IP}_{\Re}(x_1), \mathfrak{F}^{IP}_{\Re}(x_2)\big)$.
2. $\mathfrak{F}^{N}_{\Re}(x_1 \vee x_2) = \max\big(\mathfrak{F}^{N}_{\Re}(x_1), \mathfrak{F}^{N}_{\Re}(x_2)\big)$ this means that $\mathfrak{F}^{RN}_{\Re}(x_1 \vee x_2) = \max\big(\mathfrak{F}^{RN}_{\Re}(x_1), \mathfrak{F}^{RN}_{\Re}(x_2)\big)$, $\mathfrak{F}^{IN}_{\Re}(x_1 \vee x_2) = \max\big(\mathfrak{F}^{IN}_{\Re}(x_1), \mathfrak{F}^{IN}_{\Re}(x_2)\big)$.

Proof Assume that $\Re$ is a BCFI of $\mathbb{L}$ and prove that 1 and 2 hold. Let $x_1, x_2 \in \mathbb{L}$, then by 1 and 3 of the definition of BCFI, we have $\mathfrak{F}^{P}_{\Re}(x_1 \vee x_2) \geq \min\big(\mathfrak{F}^{P}_{\Re}(x_1), \mathfrak{F}^{P}_{\Re}(x_2)\big)$ that is $\mathfrak{F}^{RP}_{\Re}(x_1 \vee x_2) \geq \min\big(\mathfrak{F}^{RP}_{\Re}(x_1), \mathfrak{F}^{RP}_{\Re}(x_2)\big)$, $\mathfrak{F}^{IP}_{\Re}(x_1 \wedge x_2) \geq \min\big(\mathfrak{F}^{IP}_{\Re}(x_1), \mathfrak{F}^{IP}_{\Re}(x_2)\big)$ and $\mathfrak{F}^{N}_{\Re}(x_1 \vee x_2) \leq \max\big(\mathfrak{F}^{N}_{\Re}(x_1), \mathfrak{F}^{N}_{\Re}(x_2)\big)$ that is $\mathfrak{F}^{RN}_{\Re}(x_1 \vee x_2) \leq \max\big(\mathfrak{F}^{RN}_{\Re}(x_1), \mathfrak{F}^{RN}_{\Re}(x_2)\big)$, $\mathfrak{F}^{IN}_{\Re}(x_1 \vee x_2) \leq \max\big(\mathfrak{F}^{IN}_{\Re}(x_1), \mathfrak{F}^{IN}_{\Re}(x_2)\big)$. As $x_1 \leq x_1 \vee x_2$, and $x_2 \leq x_1 \vee x_2$ then by Theorem (3.4) we have $\mathfrak{F}^{P}_{\Re}(x_1) \geq \mathfrak{F}^{P}_{\Re}(x_1 \vee x_2)$, this means that $\mathfrak{F}^{RP}_{\Re}(x_1) \geq \mathfrak{F}^{RP}_{\Re}(x_1 \vee x_2)$, $\mathfrak{F}^{IP}_{\Re}(x_1) \geq \mathfrak{F}^{IP}_{\Re}(x_1 \vee x_2)$ and $\mathfrak{F}^{P}_{\Re}(x_2) \geq \mathfrak{F}^{P}_{\Re}(x_1 \vee x_2)$, this means that $\mathfrak{F}^{RP}_{\Re}(x_2) \geq \mathfrak{F}^{RP}_{\Re}(x_1 \vee x_2)$, $\mathfrak{F}^{IP}_{\Re}(x_2) \geq \mathfrak{F}^{IP}_{\Re}(x_1 \vee x_2)$. Hence, $\min\big(\mathfrak{F}^{RP}_{\Re}(x_1), \mathfrak{F}^{RP}_{\Re}(x_2)\big) \geq \mathfrak{F}^{RP}_{\Re}(x_1 \vee x_2)$, $\min\big(\mathfrak{F}^{IP}_{\Re}(x_1), \mathfrak{F}^{RP}_{\Re}(x_2)\big) \geq \mathfrak{F}^{IP}_{\Re}(x_1 \vee x_2)$, this implies that $\min\big(\mathfrak{F}^{P}_{\Re}(x_1), \mathfrak{F}^{RP}_{\Re}(x_2)\big) \geq \mathfrak{F}^{P}_{\Re}(x_1 \vee x_2)$. Thus, $\mathfrak{F}^{P}_{\Re}(x_1 \vee x_2) = \min\big(\mathfrak{F}^{P}_{\Re}(x_1), \mathfrak{F}^{P}_{\Re}(x_2)\big)$. Similarly, $\mathfrak{F}^{N}_{\Re}(x_1 \vee x_2) = \max\big(\mathfrak{F}^{N}_{\Re}(x_1), \mathfrak{F}^{N}_{\Re}(x_2)\big)$.

Conversely, assume that 1 and 2 hold. To show that $\Re$ is BCFI of $\mathbb{L}$. As $\mathfrak{F}^{P}_{\Re}(x_1 \vee x_2) = \min\big(\mathfrak{F}^{P}_{\Re}(x_1), \mathfrak{F}^{P}_{\Re}(x_2)\big)$ this means that $\mathfrak{F}^{RP}_{\Re}(x_1 \vee x_2) = \min\big(\mathfrak{F}^{RP}_{\Re}(x_1), \mathfrak{F}^{RP}_{\Re}(x_2)\big)$, $\mathfrak{F}^{IP}_{\Re}(x_1 \vee x_2) = \min\big(\mathfrak{F}^{IP}_{\Re}(x_1), \mathfrak{F}^{IP}_{\Re}(x_2)\big)$ and $\mathfrak{F}^{N}_{\Re}(x_1 \vee x_2) = \max\big(\mathfrak{F}^{N}_{\Re}(x_1), \mathfrak{F}^{N}_{\Re}(x_2)\big)$ this means that $\mathfrak{F}^{RN}_{\Re}(x_1 \vee x_2) = \max\big(\mathfrak{F}^{RN}_{\Re}(x_1), \mathfrak{F}^{RN}_{\Re}(x_2)\big)$, $\mathfrak{F}^{IN}_{\Re}(x_1 \vee x_2) = \max\big(\mathfrak{F}^{IN}_{\Re}(x_1), \mathfrak{F}^{IN}_{\Re}(x_2)\big)$ for any $x_1, x_2 \in \mathbb{L}$, then obviously, $\mathfrak{F}^{RP}_{\Re}(x_1 \vee x_2) \geq \min\big(\mathfrak{F}^{RP}_{\Re}(x_1), \mathfrak{F}^{RP}_{\Re}(x_2)\big)$, $\mathfrak{F}^{IP}_{\Re}(x_1 \vee x_2) \geq \min\big(\mathfrak{F}^{IP}_{\Re}(x_1), \mathfrak{F}^{IP}_{\Re}(x_2)\big)$, $\mathfrak{F}^{RN}_{\Re}(x_1 \vee x_2) \leq \max\big(\mathfrak{F}^{RN}_{\Re}(x_1), \mathfrak{F}^{RN}_{\Re}(x_2)\big)$, $\mathfrak{F}^{IN}_{\Re}(x_1 \vee x_2) \leq \max\big(\mathfrak{F}^{IN}_{\Re}(x_1), \mathfrak{F}^{IN}_{\Re}(x_2)\big)$ for any $x_1, x_2 \in \mathbb{L}$. Now to demonstrate that $\mathfrak{F}^{RP}_{\Re}(x_1 \wedge x_2) \geq \max\big(\mathfrak{F}^{RP}_{\Re}(x_1), \mathfrak{F}^{RP}_{\Re}(x_2)\big)$, $\mathfrak{F}^{IP}_{\Re}(x_1 \wedge x_2) \geq \max\big(\mathfrak{F}^{IP}_{\Re}(x_1), \mathfrak{F}^{IP}_{\Re}(x_2)\big)$, $\mathfrak{F}^{RN}_{\Re}(x_1 \wedge x_2) \leq \min\big(\mathfrak{F}^{RN}_{\Re}(x_1), \mathfrak{F}^{RN}_{\Re}(x_2)\big)$, $\mathfrak{F}^{IN}_{\Re}(x_1 \vee x_2) \leq \min\big(\mathfrak{F}^{IN}_{\Re}(x_1), \mathfrak{F}^{IN}_{\Re}(x_2)\big)$ for any $x_1, x_2 \in \mathbb{L}$.

Let $x_1, x_2 \in \mathbb{L}$ and since $x_1 \vee (x_1 \wedge x_2) = x_1$ and $x_2 \vee (x_1 \wedge x_2) = x_2$, we have $\mathfrak{F}^{RP}_{\Re}(x_1 \vee (x_1 \wedge x_2)) = \mathfrak{F}^{RP}_{\Re}(x_1)$, $\mathfrak{F}^{IP}_{\Re}(x_1 \vee (x_1 \wedge x_2)) = \mathfrak{F}^{IP}_{\Re}(x_1)$ and $\mathfrak{F}^{RP}_{\Re}(x_2 \vee (x_1 \wedge x_2)) = \mathfrak{F}^{RP}_{\Re}(x_2)$, $\mathfrak{F}^{IP}_{\Re}(x_2 \vee (x_1 \wedge x_2)) = \mathfrak{F}^{IP}_{\Re}(x_2)$. Hence, $\min\big(\mathfrak{F}^{RP}_{\Re}(x_1), \mathfrak{F}^{RP}_{\Re}(x_1 \wedge x_2)\big) = \mathfrak{F}^{RP}_{\Re}(x_1)$. $\min\big(\mathfrak{F}^{IP}_{\Re}(x_1), \mathfrak{F}^{IP}_{\Re}(x_1 \wedge x_2)\big) = \mathfrak{F}^{IP}_{\Re}(x_1)$ and $\min\big(\mathfrak{F}^{RP}_{\Re}(x_2), \mathfrak{F}^{RP}_{\Re}(x_1 \wedge x_2)\big) = \mathfrak{F}^{RP}_{\Re}(x_2)$. $\min\big(\mathfrak{F}^{IP}_{\Re}(x_2), \mathfrak{F}^{IP}_{\Re}(x_1 \wedge x_2)\big) = \mathfrak{F}^{IP}_{\Re}(x_2)$.

Consequently, $\mathfrak{F}_{\mathfrak{R}}^{RP}(x_1 \wedge x_2) \geq \mathfrak{F}_{\mathfrak{R}}^{RP}(x_1)$, $\mathfrak{F}_{\mathfrak{R}}^{IP}(x_1 \wedge x_2) \geq \mathfrak{F}_{\mathfrak{R}}^{IP}(x_1)$ and $\mathfrak{F}_{\mathfrak{R}}^{RP}(x_1 \wedge x_2) \geq \mathfrak{F}_{\mathfrak{R}}^{RP}(x_2)$, $\mathfrak{F}_{\mathfrak{R}}^{IP}(x_1 \wedge x_2) \geq \mathfrak{F}_{\mathfrak{R}}^{IP}(x_2) \Rightarrow \mathfrak{F}_{\mathfrak{R}}^{P}(x_1 \wedge x_2) \geq \mathfrak{F}_{\mathfrak{R}}^{P}(x_1)$ and $\mathfrak{F}_{\mathfrak{R}}^{P}(x_1 \wedge x_2) \geq \mathfrak{F}_{\mathfrak{R}}^{P}(x_2)$. Hence $\mathfrak{F}_{\mathfrak{R}}^{P}(x_1 \wedge x_2) \geq \max\left(\mathfrak{F}_{\mathfrak{R}}^{P}(x_1), \mathfrak{F}_{\mathfrak{R}}^{P}(x_2)\right)$ for any $x_1, x_2 \in \mathbb{L}$.

On a similar pattern, $\mathfrak{F}_{\mathfrak{R}}^{N}(x_1 \wedge x_2) \leq \min\left(\mathfrak{F}_{\mathfrak{R}}^{N}(x_1), \mathfrak{F}_{\mathfrak{R}}^{N}(x_2)\right)$ can be proved. Hence $\mathfrak{R}$ is a BCFI of $\mathbb{L}$.

3.2 Bipolar Complex Fuzzy Subsemigroup

A pair $(\hat{S}, \star)$ is identified as semigroup (SG), where $\hat{\mathrm{S}}$ is a non-empty set and $\star$ is an associative binary operation over $\hat{\mathrm{S}}$. A non-empty subset $\mathfrak{R}_{\hat{\mathrm{S}}}$ of a semigroup, $\hat{\mathrm{S}}$ is identified as a subsemigroup (S-SG) of $\hat{\mathrm{S}}$ $x_1, x_2 \in \mathfrak{R}_{\hat{\mathrm{S}}}$ implies that $x_1 x_2 \in \mathfrak{R}_{\hat{\mathrm{S}}}$.

In the semigroup notion, ideals play a significant role, but it's not quite the same as in the theory of ring. For rings, all homomorphisms are founded by ideals, whereas for semigroups, only some of them are. Consider $\mathfrak{R}_{\mathbb{I}} \neq \emptyset$ as a subset of $\hat{\mathrm{S}}$, then $\mathfrak{R}_{\mathbb{I}}$ is identified as a left ideal if $\hat{S}\mathfrak{R}_{\mathbb{I}} \subseteq \mathfrak{R}_{\mathbb{I}}\left(x \in \mathfrak{R}_{\mathbb{I}}, s \in \hat{S} \Rightarrow sx \in \mathfrak{R}_{\mathbb{I}}\right)$, right ideal if $\mathfrak{R}_{\mathbb{I}}\hat{S} \subseteq \mathfrak{R}_{\mathbb{I}}\left(x \in \mathfrak{R}_{\mathbb{I}}, s \in \hat{S} \Rightarrow xs \in \mathfrak{R}_{\mathbb{I}}\right)$ and is identified as a two-sided ideal or ideal if it is both a right and left ideal.

Rehman et al. [14] investigated the notion of BCF semigroup. A few of the important notions are demonstrated underneath.

Definition 3.3 Let us consider a BCF subset $\mathfrak{R} = \left(\mathfrak{F}_{\mathfrak{R}}^{P}, \mathfrak{F}_{\mathfrak{R}}^{N}\right) = \left(\mathfrak{F}_{\mathfrak{R}}^{RP} + \iota\mathfrak{F}_{\mathfrak{R}}^{IP}, \mathfrak{F}_{\mathfrak{R}}^{RN} + \iota\mathfrak{F}_{\mathfrak{R}}^{IN}\right)$ on $\hat{\mathrm{S}}$, then $\mathfrak{R}$ is signified as BCF S-SG (BCFS-SG) if for all $x_1, x_2 \in \hat{\mathrm{S}}$,

1. $\mathfrak{F}_{\mathfrak{R}}^{P}(x_1x_2) \geq \min\left\{\mathfrak{F}_{\mathfrak{R}}^{P}(x_1), \mathfrak{F}_{\mathfrak{R}}^{P}(x_2)\right\} \Rightarrow \mathfrak{F}_{\mathfrak{R}}^{RP}(x_1x_2) \geq \min\left\{\mathfrak{F}_{\mathfrak{R}}^{RP}(x_1), \mathfrak{F}_{\mathfrak{R}}^{RP}(x_2)\right\}$ and $\mathfrak{F}_{\mathfrak{R}}^{IP}(x_1x_2) \geq \min\left\{\mathfrak{F}_{\mathfrak{R}}^{IP}(x_1), \mathfrak{F}_{\mathfrak{R}}^{IP}(x_2)\right\}$
2. $\mathfrak{F}_{\mathfrak{R}}^{N}(x_1x_2) \leq \max\left\{\mathfrak{F}_{\mathfrak{R}}^{N}(x_1), \mathfrak{F}_{\mathfrak{R}}^{N}(x_2)\right\} \Rightarrow \mathfrak{F}_{\mathfrak{R}}^{RN}(x_1x_2) \leq \max\left\{\mathfrak{F}_{\mathfrak{R}}^{RN}(x_1), \mathfrak{F}_{\mathfrak{R}}^{RN}(x_2)\right\}$ and $\mathfrak{F}_{\mathfrak{R}}^{IN}(x_1x_2) \leq \max\left\{\mathfrak{F}_{\mathfrak{R}}^{IN}(x_1), \mathfrak{F}_{\mathfrak{R}}^{IN}(x_2)\right\}$

Example 3.3 Let $\hat{\mathrm{S}} = \{\mathfrak{e}, x_1, x_2, x_3, x_4\}$

$\cdot$	$\mathfrak{e}$	x_1	x_2	x_3	x_4
$\mathfrak{e}$	$\mathfrak{e}$	$\mathfrak{e}$	$\mathfrak{e}$	$\mathfrak{e}$	$\mathfrak{e}$
x_1	$\mathfrak{e}$	$\mathfrak{e}$	$\mathfrak{e}$	$\mathfrak{e}$	$\mathfrak{e}$
x_2	$\mathfrak{e}$	$\mathfrak{e}$	x_2	x_3	x_4
x_3	$\mathfrak{e}$	$\mathfrak{e}$	x_2	x_3	x_4
x_4	$\mathfrak{e}$	$\mathfrak{e}$	x_2	x_3	x_4

Consider a BCFS $\mathfrak{R}$ on $\hat{\mathrm{S}}$ as underneath

$$\mathfrak{R} = \left\{\begin{array}{c} (\mathfrak{e}, (0.85 + \iota 0.82, -0.17 - \iota 0.2)), (x_1, (0.65 + \iota 0.7, -0.28 - \iota 0.31)), \\ (x_2, (0.45 + \iota 0.57, -0.55 - \iota 0.25)), (x_3, (0.45 + \iota 0.57, -0.55 - \iota 0.25)), \\ (x_4, (0.45 + \iota 0.57, -0.55 - \iota 0.25)) \end{array}\right\}$$

Then, $\mathfrak{R}$ is a BCFS-SG.

The definition of BCF left and right ideals are investigated as follows.

Definition 3.4 Let $\mathfrak{R} = \left(\mathfrak{F}_{\mathfrak{R}}^{P}, \mathfrak{F}_{\mathfrak{R}}^{N}\right) = \left(\mathfrak{F}_{\mathfrak{R}}^{RP} + \iota\mathfrak{F}_{\mathfrak{R}}^{IP}, \mathfrak{F}_{\mathfrak{R}}^{RN} + \iota\mathfrak{F}_{\mathfrak{R}}^{IN}\right)$ be a BCFS over $\hat{\mathrm{S}}$, then.

1. If $\forall x_1, x_2 \in \hat{\mathrm{S}}$
 i. $\mathfrak{F}_{\mathfrak{R}}^{P}(x_1x_2) \geq \mathfrak{F}_{\mathfrak{R}}^{P}(x_2) \Rightarrow \mathfrak{F}_{\mathfrak{R}}^{RP}(x_1x_2) \geq \mathfrak{F}_{\mathfrak{R}}^{RP}(x_2)$ and $\mathfrak{F}_{\mathfrak{R}}^{IP}(x_1x_2) \geq \mathfrak{F}_{\mathfrak{R}}^{IP}(x_2)$
 ii. $\mathfrak{F}_{\mathfrak{R}}^{N}(x_1x_2) \leq \mathfrak{F}_{\mathfrak{R}}^{N}(x_2) \Rightarrow \mathfrak{F}_{\mathfrak{R}}^{RN}(x_1x_2) \leq \mathfrak{F}_{\mathfrak{R}}^{RN}(x_2)$ and $\mathfrak{F}_{\mathfrak{R}}^{IN}(x_1x_2) \leq \mathfrak{F}_{\mathfrak{R}}^{IN}(x_2)$

 Then, $\mathfrak{R}$ is signified as BCF left ideal of $\hat{\mathrm{S}}$.

2. If $\forall x_1, x_2 \in \hat{\mathrm{S}}$
 i. $\mathfrak{F}_{\mathfrak{R}}^{P}(x_1x_2) \geq \mathfrak{F}_{\mathfrak{R}}^{P}(x_1) \Rightarrow \mathfrak{F}_{\mathfrak{R}}^{RP}(x_1x_2) \geq \mathfrak{F}_{\mathfrak{R}}^{RP}(x_1)$ and $\mathfrak{F}_{\mathfrak{R}}^{IP}(x_1x_2) \geq \mathfrak{F}_{\mathfrak{R}}^{IP}(x_1)$
 ii. $\mathfrak{F}_{\mathfrak{R}}^{N}(x_1x_2) \leq \mathfrak{F}_{\mathfrak{R}}^{N}(x_1) \Rightarrow \mathfrak{F}_{\mathfrak{R}}^{RN}(x_1x_2) \leq \mathfrak{F}_{\mathfrak{R}}^{RN}(x_1)$ and $\mathfrak{F}_{\mathfrak{R}}^{IN}(x_1x_2) \leq \mathfrak{F}_{\mathfrak{R}}^{IN}(x_1)$

 Then, $\mathfrak{R}$ is signified as BCF right ideal of $\hat{\mathrm{S}}$.

2. If $\mathfrak{R}$ is both BCF left and right ideal then it is signified as BCF ideal.

Example 3.4 Reconsider an SG $\hat{\mathrm{S}}$ interpreted in example (3.3) and define a BCFS $\mathfrak{R}$ as follows.

$$\mathfrak{R} = \left\{ \begin{array}{c} (\mathfrak{e}, (0.85 + \iota 0.82, -0.55 - \iota 0.25)), (x_1, (0.65 + \iota 0.7, -0.28 - \iota 0.31)), \\ (x_2, (0.45 + \iota 0.57, -0.17 - \iota 0.2)), (x_3, (0.45 + \iota 0.57, -0.17 - \iota 0.2)), \\ (x_4, (0.45 + \iota 0.57, -0.17 - \iota 0.2)) \end{array} \right\}$$

Then $\mathfrak{R}$ is BCF left ideal, right ideal, and ideal over $\hat{\mathrm{S}}$.

Remark 3.4 In general, every BCF left ideal, right ideal, and ideal over $\hat{\mathrm{S}}$ is a BCFS-SG but the converse is not true in general.

3.3 Bipolar Complex Fuzzy Subgroups

Suppose a non-empty set $\mathbb{G}$ and $\star$ as a binary operation defined on $\mathbb{G}$, then $(\mathbb{G}, \star)$ is identified as a group if and only if.

1. $\forall x_1, x_2 \in \mathbb{G}, x_1 \star x_2 \in \mathbb{G}$
2. $\forall x_1, x_2, x_3 \in \mathbb{G}, (x_1 \star x_2) \star x_3 = x_1 \star (x_2 \star x_3)$
3. $\exists\, \mathbb{e} \in \mathbb{G}$ such that $x \star \mathbb{e} = \mathbb{e} \star x = x, \forall x \in \mathbb{G}$
4. For every $x \in \mathbb{G}$, $\exists\, x^{-1} \in \mathbb{G}$ such that $x \star x^{-1} = x^{-1} \star x = \mathbb{e}$.

A group $(\mathbb{G}, \star)$ is identified as abelian or commutative if and only if $\star$ is commutative in $\mathbb{G}$ that is $\forall x_1, x_2 \in \mathbb{G}$, $x_1 \star x_2 = x_2 \star x_1$. Let $\hat{H} \neq \emptyset$ be a subset of a group $\mathbb{G}$. Then $\hat{H}$ is identified as a subgroup of $\mathbb{G}$ if and only if $\hat{H}$ is itself a group under induced binary operation of $\mathbb{G}$ and donated as $\hat{H} \preccurlyeq \mathbb{G}$. Every group $\mathbb{G}$ has at least two subgroups that is identity group and $\mathbb{G}$ itself. A subgroup $\hat{H}$ of a group $\mathbb{G}$ is identified as a normal subgroup of $\mathbb{G}$, if and only if for each $x \in \mathbb{G}$,, $x\hat{H} = \hat{H}x$. Let $\mathcal{W}$ be a non-empty set and $\mathbb{G}$ be a group. Then normalizer of $\mathcal{W}$ is signified and demonstrated as.

$$\mathcal{N}_{\mathbb{G}}(\mathcal{W}) = \{x \in \mathbb{G} : x\mathcal{W} = \mathcal{W}x\}$$

Yang et al. [15] discussed the theory of BCF subgroups and related results. Underneath we discuss a few of them.

Definition 3.5 Let us consider a BCF subset $\mathfrak{R} = \left(\mathfrak{F}_{\mathfrak{R}}^{P}, \mathfrak{F}_{\mathfrak{R}}^{N}\right) = \left(\mathfrak{F}_{\mathfrak{R}}^{RP} + \iota\mathfrak{F}_{\mathfrak{R}}^{IP}, \mathfrak{F}_{\mathfrak{R}}^{RN} + \iota\mathfrak{F}_{\mathfrak{R}}^{IN}\right)$ of $\mathbb{G}$, then $\mathfrak{R}$ is identified as a bipolar complex fuzzy subgroup (BCF-SG) if $\forall x_1, x_2 \in \mathbb{G}$,

1. $\mathfrak{F}_{\mathfrak{R}}^{P}(x_1x_2) \geq \min\left\{\mathfrak{F}_{\mathfrak{R}}^{P}(x_1), \mathfrak{F}_{\mathfrak{R}}^{P}(x_2)\right\} \Rightarrow \mathfrak{F}_{\mathfrak{R}}^{RP}(x_1x_2) \geq \min\left\{\mathfrak{F}_{\mathfrak{R}}^{RP}(x_1), \mathfrak{F}_{\mathfrak{R}}^{RP}(x_2)\right\}$ and $\mathfrak{F}_{\mathfrak{R}}^{IP}(x_1x_2) \geq \min\left\{\mathfrak{F}_{\mathfrak{R}}^{IP}(x_1), \mathfrak{F}_{\mathfrak{R}}^{IP}(x_2)\right\}$
2. $\mathfrak{F}_{\mathfrak{R}}^{P}\left(x_1^{-1}\right) \geq \mathfrak{F}_{\mathfrak{R}}^{P}(x_1) \Rightarrow \mathfrak{F}_{\mathfrak{R}}^{RP}\left(x_1^{-1}\right) \geq \mathfrak{F}_{\mathfrak{R}}^{RP}(x_1)$ and $\mathfrak{F}_{\mathfrak{R}}^{IP}\left(x_1^{-1}\right) \geq \mathfrak{F}_{\mathfrak{R}}^{IP}(x_1)$
3. $\mathfrak{F}_{\mathfrak{R}}^{N}(x_1x_2) \leq \max\left\{\mathfrak{F}_{\mathfrak{R}}^{N}(x_1), \mathfrak{F}_{\mathfrak{R}}^{N}(x_2)\right\} \Rightarrow \mathfrak{F}_{\mathfrak{R}}^{RN}(x_1x_2) \leq \max\left\{\mathfrak{F}_{\mathfrak{R}}^{RN}(x_1), \mathfrak{F}_{\mathfrak{R}}^{RN}(x_2)\right\}$ and $\mathfrak{F}_{\mathfrak{R}}^{IN}(x_1x_2) \leq \max\left\{\mathfrak{F}_{\mathfrak{R}}^{IN}(x_1), \mathfrak{F}_{\mathfrak{R}}^{IN}(x_2)\right\}$
4. $\mathfrak{F}_{\mathfrak{R}}^{N}\left(x_1^{-1}\right) \leq \mathfrak{F}_{\mathfrak{R}}^{N}(x_1) \Rightarrow \mathfrak{F}_{\mathfrak{R}}^{RN}\left(x_1^{-1}\right) \leq \mathfrak{F}_{\mathfrak{R}}^{RN}(x_1)$ and $\mathfrak{F}_{\mathfrak{R}}^{IN}\left(x_1^{-1}\right) \leq \mathfrak{F}_{\mathfrak{R}}^{IN}(x_1)$

Or

1. $\mathfrak{F}_{\mathfrak{R}}^{P}\left(x_1^{-1}x_2\right) \geq \min\left\{\mathfrak{F}_{\mathfrak{R}}^{P}(x_1), \mathfrak{F}_{\mathfrak{R}}^{P}(x_2)\right\} \Rightarrow \mathfrak{F}_{\mathfrak{R}}^{RP}\left(x_1^{-1}x_2\right) \geq \min\left\{\mathfrak{F}_{\mathfrak{R}}^{RP}(x_1), \mathfrak{F}_{\mathfrak{R}}^{RP}(x_2)\right\}$ and $\mathfrak{F}_{\mathfrak{R}}^{IP}\left(x_1^{-1}x_2\right) \geq \min\left\{\mathfrak{F}_{\mathfrak{R}}^{IP}(x_1), \mathfrak{F}_{\mathfrak{R}}^{IP}(x_2)\right\}$
2. $\mathfrak{F}_{\mathfrak{R}}^{N}\left(x_1^{-1}x_2\right) \leq \max\left\{\mathfrak{F}_{\mathfrak{R}}^{N}(x_1), \mathfrak{F}_{\mathfrak{R}}^{N}(x_2)\right\} \Rightarrow \mathfrak{F}_{\mathfrak{R}}^{RN}\left(x_1^{-1}x_2\right) \leq \max\left\{\mathfrak{F}_{\mathfrak{R}}^{RN}(x_1), \mathfrak{F}_{\mathfrak{R}}^{RN}(x_2)\right\}$ and $\mathfrak{F}_{\mathfrak{R}}^{IN}\left(x_1^{-1}x_2\right) \leq \max\left\{\mathfrak{F}_{\mathfrak{R}}^{IN}(x_1), \mathfrak{F}_{\mathfrak{R}}^{IN}(x_2)\right\}$.

A BCF-SG $\mathfrak{R}$ of $\mathbb{G}$ is c BCF abelian subgroup of G if and only if $\forall x_1, x_2 \in \mathbb{G}$.

1. $\mathfrak{F}_{\mathfrak{R}}^{P}(x_1x_2) = \mathfrak{F}_{\mathfrak{R}}^{P}(x_2x_1) \Rightarrow \mathfrak{F}_{\mathfrak{R}}^{RP}(x_1x_2) = \mathfrak{F}_{\mathfrak{R}}^{RP}(x_2x_1)$ and $\mathfrak{F}_{\mathfrak{R}}^{IP}(x_1x_2) = \mathfrak{F}_{\mathfrak{R}}^{IP}(x_2x_1)$
2. $\mathfrak{F}_{\mathfrak{R}}^{N}(x_1x_2) = \mathfrak{F}_{\mathfrak{R}}^{N}(x_2x_1) \Rightarrow \mathfrak{F}_{\mathfrak{R}}^{RN}(x_1x_2) = \mathfrak{F}_{\mathfrak{R}}^{RN}(x_2x_1)$ and $\mathfrak{F}_{\mathfrak{R}}^{IN}(x_1x_2) = \mathfrak{F}_{\mathfrak{R}}^{IN}(x_2x_1)$

Example 3.5 Suppose $\mathbb{G} = \underset{4}{\overset{\vee}{K}} = \{1, x_1, x_2, x_1x_2\}$ as a group such that $(x_1)^2 = (x_2)^2 = (x_1x_2)^2 = 1$ and $\mathfrak{R}$ as a BCF subset of $\mathbb{G}$ interpreted as.

$$\mathfrak{R} = \left\{ \begin{array}{c} (1, (0.75 + \iota 0.68, -0.1 - \iota 0.05)), (x_1, (0.45 + \iota 0.43, -0.32 - \iota 0.56)), \\ (x_2, (0.45 + \iota 0.43, -0.32 - \iota 0.56)), (x_1x_2, (0.55 + \iota 0.7, -0.32 - \iota 0.65)) \end{array} \right\}$$

Consequently, the BCF subset $\mathfrak{R}$ of $\breve{\mathrm{G}}$ is a BCF-SG.

Underneath is the definition of the BCF normal subgroup (BCF-NSG).

Definition 3.6 A BCF-SG $\mathfrak{R} = \left(\mathfrak{F}_{\mathfrak{R}}^{P}, \mathfrak{F}_{\mathfrak{R}}^{N}\right) = \left(\mathfrak{F}_{\mathfrak{R}}^{RP} + \iota\mathfrak{F}_{\mathfrak{R}}^{IP}, \mathfrak{F}_{\mathfrak{R}}^{RN} + \iota\mathfrak{F}_{\mathfrak{R}}^{IN}\right)$ is identified as BCF-NSG of $\mathbb{G}$ if for all $x_1, x_2 \in \mathbb{G}$, the underneath hold.

1. $\mathfrak{F}_{\mathfrak{R}}^{P}\left(x_2^{-1}x_1x_2\right) \geq \mathfrak{F}_{\mathfrak{R}}^{P}(x_1) \Rightarrow \mathfrak{F}_{\mathfrak{R}}^{RP}\left(x_2^{-1}x_1x_2\right) \geq \mathfrak{F}_{\mathfrak{R}}^{RP}(x_1)$ and $\mathfrak{F}_{\mathfrak{R}}^{IP}\left(x_2^{-1}x_1x_2\right) \geq \mathfrak{F}_{\mathfrak{R}}^{IP}(x_1)$
2. $\mathfrak{F}_{\mathfrak{R}}^{N}\left(x_2^{-1}x_1x_2\right) \leq \mathfrak{F}_{\mathfrak{R}}^{N}(x_1) \Rightarrow \mathfrak{F}_{\mathfrak{R}}^{RN}\left(x_2^{-1}x_1x_2\right) \leq \mathfrak{F}_{\mathfrak{R}}^{RN}(x_1)$ and $\mathfrak{F}_{\mathfrak{R}}^{IN}\left(x_2^{-1}x_1x_2\right) \leq \mathfrak{F}_{\mathfrak{R}}^{IN}(x_1)$

The normalizer of a BCF-SG $\mathfrak{R}$ of $\mathbb{G}$ is identified and demonstrated as

$$\mathcal{N}_{\mathbb{G}}(\mathfrak{R}) = \left\{x_2 \in \mathbb{G} : \mathfrak{F}_{\mathfrak{R}}^{P}\left(x_2^{-1}x_1x_2\right) \geq \mathfrak{F}_{\mathfrak{R}}^{P}(x_1) \& \mathfrak{F}_{\mathfrak{R}}^{N}\left(x_2^{-1}x_1x_2\right) \leq \mathfrak{F}_{\mathfrak{R}}^{N}(x_1), \forall x_1 \in \mathbb{G}\right\}$$
$$= \left\{\begin{array}{l} x_2 \in \mathbb{G} : \mathfrak{F}_{\mathfrak{R}}^{RP}\left(x_2^{-1}x_1x_2\right) \geq \mathfrak{F}_{\mathfrak{R}}^{RP}(x_1), \mathfrak{F}_{\mathfrak{R}}^{IP}\left(x_2^{-1}x_1x_2\right) \geq \mathfrak{F}_{\mathfrak{R}}^{IP}(x_1), \\ \mathfrak{F}_{\mathfrak{R}}^{RN}\left(x_2^{-1}x_1x_2\right) \leq \mathfrak{F}_{\mathfrak{R}}^{RN}(x_1), \mathfrak{F}_{\mathfrak{R}}^{IN}\left(x_2^{-1}x_1x_2\right) \leq \mathfrak{F}_{\mathfrak{R}}^{IN}(x_1) \end{array}, \forall x_1 \in \mathbb{G}\right\}$$

Next, we discuss the BCF cosets.

Definition 3.7 Assume a BCF-SG $\mathfrak{R} = \left(\mathfrak{F}_{\mathfrak{R}}^{P}, \mathfrak{F}_{\mathfrak{R}}^{N}\right) = \left(\mathfrak{F}_{\mathfrak{R}}^{RP} + \iota\mathfrak{F}_{\mathfrak{R}}^{IP}, \mathfrak{F}_{\mathfrak{R}}^{RN} + \iota\mathfrak{F}_{\mathfrak{R}}^{IN}\right)$ of $\mathbb{G}$, then.

1. BCF left coset $\mathfrak{R} = \left(\mathfrak{F}_{\mathfrak{R}}^{P}, \mathfrak{F}_{\mathfrak{R}}^{N}\right) = \left(\mathfrak{F}_{\mathfrak{R}}^{RP} + \iota\mathfrak{F}_{\mathfrak{R}}^{IP}, \mathfrak{F}_{\mathfrak{R}}^{RN} + \iota\mathfrak{F}_{\mathfrak{R}}^{IN}\right)$ in $\mathbb{G}$ is interpreted by $x_1 \in \mathbb{G}$ is a bipolar complex fuzzy subset $\mathfrak{R}_{x_1}^{\mathbb{I}} = \left(\mathfrak{F}_{\mathfrak{R}_{x_1}^{\mathbb{I}}}^{P}, \mathfrak{F}_{\mathfrak{R}_{x_1}^{\mathbb{I}}}^{N}\right) = \left(\mathfrak{F}_{\mathfrak{R}_{x_1}^{\mathbb{I}}}^{RP} + \iota\mathfrak{F}_{\mathfrak{R}_{x_1}^{\mathbb{I}}}^{IP}, \mathfrak{F}_{\mathfrak{R}_{x_1}^{\mathbb{I}}}^{RN} + \iota\mathfrak{F}_{\mathfrak{R}_{x_1}^{\mathbb{I}}}^{IN}\right)$, of $\mathbb{G}$, where, $\mathfrak{F}_{\mathfrak{R}_{x_1}^{\mathbb{I}}}^{RP} : \mathbb{G} \to [0, 1]$ demonstrate $\mathfrak{F}_{\mathfrak{R}_{x_1}^{\mathbb{I}}}^{RP}(x_2) = \mathfrak{F}_{\mathfrak{R}_{x_1}^{\mathbb{I}}}^{RP}\left(x_1^{-1}x_2\right)$, $\mathfrak{F}_{\mathfrak{R}_{x_1}^{\mathbb{I}}}^{IP} : \mathbb{G} \to [0, 1]$ demonstrate $\mathfrak{F}_{\mathfrak{R}_{x_1}^{\mathbb{I}}}^{IP}(x_2) = \mathfrak{F}_{\mathfrak{R}_{x_1}^{\mathbb{I}}}^{IP}\left(x_1^{-1}x_2\right)$, $\mathfrak{F}_{\mathfrak{R}_{x_1}^{\mathbb{I}}}^{RN} : \mathbb{G} \to [-1, 0]$ designated as $\mathfrak{F}_{\mathfrak{R}_{x_1}^{\mathbb{I}}}^{RN}(x_2) = \mathfrak{F}_{\mathfrak{R}_{x_1}^{\mathbb{I}}}^{RN}\left(x_1^{-1}x_2\right)$, and $\mathfrak{F}_{\mathfrak{R}_{x_1}^{\mathbb{I}}}^{IN} : \mathbb{G} \to [-1, 0]$ designated as $\mathfrak{F}_{\mathfrak{R}_{x_1}^{\mathbb{I}}}^{IN}(x_2) = \mathfrak{F}_{\mathfrak{R}_{x_1}^{\mathbb{I}}}^{IN}\left(x_1^{-1}x_2\right)$
2. BCF right coset $\mathfrak{R} = \left(\mathfrak{F}_{\mathfrak{R}}^{P}, \mathfrak{F}_{\mathfrak{R}}^{N}\right) = \left(\mathfrak{F}_{\mathfrak{R}}^{RP} + \iota\mathfrak{F}_{\mathfrak{R}}^{IP}, \mathfrak{F}_{\mathfrak{R}}^{RN} + \iota\mathfrak{F}_{\mathfrak{R}}^{IN}\right)$ in $\mathbb{G}$ is interpreted by $x_1 \in \mathbb{G}$ is a bipolar complex fuzzy subset $\mathfrak{R}_{\mathrm{x}_1}^{\setminus} = \left(\mathfrak{F}_{\mathfrak{R}_{\mathrm{x}_1}^{\setminus}}^{\mathrm{P}}, \mathfrak{F}_{\mathfrak{R}_{\mathrm{x}_1}^{\setminus}}^{\mathrm{N}}\right) = \left(\mathfrak{F}_{\mathfrak{R}_{\mathrm{x}_1}^{\setminus}}^{\mathrm{RP}} + \iota\mathfrak{F}_{\mathfrak{R}_{\mathrm{x}_1}^{\setminus}}^{\mathrm{IP}}, \mathfrak{F}_{\mathfrak{R}_{\mathrm{x}_1}^{\setminus}}^{\mathrm{RN}} + \iota\mathfrak{F}_{\mathfrak{R}_{\mathrm{x}_1}^{\setminus}}^{\mathrm{IN}}\right)$, of $\mathbb{G}$, where, $\mathfrak{F}_{\mathfrak{R}_{\mathrm{x}_1}^{\setminus}}^{\mathrm{RP}} : \mathbb{G} \to [0, 1]$ demonstrate $\mathfrak{F}_{\mathfrak{R}_{\mathrm{x}_1}^{\setminus}}^{\mathrm{RP}}(\mathrm{x}_2) = \mathfrak{F}_{\mathfrak{R}_{\mathrm{x}_1}^{\setminus}}^{\mathrm{RP}}\left(\mathrm{x}_2\mathrm{x}_1^{-1}\right)$, $\mathfrak{F}_{\mathfrak{R}_{\mathrm{x}_1}^{\setminus}}^{\mathrm{IP}} : \mathbb{G} \to [0, 1]$ demonstrate $\mathfrak{F}_{\mathfrak{R}_{\mathrm{x}_1}^{\setminus}}^{\mathrm{IP}}(\mathrm{x}_2) = \mathfrak{F}_{\mathfrak{R}_{\mathrm{x}_1}^{\setminus}}^{\mathrm{IP}}\left(\mathrm{x}_2\mathrm{x}_1^{-1}\right)$, $\mathfrak{F}_{\mathfrak{R}_{\mathrm{x}_1}^{\setminus}}^{\mathrm{RN}} : \mathbb{G} \to [-1, 0]$ designated as $\mathfrak{F}_{\mathfrak{R}_{\mathrm{x}_1}^{\setminus}}^{\mathrm{RN}}(\mathrm{x}_2) = \mathfrak{F}_{\mathfrak{R}_{\mathrm{x}_1}^{\setminus}}^{\mathrm{RN}}\left(\mathrm{x}_2\mathrm{x}_1^{-1}\right)$, and $\mathfrak{F}_{\mathfrak{R}_{\mathrm{x}_1}^{\setminus}}^{\mathrm{IN}} : \mathbb{G} \to [-1, 0]$ designated as $\mathfrak{F}_{\mathfrak{R}_{\mathrm{x}_1}^{\setminus}}^{\mathrm{IN}}(\mathrm{x}_2) = \mathfrak{F}_{\mathfrak{R}_{\mathrm{x}_1}^{\setminus}}^{\mathrm{IN}}\left(\mathrm{x}_2\mathrm{x}_1^{-1}\right)$

3.4 Bipolar Complex Fuzzy Subsemirings

A non-empty set $\mathfrak{J}$ along two binary operations //+// and //.// is identified as semiring if $\mathfrak{J}$ is a commutative monoid under // + // and semigroup under //.// and right, left distributive laws of multiplication over addition hold. $(\mathfrak{J}, +, .)$ signifies a semiring and absorbing zero is the additive identity. A subsemiring $\mathfrak{J}_{\mathbb{S}}$ is a non-empty subset of $\mathfrak{J}$, which itself is a semiring under the induced binary operations of $\mathfrak{J}$. A left (right) ideal $\mathfrak{J}_{\mathbb{S}}$ of semiring $\mathfrak{J}$ is a non-empty subset of $\mathfrak{J}$, that is closed under $+$ and $\forall x \in \mathfrak{J}$, and $x_{\mathbb{S}} \in \mathfrak{J}_{\mathbb{S}}$, $xx_{\mathbb{S}} \in \mathfrak{J}_{\mathbb{S}}(x_{\mathbb{S}}x \in \mathfrak{J}_{\mathbb{S}})$ and $\mathfrak{J}_{\mathbb{S}}$ is identified as ideal or both-sided ideal if $\mathfrak{J}_{\mathbb{S}}$ is both left and right ideal. A bi-ideal $\mathfrak{J}_{\mathbb{S}}$ of a semiring, $\mathfrak{J}$ is a non-empty subset of $\mathfrak{J}$ which is a subsemiring of $\mathfrak{J}$ and $\mathfrak{J}_{\mathbb{S}}\mathfrak{J}\mathfrak{J}_{\mathbb{S}} \subseteq \mathfrak{J}_{\mathbb{S}}$.

Underneath we will discuss the theory of bipolar complex fuzzy semiring.

Definition 3.8 Assume a BCFS $\mathfrak{R} = \left(\mathfrak{F}^{P}_{\mathfrak{R}}, \mathfrak{F}^{N}_{\mathfrak{R}}\right) = \left(\mathfrak{F}^{RP}_{\mathfrak{R}} + \iota\mathfrak{F}^{IP}_{\mathfrak{R}}, \mathfrak{F}^{RN}_{\mathfrak{R}} + \iota\mathfrak{F}^{IN}_{\mathfrak{R}}\right)$ of a semiring $\mathfrak{J}$, then $\mathfrak{R}$ is identified as BCF subsemiring (BCFSSR) if the underneath axioms are valid.

1. $\mathfrak{F}^{P}_{\mathfrak{R}}(0) \geq \mathfrak{F}^{P}_{\mathfrak{R}}(x)$ this means that $\mathfrak{F}^{RP}_{\mathfrak{R}}(0) \geq \mathfrak{F}^{RP}_{\mathfrak{R}}(x)$ and $\mathfrak{F}^{IP}_{\mathfrak{R}}(0) \geq \mathfrak{F}^{IP}_{\mathfrak{R}}(x)$
2. $\mathfrak{F}^{N}_{\mathfrak{R}}(0) \leq \mathfrak{F}^{N}_{\mathfrak{R}}(x)$ this means that $\mathfrak{F}^{RN}_{\mathfrak{R}}(0) \leq \mathfrak{F}^{RN}_{\mathfrak{R}}(x)$ and $\mathfrak{F}^{IN}_{\mathfrak{R}}(0) \leq \mathfrak{F}^{IN}_{\mathfrak{R}}(x)$
3. $\mathfrak{F}^{P}_{\mathfrak{R}}(x_1 + x_2) \geq \min\left(\mathfrak{F}^{P}_{\mathfrak{R}}(x_1), \mathfrak{F}^{P}_{\mathfrak{R}}(x_2)\right)$ this means that $\mathfrak{F}^{RP}_{\mathfrak{R}}(x_1 + x_2) \geq \min\left(\mathfrak{F}^{RP}_{\mathfrak{R}}(x_1), \mathfrak{F}^{RP}_{\mathfrak{R}}(x_2)\right), \mathfrak{F}^{IP}_{\mathfrak{R}}(x_1 + x_2) \geq \min\left(\mathfrak{F}^{IP}_{\mathfrak{R}}(x_1), \mathfrak{F}^{IP}_{\mathfrak{R}}(x_2)\right)$
4. $\mathfrak{F}^{N}_{\mathfrak{R}}(x_1 + x_2) \leq \max\left(\mathfrak{F}^{N}_{\mathfrak{R}}(x_1), \mathfrak{F}^{N}_{\mathfrak{R}}(x_2)\right)$ this means that $\mathfrak{F}^{RN}_{\mathfrak{R}}(x_1 + x_2) \leq \max\left(\mathfrak{F}^{RN}_{\mathfrak{R}}(x_1), \mathfrak{F}^{RN}_{\mathfrak{R}}(x_2)\right), \mathfrak{F}^{IN}_{\mathfrak{R}}(x_1 + x_2) \leq \max\left(\mathfrak{F}^{IN}_{\mathfrak{R}}(x_1), \mathfrak{F}^{IN}_{\mathfrak{R}}(x_2)\right)$
5. $\mathfrak{F}^{P}_{\mathfrak{R}}(x_1x_2) \geq \min\left(\mathfrak{F}^{P}_{\mathfrak{R}}(x_1), \mathfrak{F}^{P}_{\mathfrak{R}}(x_2)\right)$ this means that $\mathfrak{F}^{RP}_{\mathfrak{R}}(x_1x_2) \geq \min\left(\mathfrak{F}^{RP}_{\mathfrak{R}}(x_1), \mathfrak{F}^{RP}_{\mathfrak{R}}(x_2)\right), \mathfrak{F}^{IP}_{\mathfrak{R}}(x_1x_2) \geq \min\left(\mathfrak{F}^{IP}_{\mathfrak{R}}(x_1), \mathfrak{F}^{IP}_{\mathfrak{R}}(x_2)\right)$
6. $\mathfrak{F}^{N}_{\mathfrak{R}}(x_1x_2) \leq \max\left(\mathfrak{F}^{N}_{\mathfrak{R}}(x_1), \mathfrak{F}^{N}_{\mathfrak{R}}(x_2)\right)$ this means that $\mathfrak{F}^{RN}_{\mathfrak{R}}(x_1x_2) \leq \max\left(\mathfrak{F}^{RN}_{\mathfrak{R}}(x_1), \mathfrak{F}^{RN}_{\mathfrak{R}}(x_2)\right), \mathfrak{F}^{IN}_{\mathfrak{R}}(x_1x_2) \leq \max\left(\mathfrak{F}^{IN}_{\mathfrak{R}}(x_1), \mathfrak{F}^{IN}_{\mathfrak{R}}(x_2)\right)$

Example 3.6 Assume that $\mathfrak{J} = \{0, x_1, x_2, x_3\}$ is a semiring with binary operation $'' + ''$ and $''.''$ interpreted as.

$+$	0	x_1	x_2	x_3
0	0	x_1	x_2	
x_1	x_1	x_1	x_2	x_3
x_2	x_2	x_2	x_2	x_3
x_3	x_3	x_3	x_3	x_2

.	0	x_1	x_2	x_3
0	0	0	0	0
x_1	0	x_1	x_1	x_1
x_2	0	x_1	x_1	x_1
x_3	0	x_1	x_1	x_1

Define a BCFS $\mathfrak{R} = \left(\mathfrak{F}^{P}_{\mathfrak{R}}, \mathfrak{F}^{N}_{\mathfrak{R}}\right) = \left(\mathfrak{F}^{RP}_{\mathfrak{R}} + \iota\mathfrak{F}^{IP}_{\mathfrak{R}}, \mathfrak{F}^{RN}_{\mathfrak{R}} + \iota\mathfrak{F}^{IN}_{\mathfrak{R}}\right)$ of $\mathfrak{J}$ as.

$\mathfrak{J}$	0	x_1	x_2	x_3
$\mathfrak{F}_{\mathfrak{R}}^{P}$	$(0.81+\iota 0.77)$	$(0.45+\iota 0.36)$	$(0.45+\iota 0.36)$	$(0.45+\iota 0.36)$
$\mathfrak{F}_{\mathfrak{R}}^{N}$	$(-0.72-\iota 0.61)$	$(-0.51-\iota 0.27)$	$(-0.51-\iota 0.27)$	$(-0.51-\iota 0.27)$

Then $\mathfrak{R}$ is a BCFSSR.

Definition 3.9 Assume a BCFS $\mathfrak{R}=\left(\mathfrak{F}_{\mathfrak{R}}^{P},\mathfrak{F}_{\mathfrak{R}}^{N}\right)=\left(\mathfrak{F}_{\mathfrak{R}}^{RP}+\iota\mathfrak{F}_{\mathfrak{R}}^{IP},\mathfrak{F}_{\mathfrak{R}}^{RN}+\iota\mathfrak{F}_{\mathfrak{R}}^{IN}\right)$ of a semiring $\mathfrak{J}$, then $\mathfrak{R}$ is identified as BCF left ideal of $\mathfrak{J}$ if the underneath axioms are valid.

1. $\mathfrak{F}_{\mathfrak{R}}^{P}(x_1+x_2)\geq\min\left(\mathfrak{F}_{\mathfrak{R}}^{P}(x_1),\mathfrak{F}_{\mathfrak{R}}^{P}(x_2)\right)$ this means that $\mathfrak{F}_{\mathfrak{R}}^{RP}(x_1+x_2)\geq\min\left(\mathfrak{F}_{\mathfrak{R}}^{RP}(x_1),\mathfrak{F}_{\mathfrak{R}}^{RP}(x_2)\right)$, $\mathfrak{F}_{\mathfrak{R}}^{IP}(x_1+x_2)\geq\min\left(\mathfrak{F}_{\mathfrak{R}}^{IP}(x_1),\mathfrak{F}_{\mathfrak{R}}^{IP}(x_2)\right)$
2. $\mathfrak{F}_{\mathfrak{R}}^{N}(x_1+x_2)\leq\max\left(\mathfrak{F}_{\mathfrak{R}}^{N}(x_1),\mathfrak{F}_{\mathfrak{R}}^{N}(x_2)\right)$ this means that $\mathfrak{F}_{\mathfrak{R}}^{RN}(x_1+x_2)\leq\max\left(\mathfrak{F}_{\mathfrak{R}}^{RN}(x_1),\mathfrak{F}_{\mathfrak{R}}^{RN}(x_2)\right)$, $\mathfrak{F}_{\mathfrak{R}}^{IN}(x_1+x_2)\leq\max\left(\mathfrak{F}_{\mathfrak{R}}^{IN}(x_1),\mathfrak{F}_{\mathfrak{R}}^{IN}(x_2)\right)$
3. $\mathfrak{F}_{\mathfrak{R}}^{P}(x_1x_2)\geq\mathfrak{F}_{\mathfrak{R}}^{P}(x_2)$ this means that $\mathfrak{F}_{\mathfrak{R}}^{RP}(x_1x_2)\geq\mathfrak{F}_{\mathfrak{R}}^{RP}(x_2)$, $\mathfrak{F}_{\mathfrak{R}}^{IP}(x_1x_2)\geq\mathfrak{F}_{\mathfrak{R}}^{IP}(x_2)$
4. $\mathfrak{F}_{\mathfrak{R}}^{N}(x_1x_2)\leq\mathfrak{F}_{\mathfrak{R}}^{N}(x_2)$ this means that $\mathfrak{F}_{\mathfrak{R}}^{RN}(x_1x_2)\leq\mathfrak{F}_{\mathfrak{R}}^{RN}(x_2)$, $\mathfrak{F}_{\mathfrak{R}}^{IN}(x_1x_2)\leq\mathfrak{F}_{\mathfrak{R}}^{IN}(x_2)$

Definition 3.10 Assume a BCFS $\mathfrak{R}=\left(\mathfrak{F}_{\mathfrak{R}}^{P},\mathfrak{F}_{\mathfrak{R}}^{N}\right)=\left(\mathfrak{F}_{\mathfrak{R}}^{RP}+\iota\mathfrak{F}_{\mathfrak{R}}^{IP},\mathfrak{F}_{\mathfrak{R}}^{RN}+\iota\mathfrak{F}_{\mathfrak{R}}^{IN}\right)$ of a semiring $\mathfrak{J}$, then $\mathfrak{R}$ is identified as BCF right ideal of $\mathfrak{J}$ if the underneath axioms are valid.

1. $\mathfrak{F}_{\mathfrak{R}}^{P}(x_1+x_2)\geq\min\left(\mathfrak{F}_{\mathfrak{R}}^{P}(x_1),\mathfrak{F}_{\mathfrak{R}}^{P}(x_2)\right)$ this means that $\mathfrak{F}_{\mathfrak{R}}^{RP}(x_1+x_2)\geq\min\left(\mathfrak{F}_{\mathfrak{R}}^{RP}(x_1),\mathfrak{F}_{\mathfrak{R}}^{RP}(x_2)\right)$, $\mathfrak{F}_{\mathfrak{R}}^{IP}(x_1+x_2)\geq\min\left(\mathfrak{F}_{\mathfrak{R}}^{IP}(x_1),\mathfrak{F}_{\mathfrak{R}}^{IP}(x_2)\right)$
2. $\mathfrak{F}_{\mathfrak{R}}^{N}(x_1+x_2)\leq\max\left(\mathfrak{F}_{\mathfrak{R}}^{N}(x_1),\mathfrak{F}_{\mathfrak{R}}^{N}(x_2)\right)$ this means that $\mathfrak{F}_{\mathfrak{R}}^{RN}(x_1+x_2)\leq\max\left(\mathfrak{F}_{\mathfrak{R}}^{RN}(x_1),\mathfrak{F}_{\mathfrak{R}}^{RN}(x_2)\right)$, $\mathfrak{F}_{\mathfrak{R}}^{IN}(x_1+x_2)\leq\max\left(\mathfrak{F}_{\mathfrak{R}}^{IN}(x_1),\mathfrak{F}_{\mathfrak{R}}^{IN}(x_2)\right)$
3. $\mathfrak{F}_{\mathfrak{R}}^{P}(x_1x_2)\geq\mathfrak{F}_{\mathfrak{R}}^{P}(x_1)$ this means that $\mathfrak{F}_{\mathfrak{R}}^{RP}(x_1x_2)\geq\mathfrak{F}_{\mathfrak{R}}^{RP}(x_1)$, $\mathfrak{F}_{\mathfrak{R}}^{IP}(x_1x_2)\geq\mathfrak{F}_{\mathfrak{R}}^{IP}(x_1)$
4. $\mathfrak{F}_{\mathfrak{R}}^{N}(x_1x_2)\leq\mathfrak{F}_{\mathfrak{R}}^{N}(x_1)$ this means that $\mathfrak{F}_{\mathfrak{R}}^{RN}(x_1x_2)\leq\mathfrak{F}_{\mathfrak{R}}^{RN}(x_1)$, $\mathfrak{F}_{\mathfrak{R}}^{IN}(x_1x_2)\leq\mathfrak{F}_{\mathfrak{R}}^{IN}(x_1)$

A BCFS $\mathfrak{R}=\left(\mathfrak{F}_{\mathfrak{R}}^{P},\mathfrak{F}_{\mathfrak{R}}^{N}\right)=\left(\mathfrak{F}_{\mathfrak{R}}^{RP}+\iota\mathfrak{F}_{\mathfrak{R}}^{IP},\mathfrak{F}_{\mathfrak{R}}^{RN}+\iota\mathfrak{F}_{\mathfrak{R}}^{IN}\right)$ is identified as BCF ideal or BCF two-sided ideal if it is both BCF right and left ideal.

Example 3.7 Let us assume a semiring $\mathfrak{J}$ of whole numbers. Now interpret a BCFS $\mathfrak{R}=\left(\mathfrak{F}_{\mathfrak{R}}^{P},\mathfrak{F}_{\mathfrak{R}}^{N}\right)=\left(\mathfrak{F}_{\mathfrak{R}}^{RP}+\iota\mathfrak{F}_{\mathfrak{R}}^{IP},\mathfrak{F}_{\mathfrak{R}}^{RN}+\iota\mathfrak{F}_{\mathfrak{R}}^{IN}\right)$ as.

$$\mathfrak{F}_{\mathfrak{R}}^{RP}(x)=\begin{cases}1 if x=0\\ 0.4 if x\in(0,0.6)\\ 0.75 if x\geq 0.6\end{cases}\quad \mathfrak{F}_{\mathfrak{R}}^{IP}(x)=\begin{cases}1 if x=0\\ 0.6 if x\in(0,0.6)\\ 0.9 if x\geq 0.6\end{cases}$$

and

$$\mathfrak{F}_{\mathfrak{R}}^{RN}(x)=\begin{cases}-1 if x=0\\ -0.4 if x\in(0,0.6)\\ -0.65 if x\geq 0.6\end{cases}\quad \mathfrak{F}_{\mathfrak{R}}^{IN}(x)=\begin{cases}-1 if x=0\\ -0.55 if x\in(0,0.6)\\ -0.8 if x\geq 0.6\end{cases}$$

Then, $\mathfrak{R}$ is BCF-I.

Following is the definition of BCF bi-ideal (BCF-BI).

Definition 3.11 Assume a BCFS $\mathfrak{R} = \left(\mathfrak{F}_{\mathfrak{R}}^{P}, \mathfrak{F}_{\mathfrak{R}}^{N}\right) = \left(\mathfrak{F}_{\mathfrak{R}}^{RP} + \iota\mathfrak{F}_{\mathfrak{R}}^{IP}, \mathfrak{F}_{\mathfrak{R}}^{RN} + \iota\mathfrak{F}_{\mathfrak{R}}^{IN}\right)$ of a semiring $\mathfrak{J}$, then $\mathfrak{R}$ is identified as BCF-BI of $\mathfrak{J}$ if the underneath axioms are valid.

1. $\mathfrak{F}_{\mathfrak{R}}^{P}(0) \geq \mathfrak{F}_{\mathfrak{R}}^{P}(x)$ this means that $\mathfrak{F}_{\mathfrak{R}}^{RP}(0) \geq \mathfrak{F}_{\mathfrak{R}}^{RP}(x)$ and $\mathfrak{F}_{\mathfrak{R}}^{IP}(0) \geq \mathfrak{F}_{\mathfrak{R}}^{IP}(x)$
2. $\mathfrak{F}_{\mathfrak{R}}^{N}(0) \leq \mathfrak{F}_{\mathfrak{R}}^{N}(x)$ this means that $\mathfrak{F}_{\mathfrak{R}}^{RN}(0) \leq \mathfrak{F}_{\mathfrak{R}}^{RN}(x)$ and $\mathfrak{F}_{\mathfrak{R}}^{IN}(0) \leq \mathfrak{F}_{\mathfrak{R}}^{IN}(x)$
3. $\mathfrak{F}_{\mathfrak{R}}^{P}(x_1 + x_2) \geq \min\left(\mathfrak{F}_{\mathfrak{R}}^{P}(x_1), \mathfrak{F}_{\mathfrak{R}}^{P}(x_2)\right)$ this means that $\mathfrak{F}_{\mathfrak{R}}^{RP}(x_1 + x_2) \geq \min\left(\mathfrak{F}_{\mathfrak{R}}^{RP}(x_1), \mathfrak{F}_{\mathfrak{R}}^{RP}(x_2)\right), \mathfrak{F}_{\mathfrak{R}}^{IP}(x_1 + x_2) \geq \min\left(\mathfrak{F}_{\mathfrak{R}}^{IP}(x_1), \mathfrak{F}_{\mathfrak{R}}^{IP}(x_2)\right)$
4. $\mathfrak{F}_{\mathfrak{R}}^{N}(x_1 + x_2) \leq \max\left(\mathfrak{F}_{\mathfrak{R}}^{N}(x_1), \mathfrak{F}_{\mathfrak{R}}^{N}(x_2)\right)$ this means that $\mathfrak{F}_{\mathfrak{R}}^{RN}(x_1 + x_2) \leq \max\left(\mathfrak{F}_{\mathfrak{R}}^{RN}(x_1), \mathfrak{F}_{\mathfrak{R}}^{RN}(x_2)\right), \mathfrak{F}_{\mathfrak{R}}^{IN}(x_1 + x_2) \leq \max\left(\mathfrak{F}_{\mathfrak{R}}^{IN}(x_1), \mathfrak{F}_{\mathfrak{R}}^{IN}(x_2)\right)$
5. $\mathfrak{F}_{\mathfrak{R}}^{P}(x_1x_2) \geq \min\left(\mathfrak{F}_{\mathfrak{R}}^{P}(x_1), \mathfrak{F}_{\mathfrak{R}}^{P}(x_2)\right)$ this means that $\mathfrak{F}_{\mathfrak{R}}^{RP}(x_1x_2) \geq \min\left(\mathfrak{F}_{\mathfrak{R}}^{RP}(x_1), \mathfrak{F}_{\mathfrak{R}}^{RP}(x_2)\right), \mathfrak{F}_{\mathfrak{R}}^{IP}(x_1x_2) \geq \min\left(\mathfrak{F}_{\mathfrak{R}}^{IP}(x_1), \mathfrak{F}_{\mathfrak{R}}^{IP}(x_2)\right)$
6. $\mathfrak{F}_{\mathfrak{R}}^{N}(x_1x_2) \leq \max\left(\mathfrak{F}_{\mathfrak{R}}^{N}(x_1), \mathfrak{F}_{\mathfrak{R}}^{N}(x_2)\right)$ this means that $\mathfrak{F}_{\mathfrak{R}}^{RN}(x_1x_2) \leq \max\left(\mathfrak{F}_{\mathfrak{R}}^{RN}(x_1), \mathfrak{F}_{\mathfrak{R}}^{RN}(x_2)\right), \mathfrak{F}_{\mathfrak{R}}^{IN}(x_1x_2) \leq \max\left(\mathfrak{F}_{\mathfrak{R}}^{IN}(x_1), \mathfrak{F}_{\mathfrak{R}}^{IN}(x_2)\right)$
7. $\mathfrak{F}_{\mathfrak{R}}^{P}(x_1x_2x_3) \geq \min\left(\mathfrak{F}_{\mathfrak{R}}^{P}(x_1), \mathfrak{F}_{\mathfrak{R}}^{P}(x_3)\right)$ this means that $\mathfrak{F}_{\mathfrak{R}}^{RP}(x_1x_2x_3) \geq \min\left(\mathfrak{F}_{\mathfrak{R}}^{RP}(x_1), \mathfrak{F}_{\mathfrak{R}}^{RP}(x_3)\right), \mathfrak{F}_{\mathfrak{R}}^{IP}(x_1x_2x_3) \geq \min\left(\mathfrak{F}_{\mathfrak{R}}^{IP}(x_1), \mathfrak{F}_{\mathfrak{R}}^{IP}(x_3)\right)$
8. $\mathfrak{F}_{\mathfrak{R}}^{N}(x_1x_2x_3) \leq \max\left(\mathfrak{F}_{\mathfrak{R}}^{N}(x_1), \mathfrak{F}_{\mathfrak{R}}^{N}(x_3)\right)$ this means that $\mathfrak{F}_{\mathfrak{R}}^{RN}(x_1x_2x_3) \leq \max\left(\mathfrak{F}_{\mathfrak{R}}^{RN}(x_1), \mathfrak{F}_{\mathfrak{R}}^{RN}(x_3)\right), \mathfrak{F}_{\mathfrak{R}}^{IN}(x_1x_2x_3) \leq \max\left(\mathfrak{F}_{\mathfrak{R}}^{IN}(x_1), \mathfrak{F}_{\mathfrak{R}}^{IN}(x_3)\right)$

3.5 Bipolar Complex Fuzzy Subrings

Let us consider a set $\mathbb{R} \neq \emptyset$ with two binary operations $'' + ''$ and $''.''$ is identified as a ring if $\mathbb{R}$ is an abelian group under $'' + ''$ and semigroup under $''.''$ and right, left distributive laws of multiplication over addition hold. $(\mathbb{R}, +, .)$ signifies a ring. If multiplicative identity (unity) exists in a ring $\mathbb{R}$, then it is signified as a ring with unity and if interpreted multiplication in $\mathbb{R}$ is commutative then, $\mathbb{R}$ is signified as a commutative ring. A division ring $\mathbb{R}$ is a ring, in which every non-zero element has multiplicative inverse in $\mathbb{R}$. A subset $\mathbb{R}_{\mathbb{S}} \neq \emptyset$ of a ring $\mathbb{R}$ is identified as subring of $\mathbb{R}$ if $\mathbb{R}_{\mathbb{S}}$ is a ring under induced operations of $\mathbb{R}$. A left (right) ideal $\mathbb{R}_{\mathbb{S}}$ of ring $\mathbb{R}$ is a non-empty subset of $\mathbb{R}$, which is subgroup of $\mathbb{R}$ under $+$ and $\forall x \in \mathbb{R}$, and $x_{\mathbb{S}} \in \mathbb{R}_{\mathbb{S}}$, $xx_{\mathbb{S}} \in \mathbb{R}_{\mathbb{S}}(x_{\mathbb{S}}x \in \mathbb{R}_{\mathbb{S}})$ and $\mathbb{R}_{\mathbb{S}}$ is identified as ideal or both-sided ideal if $\mathbb{R}_{\mathbb{S}}$ is both left and right ideal. In this subsection, $\mathbb{R}$ would identify a ring.

The notion of BCF subring is interpreted in the next definition.

Definition 3.12 Assume $\mathbb{R}$ as a ring and $\mathfrak{R} = \left(\mathfrak{F}_{\mathfrak{R}}^{P}, \mathfrak{F}_{\mathfrak{R}}^{N}\right) = \left(\mathfrak{F}_{\mathfrak{R}}^{RP} + \iota\mathfrak{F}_{\mathfrak{R}}^{IP}, \mathfrak{F}_{\mathfrak{R}}^{RN} + \iota\mathfrak{F}_{\mathfrak{R}}^{IN}\right)$ as BCF subset of $\mathbb{R}$, then $\mathfrak{R}$ is recognized as BCF subring (BCFSR) if the underneath axioms are valid.

1. $\mathfrak{F}_{\mathfrak{R}}^{P}(x_1 - x_2) \geq \min\left(\mathfrak{F}_{\mathfrak{R}}^{P}(x_1), \mathfrak{F}_{\mathfrak{R}}^{P}(x_2)\right)$ this means that $\mathfrak{F}_{\mathfrak{R}}^{RP}(x_1 - x_2) \geq \min\left(\mathfrak{F}_{\mathfrak{R}}^{RP}(x_1), \mathfrak{F}_{\mathfrak{R}}^{RP}(x_2)\right), \mathfrak{F}_{\mathfrak{R}}^{IP}(x_1 - x_2) \geq \min\left(\mathfrak{F}_{\mathfrak{R}}^{IP}(x_1), \mathfrak{F}_{\mathfrak{R}}^{IP}(x_2)\right)$
2. $\mathfrak{F}_{\mathfrak{R}}^{N}(x_1 - x_2) \leq \max\left(\mathfrak{F}_{\mathfrak{R}}^{N}(x_1), \mathfrak{F}_{\mathfrak{R}}^{N}(x_2)\right)$ this means that $\mathfrak{F}_{\mathfrak{R}}^{RN}(x_1 - x_2) \leq \max\left(\mathfrak{F}_{\mathfrak{R}}^{RN}(x_1), \mathfrak{F}_{\mathfrak{R}}^{RN}(x_2)\right), \mathfrak{F}_{\mathfrak{R}}^{IN}(x_1 - x_2) \leq \max\left(\mathfrak{F}_{\mathfrak{R}}^{IN}(x_1), \mathfrak{F}_{\mathfrak{R}}^{IN}(x_2)\right)$
3. $\mathfrak{F}_{\mathfrak{R}}^{P}(x_1x_2) \geq \min\left(\mathfrak{F}_{\mathfrak{R}}^{P}(x_1), \mathfrak{F}_{\mathfrak{R}}^{P}(x_2)\right)$ this means that $\mathfrak{F}_{\mathfrak{R}}^{RP}(x_1x_2) \geq \min\left(\mathfrak{F}_{\mathfrak{R}}^{RP}(x_1), \mathfrak{F}_{\mathfrak{R}}^{RP}(x_2)\right), \mathfrak{F}_{\mathfrak{R}}^{IP}(x_1x_2) \geq \min\left(\mathfrak{F}_{\mathfrak{R}}^{IP}(x_1), \mathfrak{F}_{\mathfrak{R}}^{IP}(x_2)\right)$

4. $\mathfrak{F}_{\mathfrak{R}}^{N}(x_1x_2) \leq \max(\mathfrak{F}_{\mathfrak{R}}^{N}(x_1), \mathfrak{F}_{\mathfrak{R}}^{N}(x_2))$ this means that $\mathfrak{F}_{\mathfrak{R}}^{RN}(x_1x_2) \leq \max(\mathfrak{F}_{\mathfrak{R}}^{RN}(x_1), \mathfrak{F}_{\mathfrak{R}}^{RN}(x_2))$, $\mathfrak{F}_{\mathfrak{R}}^{IN}(x_1x_2) \leq \max(\mathfrak{F}_{\mathfrak{R}}^{IN}(x_1), \mathfrak{F}_{\mathfrak{R}}^{IN}(x_2))$

Following is the example of BCFSR.

Example 3.8 Consider a ring $\mathbb{Z}_4 = \{\overline{0}, \overline{1}, \overline{2}, \overline{3}\}$ and define a BCF subset $\mathfrak{R} = (\mathfrak{F}_{\mathfrak{R}}^{P}, \mathfrak{F}_{\mathfrak{R}}^{N}) = (\mathfrak{F}_{\mathfrak{R}}^{RP} + \iota\mathfrak{F}_{\mathfrak{R}}^{IP}, \mathfrak{F}_{\mathfrak{R}}^{RN} + \iota\mathfrak{F}_{\mathfrak{R}}^{IN})$ as.

$$\mathfrak{R} = \left\{ \begin{array}{l} (\overline{0}, (0.88 + \iota 0.93, -0.61 - \iota 0.73)), (\overline{1}, (0.65 + \iota 0.52, -0.47 - \iota 0.51)), \\ (\overline{2}, (0.37 + \iota 0.49, -0.2 - \iota 0.18)), (\overline{3}, (0.65 + \iota 0.52, -0.47 - \iota 0.51)) \end{array} \right\}$$

Then, $\mathfrak{R}$ is a BCFSR.

Theorem 3.6 Assume a BCFSR $\mathfrak{R} = (\mathfrak{F}_{\mathfrak{R}}^{P}, \mathfrak{F}_{\mathfrak{R}}^{N}) = (\mathfrak{F}_{\mathfrak{R}}^{RP} + \iota\mathfrak{F}_{\mathfrak{R}}^{IP}, \mathfrak{F}_{\mathfrak{R}}^{RN} + \iota\mathfrak{F}_{\mathfrak{R}}^{IN})$ of a ring $\mathbb{R}$, then.

1. $\mathfrak{F}_{\mathfrak{R}}^{P}(-x) = \mathfrak{F}_{\mathfrak{R}}^{P}(x)$ i.e., $\mathfrak{F}_{\mathfrak{R}}^{RP}(-x) = \mathfrak{F}_{\mathfrak{R}}^{RP}(x)$ and $\mathfrak{F}_{\mathfrak{R}}^{IP}(-x) = \mathfrak{F}_{\mathfrak{R}}^{IP}(x)$
2. $\mathfrak{F}_{\mathfrak{R}}^{N}(-x) = \mathfrak{F}_{\mathfrak{R}}^{N}(x)$ i.e., $\mathfrak{F}_{\mathfrak{R}}^{RN}(-x) = \mathfrak{F}_{\mathfrak{R}}^{RN}(x)$ and $\mathfrak{F}_{\mathfrak{R}}^{IN}(-x) = \mathfrak{F}_{\mathfrak{R}}^{IN}(x)$
3. $\mathfrak{F}_{\mathfrak{R}}^{P}(x) \leq \mathfrak{F}_{\mathfrak{R}}^{P}(\mathfrak{e})$ i.e., $\mathfrak{F}_{\mathfrak{R}}^{RP}(x) \leq \mathfrak{F}_{\mathfrak{R}}^{RP}(\mathfrak{e})$ and $\mathfrak{F}_{\mathfrak{R}}^{IP}(x) \leq \mathfrak{F}_{\mathfrak{R}}^{IP}(\mathfrak{e}) \forall x, \mathfrak{e} \in \mathbb{R}$, where $\mathfrak{e}$ is the identity element
4. $\mathfrak{F}_{\mathfrak{R}}^{N}(x) \geq \mathfrak{F}_{\mathfrak{R}}^{N}(\mathfrak{e})$ i.e., $\mathfrak{F}_{\mathfrak{R}}^{RN}(x) \geq \mathfrak{F}_{\mathfrak{R}}^{RN}(\mathfrak{e})$ and $\mathfrak{F}_{\mathfrak{R}}^{IN}(x) \geq \mathfrak{F}_{\mathfrak{R}}^{IN}(\mathfrak{e}) \forall x, \mathfrak{e} \in \mathbb{R}$, where $\mathfrak{e}$ is the identity element

Proof 1. Assume $x \in \mathbb{R}$, then,

$$\begin{aligned} &\mathfrak{F}_{\mathfrak{R}}^{RP}(x) = \mathfrak{F}_{\mathfrak{R}}^{RP}(-(-x)) \geq \mathfrak{F}_{\mathfrak{R}}^{RP}(-x) \geq \mathfrak{F}_{\mathfrak{R}}^{RP}(x) \\ &\Rightarrow \mathfrak{F}_{\mathfrak{R}}^{RP}(x) = \mathfrak{F}_{\mathfrak{R}}^{RP}(-x) \end{aligned}$$

Similarly,

$$\mathfrak{F}_{\mathfrak{R}}^{IP}(x) = \mathfrak{F}_{\mathfrak{R}}^{IP}(-x) \Rightarrow \mathfrak{F}_{\mathfrak{R}}^{P}(x) = \mathfrak{F}_{\mathfrak{R}}^{P}(-x)$$

2. $$\begin{aligned} &\mathfrak{F}_{\mathfrak{R}}^{RN}(x) = \mathfrak{F}_{\mathfrak{R}}^{RN}(-(-x)) \leq \mathfrak{F}_{\mathfrak{R}}^{RN}(-x) \leq \mathfrak{F}_{\mathfrak{R}}^{RN}(x) \\ &\Rightarrow \mathfrak{F}_{\mathfrak{R}}^{RN}(x) = \mathfrak{F}_{\mathfrak{R}}^{RN}(-x) \end{aligned}$$

Similarly,

$$\mathfrak{F}_{\mathfrak{R}}^{IN}(x) = \mathfrak{F}_{\mathfrak{R}}^{IN}(-x) \Rightarrow \mathfrak{F}_{\mathfrak{R}}^{N}(x) = \mathfrak{F}_{\mathfrak{R}}^{N}(-x)$$

3. As

$$\mathfrak{F}_{\mathfrak{R}}^{RP}(\mathfrak{e}) = \mathfrak{F}_{\mathfrak{R}}^{RP}(x - x) \geq \min(\mathfrak{F}_{\mathfrak{R}}^{RP}(x), \mathfrak{F}_{\mathfrak{R}}^{RP}(x)) = \mathfrak{F}_{\mathfrak{R}}^{RP}(x) \Rightarrow \mathfrak{F}_{\mathfrak{R}}^{RP}(\mathfrak{e}) \geq \mathfrak{F}_{\mathfrak{R}}^{RP}(x)$$

and

$$\mathfrak{F}_{\mathfrak{R}}^{IP}(\mathfrak{e}) = \mathfrak{F}_{\mathfrak{R}}^{IP}(x - x) \geq \min\big(\mathfrak{F}_{\mathfrak{R}}^{IP}(x), \mathfrak{F}_{\mathfrak{R}}^{RP}(x)\big) = \mathfrak{F}_{\mathfrak{R}}^{IP}(x) \Rightarrow \mathfrak{F}_{\mathfrak{R}}^{IP}(\mathfrak{e}) \geq \mathfrak{F}_{\mathfrak{R}}^{IP}(x)$$
$$\Rightarrow \mathfrak{F}_{\mathfrak{R}}^{P}(\mathfrak{e}) \geq \mathfrak{F}_{\mathfrak{R}}^{P}(x)$$

4. As

$$\mathfrak{F}_{\mathfrak{R}}^{RN}(\mathfrak{e}) = \mathfrak{F}_{\mathfrak{R}}^{RN}(x - x) \leq \max\big(\mathfrak{F}_{\mathfrak{R}}^{RN}(x), \mathfrak{F}_{\mathfrak{R}}^{RN}(x)\big) = \mathfrak{F}_{\mathfrak{R}}^{RN}(x) \Rightarrow \mathfrak{F}_{\mathfrak{R}}^{RN}(\mathfrak{e}) \leq \mathfrak{F}_{\mathfrak{R}}^{RN}(x)$$

and

$$\mathfrak{F}_{\mathfrak{R}}^{IN}(\mathfrak{e}) = \mathfrak{F}_{\mathfrak{R}}^{IN}(x - x) \leq \max\big(\mathfrak{F}_{\mathfrak{R}}^{IN}(x), \mathfrak{F}_{\mathfrak{R}}^{RN}(x)\big) = \mathfrak{F}_{\mathfrak{R}}^{IN}(x) \Rightarrow \mathfrak{F}_{\mathfrak{R}}^{IN}(\mathfrak{e}) \leq \mathfrak{F}_{\mathfrak{R}}^{IN}(x)$$
$$\Rightarrow \mathfrak{F}_{\mathfrak{R}}^{N}(\mathfrak{e}) \leq \mathfrak{F}_{\mathfrak{R}}^{N}(x)$$

Theorem 3.7 Assume a BCFSR $\mathfrak{R} = \big(\mathfrak{F}_{\mathfrak{R}}^{P}, \mathfrak{F}_{\mathfrak{R}}^{N}\big) = \big(\mathfrak{F}_{\mathfrak{R}}^{RP} + \iota\mathfrak{F}_{\mathfrak{R}}^{IP}, \mathfrak{F}_{\mathfrak{R}}^{RN} + \iota\mathfrak{F}_{\mathfrak{R}}^{IN}\big)$ of $\mathbb{R}$, then $\forall x_1, x_2 \in \mathbb{R}$.

1. $\mathfrak{F}_{\mathfrak{R}}^{P}(x_1 - x_2) = 1 + \iota 1$, then $\mathfrak{F}_{\mathfrak{R}}^{P}(x_1) = \mathfrak{F}_{\mathfrak{R}}^{P}(x_2)$ i.e., $\mathfrak{F}_{\mathfrak{R}}^{RP}(x_1 - x_2) = 1$, then $\mathfrak{F}_{\mathfrak{R}}^{RP}(x_1) = \mathfrak{F}_{\mathfrak{R}}^{RP}(x_2)$ and $\mathfrak{F}_{\mathfrak{R}}^{IP}(x_1 - x_2) = 1$, then $\mathfrak{F}_{\mathfrak{R}}^{IP}(x_1) = \mathfrak{F}_{\mathfrak{R}}^{IP}(x_2)$
2. $\mathfrak{F}_{\mathfrak{R}}^{N}(x_1 - x_2) = -1 - \iota 1$, then $\mathfrak{F}_{\mathfrak{R}}^{N}(x_1) = \mathfrak{F}_{\mathfrak{R}}^{N}(x_2)$ i.e., $\mathfrak{F}_{\mathfrak{R}}^{RN}(x_1 - x_2) = -1$, then $\mathfrak{F}_{\mathfrak{R}}^{RN}(x_1) = \mathfrak{F}_{\mathfrak{R}}^{RN}(x_2)$ and $\mathfrak{F}_{\mathfrak{R}}^{IN}(x_1 - x_2) = -1$, then $\mathfrak{F}_{\mathfrak{R}}^{IN}(x_1) = \mathfrak{F}_{\mathfrak{R}}^{IN}(x_2)$

Proof 1. Assume $x_1, x_2 \in \mathbb{R}$, then,

$$\begin{aligned}
\mathfrak{F}_{\mathfrak{R}}^{RP}(x_1) &= \mathfrak{F}_{\mathfrak{R}}^{RP}(x_1 - x_2 + x_2) \geq \min\big(\mathfrak{F}_{\mathfrak{R}}^{RP}(x_1 - x_2), \mathfrak{F}_{\mathfrak{R}}^{RP}(x_2)\big)\\
&= \min\big(1, \mathfrak{F}_{\mathfrak{R}}^{RP}(x_2)\big) = \mathfrak{F}_{\mathfrak{R}}^{RP}(x_2)\\
&= \mathfrak{F}_{\mathfrak{R}}^{RP}(-x_2)\\
&= \mathfrak{F}_{\mathfrak{R}}^{RP}(-x_2 + x_1 - x_1) \geq \min\big(\mathfrak{F}_{\mathfrak{R}}^{RP}(x_1), \mathfrak{F}_{\mathfrak{R}}^{RP}(x_1 - x_2)\big)\\
&= \min\big(\mathfrak{F}_{\mathfrak{R}}^{RP}(x_1), 1\big) = \mathfrak{F}_{\mathfrak{R}}^{RP}(x_1)
\end{aligned}$$

and

$$\begin{aligned}
\mathfrak{F}_{\mathfrak{R}}^{IP}(x_1) &= \mathfrak{F}_{\mathfrak{R}}^{IP}(x_1 - x_2 + x_2) \geq \min\big(\mathfrak{F}_{\mathfrak{R}}^{IP}(x_1 - x_2), \mathfrak{F}_{\mathfrak{R}}^{IP}(x_2)\big)\\
&= \min\big(1, \mathfrak{F}_{\mathfrak{R}}^{IP}(x_2)\big) = \mathfrak{F}_{\mathfrak{R}}^{IP}(x_2)\\
&= \mathfrak{F}_{\mathfrak{R}}^{IP}(-x_2)\\
&= \mathfrak{F}_{\mathfrak{R}}^{IP}(-x_2 + x_1 - x_1) \geq \min\big(\mathfrak{F}_{\mathfrak{R}}^{IP}(x_1), \mathfrak{F}_{\mathfrak{R}}^{IP}(x_1 - x_2)\big)\\
&= \min\big(\mathfrak{F}_{\mathfrak{R}}^{IP}(x_1), 1\big) = \mathfrak{F}_{\mathfrak{R}}^{IP}(x_1)
\end{aligned}$$

This implies that $\mathfrak{F}_{\mathfrak{R}}^{P}(x_1) = \mathfrak{F}_{\mathfrak{R}}^{P}(x_2)$.

3. Assume $x_1, x_2 \in \mathbb{R}$, then,

$$\begin{aligned}
\mathfrak{F}_{\mathfrak{R}}^{RN}(x_1) &= \mathfrak{F}_{\mathfrak{R}}^{RN}(x_1 - x_2 + x_2) \leq \max\big(\mathfrak{F}_{\mathfrak{R}}^{RN}(x_1 - x_2), \mathfrak{F}_{\mathfrak{R}}^{RN}(x_2)\big)\\
&= \max\big(-1, \mathfrak{F}_{\mathfrak{R}}^{RN}(x_2)\big)
\end{aligned}$$

$$= \mathfrak{F}_{\mathfrak{R}}^{RN}(x_2) = \mathfrak{F}_{\mathfrak{R}}^{RN}(-x_2)$$
$$= \mathfrak{F}_{\mathfrak{R}}^{RN}(-x_2 + x_1 - x_1) \leq \max\left(\mathfrak{F}_{\mathfrak{R}}^{RN}(x_1), \mathfrak{F}_{\mathfrak{R}}^{RN}(x_1 - x_2)\right)$$
$$= \max\left(\mathfrak{F}_{\mathfrak{R}}^{RN}(x_1), -1\right) = \mathfrak{F}_{\mathfrak{R}}^{RN}(x_1)$$

and

$$\begin{aligned}
\mathfrak{F}_{\mathfrak{R}}^{IN}(x_1) &= \mathfrak{F}_{\mathfrak{R}}^{IN}(x_1 - x_2 + x_2) \leq \max\left(\mathfrak{F}_{\mathfrak{R}}^{IN}(x_1 - x_2), \mathfrak{F}_{\mathfrak{R}}^{IN}(x_2)\right) \\
&= \max\left(-1, \mathfrak{F}_{\mathfrak{R}}^{IN}(x_2)\right) = \mathfrak{F}_{\mathfrak{R}}^{IN}(x_2) \\
&= \mathfrak{F}_{\mathfrak{R}}^{IN}(-x_2) \\
&= \mathfrak{F}_{\mathfrak{R}}^{IN}(-x_2 + x_1 - x_1) \leq \max\left(\mathfrak{F}_{\mathfrak{R}}^{IN}(x_1), \mathfrak{F}_{\mathfrak{R}}^{IN}(x_1 - x_2)\right) \\
&= \max\left(\mathfrak{F}_{\mathfrak{R}}^{IN}(x_1), -1\right) = \mathfrak{F}_{\mathfrak{R}}^{IN}(x_1)
\end{aligned}$$

This implies that $\mathfrak{F}_{\mathfrak{R}}^{N}(x_1) = \mathfrak{F}_{\mathfrak{R}}^{N}(x_2)$.

Theorem 3.8 Assume a BCFSR $\mathfrak{R} = \left(\mathfrak{F}_{\mathfrak{R}}^{P}, \mathfrak{F}_{\mathfrak{R}}^{N}\right) = \left(\mathfrak{F}_{\mathfrak{R}}^{RP} + \iota\mathfrak{F}_{\mathfrak{R}}^{IP}, \mathfrak{F}_{\mathfrak{R}}^{RN} + \iota\mathfrak{F}_{\mathfrak{R}}^{IN}\right)$ of $\mathbb{R}$, then $\forall x_1, x_2 \in \mathbb{R}$.

1. $\mathfrak{F}_{\mathfrak{R}}^{P}\left(x_1 x_2^{-1}\right) = 0 + \iota 0$, then either $\mathfrak{F}_{\mathfrak{R}}^{P}(x_1) = 0$ or $\mathfrak{F}_{\mathfrak{R}}^{P}(x_2) = 0$ i.e., $\mathfrak{F}_{\mathfrak{R}}^{RP}\left(x_1 x_2^{-1}\right) = 0$, then either $\mathfrak{F}_{\mathfrak{R}}^{RP}(x_1) = 0$ or $\mathfrak{F}_{\mathfrak{R}}^{RP}(x_2) = 0$, and $\mathfrak{F}_{\mathfrak{R}}^{IP}\left(x_1 x_2^{-1}\right) = 0$, then either $\mathfrak{F}_{\mathfrak{R}}^{IP}(x_1) = 0$ or $\mathfrak{F}_{\mathfrak{R}}^{IP}(x_2) = 0$
2. $\mathfrak{F}_{\mathfrak{R}}^{N}\left(x_1 x_2^{-1}\right) = 0 + \iota 0$, then either $\mathfrak{F}_{\mathfrak{R}}^{N}(x_1) = 0$ or $\mathfrak{F}_{\mathfrak{R}}^{N}(x_2) = 0$ i.e., $\mathfrak{F}_{\mathfrak{R}}^{RN}\left(x_1 x_2^{-1}\right) = 0$, then either $\mathfrak{F}_{\mathfrak{R}}^{RN}(x_1) = 0$ or $\mathfrak{F}_{\mathfrak{R}}^{RN}(x_2) = 0$, and $\mathfrak{F}_{\mathfrak{R}}^{IN}\left(x_1 x_2^{-1}\right) = 0$, then either $\mathfrak{F}_{\mathfrak{R}}^{IN}(x_1) = 0$ or $\mathfrak{F}_{\mathfrak{R}}^{IN}(x_2) = 0$

Proof 1. Consider $x_1, x_2 \in \mathbb{R}$, then, we have $\mathfrak{F}_{\mathfrak{R}}^{RP}\left(x_1 x_2^{-1}\right) \geq \min\left(\mathfrak{F}_{\mathfrak{R}}^{RP}(x_1), \mathfrak{F}_{\mathfrak{R}}^{RP}(x_2)\right) \Rightarrow 0 \geq \min\left(\mathfrak{F}_{\mathfrak{R}}^{RP}(x_1), \mathfrak{F}_{\mathfrak{R}}^{RP}(x_2)\right)$. Thus, either $\mathfrak{F}_{\mathfrak{R}}^{RP}(x_1) = 0$ or $\mathfrak{F}_{\mathfrak{R}}^{RP}(x_2) = 0$. Similarly, $\mathfrak{F}_{\mathfrak{R}}^{IP}\left(x_1 x_2^{-1}\right) \geq \min\left(\mathfrak{F}_{\mathfrak{R}}^{IP}(x_1), \mathfrak{F}_{\mathfrak{R}}^{IP}(x_2)\right) \Rightarrow 0 \leq \min\left(\mathfrak{F}_{\mathfrak{R}}^{IP}(x_1), \mathfrak{F}_{\mathfrak{R}}^{IP}(x_2)\right)$. Thus, either $\mathfrak{F}_{\mathfrak{R}}^{IP}(x_1) = 0$ or $\mathfrak{F}_{\mathfrak{R}}^{IP}(x_2) = 0$. Consequently, either $\mathfrak{F}_{\mathfrak{R}}^{P}(x_1) = 0$ or $\mathfrak{F}_{\mathfrak{R}}^{P}(x_2) = 0$.

3. Consider $x_1, x_2 \in \mathbb{R}$, then, we have $\mathfrak{F}_{\mathfrak{R}}^{RN}\left(x_1 x_2^{-1}\right) \leq \max\left(\mathfrak{F}_{\mathfrak{R}}^{RN}(x_1), \mathfrak{F}_{\mathfrak{R}}^{RN}(x_2)\right) \Rightarrow 0 \leq \max\left(\mathfrak{F}_{\mathfrak{R}}^{RN}(x_1), \mathfrak{F}_{\mathfrak{R}}^{RN}(x_2)\right)$. Thus, either $\mathfrak{F}_{\mathfrak{R}}^{RN}(x_1) = 0$ or $\mathfrak{F}_{\mathfrak{R}}^{RN}(x_2) = 0$. Similarly, $\mathfrak{F}_{\mathfrak{R}}^{IN}\left(x_1 x_2^{-1}\right) \leq \max\left(\mathfrak{F}_{\mathfrak{R}}^{IN}(x_1), \mathfrak{F}_{\mathfrak{R}}^{IN}(x_2)\right) \Rightarrow 0 \leq \max\left(\mathfrak{F}_{\mathfrak{R}}^{IN}(x_1), \mathfrak{F}_{\mathfrak{R}}^{IN}(x_2)\right)$. Thus, either $\mathfrak{F}_{\mathfrak{R}}^{IN}(x_1) = 0$ or $\mathfrak{F}_{\mathfrak{R}}^{IN}(x_2) = 0$. Consequently, either $\mathfrak{F}_{\mathfrak{R}}^{N}(x_1) = 0$ or $\mathfrak{F}_{\mathfrak{R}}^{N}(x_2) = 0$.

Next, Theorem would interpret that the intersection of two BCFSRs is a BCFSR.

Theorem 3.9 Assume two BCFSRs $\mathfrak{R}_1 = \left(\mathfrak{F}_{\mathfrak{R}_1}^{P}, \mathfrak{F}_{\mathfrak{R}_1}^{N}\right) = \left(\mathfrak{F}_{\mathfrak{R}_1}^{RP} + \iota\mathfrak{F}_{\mathfrak{R}_1}^{IP}, \mathfrak{F}_{\mathfrak{R}_1}^{RN} + \iota\mathfrak{F}_{\mathfrak{R}_1}^{IN}\right)$ and $\mathfrak{R}_2 = \left(\mathfrak{F}_{\mathfrak{R}_2}^{P}, \mathfrak{F}_{\mathfrak{R}_2}^{N}\right) = \left(\mathfrak{F}_{\mathfrak{R}_2}^{RP} + \iota\mathfrak{F}_{\mathfrak{R}_2}^{IP}, \mathfrak{F}_{\mathfrak{R}_2}^{RN} + \iota\mathfrak{F}_{\mathfrak{R}_2}^{IN}\right)$ of $\mathbb{R}$, then $\mathfrak{R}_1 \cap \mathfrak{R}_2$ is a BCFSR.

Proof Consider $\mathfrak{R}_3 = \mathfrak{R}_1 \cap \mathfrak{R}_2 = \left(\mathfrak{F}_{\mathfrak{R}_3}^{P}, \mathfrak{F}_{\mathfrak{R}_3}^{N}\right) = \left(\mathfrak{F}_{\mathfrak{R}_3}^{RP} + \iota\mathfrak{F}_{\mathfrak{R}_3}^{IP}, \mathfrak{F}_{\mathfrak{R}_3}^{RN} + \iota\mathfrak{F}_{\mathfrak{R}_3}^{IN}\right)$. Now to prove that $\mathfrak{R}_3$ is a BCFSR. Let $x_1, x_2 \in \mathbb{R}$, then.

1.
$$\begin{aligned}\mathfrak{F}^{RP}_{\mathfrak{R}_3}(x_1 - x_2) &= \min\left(\mathfrak{F}^{RP}_{\mathfrak{R}_1}(x_1 - x_2), \mathfrak{F}^{RP}_{\mathfrak{R}_2}(x_1 - x_2)\right)\\ &\geq \min\left(\min\left(\mathfrak{F}^{RP}_{\mathfrak{R}_1}(x_1), \mathfrak{F}^{RP}_{\mathfrak{R}_1}(x_2)\right), \min\left(\mathfrak{F}^{RP}_{\mathfrak{R}_2}(x_1), \mathfrak{F}^{RP}_{\mathfrak{R}_1}(x_2)\right)\right)\\ &\geq \min\left(\min\left(\mathfrak{F}^{RP}_{\mathfrak{R}_1}(x_1), \mathfrak{F}^{RP}_{\mathfrak{R}_2}(x_1)\right), \min\left(\mathfrak{F}^{RP}_{\mathfrak{R}_1}(x_2), \mathfrak{F}^{RP}_{\mathfrak{R}_1}(x_2)\right)\right)\\ &= \min\left(\left(\mathfrak{F}^{RP}_{\mathfrak{R}_3}(x_1), \mathfrak{F}^{RP}_{\mathfrak{R}_3}(x_2)\right)\right)\end{aligned}$$

and

$$\begin{aligned}\mathfrak{F}^{IP}_{\mathfrak{R}_3}(x_1 - x_2) &= \min\left(\mathfrak{F}^{IP}_{\mathfrak{R}_1}(x_1 - x_2), \mathfrak{F}^{IP}_{\mathfrak{R}_2}(x_1 - x_2)\right)\\ &\geq \min\left(\min\left(\mathfrak{F}^{IP}_{\mathfrak{R}_1}(x_1), \mathfrak{F}^{IP}_{\mathfrak{R}_1}(x_2)\right), \min\left(\mathfrak{F}^{IP}_{\mathfrak{R}_2}(x_1), \mathfrak{F}^{IP}_{\mathfrak{R}_1}(x_2)\right)\right)\\ &\geq \min\left(\min\left(\mathfrak{F}^{IP}_{\mathfrak{R}_1}(x_1), \mathfrak{F}^{IP}_{\mathfrak{R}_2}(x_1)\right), \min\left(\mathfrak{F}^{IP}_{\mathfrak{R}_1}(x_2), \mathfrak{F}^{IP}_{\mathfrak{R}_1}(x_2)\right)\right)\\ &= \min\left(\left(\mathfrak{F}^{IP}_{\mathfrak{R}_3}(x_1), \mathfrak{F}^{IP}_{\mathfrak{R}_3}(x_2)\right)\right)\end{aligned}$$

This implies that $\mathfrak{F}^{P}_{\mathfrak{R}_3}(x_1 - x_2) \geq \min\left(\left(\mathfrak{F}^{P}_{\mathfrak{R}_3}(x_1), \mathfrak{F}^{P}_{\mathfrak{R}_3}(x_2)\right)\right)$.

3. Likewise we can show that $\mathfrak{F}^{N}_{\mathfrak{R}_3}(x_1 - x_2) \leq \max\left(\left(\mathfrak{F}^{N}_{\mathfrak{R}_3}(x_1), \mathfrak{F}^{N}_{\mathfrak{R}_3}(x_2)\right)\right)$

4. Now
$$\begin{aligned}\mathfrak{F}^{RP}_{\mathfrak{R}_3}(x_1 x_2) &= \min\left(\mathfrak{F}^{RP}_{\mathfrak{R}_1}(x_1 x_2), \mathfrak{F}^{RP}_{\mathfrak{R}_2}(x_1 x_2)\right)\\ &\geq \min\left(\min\left(\mathfrak{F}^{RP}_{\mathfrak{R}_1}(x_1), \mathfrak{F}^{RP}_{\mathfrak{R}_1}(x_2)\right), \min\left(\mathfrak{F}^{RP}_{\mathfrak{R}_2}(x_1), \mathfrak{F}^{RP}_{\mathfrak{R}_1}(x_2)\right)\right)\\ &\geq \min\left(\min\left(\mathfrak{F}^{RP}_{\mathfrak{R}_1}(x_1), \mathfrak{F}^{RP}_{\mathfrak{R}_2}(x_1)\right), \min\left(\mathfrak{F}^{RP}_{\mathfrak{R}_1}(x_2), \mathfrak{F}^{RP}_{\mathfrak{R}_1}(x_2)\right)\right)\\ &= \min\left(\left(\mathfrak{F}^{RP}_{\mathfrak{R}_3}(x_1), \mathfrak{F}^{RP}_{\mathfrak{R}_3}(x_2)\right)\right)\end{aligned}$$

and

$$\begin{aligned}\mathfrak{F}^{IP}_{\mathfrak{R}_3}(x_1 x_2) &= \min\left(\mathfrak{F}^{IP}_{\mathfrak{R}_1}(x_1 x_2), \mathfrak{F}^{IP}_{\mathfrak{R}_2}(x_1 x_2)\right)\\ &\geq \min\left(\min\left(\mathfrak{F}^{IP}_{\mathfrak{R}_1}(x_1), \mathfrak{F}^{IP}_{\mathfrak{R}_1}(x_2)\right), \min\left(\mathfrak{F}^{IP}_{\mathfrak{R}_2}(x_1), \mathfrak{F}^{IP}_{\mathfrak{R}_1}(x_2)\right)\right)\\ &\geq \min\left(\min\left(\mathfrak{F}^{IP}_{\mathfrak{R}_1}(x_1), \mathfrak{F}^{IP}_{\mathfrak{R}_2}(x_1)\right), \min\left(\mathfrak{F}^{IP}_{\mathfrak{R}_1}(x_2), \mathfrak{F}^{IP}_{\mathfrak{R}_1}(x_2)\right)\right)\\ &= \min\left(\left(\mathfrak{F}^{IP}_{\mathfrak{R}_3}(x_1), \mathfrak{F}^{IP}_{\mathfrak{R}_3}(x_2)\right)\right)\end{aligned}$$

This implies that $\mathfrak{F}^{P}_{\mathfrak{R}_3}(x_1 x_2) \geq \min\left(\left(\mathfrak{F}^{P}_{\mathfrak{R}_3}(x_1), \mathfrak{F}^{P}_{\mathfrak{R}_3}(x_2)\right)\right)$.

4. Likewise, we can show that $\mathfrak{F}^{N}_{\mathfrak{R}_3}(x_1 x_2) \leq \max\left(\left(\mathfrak{F}^{N}_{\mathfrak{R}_3}(x_1), \mathfrak{F}^{N}_{\mathfrak{R}_3}(x_2)\right)\right)$

This showed that $\mathfrak{R}_1 \cap \mathfrak{R}_2$ is a BCFSR.

The next Theorem will reveal that the intersection of any number of BCFSRs of $\mathbb{R}$ is a BCFSR of $\mathbb{R}$.

Theorem 3.10 The intersection of any number of BCFSRs of $\mathbb{R}$ is a BCFSR of $\mathbb{R}$.

Proof Straightforward.

In the underneath definition, we will develop the theory of BCF left ideal (BCF-LI) using a ring.

Definition 3.13 Assume $\mathbb{R}$ as a ring and $\mathfrak{R} = \left(\mathfrak{F}^{P}_{\mathfrak{R}}, \mathfrak{F}^{N}_{\mathfrak{R}}\right) = \left(\mathfrak{F}^{RP}_{\mathfrak{R}} + \iota\mathfrak{F}^{IP}_{\mathfrak{R}}, \mathfrak{F}^{RN}_{\mathfrak{R}} + \iota\mathfrak{F}^{IN}_{\mathfrak{R}}\right)$ as BCF subset of $\mathbb{R}$, then $\mathfrak{R}$ is recognized as BCF-LI if the underneath axioms are valid.

1. $\mathfrak{F}^P_{\mathfrak{R}}(x_1 - x_2) \geq \min\big(\mathfrak{F}^P_{\mathfrak{R}}(x_1), \mathfrak{F}^P_{\mathfrak{R}}(x_2)\big)$ this means that $\mathfrak{F}^{RP}_{\mathfrak{R}}(x_1 - x_2) \geq \min\big(\mathfrak{F}^{RP}_{\mathfrak{R}}(x_1), \mathfrak{F}^{RP}_{\mathfrak{R}}(x_2)\big), \mathfrak{F}^{IP}_{\mathfrak{R}}(x_1 - x_2) \geq \min\big(\mathfrak{F}^{IP}_{\mathfrak{R}}(x_1), \mathfrak{F}^{IP}_{\mathfrak{R}}(x_2)\big)$
2. $\mathfrak{F}^P_{\mathfrak{R}}(x_1x_2) \geq \mathfrak{F}^P_{\mathfrak{R}}(x_1)$ this means that $\mathfrak{F}^{RP}_{\mathfrak{R}}(x_1x_2) \geq \mathfrak{F}^{RP}_{\mathfrak{R}}(x_1), \mathfrak{F}^{IP}_{\mathfrak{R}}(x_1x_2) \geq \mathfrak{F}^{IP}_{\mathfrak{R}}(x_1)$
3. $\mathfrak{F}^N_{\mathfrak{R}}(x_1 - x_2) \leq \max\big(\mathfrak{F}^N_{\mathfrak{R}}(x_1), \mathfrak{F}^N_{\mathfrak{R}}(x_2)\big)$ this means that $\mathfrak{F}^{RN}_{\mathfrak{R}}(x_1 - x_2) \leq \max\big(\mathfrak{F}^{RN}_{\mathfrak{R}}(x_1), \mathfrak{F}^{RN}_{\mathfrak{R}}(x_2)\big), \mathfrak{F}^{IN}_{\mathfrak{R}}(x_1 - x_2) \leq \max\big(\mathfrak{F}^{IN}_{\mathfrak{R}}(x_1), \mathfrak{F}^{IN}_{\mathfrak{R}}(x_2)\big)$
4. $\mathfrak{F}^N_{\mathfrak{R}}(x_1x_2) \leq \mathfrak{F}^N_{\mathfrak{R}}(x_1)$ this means that $\mathfrak{F}^{RN}_{\mathfrak{R}}(x_1x_2) \leq \mathfrak{F}^{RN}_{\mathfrak{R}}(x_1), \mathfrak{F}^{IN}_{\mathfrak{R}}(x_1x_2) \leq \mathfrak{F}^{IN}_{\mathfrak{R}}(x_1)$.

Next is the definition of BCF right ideal (BCF-RI) in ring theory.

Definition 3.14 Assume $\mathbb{R}$ as a ring and $\mathfrak{R} = \big(\mathfrak{F}^P_{\mathfrak{R}}, \mathfrak{F}^N_{\mathfrak{R}}\big) = \big(\mathfrak{F}^{RP}_{\mathfrak{R}} + \iota\mathfrak{F}^{IP}_{\mathfrak{R}}, \mathfrak{F}^{RN}_{\mathfrak{R}} + \iota\mathfrak{F}^{IN}_{\mathfrak{R}}\big)$ as BCF subset of $\mathbb{R}$, then $\mathfrak{R}$ is recognized as BCF-LI if the underneath axioms are valid.

1. $\mathfrak{F}^P_{\mathfrak{R}}(x_1 - x_2) \geq \min\big(\mathfrak{F}^P_{\mathfrak{R}}(x_1), \mathfrak{F}^P_{\mathfrak{R}}(x_2)\big)$ this means that $\mathfrak{F}^{RP}_{\mathfrak{R}}(x_1 - x_2) \geq \min\big(\mathfrak{F}^{RP}_{\mathfrak{R}}(x_1), \mathfrak{F}^{RP}_{\mathfrak{R}}(x_2)\big), \mathfrak{F}^{IP}_{\mathfrak{R}}(x_1 - x_2) \geq \min\big(\mathfrak{F}^{IP}_{\mathfrak{R}}(x_1), \mathfrak{F}^{IP}_{\mathfrak{R}}(x_2)\big)$
2. $\mathfrak{F}^P_{\mathfrak{R}}(x_1x_2) \geq \mathfrak{F}^P_{\mathfrak{R}}(x_2)$ this means that $\mathfrak{F}^{RP}_{\mathfrak{R}}(x_1x_2) \geq \mathfrak{F}^{RP}_{\mathfrak{R}}(x_2), \mathfrak{F}^{IP}_{\mathfrak{R}}(x_1x_2) \geq \mathfrak{F}^{IP}_{\mathfrak{R}}(x_2)$
3. $\mathfrak{F}^N_{\mathfrak{R}}(x_1 - x_2) \leq \max\big(\mathfrak{F}^N_{\mathfrak{R}}(x_1), \mathfrak{F}^N_{\mathfrak{R}}(x_2)\big)$ this means that $\mathfrak{F}^{RN}_{\mathfrak{R}}(x_1 - x_2) \leq \max\big(\mathfrak{F}^{RN}_{\mathfrak{R}}(x_1), \mathfrak{F}^{RN}_{\mathfrak{R}}(x_2)\big), \mathfrak{F}^{IN}_{\mathfrak{R}}(x_1 - x_2) \leq \max\big(\mathfrak{F}^{IN}_{\mathfrak{R}}(x_1), \mathfrak{F}^{IN}_{\mathfrak{R}}(x_2)\big)$
4. $\mathfrak{F}^N_{\mathfrak{R}}(x_1x_2) \leq \mathfrak{F}^N_{\mathfrak{R}}(x_2)$ this means that $\mathfrak{F}^{RN}_{\mathfrak{R}}(x_1x_2) \leq \mathfrak{F}^{RN}_{\mathfrak{R}}(x_2), \mathfrak{F}^{IN}_{\mathfrak{R}}(x_1x_2) \leq \mathfrak{F}^{IN}_{\mathfrak{R}}(x_2)$.

Both BCF-LI and BCF-RI would be identified as BCF two-sided ideal or ideal (BCF-I) of $\mathbb{R}$.

The underneath Theorem tells us that the intersection of two BCF-Is is a BCF-I.

Theorem 3.11 Assume two BCF-Is $\mathfrak{R}_1 = \big(\mathfrak{F}^P_{\mathfrak{R}_1}, \mathfrak{F}^N_{\mathfrak{R}_1}\big) = \big(\mathfrak{F}^{RP}_{\mathfrak{R}_1} + \iota\mathfrak{F}^{IP}_{\mathfrak{R}_1}, \mathfrak{F}^{RN}_{\mathfrak{R}_1} + \iota\mathfrak{F}^{IN}_{\mathfrak{R}_1}\big)$ and $\mathfrak{R}_2 = \big(\mathfrak{F}^P_{\mathfrak{R}_2}, \mathfrak{F}^N_{\mathfrak{R}_2}\big) = \big(\mathfrak{F}^{RP}_{\mathfrak{R}_2} + \iota\mathfrak{F}^{IP}_{\mathfrak{R}_2}, \mathfrak{F}^{RN}_{\mathfrak{R}_2} + \iota\mathfrak{F}^{IN}_{\mathfrak{R}_2}\big)$ of $\mathbb{R}$, then $\mathfrak{R}_1 \cap \mathfrak{R}_2$ is a BCF-I.

Proof Omitted.

3.6 Other Related Algebraic Structures

A non-empty set $\hat{\mathrm{S}}$ along with a binary operation $\star$ is identified as a semilattice iff $\forall x_1, x_2, x_2 \in \hat{\mathrm{S}}$, we have.

1. $x_1 \star \mathrm{x}_2 \in \hat{\mathrm{S}}$
2. $x_1 \star \mathrm{x}_1 = \mathrm{x}_1$
3. $x_1 \star \mathrm{x}_2 = \mathrm{x}_2 \star \mathrm{x}_1$
4. $x_1 \star (\mathrm{x}_2 \star \mathrm{x}_3) = (\mathrm{x}_1 \star \mathrm{x}_2) \star \mathrm{x}_3$

We can also say, an idempotent commutative semigroup is identified as semilattice. A subsemilattice of a lattice $\hat{\mathrm{S}}$ is a subset of $\hat{\mathrm{S}}$ that is closed under the same binary operation. A lattice $\mathbb{L}$ is identified as a distributive lattice if and only if it holds the distributive laws. In this section, $\mathfrak{R}(\mathbb{E})$ would be identified as a collection of BCFSs.

In the next Theorem, we will reveal that $\mathfrak{R}(\mathbb{E})$ is semilattice with the binary operation $\cup$ and $\cap$.

Theorem 3.12 $(\mathfrak{R}(\mathbb{E}), \cup)$ and $(\mathfrak{R}(\mathbb{E}), \cap)$ is semilattice i.e. for all $\mathfrak{R}_1, \mathfrak{R}_2, \mathfrak{R}_3 \in \mathfrak{R}(\mathbb{E})$ the following axioms hold:

1. $\mathfrak{R}_1 \cup \mathfrak{R}_2 \in \mathfrak{R}(\mathbb{E})$, and $\mathfrak{R}_1 \cap \mathfrak{R}_2 \in \mathfrak{R}(\mathbb{E})$
2. $\mathfrak{R}_1 \cup \mathfrak{R}_1 = \mathfrak{R}_1$, and $\mathfrak{R}_1 \cap \mathfrak{R}_1 = \mathfrak{R}_1$,
3. $\mathfrak{R}_1 \cup \mathfrak{R}_2 = \mathfrak{R}_2 \cup \mathfrak{R}_1$ and $\mathfrak{R}_1 \cap \mathfrak{R}_2 = \mathfrak{R}_2 \cap \mathfrak{R}_1$,
4. $\mathfrak{R}_1 \cup (\mathfrak{R}_2 \cup \mathfrak{R}_3) = (\mathfrak{R}_1 \cup \mathfrak{R}_2) \cup \mathfrak{R}_3$ and $\mathfrak{R}_1 \cap (\mathfrak{R}_2 \cap \mathfrak{R}_3) = (\mathfrak{R}_1 \cap \mathfrak{R}_2) \cap \mathfrak{R}_3$.

Proof 1. It is obvious. The remaining axioms 2, 3, and 4 are proved in Proposition (2.1).

The following Theorem will show that $\mathfrak{R}(\mathbb{E})$ is a lattice with two binary operations $\cup$ and $\cap$.

Theorem 3.13 $(\mathfrak{R}(\mathbb{E}), \cup, \cap)$ is lattice i.e., for all $\mathfrak{R}_1, \mathfrak{R}_2, \mathfrak{R}_3 \in \mathfrak{R}(\mathbb{E})$ the following axioms hold:

1. $\mathfrak{R}_1 \cup \mathfrak{R}_1 = \mathfrak{R}_1$, and $\mathfrak{R}_1 \cap \mathfrak{R}_1 = \mathfrak{R}_1$,
2. $\mathfrak{R}_1 \cup \mathfrak{R}_2 = \mathfrak{R}_2 \cup \mathfrak{R}_1$ and $\mathfrak{R}_1 \cap \mathfrak{R}_2 = \mathfrak{R}_2 \cap \mathfrak{R}_1$,
3. $\mathfrak{R}_1 \cup (\mathfrak{R}_2 \cup \mathfrak{R}_3) = (\mathfrak{R}_1 \cup \mathfrak{R}_2) \cup \mathfrak{R}_3$ and $\mathfrak{R}_1 \cap (\mathfrak{R}_2 \cap \mathfrak{R}_3) = (\mathfrak{R}_1 \cap \mathfrak{R}_2) \cap \mathfrak{R}_3$,
4. $\mathfrak{R}_1 \cup (\mathfrak{R}_1 \cap \mathfrak{R}_2) = \mathfrak{R}_1$ and $\mathfrak{R}_1 \cap (\mathfrak{R}_1 \cup \mathfrak{R}_2) = \mathfrak{R}_1$.

Proof 1, 2, and 3 are already proved in Proposition (2.1).

4. Suppose $\mathfrak{R}_1 = \{(e, \mathfrak{F}^P_{\mathfrak{R}_1}(e), \mathfrak{F}^N_{\mathfrak{R}_1}(e)) | e \in \mathbb{E}\}$, $\mathfrak{R}_2 = \{(e, \mathfrak{F}^R_{\mathfrak{R}_2}(e), \mathfrak{F}^P_{\mathfrak{R}_2}(e)) | e \in \mathbb{E}\} \in \mathfrak{R}(\mathbb{E})$, then first we will prove that $\mathfrak{R}_1 \cup (\mathfrak{R}_1 \cap \mathfrak{R}_2) = \mathfrak{R}_1$

$$\mathfrak{R}_1 \cup (\mathfrak{R}_1 \cap \mathfrak{R}_2) = \left\{\left(e, \mathfrak{F}^P_{\mathfrak{R}_1 \cup (\mathfrak{R}_1 \cap \mathfrak{R}_2)}(e), \mathfrak{F}^N_{\mathfrak{R}_1 \cup (\mathfrak{R}_1 \cap \mathfrak{R}_2)}(e)\right) | e \in \mathbb{E}\right\}$$
$$= \left\{\left(\begin{array}{c} e, \max\left(\mathfrak{F}^{RP}_{\mathfrak{R}_1}(e), \min\left(\mathfrak{F}^{RP}_{\mathfrak{R}_1}(e), \mathfrak{F}^{RP}_{\mathfrak{R}_2}(e)\right)\right) + \iota \max\left(\mathfrak{F}^{IP}_{\mathfrak{R}_1}(e), \min\left(\mathfrak{F}^{IP}_{\mathfrak{R}_1}(e), \mathfrak{F}^{IP}_{\mathfrak{R}_2}(e)\right)\right), \\ \min\left(\mathfrak{F}^{RN}_{\mathfrak{R}_1}(e), \max\left(\mathfrak{F}^{RN}_{\mathfrak{R}_1}(e), \mathfrak{F}^{RN}_{\mathfrak{R}_2}(e)\right)\right) + \iota \min\left(\mathfrak{F}^{IN}_{\mathfrak{R}_1}(e), \max\left(\mathfrak{F}^{IN}_{\mathfrak{R}_1}(e), \mathfrak{F}^{IN}_{\mathfrak{R}_2}(e)\right)\right) \end{array}\right) | e \in \mathbb{E}\right\}$$

Since $\min(\mathfrak{F}^{RP}_{\mathfrak{R}_1}(e), \mathfrak{F}^{RP}_{\mathfrak{R}_2}(e)) \le \mathfrak{F}^{RP}_{\mathfrak{R}_1}(e)$ then $\max(\mathfrak{F}^{RP}_{\mathfrak{R}_1}(e), \min(\mathfrak{F}^{RP}_{\mathfrak{R}_1}(e), \mathfrak{F}^{RP}_{\mathfrak{R}_2}(e))) = \mathfrak{F}^{RP}_{\mathfrak{R}_1}(e)$, $\min(\mathfrak{F}^{IP}_{\mathfrak{R}_1}(e), \mathfrak{F}^{IP}_{\mathfrak{R}_2}(e)) \le \mathfrak{F}^{IP}_{\mathfrak{R}_1}(e)$ then $\max(\mathfrak{F}^{IP}_{\mathfrak{R}_1}(e), \min(\mathfrak{F}^{IP}_{\mathfrak{R}_1}(e), \mathfrak{F}^{IP}_{\mathfrak{R}_2}(e))) = \mathfrak{F}^{IP}_{\mathfrak{R}_1}(e)$, $\mathfrak{F}^{RN}_{\mathfrak{R}_1}(e) \le \max(\mathfrak{F}^{RN}_{\mathfrak{R}_1}(e), \mathfrak{F}^{RN}_{\mathfrak{R}_2}(e))$ then $\min(\mathfrak{F}^{RN}_{\mathfrak{R}_1}(e), \max(\mathfrak{F}^{RN}_{\mathfrak{R}_1}(e), \mathfrak{F}^{RN}_{\mathfrak{R}_2}(e))) = \mathfrak{F}^{RN}_{\mathfrak{R}_1}(e)$, and $\mathfrak{F}^{IN}_{\mathfrak{R}_1}(e) \le \max(\mathfrak{F}^{IN}_{\mathfrak{R}_1}(e), \mathfrak{F}^{IN}_{\mathfrak{R}_2}(e))$ then $\min(\mathfrak{F}^{IN}_{\mathfrak{R}_1}(e), \max(\mathfrak{F}^{IN}_{\mathfrak{R}_1}(e), \mathfrak{F}^{IN}_{\mathfrak{R}_2}(e))) = \mathfrak{F}^{IN}_{\mathfrak{R}_1}(e)$. Hence we obtain

$$\mathfrak{R}_1 \cup (\mathfrak{R}_1 \cap \mathfrak{R}_2) = \{(e, (\mathfrak{F}^{RP}_{\mathfrak{R}_1}(e) + \iota \mathfrak{F}^{IP}_{\mathfrak{R}_1}(e), \mathfrak{F}^{RN}_{\mathfrak{R}_1}(e) + \iota \mathfrak{F}^{IN}_{\mathfrak{R}_1}(e))) | e \in \mathbb{E}\} = \mathfrak{R}_1.$$

Next, we will prove that $\mathfrak{R}_1 \cap (\mathfrak{R}_1 \cup \mathfrak{R}_2) = \mathfrak{R}_1$. By definition (13), we obtain

$$\mathfrak{R}_1 \cap (\mathfrak{R}_1 \cup \mathfrak{R}_2) = \left\{\left(e, \mathfrak{F}^P_{\mathfrak{R}_1 \cap (\mathfrak{R}_1 \cup \mathfrak{R}_2)}(e), \mathfrak{F}^N_{\mathfrak{R}_1 \cap (\mathfrak{R}_1 \cup \mathfrak{R}_2)}(e)\right) | e \in \mathbb{E}\right\}$$
$$= \left\{\left(\begin{array}{c} e, \min\left(\mathfrak{F}^{RP}_{\mathfrak{R}_1}(e), \max\left(\mathfrak{F}^{RP}_{\mathfrak{R}_1}(e), \mathfrak{F}^{RP}_{\mathfrak{R}_2}(e)\right)\right) + \iota \min\left(\mathfrak{F}^{IP}_{\mathfrak{R}_1}(e), \max\left(\mathfrak{F}^{IP}_{\mathfrak{R}_1}(e), \mathfrak{F}^{IP}_{\mathfrak{R}_2}(e)\right)\right), \\ \max\left(\mathfrak{F}^{RN}_{\mathfrak{R}_1}(e), \min\left(\mathfrak{F}^{RN}_{\mathfrak{R}_1}(e), \mathfrak{F}^{RN}_{\mathfrak{R}_2}(e)\right)\right) + \iota \max\left(\mathfrak{F}^{IN}_{\mathfrak{R}_1}(e), \min\left(\mathfrak{F}^{IN}_{\mathfrak{R}_1}(e), \mathfrak{F}^{IN}_{\mathfrak{R}_2}(e)\right)\right) \end{array}\right) | e \in \mathbb{E}\right\}$$

Since $\mathfrak{F}^{RP}_{\Re_1}(e) \le \max(\mathfrak{F}^{RP}_{\Re_1}(e), \mathfrak{F}^{RP}_{\Re_2}(e))$ then $\min(\mathfrak{F}^{RP}_{\Re_1}(e), \max(\mathfrak{F}^{RP}_{\Re_1}(e), \mathfrak{F}^{RP}_{\Re_2}(e))) = \mathfrak{F}^{RP}_{\Re_1}(e)$, $\mathfrak{F}^{IP}_{\Re_1}(e) \le \max(\mathfrak{F}^{IP}_{\Re_1}(e), \mathfrak{F}^{IP}_{\Re_2}(e))$ then $\min(\mathfrak{F}^{IP}_{\Re_1}(e), \max(\mathfrak{F}^{IP}_{\Re_1}(e), \mathfrak{F}^{IP}_{\Re_2}(e))) = \mathfrak{F}^{IP}_{\Re_1}(e)$, $\min(\mathfrak{F}^{RN}_{\Re_1}(e), \mathfrak{F}^{RN}_{\Re_2}(e)) \le \mathfrak{F}^{RN}_{\Re_1}(e)$ then $\max(\mathfrak{F}^{RN}_{\Re_1}(e), \min(\mathfrak{F}^{RN}_{\Re_1}(e), \mathfrak{F}^{RN}_{\Re_2}(e))) = \mathfrak{F}^{RN}_{\Re_1}(e)$, and $\min(\mathfrak{F}^{IN}_{\Re_1}(e), \mathfrak{F}^{IN}_{\Re_2}(e)) \le \mathfrak{F}^{IN}_{\Re_1}(e)$ then $\max(\mathfrak{F}^{IN}_{\Re_1}(e), \min(\mathfrak{F}^{IN}_{\Re_1}(e), \mathfrak{F}^{IN}_{\Re_2}(e))) = \mathfrak{F}^{IN}_{\Re_1}(e)$. Hence we obtain

$$\Re_1 \cap (\Re_1 \cup \Re_2) = \left\{\left(e, \left(\mathfrak{F}^{RP}_{\Re_1}(e) + \iota\mathfrak{F}^{IP}_{\Re_1}(e), \mathfrak{F}^{RN}_{\Re_1}(e) + \iota\mathfrak{F}^{IN}_{\Re_1}(e)\right)\right) | e \in \mathbb{E}\right\} = \Re_1.$$

Proposition 3.1. $(\Re(\mathbb{E}), \cup, \cap)$ is a distributive lattice.

Proof From Theorem (3.13), we have that $(\Re(\mathbb{E}), \cup, \cap)$ is a lattice and from Proposition (2.1), we have that $(\Re(\mathbb{E}), \cup, \cap)$ holds the distributive laws.

Theorem 3.14 $(\Re(\mathbb{E}), \cup, \cap, \emptyset_{\mathbb{E}}, \wp)$ is a bounded lattice.

Proof In Theorem (3.13) it is proved that $(\Re(\mathbb{E}), \cup, \cap, \emptyset_{\mathbb{E}}, \wp)$ is lattice. So $(\Re(\mathbb{E}), \cup, \cap, \emptyset_{\mathbb{E}}, \wp)$ is a bounded lattice with the first element $\emptyset_{\mathbb{E}}$ and last element $\wp$.

Theorem 3.15 $(\Re(\mathbb{E}), \cup, \cap, \emptyset_{\mathbb{E}}, \wp)$ is a Boolean algebra.

Proof In Theorem (3.14) it is proved that $(\Re(\mathbb{E}), \cup, \cap, \emptyset_{\mathbb{E}}, \wp)$ is bounded lattice, distributive laws hold by proposition (2.1). Let us take the last element $\emptyset_{\mathbb{E}} = (0.0 + \iota 0.0, -0.0 - \iota 0.0)$ and first element $\wp = (1.0 + \iota 1.0, -1.0 - \iota 1.0)$ then, we can find the complement of every element i.e., for every $\Re \in \Re(\mathbb{E})$, we can find $\Re^c$ which holds the underneath axioms.

1. $\Re \cup \Re^c = \wp$,
2. $\Re \cap \Re^c = \emptyset_{\mathbb{E}}$

i.e. $\Re(\mathbb{E})$ is a complemented lattice. Hence $(\Re(\mathbb{E}), \cup, \cap, \emptyset_{\mathbb{E}}, \wp)$ is a Boolean algebra.

Theorem 3.16 $(\Re(\mathbb{E}), \cup, \cap, \emptyset_{\mathbb{E}}, \wp, c)$ is De-Morgan's algebra.

Proof Since $(\Re(\mathbb{E}), \cup, \cap, \emptyset_{\mathbb{E}}, \wp, c)$ is bounded distributive lattice by Theorem (3.15), by proposition (2.2) we have that the De-Morgan's laws hold, and by proposition (2.1), we have that for all BCFSs $\Re$, $(\Re^c)^c = \Re$. So, this implies that $(\Re(\mathbb{E}), \cup, \cap, \emptyset_{\mathbb{E}}, \wp, c)$ is De-Morgan's algebra.

Theorem 3.17 $(\Re(\mathbb{E}), \cup, \cap, \emptyset_{\mathbb{E}}, \wp, c)$ is Kleene algebra.

Proof It is clear that for all $\Re_1, \Re_2 \in \Re(\mathbb{E})$ $\Re_1 \cap \Re_1^c \subseteq \Re_2 \cup \Re_2{}^c$ and by Theorem (3.16) $(\Re(\mathbb{E}), \cup, \cap, \emptyset_{\mathbb{E}}, \wp, c)$ is De-Morgan's algebra. Hence $(\Re(\mathbb{E}), \cup, \cap, \emptyset_{\mathbb{E}}, \wp, c)$ is a Kleene algebra.

Chapter 4
Distance Measures for Bipolar Complex Fuzzy Sets and Their Application in Automated Trading Systems

In several disciplines, including mathematics, statistics, and computer science, distance measures are used to quantify the differences between objects. Numerous applications, such as clustering, classification, data analysis, pattern recognition, and machine learning, depend greatly on these measures. The dissimilarity between two objects in a dataset is numerically represented by distance measures. It is possible to use the distance between items as a gauge for how "far apart" or unlike they are from one another. There are several well-known distance measures such as Hamming distance, Euclidean distance, Minkowski distance, Hausdorff distance that are frequently used in various situations. The type of distance measure chosen will rely on the nature of information, the scope of the issue, and the analysis or applications are particular goals.

The term "automated trading systems" also known as "algorithmic trading" or "black-box trading," refers to computer programs that are created to carry out buy and sell orders in the financial markets in accordance with a predetermined set of guidelines. These systems employ sophisticated algorithms to instantly assess a massive quantity of market data, including price changes, trade activity, and other pertinent factors. They may be created to trade a variety of financial items, such as derivatives, currencies, commodities, and stocks. By automatically placing trades when certain criteria are satisfied, frequently at very high speeds and frequencies, automated trading system tries to eliminate the emotional and impulsive components of trading. Institutional investors, hedge funds, and individual traders all use this technology in an effort to take advantage of market opportunities and effectively manage their portfolios. With the use of automated trading system, a variety of trading approaches are possible, including trend-following, mean-reversion, arbitrage, and statistical analysis. Although they could improve productivity and decrease human error, it's important to be aware that they also come with hazards, especially in fast shifting market settings where unforeseen circumstances might produce unexpected results. Therefore, for Automated Trading Systems to be successful and safe, good design, thorough testing, and continuing monitoring are crucial.

U. ur Rehman and T. Mahmood, *Bipolar Complex Fuzzy Sets*, Studies in Fuzziness and Soft Computing 439, https://doi.org/10.1007/978-3-032-07388-4_4

In Sect. 4.1. of this chapter, we initiate a few distance measures among BCFSs relying on the well-known Hamming and Hausdorff distance such as generalized BCF normalized Hamming distance measure (GBCFNHDM), generalized BCF normalized Hausdroff distance measure (GBCFNHDDM), and generalized hybrid BCF normalized distance measure (GHBCFNDM) and their particular cases. In Sect. 4.2 of this chapter, we consider weight in the originated distance measures of Sect. 4.1 to investigate the weighted distance measures. In Sect. 4.3. we inaugurate continuous distance measures between BCFSs. Section 4.4 contains the VIKOR technique relying on the initiated distance measures in the structure of BCF information and in Sect. 4.5, we devise an application (automated trading system) by employing the VIKOR approach in the setting of BCF information. In Sect. 4.6, we do the comparison of initiated work with a few prevailing theories for advantages and supremacy of the initiated work.

4.1 Distance Measures for BCFSs

We interpret a few distance measures among BCFSs relying on the well-known Hamming and Hausdorff distances such as generalized GBCFNHDM, GBCFNHDDM, and GHBCFNDM and their particular cases.

Definition 4.1 Consider two BCFSs $\mathfrak{R}_1 = \left\{e_\sigma, \left(\mathfrak{F}^{P}_{\mathfrak{R}_1}(e_\sigma), \mathfrak{F}^{N}_{\mathfrak{R}_1}(e_\sigma)\right) | e_\sigma \in \mathbb{E}, \sigma = 1, 2, .., n\right\} = \left\{e_\sigma, \left(\mathfrak{F}^{RP}_{\mathfrak{R}_1}(e_\sigma) + \iota\mathfrak{F}^{IP}_{\mathfrak{R}_1}(e_\sigma), \mathfrak{F}^{RN}_{\mathfrak{R}_1}(e_\sigma) + \iota\mathfrak{F}^{IN}_{\mathfrak{R}_1}(e_\sigma)\right) | e_\sigma \in \mathbb{E}, \sigma = 1, 2, .., n\right\}$ and $\mathfrak{R}_2 = \left\{e, \left(\mathfrak{F}^{P}_{\mathfrak{R}_2}(e_\sigma), \mathfrak{F}^{N}_{\mathfrak{R}_2}(e_\sigma)\right) | e \in \mathbb{E}, \sigma = 1, 2, .., n\right\} = \left\{e_\sigma, \left(\mathfrak{F}^{RP}_{\mathfrak{R}_2}(e_\sigma) + \iota\mathfrak{F}^{IP}_{\mathfrak{R}_2}(e_\sigma), \mathfrak{F}^{RN}_{\mathfrak{R}_2}(e_\sigma) + \iota\mathfrak{F}^{IN}_{\mathfrak{R}_2}(e_\sigma)\right) | e_\sigma \in \mathbb{E}, \sigma = 1, 2, .., n\right\}$ over $\mathbb{E}$. Then the GBCFNHDM between $\mathfrak{R}_1$ and $\mathfrak{R}_2$ is denoted and defined as.

$$d^1_{BCFS}(\mathfrak{R}_1, \mathfrak{R}_2) = \left(\frac{1}{4n}\sum_{\sigma=1}^{n}\left(\begin{array}{c} \left|\left(\mathfrak{F}^{RP}_{\mathfrak{R}_1}(e_\sigma)\right)^2 - \left(\mathfrak{F}^{RP}_{\mathfrak{R}_2}(e_\sigma)\right)^2\right| + \left|\left(\mathfrak{F}^{IP}_{\mathfrak{R}_1}(e_\sigma)\right)^2 - \left(\mathfrak{F}^{IP}_{\mathfrak{R}_2}(e_\sigma)\right)^2\right| \\ + \left|\left(\mathfrak{F}^{RN}_{\mathfrak{R}_1}(e_\sigma)\right)^2 - \left(\mathfrak{F}^{RN}_{\mathfrak{R}_2}(e_\sigma)\right)^2\right| + \left|\left(\mathfrak{F}^{IN}_{\mathfrak{R}_1}(e_\sigma)\right)^2 - \left(\mathfrak{F}^{IN}_{\mathfrak{R}_2}(e_\sigma)\right)^2\right| \end{array}\right)\right)^{\frac{1}{\lambda}}, \lambda \geq 1$$

Next, Theorem interpret that d^1_{BCFS} satisfies the necessary properties of distance measure.

Theorem 4.1 The GBCFNHDM satisfies the underneath properties.

1. $0 \leq d^1_{BCFS}(\mathfrak{R}_1, \mathfrak{R}_2) \leq 1$
2. $d^1_{BCFS}(\mathfrak{R}_1, \mathfrak{R}_2) = 0 \Leftrightarrow \mathfrak{R}_1 = \mathfrak{R}_2$
3. $d^1_{BCFS}(\mathfrak{R}_1, \mathfrak{R}_2) = d^1_{BCFS}(\mathfrak{R}_2, \mathfrak{R}_1)$
4. If $\mathfrak{R}_1 \subset \mathfrak{R}_2 \subset \mathfrak{R}_3$, then $d^1_{BCFS}(\mathfrak{R}_1, \mathfrak{R}_2) \leq d^1_{BCFS}(\mathfrak{R}_1, \mathfrak{R}_3)$ and $d^1_{BCFS}(\mathfrak{R}_2, \mathfrak{R}_3) \leq d^1_{BCFS}(\mathfrak{R}_1, \mathfrak{R}_3)$

Proof: 1. As $\mathfrak{F}^{RP}_{\mathfrak{R}_1}(e_\sigma), \mathfrak{F}^{IP}_{\mathfrak{R}_1}(e_\sigma), \mathfrak{F}^{RP}_{\mathfrak{R}_2}(e_\sigma), \mathfrak{F}^{IP}_{\mathfrak{R}_2}(e_\sigma) \in [0, 1]$ and $\mathfrak{F}^{RN}_{\mathfrak{R}_1}(e_\sigma), \mathfrak{F}^{IN}_{\mathfrak{R}_1}(e_\sigma), \mathfrak{F}^{RN}_{\mathfrak{R}_2}(e_\sigma), \mathfrak{F}^{IN}_{\mathfrak{R}_2}(e_\sigma) \in [-1, 0]$ for all σ. Then

$\left(\mathfrak{F}_{\mathfrak{R}_1}^{RP}(e_\sigma)\right)^2$, $\left(\mathfrak{F}_{\mathfrak{R}_1}^{IP}(e_\sigma)\right)^2$, $\left(\mathfrak{F}_{\mathfrak{R}_2}^{RP}(e_\sigma)\right)^2$, $\left(\mathfrak{F}_{\mathfrak{R}_2}^{IP}(e_\sigma)\right)^2 \in [0, 1]$ and $\left(\mathfrak{F}_{\mathfrak{R}_1}^{RN}(e_\sigma)\right)^2$ $\left(\mathfrak{F}_{\mathfrak{R}_1}^{IN}(e_\sigma)\right)^2$ $\left(\mathfrak{F}_{\mathfrak{R}_2}^{RN}(e_\sigma)\right)^2$, $\left(\mathfrak{F}_{\mathfrak{R}_2}^{IN}(e_\sigma)\right)^2 \in [0, 1]$ for all σ. $\Rightarrow$ $\left|\left(\mathfrak{F}_{\mathfrak{R}_1}^{RP}(e_\sigma)\right)^2 - \left(\mathfrak{F}_{\mathfrak{R}_2}^{RP}(e_\sigma)\right)^2\right|$, $\left|\left(\mathfrak{F}_{\mathfrak{R}_1}^{IP}(e_\sigma)\right)^2 - \left(\mathfrak{F}_{\mathfrak{R}_2}^{IP}(e_\sigma)\right)^2\right|$, $\left|\left(\mathfrak{F}_{\mathfrak{R}_1}^{RN}(e_\sigma)\right)^2 - \left(\mathfrak{F}_{\mathfrak{R}_2}^{RN}(e_\sigma)\right)^2\right|$, $\left|\left(\mathfrak{F}_{\mathfrak{R}_1}^{IN}(e_\sigma)\right)^2 - \left(\mathfrak{F}_{\mathfrak{R}_2}^{IN}(e_\sigma)\right)^2\right| \in [0, 1]\ \forall\sigma$. Consequently, $\forall\sigma$

$$\left(\begin{array}{c}\left|\left(\mathfrak{F}_{\mathfrak{R}_1}^{RP}(e_\sigma)\right)^2 - \left(\mathfrak{F}_{\mathfrak{R}_2}^{RP}(e_\sigma)\right)^2\right| + \left|\left(\mathfrak{F}_{\mathfrak{R}_1}^{IP}(e_\sigma)\right)^2 - \left(\mathfrak{F}_{\mathfrak{R}_2}^{IP}(e_\sigma)\right)^2\right| \\ +\left|\left(\mathfrak{F}_{\mathfrak{R}_1}^{RN}(e_\sigma)\right)^2 - \left(\mathfrak{F}_{\mathfrak{R}_2}^{RN}(e_\sigma)\right)^2\right| + \left|\left(\mathfrak{F}_{\mathfrak{R}_1}^{IN}(e_\sigma)\right)^2 - \left(\mathfrak{F}_{\mathfrak{R}_2}^{IN}(e_\sigma)\right)^2\right|\end{array}\right) \in [0, 4]$$

$$0 \le \left(\begin{array}{c}\left|\left(\mathfrak{F}_{\mathfrak{R}_1}^{RP}(e_\sigma)\right)^2 - \left(\mathfrak{F}_{\mathfrak{R}_2}^{RP}(e_\sigma)\right)^2\right| + \left|\left(\mathfrak{F}_{\mathfrak{R}_1}^{IP}(e_\sigma)\right)^2 - \left(\mathfrak{F}_{\mathfrak{R}_2}^{IP}(e_\sigma)\right)^2\right| \\ +\left|\left(\mathfrak{F}_{\mathfrak{R}_1}^{RN}(e_\sigma)\right)^2 - \left(\mathfrak{F}_{\mathfrak{R}_2}^{RN}(e_\sigma)\right)^2\right| + \left|\left(\mathfrak{F}_{\mathfrak{R}_1}^{IN}(e_\sigma)\right)^2 - \left(\mathfrak{F}_{\mathfrak{R}_2}^{IN}(e_\sigma)\right)^2\right|\end{array}\right) \le 4$$

Thus,

$$0 \le \sum_{\sigma=1}^{n}\left(\begin{array}{c}\left|\left(\mathfrak{F}_{\mathfrak{R}_1}^{RP}(e_\sigma)\right)^2 - \left(\mathfrak{F}_{\mathfrak{R}_2}^{RP}(e_\sigma)\right)^2\right| + \left|\left(\mathfrak{F}_{\mathfrak{R}_1}^{IP}(e_\sigma)\right)^2 - \left(\mathfrak{F}_{\mathfrak{R}_2}^{IP}(e_\sigma)\right)^2\right| \\ +\left|\left(\mathfrak{F}_{\mathfrak{R}_1}^{RN}(e_\sigma)\right)^2 - \left(\mathfrak{F}_{\mathfrak{R}_2}^{RN}(e_\sigma)\right)^2\right| + \left|\left(\mathfrak{F}_{\mathfrak{R}_1}^{IN}(e_\sigma)\right)^2 - \left(\mathfrak{F}_{\mathfrak{R}_2}^{IN}(e_\sigma)\right)^2\right|\end{array}\right) \le 4n$$

$$0 \le \frac{1}{4n}\sum_{\sigma=1}^{n}\left(\begin{array}{c}\left|\left(\mathfrak{F}_{\mathfrak{R}_1}^{RP}(e_\sigma)\right)^2 - \left(\mathfrak{F}_{\mathfrak{R}_2}^{RP}(e_\sigma)\right)^2\right| + \left|\left(\mathfrak{F}_{\mathfrak{R}_1}^{IP}(e_\sigma)\right)^2 - \left(\mathfrak{F}_{\mathfrak{R}_2}^{IP}(e_\sigma)\right)^2\right| \\ +\left|\left(\mathfrak{F}_{\mathfrak{R}_1}^{RN}(e_\sigma)\right)^2 - \left(\mathfrak{F}_{\mathfrak{R}_2}^{RN}(e_\sigma)\right)^2\right| + \left|\left(\mathfrak{F}_{\mathfrak{R}_1}^{IN}(e_\sigma)\right)^2 - \left(\mathfrak{F}_{\mathfrak{R}_2}^{IN}(e_\sigma)\right)^2\right|\end{array}\right) \le 1$$

Since $\lambda \ge 1$, then

$$0 \le \left(\frac{1}{4n}\sum_{\sigma=1}^{n}\left(\begin{array}{c}\left|\left(\mathfrak{F}_{\mathfrak{R}_1}^{RP}(e_\sigma)\right)^2 - \left(\mathfrak{F}_{\mathfrak{R}_2}^{RP}(e_\sigma)\right)^2\right| + \left|\left(\mathfrak{F}_{\mathfrak{R}_1}^{IP}(e_\sigma)\right)^2 - \left(\mathfrak{F}_{\mathfrak{R}_2}^{IP}(e_\sigma)\right)^2\right| \\ +\left|\left(\mathfrak{F}_{\mathfrak{R}_1}^{RN}(e_\sigma)\right)^2 - \left(\mathfrak{F}_{\mathfrak{R}_2}^{RN}(e_\sigma)\right)^2\right| + \left|\left(\mathfrak{F}_{\mathfrak{R}_1}^{IN}(e_\sigma)\right)^2 - \left(\mathfrak{F}_{\mathfrak{R}_2}^{IN}(e_\sigma)\right)^2\right|\end{array}\right)\right)^{\frac{1}{\lambda}} \le 1$$

$$\Rightarrow 0 \le d_{BCFS}^{1}(\mathfrak{R}_1, \mathfrak{R}_2) \le 1.$$

2. Let $d_{BCFS}^{1}(\mathfrak{R}_1, \mathfrak{R}_2) = 0$. Then to show that $\mathfrak{R}_1 = \mathfrak{R}_2$. as $d_{BCFS}^{1}(\mathfrak{R}_1, \mathfrak{R}_2) = 0$, then

$$\left(\frac{1}{4n}\sum_{\sigma=1}^{n}\left(\begin{array}{c}\left|\left(\mathfrak{F}_{\mathfrak{R}_1}^{RP}(e_\sigma)\right)^2 - \left(\mathfrak{F}_{\mathfrak{R}_2}^{RP}(e_\sigma)\right)^2\right| + \left|\left(\mathfrak{F}_{\mathfrak{R}_1}^{IP}(e_\sigma)\right)^2 - \left(\mathfrak{F}_{\mathfrak{R}_2}^{IP}(e_\sigma)\right)^2\right| \\ +\left|\left(\mathfrak{F}_{\mathfrak{R}_1}^{RN}(e_\sigma)\right)^2 - \left(\mathfrak{F}_{\mathfrak{R}_2}^{RN}(e_\sigma)\right)^2\right| + \left|\left(\mathfrak{F}_{\mathfrak{R}_1}^{IN}(e_\sigma)\right)^2 - \left(\mathfrak{F}_{\mathfrak{R}_2}^{IN}(e_\sigma)\right)^2\right|\end{array}\right)\right)^{\frac{1}{\lambda}} = 0$$

$$\frac{1}{4n}\sum_{\sigma=1}^{n}\left(\begin{array}{c}\left|\left(\mathfrak{F}_{\mathfrak{R}_1}^{RP}(e_\sigma)\right)^2 - \left(\mathfrak{F}_{\mathfrak{R}_2}^{RP}(e_\sigma)\right)^2\right| + \left|\left(\mathfrak{F}_{\mathfrak{R}_1}^{IP}(e_\sigma)\right)^2 - \left(\mathfrak{F}_{\mathfrak{R}_2}^{IP}(e_\sigma)\right)^2\right| \\ +\left|\left(\mathfrak{F}_{\mathfrak{R}_1}^{RN}(e_\sigma)\right)^2 - \left(\mathfrak{F}_{\mathfrak{R}_2}^{RN}(e_\sigma)\right)^2\right| + \left|\left(\mathfrak{F}_{\mathfrak{R}_1}^{IN}(e_\sigma)\right)^2 - \left(\mathfrak{F}_{\mathfrak{R}_2}^{IN}(e_\sigma)\right)^2\right|\end{array}\right) = 0$$

$$\sum_{\sigma=1}^{n}\left(\begin{array}{c}\left|\left(\mathfrak{F}_{\mathfrak{R}_1}^{RP}(e_\sigma)\right)^2 - \left(\mathfrak{F}_{\mathfrak{R}_2}^{RP}(e_\sigma)\right)^2\right| + \left|\left(\mathfrak{F}_{\mathfrak{R}_1}^{IP}(e_\sigma)\right)^2 - \left(\mathfrak{F}_{\mathfrak{R}_2}^{IP}(e_\sigma)\right)^2\right| \\ +\left|\left(\mathfrak{F}_{\mathfrak{R}_1}^{RN}(e_\sigma)\right)^2 - \left(\mathfrak{F}_{\mathfrak{R}_2}^{RN}(e_\sigma)\right)^2\right| + \left|\left(\mathfrak{F}_{\mathfrak{R}_1}^{IN}(e_\sigma)\right)^2 - \left(\mathfrak{F}_{\mathfrak{R}_2}^{IN}(e_\sigma)\right)^2\right|\end{array}\right) = 0$$

$$\left(\begin{array}{c}\left|\left(\mathfrak{F}_{\mathfrak{R}_1}^{\mathrm{RP}}(\mathrm{e}_\sigma)\right)^2-\left(\mathfrak{F}_{\mathfrak{R}_2}^{\mathrm{RP}}(\mathrm{e}_\sigma)\right)^2\right|+\left|\left(\mathfrak{F}_{\mathfrak{R}_1}^{\mathrm{IP}}(\mathrm{e}_\sigma)\right)^2-\left(\mathfrak{F}_{\mathfrak{R}_2}^{\mathrm{IP}}(\mathrm{e}_\sigma)\right)^2\right| \\ +\left|\left(\mathfrak{F}_{\mathfrak{R}_1}^{\mathrm{RN}}(\mathrm{e}_\sigma)\right)^2-\left(\mathfrak{F}_{\mathfrak{R}_2}^{\mathrm{RN}}(\mathrm{e}_\sigma)\right)^2\right|+\left|\left(\mathfrak{F}_{\mathfrak{R}_1}^{\mathrm{IN}}(\mathrm{e}_\sigma)\right)^2-\left(\mathfrak{F}_{\mathfrak{R}_2}^{\mathrm{IN}}(\mathrm{e}_\sigma)\right)^2\right|\end{array}\right)=0\forall\sigma$$

$\Rightarrow \forall\sigma$, $\left|\left(\mathfrak{F}_{\mathfrak{R}_1}^{\mathrm{RP}}(\mathrm{e}_\sigma)\right)^2-\left(\mathfrak{F}_{\mathfrak{R}_2}^{\mathrm{RP}}(\mathrm{e}_\sigma)\right)^2\right| = 0$, $\left|\left(\mathfrak{F}_{\mathfrak{R}_1}^{\mathrm{IP}}(\mathrm{e}_\sigma)\right)^2-\left(\mathfrak{F}_{\mathfrak{R}_2}^{\mathrm{IP}}(\mathrm{e}_\sigma)\right)^2\right| = 0$, $\left|\left(\mathfrak{F}_{\mathfrak{R}_1}^{\mathrm{RN}}(\mathrm{e}_\sigma)\right)^2-\left(\mathfrak{F}_{\mathfrak{R}_2}^{\mathrm{RN}}(\mathrm{e}_\sigma)\right)^2\right| = 0$, $\left|\left(\mathfrak{F}_{\mathfrak{R}_1}^{\mathrm{RN}}(\mathrm{e}_\sigma)\right)^2-\left(\mathfrak{F}_{\mathfrak{R}_2}^{\mathrm{RN}}(\mathrm{e}_\sigma)\right)^2\right| = 0$. Thus, $\mathfrak{F}_{\mathfrak{R}_1}^{\mathrm{RP}}(\mathrm{e}_\sigma) = \mathfrak{F}_{\mathfrak{R}_2}^{\mathrm{RP}}(\mathrm{e}_\sigma)$, $\mathfrak{F}_{\mathfrak{R}_1}^{\mathrm{IP}}(\mathrm{e}_\sigma) = \mathfrak{F}_{\mathfrak{R}_2}^{\mathrm{IP}}(\mathrm{e}_\sigma)$, $\mathfrak{F}_{\mathfrak{R}_1}^{\mathrm{RN}}(\mathrm{e}_\sigma) = \mathfrak{F}_{\mathfrak{R}_2}^{\mathrm{RN}}(\mathrm{e}_\sigma)$, $\mathfrak{F}_{\mathfrak{R}_1}^{\mathrm{IN}}(\mathrm{e}_\sigma) = \mathfrak{F}_{\mathfrak{R}_2}^{\mathrm{IN}}(\mathrm{e}_\sigma)$ $\forall\sigma$. Consequently, $\mathfrak{R}_1 = \mathfrak{R}_2$.

Conversely, Let $\mathfrak{R}_1 = \mathfrak{R}_2$. To show that $d_{BCFS}^1(\mathfrak{R}_1, \mathfrak{R}_2) = 0$. As $\mathfrak{R}_1 = \mathfrak{R}_2$, then $\mathfrak{F}_{\mathfrak{R}_1}^{\mathrm{RP}}(\mathrm{e}_\sigma) = \mathfrak{F}_{\mathfrak{R}_2}^{\mathrm{RP}}(\mathrm{e}_\sigma)$, $\mathfrak{F}_{\mathfrak{R}_1}^{\mathrm{IP}}(\mathrm{e}_\sigma) = \mathfrak{F}_{\mathfrak{R}_2}^{\mathrm{IP}}(\mathrm{e}_\sigma)$, $\mathfrak{F}_{\mathfrak{R}_1}^{\mathrm{RN}}(\mathrm{e}_\sigma) = \mathfrak{F}_{\mathfrak{R}_2}^{\mathrm{RN}}(\mathrm{e}_\sigma)$, $\mathfrak{F}_{\mathfrak{R}_1}^{\mathrm{IN}}(\mathrm{e}_\sigma) = \mathfrak{F}_{\mathfrak{R}_2}^{\mathrm{IN}}(\mathrm{e}_\sigma)$ $\forall\sigma$. Then $\forall\sigma$

$$d_{BCFS}^1(\mathfrak{R}_1,\mathfrak{R}_2)=\left(\frac{1}{4n}\sum_{\sigma=1}^{n}\left(\begin{array}{c}\left|\left(\mathfrak{F}_{\mathfrak{R}_1}^{\mathrm{RP}}(\mathrm{e}_\sigma)\right)^2-\left(\mathfrak{F}_{\mathfrak{R}_1}^{\mathrm{RP}}(\mathrm{e}_\sigma)\right)^2\right|+\left|\left(\mathfrak{F}_{\mathfrak{R}_1}^{\mathrm{IP}}(\mathrm{e}_\sigma)\right)^2-\left(\mathfrak{F}_{\mathfrak{R}_1}^{\mathrm{IP}}(\mathrm{e}_\sigma)\right)^2\right| \\ +\left|\left(\mathfrak{F}_{\mathfrak{R}_1}^{\mathrm{RN}}(\mathrm{e}_\sigma)\right)^2-\left(\mathfrak{F}_{\mathfrak{R}_1}^{\mathrm{RN}}(\mathrm{e}_\sigma)\right)^2\right|+\left|\left(\mathfrak{F}_{\mathfrak{R}_1}^{\mathrm{IN}}(\mathrm{e}_\sigma)\right)^2-\left(\mathfrak{F}_{\mathfrak{R}_1}^{\mathrm{IN}}(\mathrm{e}_\sigma)\right)^2\right|\end{array}\right)\right)^{\frac{1}{\lambda}}=0$$

3. As

$$d_{BCFS}^1(\mathfrak{R}_1,\mathfrak{R}_2)=\left(\frac{1}{4n}\sum_{\sigma=1}^{n}\left(\begin{array}{c}\left|\left(\mathfrak{F}_{\mathfrak{R}_1}^{\mathrm{RP}}(\mathrm{e}_\sigma)\right)^2-\left(\mathfrak{F}_{\mathfrak{R}_2}^{\mathrm{RP}}(\mathrm{e}_\sigma)\right)^2\right|+\left|\left(\mathfrak{F}_{\mathfrak{R}_1}^{\mathrm{IP}}(\mathrm{e}_\sigma)\right)^2-\left(\mathfrak{F}_{\mathfrak{R}_2}^{\mathrm{IP}}(\mathrm{e}_\sigma)\right)^2\right| \\ +\left|\left(\mathfrak{F}_{\mathfrak{R}_1}^{\mathrm{RN}}(\mathrm{e}_\sigma)\right)^2-\left(\mathfrak{F}_{\mathfrak{R}_2}^{\mathrm{RN}}(\mathrm{e}_\sigma)\right)^2\right|+\left|\left(\mathfrak{F}_{\mathfrak{R}_1}^{\mathrm{IN}}(\mathrm{e}_\sigma)\right)^2-\left(\mathfrak{F}_{\mathfrak{R}_2}^{\mathrm{IN}}(\mathrm{e}_\sigma)\right)^2\right|\end{array}\right)\right)^{\frac{1}{\lambda}}$$

$$=\left(\frac{1}{4n}\sum_{\sigma=1}^{n}\left(\begin{array}{c}\left|\left(\mathfrak{F}_{\mathfrak{R}_2}^{\mathrm{RP}}(\mathrm{e}_\sigma)\right)^2-\left(\mathfrak{F}_{\mathfrak{R}_1}^{\mathrm{RP}}(\mathrm{e}_\sigma)\right)^2\right|+\left|\left(\mathfrak{F}_{\mathfrak{R}_2}^{\mathrm{IP}}(\mathrm{e}_\sigma)\right)^2-\left(\mathfrak{F}_{\mathfrak{R}_1}^{\mathrm{IP}}(\mathrm{e}_\sigma)\right)^2\right| \\ +\left|\left(\mathfrak{F}_{\mathfrak{R}_2}^{\mathrm{RN}}(\mathrm{e}_\sigma)\right)^2-\left(\mathfrak{F}_{\mathfrak{R}_1}^{\mathrm{RN}}(\mathrm{e}_\sigma)\right)^2\right|+\left|\left(\mathfrak{F}_{\mathfrak{R}_2}^{\mathrm{IN}}(\mathrm{e}_\sigma)\right)^2-\left(\mathfrak{F}_{\mathfrak{R}_1}^{\mathrm{IN}}(\mathrm{e}_\sigma)\right)^2\right|\end{array}\right)\right)^{\frac{1}{\lambda}}=d_{BCFS}^1(\mathfrak{R}_2,\mathfrak{R}_1)$$

4. Assume that $\mathfrak{R}_1 \subset \mathfrak{R}_2 \subset \mathfrak{R}_3$, then $\mathfrak{F}_{\mathfrak{R}_1}^{\mathrm{RP}}(\mathrm{e}_\sigma) \le \mathfrak{F}_{\mathfrak{R}_2}^{\mathrm{RP}}(\mathrm{e}_\sigma) \le \mathfrak{F}_{\mathfrak{R}_3}^{\mathrm{RP}}(\mathrm{e}_\sigma)$, $\mathfrak{F}_{\mathfrak{R}_1}^{\mathrm{IP}}(\mathrm{e}_\sigma) \le \mathfrak{F}_{\mathfrak{R}_2}^{\mathrm{IP}}(\mathrm{e}_\sigma) \le \mathfrak{F}_{\mathfrak{R}_3}^{\mathrm{IP}}(\mathrm{e}_\sigma)$, $\mathfrak{F}_{\mathfrak{R}_1}^{\mathrm{RN}}(\mathrm{e}_\sigma) \le \mathfrak{F}_{\mathfrak{R}_2}^{\mathrm{RN}}(\mathrm{e}_\sigma) \le \mathfrak{F}_{\mathfrak{R}_3}^{\mathrm{RN}}(\mathrm{e}_\sigma)$, $\mathfrak{F}_{\mathfrak{R}_1}^{\mathrm{IN}}(\mathrm{e}_\sigma) \le \mathfrak{F}_{\mathfrak{R}_2}^{\mathrm{IN}}(\mathrm{e}_\sigma) \le \mathfrak{F}_{\mathfrak{R}_3}^{\mathrm{IN}}(\mathrm{e}_\sigma)$ $\forall\sigma$. Thus, we have the inequalities

$$\left|\left(\mathfrak{F}_{\mathfrak{R}_1}^{\mathrm{RP}}(\mathrm{e}_\sigma)\right)^2-\left(\mathfrak{F}_{\mathfrak{R}_2}^{\mathrm{RP}}(\mathrm{e}_\sigma)\right)^2\right| \le \left|\left(\mathfrak{F}_{\mathfrak{R}_1}^{\mathrm{RP}}(\mathrm{e}_\sigma)\right)^2-\left(\mathfrak{F}_{\mathfrak{R}_3}^{\mathrm{RP}}(\mathrm{e}_\sigma)\right)^2\right|$$

$$\left|\left(\mathfrak{F}_{\mathfrak{R}_1}^{\mathrm{IP}}(\mathrm{e}_\sigma)\right)^2-\left(\mathfrak{F}_{\mathfrak{R}_2}^{\mathrm{IP}}(\mathrm{e}_\sigma)\right)^2\right| \le \left|\left(\mathfrak{F}_{\mathfrak{R}_1}^{\mathrm{IP}}(\mathrm{e}_\sigma)\right)^2-\left(\mathfrak{F}_{\mathfrak{R}_3}^{\mathrm{IP}}(\mathrm{e}_\sigma)\right)^2\right|$$

And

$$\left|\left(\mathfrak{F}_{\mathfrak{R}_1}^{\mathrm{RN}}(\mathrm{e}_\sigma)\right)^2-\left(\mathfrak{F}_{\mathfrak{R}_2}^{\mathrm{RN}}(\mathrm{e}_\sigma)\right)^2\right| \le \left|\left(\mathfrak{F}_{\mathfrak{R}_1}^{\mathrm{RN}}(\mathrm{e}_\sigma)\right)^2-\left(\mathfrak{F}_{\mathfrak{R}_3}^{\mathrm{RN}}(\mathrm{e}_\sigma)\right)^2\right|$$

$$\left|\left(\mathfrak{F}_{\mathfrak{R}_1}^{\mathrm{IN}}(\mathrm{e}_\sigma)\right)^2-\left(\mathfrak{F}_{\mathfrak{R}_2}^{\mathrm{IN}}(\mathrm{e}_\sigma)\right)^2\right| \le \left|\left(\mathfrak{F}_{\mathfrak{R}_1}^{\mathrm{IN}}(\mathrm{e}_\sigma)\right)^2-\left(\mathfrak{F}_{\mathfrak{R}_3}^{\mathrm{IN}}(\mathrm{e}_\sigma)\right)^2\right|$$

Thus, $\forall\sigma$

$$\left(\begin{array}{c}\left|\left(\mathfrak{F}^{RP}_{\mathfrak{R}_1}(\mathrm{e}_\sigma)\right)^2-\left(\mathfrak{F}^{RP}_{\mathfrak{R}_2}(\mathrm{e}_\sigma)\right)^2\right|+\left|\left(\mathfrak{F}^{IP}_{\mathfrak{R}_1}(\mathrm{e}_\sigma)\right)^2-\left(\mathfrak{F}^{IP}_{\mathfrak{R}_2}(\mathrm{e}_\sigma)\right)^2\right|\\+\left|\left(\mathfrak{F}^{RN}_{\mathfrak{R}_1}(\mathrm{e}_\sigma)\right)^2-\left(\mathfrak{F}^{RN}_{\mathfrak{R}_2}(\mathrm{e}_\sigma)\right)^2\right|+\left|\left(\mathfrak{F}^{RN}_{\mathfrak{R}_1}(\mathrm{e}_\sigma)\right)^2-\left(\mathfrak{F}^{RN}_{\mathfrak{R}_2}(\mathrm{e}_\sigma)\right)^2\right|\end{array}\right)$$

$$\leq\left(\begin{array}{c}\left|\left(\mathfrak{F}^{RP}_{\mathfrak{R}_1}(\mathrm{e}_\sigma)\right)^2-\left(\mathfrak{F}^{RP}_{\mathfrak{R}_3}(\mathrm{e}_\sigma)\right)^2\right|+\left|\left(\mathfrak{F}^{IP}_{\mathfrak{R}_1}(\mathrm{e}_\sigma)\right)^2-\left(\mathfrak{F}^{IP}_{\mathfrak{R}_3}(\mathrm{e}_\sigma)\right)^2\right|\\+\left|\left(\mathfrak{F}^{RN}_{\mathfrak{R}_1}(\mathrm{e}_\sigma)\right)^2-\left(\mathfrak{F}^{RN}_{\mathfrak{R}_3}(\mathrm{e}_\sigma)\right)^2\right|+\left|\left(\mathfrak{F}^{RN}_{\mathfrak{R}_1}(\mathrm{e}_\sigma)\right)^2-\left(\mathfrak{F}^{RN}_{\mathfrak{R}_3}(\mathrm{e}_\sigma)\right)^2\right|\end{array}\right)$$

$$\left(\frac{1}{4n}\sum_{\sigma=1}^{n}\left(\begin{array}{c}\left|\left(\mathfrak{F}^{RP}_{\mathfrak{R}_1}(\mathrm{e}_\sigma)\right)^2-\left(\mathfrak{F}^{RP}_{\mathfrak{R}_2}(\mathrm{e}_\sigma)\right)^2\right|+\left|\left(\mathfrak{F}^{IP}_{\mathfrak{R}_1}(\mathrm{e}_\sigma)\right)^2-\left(\mathfrak{F}^{IP}_{\mathfrak{R}_2}(\mathrm{e}_\sigma)\right)^2\right|\\+\left|\left(\mathfrak{F}^{RN}_{\mathfrak{R}_1}(\mathrm{e}_\sigma)\right)^2-\left(\mathfrak{F}^{RN}_{\mathfrak{R}_2}(\mathrm{e}_\sigma)\right)^2\right|+\left|\left(\mathfrak{F}^{RN}_{\mathfrak{R}_1}(\mathrm{e}_\sigma)\right)^2-\left(\mathfrak{F}^{RN}_{\mathfrak{R}_2}(\mathrm{e}_\sigma)\right)^2\right|\end{array}\right)\right)^{\frac{1}{\lambda}}$$

$$\leq\left(\frac{1}{4n}\sum_{\sigma=1}^{n}\left(\begin{array}{c}\left|\left(\mathfrak{F}^{RP}_{\mathfrak{R}_1}(\mathrm{e}_\sigma)\right)^2-\left(\mathfrak{F}^{RP}_{\mathfrak{R}_3}(\mathrm{e}_\sigma)\right)^2\right|+\left|\left(\mathfrak{F}^{IP}_{\mathfrak{R}_1}(\mathrm{e}_\sigma)\right)^2-\left(\mathfrak{F}^{IP}_{\mathfrak{R}_3}(\mathrm{e}_\sigma)\right)^2\right|\\+\left|\left(\mathfrak{F}^{RN}_{\mathfrak{R}_1}(\mathrm{e}_\sigma)\right)^2-\left(\mathfrak{F}^{RN}_{\mathfrak{R}_3}(\mathrm{e}_\sigma)\right)^2\right|+\left|\left(\mathfrak{F}^{RN}_{\mathfrak{R}_1}(\mathrm{e}_\sigma)\right)^2-\left(\mathfrak{F}^{RN}_{\mathfrak{R}_3}(\mathrm{e}_\sigma)\right)^2\right|\end{array}\right)\right)^{\frac{1}{\lambda}}$$

$$\Rightarrow d^1_{BCFS}(\mathfrak{R}_1,\mathfrak{R}_2)\leq d^1_{BCFS}(\mathfrak{R}_1,\mathfrak{R}_3).$$

By the same procedure, $d^1_{BCFS}(\mathfrak{R}_2,\mathfrak{R}_3)\leq d^1_{BCFS}(\mathfrak{R}_1,\mathfrak{R}_3)$ can be proved.

Particular Cases

Underneath are the particular cases of GBCFNHDM.

i. By considering $\lambda = 1,$, the GBCFNHDM would decrease to BCF normalized Hamming distance measure (BCFNHDM)

$$d^2_{BCFS}(\mathfrak{R}_1,\mathfrak{R}_2)=\frac{1}{4n}\sum_{\sigma=1}^{n}\left(\begin{array}{c}\left|\left(\mathfrak{F}^{RP}_{\mathfrak{R}_1}(\mathrm{e}_\sigma)\right)^2-\left(\mathfrak{F}^{RP}_{\mathfrak{R}_2}(\mathrm{e}_\sigma)\right)^2\right|+\left|\left(\mathfrak{F}^{IP}_{\mathfrak{R}_1}(\mathrm{e}_\sigma)\right)^2-\left(\mathfrak{F}^{IP}_{\mathfrak{R}_2}(\mathrm{e}_\sigma)\right)^2\right|\\+\left|\left(\mathfrak{F}^{RN}_{\mathfrak{R}_1}(\mathrm{e}_\sigma)\right)^2-\left(\mathfrak{F}^{RN}_{\mathfrak{R}_2}(\mathrm{e}_\sigma)\right)^2\right|+\left|\left(\mathfrak{F}^{IN}_{\mathfrak{R}_1}(\mathrm{e}_\sigma)\right)^2-\left(\mathfrak{F}^{IN}_{\mathfrak{R}_2}(\mathrm{e}_\sigma)\right)^2\right|\end{array}\right)$$

ii. By considering $\lambda = 2,$, the GBCFNHDM would decrease to BCF normalized Euclidean distance measure (BCFNEDM)

$$d^3_{BCFS}(\mathfrak{R}_1,\mathfrak{R}_2)=\sqrt{\frac{1}{4n}\sum_{\sigma=1}^{n}\left(\begin{array}{c}\left|\left(\mathfrak{F}^{RP}_{\mathfrak{R}_1}(\mathrm{e}_\sigma)\right)^2-\left(\mathfrak{F}^{RP}_{\mathfrak{R}_2}(\mathrm{e}_\sigma)\right)^2\right|+\left|\left(\mathfrak{F}^{IP}_{\mathfrak{R}_1}(\mathrm{e}_\sigma)\right)^2-\left(\mathfrak{F}^{IP}_{\mathfrak{R}_2}(\mathrm{e}_\sigma)\right)^2\right|\\+\left|\left(\mathfrak{F}^{RN}_{\mathfrak{R}_1}(\mathrm{e}_\sigma)\right)^2-\left(\mathfrak{F}^{RN}_{\mathfrak{R}_2}(\mathrm{e}_\sigma)\right)^2\right|+\left|\left(\mathfrak{F}^{IN}_{\mathfrak{R}_1}(\mathrm{e}_\sigma)\right)^2-\left(\mathfrak{F}^{IN}_{\mathfrak{R}_2}(\mathrm{e}_\sigma)\right)^2\right|\end{array}\right)}$$

The Hausdorff distance investigated by Nadler [16] in 1978, is the maximum distance of a set to the nearest point in the other set. Formally, let U_1 and U_2 be two finite sets. Then the Hausdorff distance is identified and interpreted as

$$\mathcal{H}(\mathrm{U}_1,\mathrm{U}_2)=\max\{\mathrm{h}(\mathrm{U}_1,\mathrm{U}_2),\mathrm{h}(\mathrm{U}_2,\mathrm{U}_1)\}$$

where

$$h(\mathrm{U}_1,\mathrm{U}_2)=\max_{u_1\in U_1}\min_{u_2\in U_2}d(u_1,u_2)$$

Noted that d is the metric among U_1 and U_2. The directed Hausdorff distances are the two distances $h(\mathrm{U}_1,\mathrm{U}_2)$ and $h(\mathrm{U}_2,\mathrm{U}_1)$. According to the function $h(\mathrm{U}_1,\mathrm{U}_2)$ (the directed Hausdorff distance from U_1 to U_2), each element of U_1 is ranked according to how far away the nearest element of U_1 is, and the element with the highest ranking determines the distance's value. $h(U_1, U_2)$ and $h(U_2, U_1)$ can typically have distinct values (the directed distances are not symmetric). If $U_1 = \{u_1\}$ and $U_2 = \{u_2\}$, then the Hausdorff distance reduces to $d(u_1, u_2)$. Underneath, we define GBCFNHDDM between BCFSs.

Definition 4.2 Consider two BCFSs $\mathfrak{R}_1 = \left\{e_\sigma, \left(\mathfrak{F}^{P}_{\mathfrak{R}_1}(e_\sigma), \mathfrak{F}^{N}_{\mathfrak{R}_1}(e_\sigma)\right) | e_\sigma \in \mathbb{E}, \sigma = 1, 2, .., n\right\} = \left\{e_\sigma, \left(\mathfrak{F}^{RP}_{\mathfrak{R}_1}(e_\sigma) + \iota\mathfrak{F}^{IP}_{\mathfrak{R}_1}(e_\sigma), \mathfrak{F}^{RN}_{\mathfrak{R}_1}(e_\sigma) + \iota\mathfrak{F}^{IN}_{\mathfrak{R}_1}(e_\sigma)\right) | e_\sigma \in \mathbb{E}, \sigma = 1, 2, .., n\right\}$ and $\mathfrak{R}_2 = \left\{e, \left(\mathfrak{F}^{P}_{\mathfrak{R}_2}(e_\sigma), \mathfrak{F}^{N}_{\mathfrak{R}_2}(e_\sigma)\right) | e \in \mathbb{E}, \sigma = 1, 2, .., n\right\} = \left\{e_\sigma, \left(\mathfrak{F}^{RP}_{\mathfrak{R}_2}(e_\sigma) + \iota\mathfrak{F}^{IP}_{\mathfrak{R}_2}(e_\sigma), \mathfrak{F}^{RN}_{\mathfrak{R}_2}(e_\sigma) + \iota\mathfrak{F}^{IN}_{\mathfrak{R}_2}(e_\sigma)\right) | e_\sigma \in \mathbb{E}, \sigma = 1, 2, .., n\right\}$ over $\mathbb{E}$. Then the GBCFNHDDM between $\mathfrak{R}_1$ and $\mathfrak{R}_2$ is inaugurated as.

$$d^4_{BCFS}(\mathfrak{R}_1, \mathfrak{R}_2) = \left(\frac{1}{n}\sum_{\sigma=1}^{n} \max\left(\begin{array}{c} \left|\left(\mathfrak{F}^{RP}_{\mathfrak{R}_1}(e_\sigma)\right)^2 - \left(\mathfrak{F}^{RP}_{\mathfrak{R}_2}(e_\sigma)\right)^2\right|, \left|\left(\mathfrak{F}^{IP}_{\mathfrak{R}_1}(e_\sigma)\right)^2 - \left(\mathfrak{F}^{IP}_{\mathfrak{R}_2}(e_\sigma)\right)^2\right| \\ , \left|\left(\mathfrak{F}^{RN}_{\mathfrak{R}_1}(e_\sigma)\right)^2 - \left(\mathfrak{F}^{RN}_{\mathfrak{R}_2}(e_\sigma)\right)^2\right|, \left|\left(\mathfrak{F}^{IN}_{\mathfrak{R}_1}(e_\sigma)\right)^2 - \left(\mathfrak{F}^{IN}_{\mathfrak{R}_2}(e_\sigma)\right)^2\right| \end{array}\right)\right)^{\frac{1}{\lambda}}, \quad \lambda \geq 1$$

Theorem 4.2 The GBCFNHDDM satisfies the underneath properties.

1. $0 \leq d^4_{BCFS}(\mathfrak{R}_1, \mathfrak{R}_2) \leq 1$
2. $d^4_{BCFS}(\mathfrak{R}_1, \mathfrak{R}_2) = 0 \Leftrightarrow \mathfrak{R}_1 = \mathfrak{R}_2$
3. $d^4_{BCFS}(\mathfrak{R}_1, \mathfrak{R}_2) = d^4_{BCFS}(\mathfrak{R}_2, \mathfrak{R}_1)$
4. If $\mathfrak{R}_1 \subset \mathfrak{R}_2 \subset \mathfrak{R}_3$, then $d^4_{BCFS}(\mathfrak{R}_1, \mathfrak{R}_2) \leq d^4_{BCFS}(\mathfrak{R}_1, \mathfrak{R}_3)$ and $d^4_{BCFS}(\mathfrak{R}_2, \mathfrak{R}_3) \leq d^4_{BCFS}(\mathfrak{R}_1, \mathfrak{R}_3)$

Proof: 1. As $\mathfrak{F}^{RP}_{\mathfrak{R}_1}(e_\sigma), \mathfrak{F}^{IP}_{\mathfrak{R}_1}(e_\sigma), \mathfrak{F}^{RP}_{\mathfrak{R}_2}(e_\sigma), \mathfrak{F}^{IP}_{\mathfrak{R}_2}(e_\sigma) \in [0, 1]$ and $\mathfrak{F}^{RN}_{\mathfrak{R}_1}(e_\sigma), \mathfrak{F}^{IN}_{\mathfrak{R}_1}(e_\sigma), \mathfrak{F}^{RN}_{\mathfrak{R}_2}(e_\sigma), \mathfrak{F}^{IN}_{\mathfrak{R}_2}(e_\sigma) \in [-1, 0]$ for all σ. Then $\left(\mathfrak{F}^{RP}_{\mathfrak{R}_1}(e_\sigma)\right)^2$ $\left(\mathfrak{F}^{IP}_{\mathfrak{R}_1}(e_\sigma)\right)^2$, $\left(\mathfrak{F}^{RP}_{\mathfrak{R}_2}(e_\sigma)\right)^2$ $\left(\mathfrak{F}^{IP}_{\mathfrak{R}_2}(e_\sigma)\right)^2 \in [0, 1]$ and $\left(\mathfrak{F}^{RN}_{\mathfrak{R}_1}(e_\sigma)\right)^2$, $\left(\mathfrak{F}^{IN}_{\mathfrak{R}_1}(e_\sigma)\right)^2$ $\left(\mathfrak{F}^{RN}_{\mathfrak{R}_2}(e_\sigma)\right)^2$, $\left(\mathfrak{F}^{IN}_{\mathfrak{R}_2}(e_\sigma)\right)^2 \in [0, 1]$ for all σ. $\Rightarrow$ $\left|\left(\mathfrak{F}^{RP}_{\mathfrak{R}_1}(e_\sigma)\right)^2 - \left(\mathfrak{F}^{RP}_{\mathfrak{R}_2}(e_\sigma)\right)^2\right|$, $\left|\left(\mathfrak{F}^{IP}_{\mathfrak{R}_1}(e_\sigma)\right)^2 - \left(\mathfrak{F}^{IP}_{\mathfrak{R}_2}(e_\sigma)\right)^2\right|$, $\left|\left(\mathfrak{F}^{RN}_{\mathfrak{R}_1}(e_\sigma)\right)^2 - \left(\mathfrak{F}^{RN}_{\mathfrak{R}_2}(e_\sigma)\right)^2\right|$, $\left|\left(\mathfrak{F}^{IN}_{\mathfrak{R}_1}(e_\sigma)\right)^2 - \left(\mathfrak{F}^{IN}_{\mathfrak{R}_2}(e_\sigma)\right)^2\right| \in [0, 1]\ \forall\sigma$. Consequently, $\forall\sigma$

$$\max\left(\begin{array}{c} \left|\left(\mathfrak{F}^{RP}_{\mathfrak{R}_1}(e_\sigma)\right)^2 - \left(\mathfrak{F}^{RP}_{\mathfrak{R}_2}(e_\sigma)\right)^2\right|, \left|\left(\mathfrak{F}^{IP}_{\mathfrak{R}_1}(e_\sigma)\right)^2 - \left(\mathfrak{F}^{IP}_{\mathfrak{R}_2}(e_\sigma)\right)^2\right| \\ , \left|\left(\mathfrak{F}^{RN}_{\mathfrak{R}_1}(e_\sigma a)\right)^2 - \left(\mathfrak{F}^{RN}_{\mathfrak{R}_2}(e_\sigma)\right)^2\right|, \left|\left(\mathfrak{F}^{IN}_{\mathfrak{R}_1}(e_\sigma)\right)^2 - \left(\mathfrak{F}^{IN}_{\mathfrak{R}_2}(e_\sigma)\right)^2\right| \end{array}\right) \in [0, 1]$$

$$0 \leq \max\left(\begin{array}{c} \left|\left(\mathfrak{F}^{RP}_{\mathfrak{R}_1}(e_\sigma)\right)^2 - \left(\mathfrak{F}^{RP}_{\mathfrak{R}_2}(e_\sigma)\right)^2\right|, \left|\left(\mathfrak{F}^{IP}_{\mathfrak{R}_1}(e_\sigma)\right)^2 - \left(\mathfrak{F}^{IP}_{\mathfrak{R}_2}(e_\sigma)\right)^2\right| \\ , \left|\left(\mathfrak{F}^{RN}_{\mathfrak{R}_1}(e_\sigma)\right)^2 - \left(\mathfrak{F}^{RN}_{\mathfrak{R}_2}(e_\sigma)\right)^2\right|, \left|\left(\mathfrak{F}^{IN}_{\mathfrak{R}_1}(e_\sigma)\right)^2 - \left(\mathfrak{F}^{IN}_{\mathfrak{R}_2}(e_\sigma)\right)^2\right| \end{array}\right) \leq 1$$

Thus,

$$0 \leq \sum_{\sigma=1}^{n} \max\left(\begin{array}{c} \left|\left(\mathfrak{F}^{RP}_{\mathfrak{R}_1}(e_\sigma)\right)^2 - \left(\mathfrak{F}^{RP}_{\mathfrak{R}_2}(e_\sigma)\right)^2\right|, \left|\left(\mathfrak{F}^{IP}_{\mathfrak{R}_1}(e_\sigma)\right)^2 - \left(\mathfrak{F}^{IP}_{\mathfrak{R}_2}(e_\sigma)\right)^2\right| \\ , \left|\left(\mathfrak{F}^{RN}_{\mathfrak{R}_1}(e_\sigma)\right)^2 - \left(\mathfrak{F}^{RN}_{\mathfrak{R}_2}(e_\sigma)\right)^2\right|, \left|\left(\mathfrak{F}^{IN}_{\mathfrak{R}_1}(e_\sigma)\right)^2 - \left(\mathfrak{F}^{IN}_{\mathfrak{R}_2}(e_\sigma)\right)^2\right| \end{array}\right) \leq n$$

$$0 \le \frac{1}{n}\sum_{\sigma=1}^{n} \max\begin{pmatrix} \left|\left(\mathfrak{F}^{RP}_{\mathfrak{R}_1}(e_\sigma)\right)^2-\left(\mathfrak{F}^{RP}_{\mathfrak{R}_2}(e_\sigma)\right)^2\right|,\left|\left(\mathfrak{F}^{IP}_{\mathfrak{R}_1}(e_\sigma)\right)^2-\left(\mathfrak{F}^{IP}_{\mathfrak{R}_2}(e_\sigma)\right)^2\right| \\ ,\left|\left(\mathfrak{F}^{RN}_{\mathfrak{R}_1}(e_\sigma)\right)^2-\left(\mathfrak{F}^{RN}_{\mathfrak{R}_2}(e_\sigma)\right)^2\right|,\left|\left(\mathfrak{F}^{IN}_{\mathfrak{R}_1}(e_\sigma)\right)^2-\left(\mathfrak{F}^{IN}_{\mathfrak{R}_2}(e_\sigma)\right)^2\right| \end{pmatrix} \le 1$$

Since $\lambda \ge 1$, then

$$0 \le \left(\frac{1}{n}\sum_{\sigma=1}^{n} \max\begin{pmatrix} \left|\left(\mathfrak{F}^{RP}_{\mathfrak{R}_1}(e_\sigma)\right)^2-\left(\mathfrak{F}^{RP}_{\mathfrak{R}_2}(e_\sigma)\right)^2\right|,\left|\left(\mathfrak{F}^{IP}_{\mathfrak{R}_1}(e_\sigma)\right)^2-\left(\mathfrak{F}^{IP}_{\mathfrak{R}_2}(e_\sigma)\right)^2\right| \\ ,\left|\left(\mathfrak{F}^{RN}_{\mathfrak{R}_1}(e_\sigma)\right)^2-\left(\mathfrak{F}^{RN}_{\mathfrak{R}_2}(e_\sigma)\right)^2\right|,\left|\left(\mathfrak{F}^{IN}_{\mathfrak{R}_1}(e_\sigma)\right)^2-\left(\mathfrak{F}^{IN}_{\mathfrak{R}_2}(e_\sigma)\right)^2\right| \end{pmatrix}\right)^{\frac{1}{\lambda}} \le 1$$

$$\Rightarrow 0 \le d^4_{BCFS}(\mathfrak{R}_1,\mathfrak{R}_2) \le 1.$$

2 Let $d^4_{BCFS}(\mathfrak{R}_1,\mathfrak{R}_2)=0$. Then to show that $\mathfrak{R}_1=\mathfrak{R}_2$. as $d^4_{BCFS}(\mathfrak{R}_1,\mathfrak{R}_2)=0$, then

$$\left(\frac{1}{n}\sum_{\sigma=1}^{n} \max\begin{pmatrix} \left|\left(\mathfrak{F}^{RP}_{\mathfrak{R}_1}(e_\sigma)\right)^2-\left(\mathfrak{F}^{RP}_{\mathfrak{R}_2}(e_\sigma)\right)^2\right|,\left|\left(\mathfrak{F}^{IP}_{\mathfrak{R}_1}(e_\sigma)\right)^2-\left(\mathfrak{F}^{IP}_{\mathfrak{R}_2}(e_\sigma)\right)^2\right| \\ ,\left|\left(\mathfrak{F}^{RN}_{\mathfrak{R}_1}(e_\sigma)\right)^2-\left(\mathfrak{F}^{RN}_{\mathfrak{R}_2}(e_\sigma)\right)^2\right|,\left|\left(\mathfrak{F}^{IN}_{\mathfrak{R}_1}(e_\sigma)\right)^2-\left(\mathfrak{F}^{IN}_{\mathfrak{R}_2}(e_\sigma)\right)^2\right| \end{pmatrix}\right)^{\frac{1}{\lambda}} = 0$$

$$\frac{1}{n}\sum_{\sigma=1}^{n} \max\begin{pmatrix} \left|\left(\mathfrak{F}^{RP}_{\mathfrak{R}_1}(e_\sigma)\right)^2-\left(\mathfrak{F}^{RP}_{\mathfrak{R}_2}(e_\sigma)\right)^2\right|,\left|\left(\mathfrak{F}^{IP}_{\mathfrak{R}_1}(e_\sigma)\right)^2-\left(\mathfrak{F}^{IP}_{\mathfrak{R}_2}(e_\sigma)\right)^2\right| \\ ,\left|\left(\mathfrak{F}^{RN}_{\mathfrak{R}_1}(e_\sigma)\right)^2-\left(\mathfrak{F}^{RN}_{\mathfrak{R}_2}(e_\sigma)\right)^2\right|,\left|\left(\mathfrak{F}^{IN}_{\mathfrak{R}_1}(e_\sigma)\right)^2-\left(\mathfrak{F}^{IN}_{\mathfrak{R}_2}(e_\sigma)\right)^2\right| \end{pmatrix} = 0$$

$$\sum_{\sigma=1}^{n} \max\begin{pmatrix} \left|\left(\mathfrak{F}^{RP}_{\mathfrak{R}_1}(e_\sigma)\right)^2-\left(\mathfrak{F}^{RP}_{\mathfrak{R}_2}(e_\sigma)\right)^2\right|,\left|\left(\mathfrak{F}^{IP}_{\mathfrak{R}_1}(e_\sigma)\right)^2-\left(\mathfrak{F}^{IP}_{\mathfrak{R}_2}(e_\sigma)\right)^2\right| \\ ,\left|\left(\mathfrak{F}^{RN}_{\mathfrak{R}_1}(e_\sigma)\right)^2-\left(\mathfrak{F}^{RN}_{\mathfrak{R}_2}(e_\sigma)\right)^2\right|,\left|\left(\mathfrak{F}^{IN}_{\mathfrak{R}_1}(e_\sigma)\right)^2-\left(\mathfrak{F}^{IN}_{\mathfrak{R}_2}(e_\sigma)\right)^2\right| \end{pmatrix} = 0$$

$$\max\begin{pmatrix} \left|\left(\mathfrak{F}^{RP}_{\mathfrak{R}_1}(e_\sigma)\right)^2-\left(\mathfrak{F}^{RP}_{\mathfrak{R}_2}(e_\sigma)\right)^2\right|,\left|\left(\mathfrak{F}^{IP}_{\mathfrak{R}_1}(e_\sigma)\right)^2-\left(\mathfrak{F}^{IP}_{\mathfrak{R}_2}(e_\sigma)\right)^2\right| \\ ,\left|\left(\mathfrak{F}^{RN}_{\mathfrak{R}_1}(e_\sigma)\right)^2-\left(\mathfrak{F}^{RN}_{\mathfrak{R}_2}(e_\sigma)\right)^2\right|,\left|\left(\mathfrak{F}^{IN}_{\mathfrak{R}_1}(e_\sigma)\right)^2-\left(\mathfrak{F}^{IN}_{\mathfrak{R}_2}(e_\sigma)\right)^2\right| \end{pmatrix} = 0\forall\sigma$$

The above inequality holds only when $\forall\sigma$, $\left|\left(\mathfrak{F}^{RP}_{\mathfrak{R}_1}(e_\sigma)\right)^2-\left(\mathfrak{F}^{RP}_{\mathfrak{R}_2}(e_\sigma)\right)^2\right| = 0$, $\left|\left(\mathfrak{F}^{IP}_{\mathfrak{R}_1}(e_\sigma)\right)^2-\left(\mathfrak{F}^{IP}_{\mathfrak{R}_2}(e_\sigma)\right)^2\right| = 0$, $\left|\left(\mathfrak{F}^{RN}_{\mathfrak{R}_1}(e_\sigma)\right)^2-\left(\mathfrak{F}^{RN}_{\mathfrak{R}_2}(e_\sigma)\right)^2\right| = 0$, $\left|\left(\mathfrak{F}^{RN}_{\mathfrak{R}_1}(e_\sigma)\right)^2-\left(\mathfrak{F}^{RN}_{\mathfrak{R}_2}(e_\sigma)\right)^2\right| = 0$. Thus, $\mathfrak{F}^{RP}_{\mathfrak{R}_1}(e_\sigma)=\mathfrak{F}^{RP}_{\mathfrak{R}_2}(e_\sigma)$, $\mathfrak{F}^{IP}_{\mathfrak{R}_1}(e_\sigma)=\mathfrak{F}^{IP}_{\mathfrak{R}_2}(e_\sigma)$, $\mathfrak{F}^{RN}_{\mathfrak{R}_1}(e_\sigma)=\mathfrak{F}^{RN}_{\mathfrak{R}_2}(e_\sigma)$, $\mathfrak{F}^{IN}_{\mathfrak{R}_1}(e_\sigma)=\mathfrak{F}^{IN}_{\mathfrak{R}_2}(e_\sigma)$ $\forall\sigma$. Consequently, $\mathfrak{R}_1=\mathfrak{R}_2$.

Conversely, Let $\mathfrak{R}_1=\mathfrak{R}_2$. To show that $d^4_{BCFS}(\mathfrak{R}_1,\mathfrak{R}_2)=0$. As $\mathfrak{R}_1=\mathfrak{R}_2$, then $\mathfrak{F}^{RP}_{\mathfrak{R}_1}(e_\sigma)=\mathfrak{F}^{RP}_{\mathfrak{R}_2}(e_\sigma)$, $\mathfrak{F}^{IP}_{\mathfrak{R}_1}(e_\sigma)=\mathfrak{F}^{IP}_{\mathfrak{R}_2}(e_\sigma)$, $\mathfrak{F}^{RN}_{\mathfrak{R}_1}(e_\sigma)=\mathfrak{F}^{RN}_{\mathfrak{R}_2}(e_\sigma)$, $\mathfrak{F}^{IN}_{\mathfrak{R}_1}(e_\sigma)=\mathfrak{F}^{IN}_{\mathfrak{R}_2}(e_\sigma)$ $\forall\sigma$. Then $\forall\sigma$

$$d^4_{BCFS}(\mathfrak{R}_1,\mathfrak{R}_2) = \left(\frac{1}{n}\sum_{\sigma=1}^{n} \max\begin{pmatrix} \left|\left(\mathfrak{F}^{RP}_{\mathfrak{R}_1}(e_\sigma)\right)^2-\left(\mathfrak{F}^{RP}_{\mathfrak{R}_1}(e_\sigma)\right)^2\right|,\left|\left(\mathfrak{F}^{IP}_{\mathfrak{R}_1}(e_\sigma)\right)^2-\left(\mathfrak{F}^{IP}_{\mathfrak{R}_1}(e_\sigma)\right)^2\right| \\ ,\left|\left(\mathfrak{F}^{RN}_{\mathfrak{R}_1}(e_\sigma)\right)^2-\left(\mathfrak{F}^{RN}_{\mathfrak{R}_1}(e_\sigma)\right)^2\right|,\left|\left(\mathfrak{F}^{IN}_{\mathfrak{R}_1}(e_\sigma)\right)^2-\left(\mathfrak{F}^{IN}_{\mathfrak{R}_1}(e_\sigma)\right)^2\right| \end{pmatrix}\right)^{\frac{1}{\lambda}} = 0$$

3 As

$$d^4_{BCFS}(\Re_1, \Re_2) = \left(\frac{1}{n}\sum_{\sigma=1}^{n} \max\left(\begin{array}{l} \left|\left(\mathfrak{F}^{RP}_{\Re_1}(e_\sigma)\right)^2 - \left(\mathfrak{F}^{RP}_{\Re_2}(e_\sigma)\right)^2\right|, \left|\left(\mathfrak{F}^{IP}_{\Re_1}(e_\sigma)\right)^2 - \left(\mathfrak{F}^{IP}_{\Re_2}(e_\sigma)\right)^2\right| \\ , \left|\left(\mathfrak{F}^{RN}_{\Re_1}(e_\sigma)\right)^2 - \left(\mathfrak{F}^{RN}_{\Re_2}(e_\sigma)\right)^2\right|, \left|\left(\mathfrak{F}^{IN}_{\Re_1}(e_\sigma)\right)^2 - \left(\mathfrak{F}^{IN}_{\Re_2}(e_\sigma)\right)^2\right| \end{array}\right)\right)^{\frac{1}{\lambda}}$$

$$= \left(\frac{1}{n}\sum_{\sigma=1}^{n} \max\left(\begin{array}{l} \left|\left(\mathfrak{F}^{RP}_{\Re_2}(e_\sigma)\right)^2 - \left(\mathfrak{F}^{RP}_{\Re_2}(e_\sigma)\right)^2\right|, \left|\left(\mathfrak{F}^{IP}_{\Re_2}(e_\sigma)\right)^2 - \left(\mathfrak{F}^{IP}_{\Re_1}(e_\sigma)\right)^2\right| \\ , \left|\left(\mathfrak{F}^{RN}_{\Re_2}(e_\sigma)\right)^2 - \left(\mathfrak{F}^{RN}_{\Re_2}(e_\sigma)\right)^2\right|, \left|(\mathfrak{F}^{IN}_{2}(e_\sigma))^2 - \left(\mathfrak{F}^{IN}_{\Re_1}(e_\sigma)\right)^2\right| \end{array}\right)\right)^{\frac{1}{\lambda}} = d^4_{BCFS}(\Re_2, \Re_1)$$

4 Assume that $\Re_1 \subset \Re_2 \subset \Re_3$ then $\mathfrak{F}^{RP}_{\Re_1}(e_\sigma) \leq \mathfrak{F}^{RP}_{\Re_2}(e_\sigma) \leq \mathfrak{F}^{RP}_{\Re_3}(e_\sigma)$, $\mathfrak{F}^{IP}_{\Re_1}(e_\sigma) \leq \mathfrak{F}^{IP}_{\Re_2}(e_\sigma) \leq \mathfrak{F}^{IP}_{\Re_3}(e_\sigma)$, $\mathfrak{F}^{RN}_{\Re_1}(e_\sigma) \leq \mathfrak{F}^{RN}_{\Re_2}(e_\sigma) \leq \mathfrak{F}^{RN}_{\Re_3}(e_\sigma)$, $\mathfrak{F}^{IN}_{\Re_1}(e_\sigma) \leq \mathfrak{F}^{IN}_{\Re_2}(e_\sigma) \leq \mathfrak{F}^{IN}_{\Re_3}(e_\sigma)$ $\forall\sigma$ Thus, we have the inequalities

$$\left|\left(\mathfrak{F}^{RP}_{\Re_1}(e_\sigma)\right)^2 - \left(\mathfrak{F}^{RP}_{\Re_2}(e_\sigma)\right)^2\right| \leq \left|\left(\mathfrak{F}^{RP}_{\Re_1}(e_\sigma)\right)^2 - \left(\mathfrak{F}^{RP}_{\Re_3}(e_\sigma)\right)^2\right|,$$

$$\left|\left(\mathfrak{F}^{IP}_{\Re_1}(e_\sigma)\right)^2 - \left(\mathfrak{F}^{IP}_{\Re_2}(e_\sigma)\right)^2\right| \leq \left|\left(\mathfrak{F}^{IP}_{\Re_1}(e_\sigma)\right)^2 - \left(\mathfrak{F}^{IP}_{\Re_3}(e_\sigma)\right)^2\right|$$

And

$$\left|\left(\mathfrak{F}^{RN}_{\Re_1}(e_\sigma)\right)^2 - \left(\mathfrak{F}^{RN}_{\Re_2}(e_\sigma)\right)^2\right| \leq \left|\left(\mathfrak{F}^{RN}_{\Re_1}(e_\sigma)\right)^2 - \left(\mathfrak{F}^{RN}_{\Re_3}(e_\sigma)\right)^2\right|$$

$$\left|\left(\mathfrak{F}^{IN}_{\Re_1}(e_\sigma)\right)^2 - \left(\mathfrak{F}^{IN}_{\Re_2}(e_\sigma)\right)^2\right| \leq \left|\left(\mathfrak{F}^{IN}_{\Re_1}(e_\sigma)\right)^2 - \left(\mathfrak{F}^{IN}_{\Re_3}(e_\sigma)\right)^2\right|$$

Thus, $\forall\sigma$

$$\max\left(\begin{array}{l} \left|\left(\mathfrak{F}^{RP}_{\Re_1}(e_\sigma)\right)^2 - \left(\mathfrak{F}^{RP}_{\Re_2}(e_\sigma)\right)^2\right|, \left|\left(\mathfrak{F}^{IP}_{\Re_1}(e_\sigma)\right)^2 - \left(\mathfrak{F}^{IP}_{\Re_2}(e_\sigma)\right)^2\right|, \\ \left|\left(\mathfrak{F}^{RN}_{\Re_1}(e_\sigma)\right)^2 - \left(\mathfrak{F}^{RN}_{\Re_2}(e_\sigma)\right)^2\right|, \left|\left(\mathfrak{F}^{RN}_{\Re_1}(e_\sigma)\right)^2 - \left(\mathfrak{F}^{RN}_{\Re_2}(e_\sigma)\right)^2\right| \end{array}\right)$$

$$\leq \max\left(\begin{array}{l} \left|\left(\mathfrak{F}^{RP}_{\Re_1}(e_\sigma)\right)^2 - \left(\mathfrak{F}^{RP}_{\Re_3}(e_\sigma)\right)^2\right|, \left|\left(\mathfrak{F}^{IP}_{\Re_1}(e_\sigma)\right)^2 - \left(\mathfrak{F}^{IP}_{\Re_3}(e_\sigma)\right)^2\right| \\ , \left|\left(\mathfrak{F}^{RN}_{\Re_1}(e_\sigma)\right)^2 - \left(\mathfrak{F}^{RN}_{\Re_3}(e_\sigma)\right)^2\right|, \left|\left(\mathfrak{F}^{RN}_{\Re_1}(e_\sigma)\right)^2 - \left(\mathfrak{F}^{RN}_{\Re_3}(e_\sigma)\right)^2\right| \end{array}\right)$$

$$\left(\frac{1}{n}\sum_{\sigma=1}^{n} \max\left(\begin{array}{l} \left|\left(\mathfrak{F}^{RP}_{\Re_1}(e_\sigma)\right)^2 - \left(\mathfrak{F}^{RP}_{\Re_2}(e_\sigma)\right)^2\right|, \left|\left(\mathfrak{F}^{IP}_{\Re_1}(e_\sigma)\right)^2 - \left(\mathfrak{F}^{IP}_{\Re_2}(e_\sigma)\right)^2\right|, \\ \left|\left(\mathfrak{F}^{RN}_{\Re_1}(e_\sigma)\right)^2 - \left(\mathfrak{F}^{RN}_{\Re_2}(e_\sigma)\right)^2\right|, \left|\left(\mathfrak{F}^{RN}_{\Re_1}(e_\sigma)\right)^2 - \left(\mathfrak{F}^{RN}_{\Re_2}(e_\sigma)\right)^2\right| \end{array}\right)\right)^{\frac{1}{\lambda}}$$

$$\leq \left(\frac{1}{n}\sum_{\sigma=1}^{n} \max\left(\begin{array}{l} \left|\left(\mathfrak{F}^{RP}_{\Re_1}(e_\sigma)\right)^2 - \left(\mathfrak{F}^{RP}_{\Re_3}(e_\sigma)\right)^2\right|, \left|\left(\mathfrak{F}^{IP}_{\Re_1}(e_\sigma)\right)^2 - \left(\mathfrak{F}^{IP}_{\Re_3}(e_\sigma)\right)^2\right| \\ , \left|\left(\mathfrak{F}^{RN}_{\Re_1}(e_\sigma)\right)^2 - \left(\mathfrak{F}^{RN}_{\Re_3}(e_\sigma)\right)^2\right|, \left|\left(\mathfrak{F}^{RN}_{\Re_1}(e_\sigma)\right)^2 - \left(\mathfrak{F}^{RN}_{\Re_3}(e_\sigma)\right)^2\right| \end{array}\right)\right)^{\frac{1}{\lambda}}$$

$$\Rightarrow d^4_{BCFS}(\Re_1, \Re_2) \leq d^4_{BCFS}(\Re_1, \Re_3).$$

By the same procedure, $d^4_{BCFS}(\Re_2, \Re_3) \leq d^4_{BCFS}(\Re_1, \Re_3)$ can be proved.

Particular Cases

Underneath are the particular cases of GBCFNHDDM.

i. By considering $\lambda = 1$, , the GBCFNHDDM would decrease to BCF normalized Hamming-Hausdroff distance measure (BCFNHHDDM)

$$d_{BCFS}^{5}(\Re_1, \Re_2) = \frac{1}{n}\sum_{\sigma=1}^{n} \max\left(\begin{array}{c} \left|\left(\mathfrak{F}_{\Re_1}^{RP}(e_\sigma)\right)^2 - \left(\mathfrak{F}_{\Re_2}^{RP}(e_\sigma)\right)^2\right|, \left|\left(\mathfrak{F}_{\Re_1}^{IP}(e_\sigma)\right)^2 - \left(\mathfrak{F}_{\Re_2}^{IP}(e_\sigma)\right)^2\right| \\ , \left|\left(\mathfrak{F}_{\Re_1}^{RN}(e_\sigma)\right)^2 - \left(\mathfrak{F}_{\Re_2}^{RN}(e_\sigma)\right)^2\right|, \left|\left(\mathfrak{F}_{\Re_1}^{IN}(e_\sigma)\right)^2 - \left(\mathfrak{F}_{\Re_2}^{IN}(e_\sigma)\right)^2\right| \end{array}\right)$$

ii. By considering $\lambda = 2$, , the GBCFNHDDM would decrease to BCF normalized Euclidean-Hausdroff distance measure (BCFNEHDDM)

$$d_{BCFS}^{6}(\Re_1, \Re_2) = \sqrt{\frac{1}{n}\sum_{\sigma=1}^{n} \max\left(\begin{array}{c} \left|\left(\mathfrak{F}_{\Re_1}^{RP}(e_\sigma)\right)^2 - \left(\mathfrak{F}_{\Re_2}^{RP}(e_\sigma)\right)^2\right|, \left|\left(\mathfrak{F}_{\Re_1}^{IP}(e_\sigma)\right)^2 - \left(\mathfrak{F}_{\Re_2}^{IP}(e_\sigma)\right)^2\right| \\ , \left|\left(\mathfrak{F}_{\Re_1}^{RN}(e_\sigma)\right)^2 - \left(\mathfrak{F}_{\Re_2}^{RN}(e_\sigma)\right)^2\right|, \left|\left(\mathfrak{F}_{\Re_1}^{IN}(e_\sigma)\right)^2 - \left(\mathfrak{F}_{\Re_2}^{IN}(e_\sigma)\right)^2\right| \end{array}\right)}$$

To get the advantage of both GBCFNHDM and GBCFNHDDM at a time, we fuse GBCFNHDM and GBCFNHDDM, to develop GHBCFNDM as follows.

Definition 4.3 Consider two BCFSs $\Re_1 = \left\{e_\sigma, \left(\mathfrak{F}_{\Re_1}^{P}(e_\sigma), \mathfrak{F}_{\Re_1}^{N}(e_\sigma)\right) | e_\sigma \in \mathbb{E}, \sigma = 1, 2, .., n\right\} = \left\{e_\sigma, \left(\mathfrak{F}_{\Re_1}^{RP}(e_\sigma) + \iota\mathfrak{F}_{\Re_1}^{IP}(e_\sigma), \mathfrak{F}_{\Re_1}^{RN}(e_\sigma) + \iota\mathfrak{F}_{\Re_1}^{IN}(e_\sigma)\right) | e_\sigma \in \mathbb{E}, \sigma = 1, 2, .., n\right\}$ and $\Re_2 = \left\{e, \left(\mathfrak{F}_{\Re_2}^{P}(e_\sigma), \mathfrak{F}_{\Re_2}^{N}(e_\sigma)\right) | e \in \mathbb{E}, \sigma = 1, 2, .., n\right\} = \left\{e_\sigma, \left(\mathfrak{F}_{\Re_2}^{RP}(e_\sigma) + \iota\mathfrak{F}_{\Re_2}^{IP}(e_\sigma), \mathfrak{F}_{\Re_2}^{RN}(e_\sigma) + \iota\mathfrak{F}_{\Re_2}^{IN}(e_\sigma)\right) | e_\sigma \in \mathbb{E}, \sigma = 1, 2, .., n\right\}$ over $\mathbb{E}$. Then the GHBCFNDM between $\Re_1$ and $\Re_2$ is inaugurated as.

$$d_{BCFS}^{7}(\Re_1, \Re_2) = \left(\frac{1}{2n}\sum_{\sigma=1}^{n}\left(\begin{array}{c} \frac{1}{4}\left(\begin{array}{c} \left|\left(\mathfrak{F}_{\Re_1}^{RP}(e_\sigma)\right)^2 - \left(\mathfrak{F}_{\Re_2}^{RP}(e_\sigma)\right)^2\right| + \left|\left(\mathfrak{F}_{\Re_1}^{IP}(e_\sigma)\right)^2 - \left(\mathfrak{F}_{\Re_2}^{IP}(e_\sigma)\right)^2\right| \\ + \left|\left(\mathfrak{F}_{\Re_1}^{RN}(e_\sigma)\right)^2 - \left(\mathfrak{F}_{\Re_2}^{RN}(e_\sigma)\right)^2\right| + \left|\left(\mathfrak{F}_{\Re_1}^{IN}(e_\sigma)\right)^2 - \left(\mathfrak{F}_{\Re_2}^{IN}(e_\sigma)\right)^2\right| \end{array}\right) \\ + \max\left(\begin{array}{c} \left|\left(\mathfrak{F}_{\Re_1}^{RP}(e_\sigma)\right)^2 - \left(\mathfrak{F}_{\Re_2}^{RP}(e_\sigma)\right)^2\right|, \left|\left(\mathfrak{F}_{\Re_1}^{IP}(e_\sigma)\right)^2 - \left(\mathfrak{F}_{\Re_2}^{IP}(e_\sigma)\right)^2\right| \\ \left|\left(\mathfrak{F}_{\Re_1}^{RN}(e_\sigma)\right)^2 - \left(\mathfrak{F}_{\Re_2}^{RN}(e_\sigma)\right)^2\right|, \left|\left(\mathfrak{F}_{\Re_1}^{IN}(e_\sigma)\right)^2 - \left(\mathfrak{F}_{\Re_2}^{IN}(e_\sigma)\right)^2\right| \end{array}\right) \end{array}\right)\right), \quad \lambda \geq 1$$

Theorem 4.3 The GHBCFNDM satisfies the underneath properties.

1. $0 \leq d_{BCFS}^{7}(\Re_1, \Re_2) \leq 1$
2. $d_{BCFS}^{7}(\Re_1, \Re_2) = 0 \Leftrightarrow \Re_1 = \Re_2$
3. $d_{BCFS}^{7}(\Re_1, \Re_2) = d_{BCFS}^{7}(\Re_2, \Re_1)$
4. If $\Re_1 \subset \Re_2 \subset \Re_3$, then $d_{BCFS}^{7}(\Re_1, \Re_2) \leq d_{BCFS}^{7}(\Re_1, \Re_3)$ and $d_{BCFS}^{7}(\Re_2, \Re_3) \leq d_{BCFS}^{7}(\Re_1, \Re_3)$

Proof: Correspondingly, Theorem (4.1).

4.2 Weighted Distance Measure for BCFSs

For the situations, where the weight of each element is involved, we initiate underneath weighted distance measures.

Definition 4.4 Consider two BCFSs $\mathfrak{R}_1 = \left\{e_\sigma, \left(\mathfrak{F}^{P}_{\mathfrak{R}_1}(e_\sigma), \mathfrak{F}^{N}_{\mathfrak{R}_1}(e_\sigma)\right) | e_\sigma \in \mathbb{E}, \sigma = 1, 2, .., n\right\} = \left\{e_\sigma, \left(\mathfrak{F}^{RP}_{\mathfrak{R}_1}(e_\sigma) + \iota\mathfrak{F}^{IP}_{\mathfrak{R}_1}(e_\sigma), \mathfrak{F}^{RN}_{\mathfrak{R}_1}(e_\sigma) + \iota\mathfrak{F}^{IN}_{\mathfrak{R}_1}(e_\sigma)\right) | e_\sigma \in \mathbb{E}, \sigma = 1, 2, .., n\right\}$ and $\mathfrak{R}_2 = \left\{e, \left(\mathfrak{F}^{P}_{\mathfrak{R}_2}(e_\sigma), \mathfrak{F}^{N}_{\mathfrak{R}_2}(e_\sigma)\right) | e \in \mathbb{E}, \sigma = 1, 2, .., n\right\} = \left\{e_\sigma, \left(\mathfrak{F}^{RP}_{\mathfrak{R}_2}(e_\sigma) + \iota\mathfrak{F}^{IP}_{\mathfrak{R}_2}(e_\sigma), \mathfrak{F}^{RN}_{\mathfrak{R}_2}(e_\sigma) + \iota\mathfrak{F}^{IN}_{\mathfrak{R}_2}(e_\sigma)\right) | e_\sigma \in \mathbb{E}, \sigma = 1, 2, .., n\right\}$ over $\mathbb{E}$. Then the GBCFWNHDM between $\mathfrak{R}_1$ and $\mathfrak{R}_2$ is inaugurated as.

$$d^{8}_{BCFS}(\mathfrak{R}_1, \mathfrak{R}_2) = \left(\frac{1}{4}\sum_{\sigma=1}^{n}\omega_\sigma\left(\begin{array}{c}\left|\left(\mathfrak{F}^{RP}_{\mathfrak{R}_1}(e_\sigma)\right)^2 - \left(\mathfrak{F}^{RP}_{\mathfrak{R}_2}(e_\sigma)\right)^2\right| + \left|\left(\mathfrak{F}^{IP}_{\mathfrak{R}_1}(e_\sigma)\right)^2 - \left(\mathfrak{F}^{IP}_{\mathfrak{R}_2}(e_\sigma)\right)^2\right| \\ + \left|\left(\mathfrak{F}^{RN}_{\mathfrak{R}_1}(e_\sigma)\right)^2 - \left(\mathfrak{F}^{RN}_{\mathfrak{R}_2}(e_\sigma)\right)^2\right| + \left|\left(\mathfrak{F}^{IN}_{\mathfrak{R}_1}(e_\sigma)\right)^2 - \left(\mathfrak{F}^{IN}_{\mathfrak{R}_2}(e_\sigma)\right)^2\right|\end{array}\right)\right)^{\frac{1}{\lambda}}, \quad \lambda \geq 1$$

Observed that $\omega = \left(\omega_1, \omega_2, \ldots, \omega_y\right)^T$ is a weight vector and $0 \leq \omega_\sigma \leq 1$ for all σ with $\sum_{\sigma=1}^{y}\omega_\sigma = 1$.

The underneath Theorem would show that d^{8}_{BCFS} satisfies the necessary conditions of distance measures.

Theorem 4.4 The GBCFWNHDM satisfies the underneath properties.

1. $0 \leq d^{8}_{BCFS}(\mathfrak{R}_1, \mathfrak{R}_2) \leq 1$
2. $d^{8}_{BCFS}(\mathfrak{R}_1, \mathfrak{R}_2) = 0 \Leftrightarrow \mathfrak{R}_1 = \mathfrak{R}_2$
3. $d^{8}_{BCFS}(\mathfrak{R}_1, \mathfrak{R}_2) = 0 \Leftrightarrow \mathfrak{R}_1 = \mathfrak{R}_2$
4. If $\mathfrak{R}_1 \subset \mathfrak{R}_2 \subset \mathfrak{R}_3$, then $d^{8}_{BCFS}(\mathfrak{R}_1, \mathfrak{R}_2) \leq d^{8}_{BCFS}(\mathfrak{R}_1, \mathfrak{R}_3)$ and $d^{8}_{BCFS}(\mathfrak{R}_2, \mathfrak{R}_3) \leq d^{8}_{BCFS}(\mathfrak{R}_1, \mathfrak{R}_3)$

Proof: 1. As $\mathfrak{F}^{RP}_{\mathfrak{R}_1}(e_\sigma), \mathfrak{F}^{IP}_{\mathfrak{R}_1}(e_\sigma), \mathfrak{F}^{RP}_{\mathfrak{R}_2}(e_\sigma), \mathfrak{F}^{IP}_{\mathfrak{R}_2}(e_\sigma) \in [0, 1]$ and $\mathfrak{F}^{RN}_{\mathfrak{R}_1}(e_\sigma), \mathfrak{F}^{IN}_{\mathfrak{R}_1}(e_\sigma), \mathfrak{F}^{RN}_{\mathfrak{R}_2}(e_\sigma), \mathfrak{F}^{IN}_{\mathfrak{R}_2}(e_\sigma) \in [-1, 0]$ for all σ. Then $\left(\mathfrak{F}^{RP}_{\mathfrak{R}_1}(e_\sigma)\right)^2 \left(\mathfrak{F}^{IP}_{\mathfrak{R}_1}(e_\sigma)\right)^2 \left(\mathfrak{F}^{RP}_{\mathfrak{R}_2}(e_\sigma)\right)^2, \left(\mathfrak{F}^{IP}_{\mathfrak{R}_2}(e_\sigma)\right)^2 \in [0, 1]$ and $\left(\mathfrak{F}^{RN}_{\mathfrak{R}_1}(e_\sigma)\right)^2 \left(\mathfrak{F}^{IN}_{\mathfrak{R}_1}(e_\sigma)\right)^2 \left(\mathfrak{F}^{RN}_{\mathfrak{R}_2}(e_\sigma)\right)^2, \left(\mathfrak{F}^{IN}_{\mathfrak{R}_2}(e_\sigma)\right)^2 \in [0, 1]$ for all σ. $\Rightarrow \left|\left(\mathfrak{F}^{RP}_{\mathfrak{R}_1}(e_\sigma)\right)^2 - \left(\mathfrak{F}^{RP}_{\mathfrak{R}_2}(e_\sigma)\right)^2\right|, \left|\left(\mathfrak{F}^{IP}_{\mathfrak{R}_1}(e_\sigma)\right)^2 - \left(\mathfrak{F}^{IP}_{\mathfrak{R}_2}(e_\sigma)\right)^2\right|, \left|\left(\mathfrak{F}^{RN}_{\mathfrak{R}_1}(e_\sigma)\right)^2 - \left(\mathfrak{F}^{RN}_{\mathfrak{R}_2}(e_\sigma)\right)^2\right|, \left|\left(\mathfrak{F}^{IN}_{\mathfrak{R}_1}(e_\sigma)\right)^2 - \left(\mathfrak{F}^{IN}_{\mathfrak{R}_2}(e_\sigma)\right)^2\right| \in [0, 1]\ \forall\sigma$. Consequently, $\forall\sigma$

$$\left(\begin{array}{c}\left|\left(\mathfrak{F}^{RP}_{\mathfrak{R}_1}(e_\sigma)\right)^2 - \left(\mathfrak{F}^{RP}_{\mathfrak{R}_2}(e_\sigma)\right)^2\right| + \left|\left(\mathfrak{F}^{IP}_{\mathfrak{R}_1}(e_\sigma)\right)^2 - \left(\mathfrak{F}^{IP}_{\mathfrak{R}_2}(e_\sigma)\right)^2\right| \\ + \left|\left(\mathfrak{F}^{RN}_{\mathfrak{R}_1}(e_\sigma)\right)^2 - \left(\mathfrak{F}^{RN}_{\mathfrak{R}_2}(e_\sigma)\right)^2\right| + \left|\left(\mathfrak{F}^{IN}_{\mathfrak{R}_1}(e_\sigma)\right)^2 - \left(\mathfrak{F}^{IN}_{\mathfrak{R}_2}(e_\sigma)\right)^2\right|\end{array}\right) \in [0, 4]$$

$$0 \leq \left(\begin{array}{c}\left|\left(\mathfrak{F}^{RP}_{\mathfrak{R}_1}(e_\sigma)\right)^2 - \left(\mathfrak{F}^{RP}_{\mathfrak{R}_2}(e_\sigma)\right)^2\right| + \left|\left(\mathfrak{F}^{IP}_{\mathfrak{R}_1}(e_\sigma)\right)^2 - \left(\mathfrak{F}^{IP}_{\mathfrak{R}_2}(e_\sigma)\right)^2\right| \\ + \left|\left(\mathfrak{F}^{RN}_{\mathfrak{R}_1}(e_\sigma)\right)^2 - \left(\mathfrak{F}^{RN}_{\mathfrak{R}_2}(e_\sigma)\right)^2\right| + \left|\left(\mathfrak{F}^{IN}_{\mathfrak{R}_1}(e_\sigma)\right)^2 - \left(\mathfrak{F}^{IN}_{\mathfrak{R}_2}(e_\sigma)\right)^2\right|\end{array}\right) \leq 4$$

$$0 \leq \omega_\sigma\left(\begin{array}{c}\left|\left(\mathfrak{F}^{RP}_{\mathfrak{R}_1}(e_\sigma)\right)^2 - \left(\mathfrak{F}^{RP}_{\mathfrak{R}_2}(e_\sigma)\right)^2\right| + \left|\left(\mathfrak{F}^{IP}_{\mathfrak{R}_1}(e_\sigma)\right)^2 - \left(\mathfrak{F}^{IP}_{\mathfrak{R}_2}(e_\sigma)\right)^2\right| \\ + \left|\left(\mathfrak{F}^{RN}_{\mathfrak{R}_1}(e_\sigma)\right)^2 - \left(\mathfrak{F}^{RN}_{\mathfrak{R}_2}(e_\sigma)\right)^2\right| + \left|\left(\mathfrak{F}^{IN}_{\mathfrak{R}_1}(e_\sigma)\right)^2 - \left(\mathfrak{F}^{IN}_{\mathfrak{R}_2}(e_\sigma)\right)^2\right|\end{array}\right) \leq 4\omega_\sigma$$

where $\omega_\sigma \in [0, 1], \forall\sigma$, and $\sum_{\sigma=1}^{y}\omega_\sigma = 1$. Thus,

$$0 \le \sum_{\sigma=1}^{n} \omega_\sigma \begin{pmatrix} \left|(\mathfrak{F}^{RP}_{\Re_1}(e_\sigma))^2-(\mathfrak{F}^{RP}_{\Re_2}(e_\sigma))^2\right| + \left|(\mathfrak{F}^{IP}_{\Re_1}(e_\sigma))^2-(\mathfrak{F}^{IP}_{\Re_2}(e_\sigma))^2\right| \\ +\left|(\mathfrak{F}^{RN}_{\Re_1}(e_\sigma))^2-(\mathfrak{F}^{RN}_{\Re_2}(e_\sigma))^2\right| + \left|(\mathfrak{F}^{IN}_{\Re_1}(e_\sigma))^2-(\mathfrak{F}^{IN}_{\Re_2}(e_\sigma))^2\right| \end{pmatrix} \le 4$$

$$0 \le \frac{1}{4}\sum_{\sigma=1}^{n} \omega_\sigma \begin{pmatrix} \left|(\mathfrak{F}^{RP}_{\Re_1}(e_\sigma))^2-(\mathfrak{F}^{RP}_{\Re_2}(e_\sigma))^2\right| + \left|(\mathfrak{F}^{IP}_{\Re_1}(e_\sigma))^2-(\mathfrak{F}^{IP}_{\Re_2}(e_\sigma))^2\right| \\ +\left|(\mathfrak{F}^{RN}_{\Re_1}(e_\sigma))^2-(\mathfrak{F}^{RN}_{\Re_2}(e_\sigma))^2\right| + \left|(\mathfrak{F}^{IN}_{\Re_1}(e_\sigma))^2-(\mathfrak{F}^{IN}_{\Re_2}(e_\sigma))^2\right| \end{pmatrix} \le 1$$

Since $\lambda \ge 1$, then

$$0 \le \left(\frac{1}{4}\sum_{\sigma=1}^{n} \omega_\sigma \begin{pmatrix} \left|(\mathfrak{F}^{RP}_{\Re_1}(e_\sigma))^2-(\mathfrak{F}^{RP}_{\Re_2}(e_\sigma))^2\right| + \left|(\mathfrak{F}^{IP}_{\Re_1}(e_\sigma))^2-(\mathfrak{F}^{IP}_{\Re_2}(e_\sigma))^2\right| \\ +\left|(\mathfrak{F}^{RN}_{\Re_1}(e_\sigma))^2-(\mathfrak{F}^{RN}_{\Re_2}(e_\sigma))^2\right| + \left|(\mathfrak{F}^{IN}_{\Re_1}(e_\sigma))^2-(\mathfrak{F}^{IN}_{\Re_2}(e_\sigma))^2\right| \end{pmatrix}\right)^{\frac{1}{\lambda}} \le 1$$

$$\Rightarrow 0 \le d^8_{BCFS}(\Re_1, \Re_2) \le 1.$$

2 Let $d^8_{BCFS}(\Re_1, \Re_2) = 0$. Then to show that $\Re_1 = \Re_2$. As $d^8_{BCFS}(\Re_1, \Re_2) = 0$, then

$$\left(\frac{1}{4}\sum_{\sigma=1}^{n} \omega_\sigma \begin{pmatrix} \left|(\mathfrak{F}^{RP}_{\Re_1}(e_\sigma))^2-(\mathfrak{F}^{RP}_{\Re_2}(e_\sigma))^2\right| + \left|(\mathfrak{F}^{IP}_{\Re_1}(e_\sigma))^2-(\mathfrak{F}^{IP}_{\Re_2}(e_\sigma))^2\right| \\ +\left|(\mathfrak{F}^{RN}_{\Re_1}(e_\sigma))^2-(\mathfrak{F}^{RN}_{\Re_2}(e_\sigma))^2\right| + \left|(\mathfrak{F}^{IN}_{\Re_1}(e_\sigma))^2-(\mathfrak{F}^{IN}_{\Re_2}(e_\sigma))^2\right| \end{pmatrix}\right)^{\frac{1}{\lambda}} = 0$$

$$\frac{1}{4}\sum_{\sigma=1}^{n} \omega_\sigma \begin{pmatrix} \left|(\mathfrak{F}^{RP}_{\Re_1}(e_\sigma))^2-(\mathfrak{F}^{RP}_{\Re_2}(e_\sigma))^2\right| + \left|(\mathfrak{F}^{IP}_{\Re_1}(e_\sigma))^2-(\mathfrak{F}^{IP}_{\Re_2}(e_\sigma))^2\right| \\ +\left|(\mathfrak{F}^{RN}_{\Re_1}(e_\sigma))^2-(\mathfrak{F}^{RN}_{\Re_2}(e_\sigma))^2\right| + \left|(\mathfrak{F}^{IN}_{\Re_1}(e_\sigma))^2-(\mathfrak{F}^{IN}_{\Re_2}(e_\sigma))^2\right| \end{pmatrix} = 0$$

$$\sum_{\sigma=1}^{n} \omega_\sigma \begin{pmatrix} \left|(\mathfrak{F}^{RP}_{\Re_1}(e_\sigma))^2-(\mathfrak{F}^{RP}_{\Re_2}(e_\sigma))^2\right| + \left|(\mathfrak{F}^{IP}_{\Re_1}(e_\sigma))^2-(\mathfrak{F}^{IP}_{\Re_2}(e_\sigma))^2\right| \\ +\left|(\mathfrak{F}^{RN}_{\Re_1}(e_\sigma))^2-(\mathfrak{F}^{RN}_{\Re_2}(e_\sigma))^2\right| + \left|(\mathfrak{F}^{IN}_{\Re_1}(e_\sigma))^2-(\mathfrak{F}^{IN}_{\Re_2}(e_\sigma))^2\right| \end{pmatrix} = 0$$

$$\omega_\sigma \begin{pmatrix} \left|(\mathfrak{F}^{RP}_{\Re_1}(e_\sigma))^2-(\mathfrak{F}^{RP}_{\Re_2}(e_\sigma))^2\right| + \left|(\mathfrak{F}^{IP}_{\Re_1}(e_\sigma))^2-(\mathfrak{F}^{IP}_{\Re_2}(e_\sigma))^2\right| \\ +\left|(\mathfrak{F}^{RN}_{\Re_1}(e_\sigma))^2-(\mathfrak{F}^{RN}_{\Re_2}(e_\sigma))^2\right| + \left|(\mathfrak{F}^{IN}_{\Re_1}(e_\sigma))^2-(\mathfrak{F}^{IN}_{\Re_2}(e_\sigma))^2\right| \end{pmatrix} = 0 \forall\sigma$$

$\Rightarrow \forall\sigma$, $\left|(\mathfrak{F}^{RP}_{\Re_1}(e_\sigma))^2-(\mathfrak{F}^{RP}_{\Re_2}(e_\sigma))^2\right| = 0$, $\left|(\mathfrak{F}^{IP}_{\Re_1}(e_\sigma))^2-(\mathfrak{F}^{IP}_{\Re_2}(e_\sigma))^2\right| = 0$, $\left|(\mathfrak{F}^{RN}_{\Re_1}(e_\sigma))^2-(\mathfrak{F}^{RN}_{\Re_2}(e_\sigma))^2\right| = 0$, $\left|(\mathfrak{F}^{RN}_{\Re_1}(e_\sigma))^2-(\mathfrak{F}^{RN}_{\Re_2}(e_\sigma))^2\right| = 0$. Thus, $\mathfrak{F}^{RP}_{\Re_1}(e_\sigma) = \mathfrak{F}^{RP}_{\Re_2}(e_\sigma)$, $\mathfrak{F}^{IP}_{\Re_1}(e_\sigma) = \mathfrak{F}^{IP}_{\Re_2}(e_\sigma)$, $\mathfrak{F}^{RN}_{\Re_1}(e_\sigma) = \mathfrak{F}^{RN}_{\Re_2}(e_\sigma)$, $\mathfrak{F}^{IN}_{\Re_1}(e_\sigma) = \mathfrak{F}^{IN}_{\Re_2}(e_\sigma)$ $\forall\sigma$. Consequently, $\Re_1 = \Re_2$.

Conversely, Let $\Re_1 = \Re_2$. To show that $d^8_{BCFS}(\Re_1, \Re_2) = 0$. As $\Re_1 = \Re_2$, then $\mathfrak{F}^{RP}_{\Re_1}(e_\sigma) = \mathfrak{F}^{RP}_{\Re_2}(e_\sigma)$, $\mathfrak{F}^{IP}_{\Re_1}(e_\sigma) = \mathfrak{F}^{IP}_{\Re_2}(e_\sigma)$, $\mathfrak{F}^{RN}_{\Re_1}(e_\sigma) = \mathfrak{F}^{RN}_{\Re_2}(e_\sigma)$, $\mathfrak{F}^{IN}_{\Re_1}(e_\sigma) = \mathfrak{F}^{IN}_{\Re_2}(e_\sigma)$ $\forall\sigma$. Then $\forall\sigma$

$$d^8_{BCFS}(\Re_1, \Re_2) = \left(\frac{1}{4}\sum_{\sigma=1}^{n} \omega_\sigma \begin{pmatrix} \left|(\mathfrak{F}^{RP}_{\Re_1}(e_\sigma))^2-(\mathfrak{F}^{RP}_{\Re_1}(e_\sigma))^2\right| + \left|(\mathfrak{F}^{IP}_{\Re_1}(e_\sigma))^2-(\mathfrak{F}^{IP}_{\Re_1}(e_\sigma))^2\right| \\ +\left|(\mathfrak{F}^{RN}_{\Re_1}(e_\sigma))^2-(\mathfrak{F}^{RN}_{\Re_1}(e_\sigma))^2\right| + \left|(\mathfrak{F}^{IN}_{\Re_1}(e_\sigma))^2-(\mathfrak{F}^{IN}_{\Re_1}(e_\sigma))^2\right| \end{pmatrix}\right)^{\frac{1}{\lambda}} = 0$$

3 As

$$d_{BCFS}^{8}(\mathfrak{R}_1,\mathfrak{R}_2)=\left(\frac{1}{4}\sum_{\sigma=1}^{n}\omega_\sigma\left(\begin{array}{c}\left|\left(\mathfrak{F}_{\mathfrak{R}_1}^{\mathrm{RP}}(\mathrm{e}_\sigma)\right)^2-\left(\mathfrak{F}_{\mathfrak{R}_2}^{\mathrm{RP}}(\mathrm{e}_\sigma)\right)^2\right|+\left|\left(\mathfrak{F}_{\mathfrak{R}_1}^{\mathrm{IP}}(\mathrm{e}_\sigma)\right)^2-\left(\mathfrak{F}_{\mathfrak{R}_2}^{\mathrm{IP}}(\mathrm{e}_\sigma)\right)^2\right|\\+\left|\left(\mathfrak{F}_{\mathfrak{R}_1}^{\mathrm{RN}}(\mathrm{e}_\sigma)\right)^2-\left(\mathfrak{F}_{\mathfrak{R}_2}^{\mathrm{RN}}(\mathrm{e}_\sigma)\right)^2\right|+\left|\left(\mathfrak{F}_{\mathfrak{R}_1}^{\mathrm{IN}}(\mathrm{e}_\sigma)\right)^2-\left(\mathfrak{F}_{\mathfrak{R}_2}^{\mathrm{IN}}(\mathrm{e}_\sigma)\right)^2\right|\end{array}\right)\right)^{\frac{1}{\lambda}}$$

$$=\left(\frac{1}{4}\sum_{\sigma=1}^{n}\omega_\sigma\left(\begin{array}{c}\left|\left(\mathfrak{F}_{\mathfrak{R}_2}^{\mathrm{RP}}(\mathrm{e}_\sigma)\right)^2-\left(\mathfrak{F}_{\mathfrak{R}_1}^{\mathrm{RP}}(\mathrm{e}_\sigma)\right)^2\right|+\left|\left(\mathfrak{F}_{\mathfrak{R}_2}^{\mathrm{IP}}(\mathrm{e}_\sigma)\right)^2-\left(\mathfrak{F}_{\mathfrak{R}_1}^{\mathrm{IP}}(\mathrm{e}_\sigma)\right)^2\right|\\+\left|\left(\mathfrak{F}_{\mathfrak{R}_2}^{\mathrm{RN}}(\mathrm{e}_\sigma)\right)^2-\left(\mathfrak{F}_{\mathfrak{R}_1}^{\mathrm{RN}}(\mathrm{e}_\sigma)\right)^2\right|+\left|\left(\mathfrak{F}_{\mathfrak{R}_2}^{\mathrm{IN}}(\mathrm{e}_\sigma)\right)^2-\left(\mathfrak{F}_{\mathfrak{R}_1}^{\mathrm{IN}}(\mathrm{e}_\sigma)\right)^2\right|\end{array}\right)\right)^{\frac{1}{\lambda}}$$

$$=d_{BCFS}^{8}(\mathfrak{R}_2,\mathfrak{R}_1)$$

4 Assume that $\mathfrak{R}_1\subset\mathfrak{R}_2\subset\mathfrak{R}_3$, then $\mathfrak{F}_{\mathfrak{R}_1}^{\mathrm{RP}}(\mathrm{e}_\sigma)\le\mathfrak{F}_{\mathfrak{R}_2}^{\mathrm{RP}}(\mathrm{e}_\sigma)\le\mathfrak{F}_{\mathfrak{R}_3}^{\mathrm{RP}}(\mathrm{e}_\sigma)$, $\mathfrak{F}_{\mathfrak{R}_1}^{\mathrm{IP}}(\mathrm{e}_\sigma)\le\mathfrak{F}_{\mathfrak{R}_2}^{\mathrm{IP}}(\mathrm{e}_\sigma)\le\mathfrak{F}_{\mathfrak{R}_3}^{\mathrm{IP}}(\mathrm{e}_\sigma)$, $\mathfrak{F}_{\mathfrak{R}_1}^{\mathrm{RN}}(\mathrm{e}_\sigma)\le\mathfrak{F}_{\mathfrak{R}_2}^{\mathrm{RN}}(\mathrm{e}_\sigma)\le\mathfrak{F}_{\mathfrak{R}_3}^{\mathrm{RN}}(\mathrm{e}_\sigma)$, $\mathfrak{F}_{\mathfrak{R}_1}^{\mathrm{IN}}(\mathrm{e}_\sigma)\le\mathfrak{F}_{\mathfrak{R}_2}^{\mathrm{IN}}(\mathrm{e}_\sigma)\le\mathfrak{F}_{\mathfrak{R}_3}^{\mathrm{IN}}(\mathrm{e}_\sigma)$ $\forall\sigma$. Thus, we have the inequalities

$$\left|\left(\mathfrak{F}_{\mathfrak{R}_1}^{\mathrm{RP}}(\mathrm{e}_\sigma)\right)^2-\left(\mathfrak{F}_{\mathfrak{R}_2}^{\mathrm{RP}}(\mathrm{e}_\sigma)\right)^2\right|\le\left|\left(\mathfrak{F}_{\mathfrak{R}_1}^{\mathrm{RP}}(\mathrm{e}_\sigma)\right)^2-\left(\mathfrak{F}_{\mathfrak{R}_3}^{\mathrm{RP}}(\mathrm{e}_\sigma)\right)^2\right|$$

$$\left|\left(\mathfrak{F}_{\mathfrak{R}_1}^{\mathrm{IP}}(\mathrm{e}_\sigma)\right)^2-\left(\mathfrak{F}_{\mathfrak{R}_2}^{\mathrm{IP}}(\mathrm{e}_\sigma)\right)^2\right|\le\left|\left(\mathfrak{F}_{\mathfrak{R}_1}^{\mathrm{IP}}(\mathrm{e}_\sigma)\right)^2-\left(\mathfrak{F}_{\mathfrak{R}_3}^{\mathrm{IP}}(\mathrm{e}_\sigma)\right)^2\right|$$

And

$$\left|\left(\mathfrak{F}_{\mathfrak{R}_1}^{\mathrm{RN}}(\mathrm{e}_\sigma)\right)^2-\left(\mathfrak{F}_{\mathfrak{R}_2}^{\mathrm{RN}}(\mathrm{e}_\sigma)\right)^2\right|\le\left|\left(\mathfrak{F}_{\mathfrak{R}_1}^{\mathrm{RN}}(\mathrm{e}_\sigma)\right)^2-\left(\mathfrak{F}_{\mathfrak{R}_3}^{\mathrm{RN}}(\mathrm{e}_\sigma)\right)^2\right|$$

$$\left|\left(\mathfrak{F}_{\mathfrak{R}_1}^{\mathrm{IN}}(\mathrm{e}_\sigma)\right)^2-\left(\mathfrak{F}_{\mathfrak{R}_2}^{\mathrm{IN}}(\mathrm{e}_\sigma)\right)^2\right|\le\left|\left(\mathfrak{F}_{\mathfrak{R}_1}^{\mathrm{IN}}(\mathrm{e}_\sigma)\right)^2-\left(\mathfrak{F}_{\mathfrak{R}_3}^{\mathrm{IN}}(\mathrm{e}_\sigma)\right)^2\right|$$

Thus, $\forall\sigma$

$$\omega_\sigma\left(\begin{array}{c}\left|\left(\mathfrak{F}_{\mathfrak{R}_1}^{\mathrm{RP}}(\mathrm{e}_\sigma)\right)^2-\left(\mathfrak{F}_{\mathfrak{R}_2}^{\mathrm{RP}}(\mathrm{e}_\sigma)\right)^2\right|+\left|\left(\mathfrak{F}_{\mathfrak{R}_1}^{\mathrm{IP}}(\mathrm{e}_\sigma)\right)^2-\left(\mathfrak{F}_{\mathfrak{R}_2}^{\mathrm{IP}}(\mathrm{e}_\sigma)\right)^2\right|\\+\left|\left(\mathfrak{F}_{\mathfrak{R}_1}^{\mathrm{RN}}(\mathrm{e}_\sigma)\right)^2-\left(\mathfrak{F}_{\mathfrak{R}_2}^{\mathrm{RN}}(\mathrm{e}_\sigma)\right)^2\right|+\left|\left(\mathfrak{F}_{\mathfrak{R}_1}^{\mathrm{RN}}(\mathrm{e}_\sigma)\right)^2-\left(\mathfrak{F}_{\mathfrak{R}_2}^{\mathrm{RN}}(\mathrm{e}_\sigma)\right)^2\right|\end{array}\right)$$

$$\le\omega_\sigma\left(\begin{array}{c}\left|\left(\mathfrak{F}_{\mathfrak{R}_1}^{\mathrm{RP}}(\mathrm{e}_\sigma)\right)^2-\left(\mathfrak{F}_{\mathfrak{R}_3}^{\mathrm{RP}}(\mathrm{e}_\sigma)\right)^2\right|+\left|\left(\mathfrak{F}_{\mathfrak{R}_1}^{\mathrm{IP}}(\mathrm{e}_\sigma)\right)^2-\left(\mathfrak{F}_{\mathfrak{R}_3}^{\mathrm{IP}}(\mathrm{e}_\sigma)\right)^2\right|\\+\left|\left(\mathfrak{F}_{\mathfrak{R}_1}^{\mathrm{RN}}(\mathrm{e}_\sigma)\right)^2-\left(\mathfrak{F}_{\mathfrak{R}_3}^{\mathrm{RN}}(\mathrm{e}_\sigma)\right)^2\right|+\left|\left(\mathfrak{F}_{\mathfrak{R}_1}^{\mathrm{RN}}(\mathrm{e}_\sigma)\right)^2-\left(\mathfrak{F}_{\mathfrak{R}_3}^{\mathrm{RN}}(\mathrm{e}_\sigma)\right)^2\right|\end{array}\right)$$

$$\left(\frac{1}{4}\sum_{\sigma=1}^{n}\omega_\sigma\left(\begin{array}{c}\left|\left(\mathfrak{F}_{\mathfrak{R}_1}^{\mathrm{RP}}(\mathrm{e}_\sigma)\right)^2-\left(\mathfrak{F}_{\mathfrak{R}_2}^{\mathrm{RP}}(\mathrm{e}_\sigma)\right)^2\right|+\left|\left(\mathfrak{F}_{\mathfrak{R}_1}^{\mathrm{IP}}(\mathrm{e}_\sigma)\right)^2-\left(\mathfrak{F}_{\mathfrak{R}_2}^{\mathrm{IP}}(\mathrm{e}_\sigma)\right)^2\right|\\+\left|\left(\mathfrak{F}_{\mathfrak{R}_1}^{\mathrm{RN}}(\mathrm{e}_\sigma)\right)^2-\left(\mathfrak{F}_{\mathfrak{R}_2}^{\mathrm{RN}}(\mathrm{e}_\sigma)\right)^2\right|+\left|\left(\mathfrak{F}_{\mathfrak{R}_1}^{\mathrm{RN}}(\mathrm{e}_\sigma)\right)^2-\left(\mathfrak{F}_{\mathfrak{R}_2}^{\mathrm{RN}}(\mathrm{e}_\sigma)\right)^2\right|\end{array}\right)\right)^{\frac{1}{\lambda}}$$

$$\le\left(\frac{1}{4}\sum_{\sigma=1}^{n}\omega_\sigma\left(\begin{array}{c}\left|\left(\mathfrak{F}_{\mathfrak{R}_1}^{\mathrm{RP}}(\mathrm{e}_\sigma)\right)^2-\left(\mathfrak{F}_{\mathfrak{R}_3}^{\mathrm{RP}}(\mathrm{e}_\sigma)\right)^2\right|+\left|\left(\mathfrak{F}_{\mathfrak{R}_1}^{\mathrm{IP}}(\mathrm{e}_\sigma)\right)^2-\left(\mathfrak{F}_{\mathfrak{R}_3}^{\mathrm{IP}}(\mathrm{e}_\sigma)\right)^2\right|\\+\left|\left(\mathfrak{F}_{\mathfrak{R}_1}^{\mathrm{RN}}(\mathrm{e}_\sigma)\right)^2-\left(\mathfrak{F}_{\mathfrak{R}_3}^{\mathrm{RN}}(\mathrm{e}_\sigma)\right)^2\right|+\left|\left(\mathfrak{F}_{\mathfrak{R}_1}^{\mathrm{RN}}(\mathrm{e}_\sigma)\right)^2-\left(\mathfrak{F}_{\mathfrak{R}_3}^{\mathrm{RN}}(\mathrm{e}_\sigma)\right)^2\right|\end{array}\right)\right)^{\frac{1}{\lambda}}$$

$$\Rightarrow d_{BCFS}^{8}(\mathfrak{R}_1,\mathfrak{R}_2)\le d_{BCFS}^{8}(\mathfrak{R}_1,\mathfrak{R}_3).$$

By the same procedure, $d_{BCFS}^{8}(\mathfrak{R}_2,\mathfrak{R}_3)\le d_{BCFS}^{8}(\mathfrak{R}_1,\mathfrak{R}_3)$ can be proved.

Particular cases

Underneath are the particular cases of GBCFWNHDM.

i. By considering $\lambda = 1,$, the GBCFWNHDM would decrease to BCF weighted normalized Hamming distance measure (BCFWNHDM)

$$d^{9}_{BCFS}(\mathfrak{R}_1,\mathfrak{R}_2)=\frac{1}{4}\sum_{\sigma=1}^{n}\omega_\sigma\left(\begin{array}{c}\left|\left(\mathfrak{F}^{RP}_{\mathfrak{R}_1}(e_\sigma)\right)^2-\left(\mathfrak{F}^{RP}_{\mathfrak{R}_2}(e_\sigma)\right)^2\right|+\left|\left(\mathfrak{F}^{IP}_{\mathfrak{R}_1}(e_\sigma)\right)^2-\left(\mathfrak{F}^{IP}_{\mathfrak{R}_2}(e_\sigma)\right)^2\right|\\+\left|\left(\mathfrak{F}^{RN}_{\mathfrak{R}_1}(e_\sigma)\right)^2-\left(\mathfrak{F}^{RN}_{\mathfrak{R}_2}(e_\sigma)\right)^2\right|+\left|\left(\mathfrak{F}^{IN}_{\mathfrak{R}_1}(e_\sigma)\right)^2-\left(\mathfrak{F}^{IN}_{\mathfrak{R}_2}(e_\sigma)\right)^2\right|\end{array}\right)$$

ii. By considering $\lambda = 2,$, the GBCFWNHDM would decrease to BCF weighted normalized Euclidean distance measure (BCFWNEDM)

$$d^{10}_{BCFS}(\mathfrak{R}_1,\mathfrak{R}_2)=\sqrt{\frac{1}{4}\sum_{\sigma=1}^{n}\omega_\sigma\left(\begin{array}{c}\left|\left(\mathfrak{F}^{RP}_{\mathfrak{R}_1}(e_\sigma)\right)^2-\left(\mathfrak{F}^{RP}_{\mathfrak{R}_2}(e_\sigma)\right)^2\right|+\left|\left(\mathfrak{F}^{IP}_{\mathfrak{R}_1}(e_\sigma)\right)^2-\left(\mathfrak{F}^{IP}_{\mathfrak{R}_2}(e_\sigma)\right)^2\right|\\+\left|\left(\mathfrak{F}^{RN}_{\mathfrak{R}_1}(e_\sigma)\right)^2-\left(\mathfrak{F}^{RN}_{\mathfrak{R}_2}(e_\sigma)\right)^2\right|+\left|\left(\mathfrak{F}^{IN}_{\mathfrak{R}_1}(e_\sigma)\right)^2-\left(\mathfrak{F}^{IN}_{\mathfrak{R}_2}(e_\sigma)\right)^2\right|\end{array}\right)}$$

Definition 4.5 Let two BCFSs $\mathfrak{R}_1 = \left\{e_\sigma, \left(\mathfrak{F}^{P}_{\mathfrak{R}_1}(e_\sigma), \mathfrak{F}^{N}_{\mathfrak{R}_1}(e_\sigma)\right)|e_\sigma \in \mathbb{E}, \sigma = 1, 2, .., n\right\} = \left\{e_\sigma, \left(\mathfrak{F}^{RP}_{\mathfrak{R}_1}(e_\sigma) + \iota\mathfrak{F}^{IP}_{\mathfrak{R}_1}(e_\sigma), \mathfrak{F}^{RN}_{\mathfrak{R}_1}(e_\sigma) + \iota\mathfrak{F}^{IN}_{\mathfrak{R}_1}(e_\sigma)\right)|e_\sigma \in \mathbb{E}, \sigma = 1, 2, .., n\right\}$ and $\mathfrak{R}_2 = \left\{e, \left(\mathfrak{F}^{P}_{\mathfrak{R}_2}(e_\sigma), \mathfrak{F}^{N}_{\mathfrak{R}_2}(e_\sigma)\right)|e \in \mathbb{E}, \sigma = 1, 2, .., n\right\} = \left\{e_\sigma, \left(\mathfrak{F}^{RP}_{\mathfrak{R}_2}(e_\sigma) + \iota\mathfrak{F}^{IP}_{\mathfrak{R}_2}(e_\sigma), \mathfrak{F}^{RN}_{\mathfrak{R}_2}(e_\sigma) + \iota\mathfrak{F}^{IN}_{\mathfrak{R}_2}(e_\sigma)\right)|e_\sigma \in \mathbb{E}, \sigma = 1, 2, .., n\right\}$ over $\mathbb{E}$. Then the GBCFWNHDDM between $\mathfrak{R}_1$ and $\mathfrak{R}_2$ is inaugurated as.

$$d^{11}_{BCFS}(\mathfrak{R}_1,\mathfrak{R}_2)=\left(\sum_{\sigma=1}^{n}\omega_\sigma\max\left(\begin{array}{c}\left|\left(\mathfrak{F}^{RP}_{\mathfrak{R}_1}(e_\sigma)\right)^2-\left(\mathfrak{F}^{RP}_{\mathfrak{R}_2}(e_\sigma)\right)^2\right|,\left|\left(\mathfrak{F}^{IP}_{\mathfrak{R}_1}(e_\sigma)\right)^2-\left(\mathfrak{F}^{IP}_{\mathfrak{R}_2}(e_\sigma)\right)^2\right|\\,\left|\left(\mathfrak{F}^{RN}_{\mathfrak{R}_1}(e_\sigma)\right)^2-\left(\mathfrak{F}^{RN}_{\mathfrak{R}_2}(e_\sigma)\right)^2\right|,\left|\left(\mathfrak{F}^{IN}_{\mathfrak{R}_1}(e_\sigma)\right)^2-\left(\mathfrak{F}^{IN}_{\mathfrak{R}_2}(e_\sigma)\right)^2\right|\end{array}\right)\right)^{\frac{1}{\lambda}},\quad \lambda\geq 1$$

Observed that $\omega = \left(\omega_{1,}\omega_2, \ldots, \omega_y\right)^T$ is a weight vector and $0 \leq \omega_\sigma \leq 1$ for all σ with $\sum_{\sigma=1}^{y} \omega_\sigma = 1$.

Theorem 4.5 The GBCFWNHDDM satisfies the underneath properties.

1. $0 \leq d^{11}_{BCFS}(\mathfrak{R}_1, \mathfrak{R}_2) \leq 1$
2. $d^{11}_{BCFS}(\mathfrak{R}_1, \mathfrak{R}_2) = 0 \Leftrightarrow \mathfrak{R}_1 = \mathfrak{R}_2$
3. $d^{11}_{BCFS}(\mathfrak{R}_1, \mathfrak{R}_2) = d^{11}_{BCFS}(\mathfrak{R}_2, \mathfrak{R}_1)$
4. If $\mathfrak{R}_1 \subset \mathfrak{R}_2 \subset \mathfrak{R}_3$, then $d^{11}_{BCFS}(\mathfrak{R}_1, \mathfrak{R}_2) \leq d^{11}_{BCFS}(\mathfrak{R}_1, \mathfrak{R}_3)$ and $d^{11}_{BCFS}(\mathfrak{R}_2, \mathfrak{R}_3) \leq d^{11}_{BCFS}(\mathfrak{R}_1, \mathfrak{R}_3)$

Proof: Omitted.

Particular cases.

Underneath are the particular cases of GBCFWNHDDM.

i. By considering $\lambda = 1,$, the GBCFWNHDDM would decrease to BCF weighted normalized Hamming-Hausdroff distance measure (BCFWNHHDDM)

$$d_{BCFS}^{12}(\Re_1,\Re_2)=\sum_{\sigma=1}^{n}\omega_\sigma \max\left(\begin{array}{c}\left|\left(\mathfrak{F}_{\Re_1}^{RP}(e_\sigma)\right)^2-\left(\mathfrak{F}_{\Re_2}^{RP}(e_\sigma)\right)^2\right|,\left|\left(\mathfrak{F}_{\Re_1}^{IP}(e_\sigma)\right)^2-\left(\mathfrak{F}_{\Re_2}^{IP}(e_\sigma)\right)^2\right|\\,\left|\left(\mathfrak{F}_{\Re_1}^{RN}(e_\sigma)\right)^2-\left(\mathfrak{F}_{\Re_2}^{RN}(e_\sigma)\right)^2\right|,\left|\left(\mathfrak{F}_{\Re_1}^{IN}(e_\sigma)\right)^2-\left(\mathfrak{F}_{\Re_2}^{IN}(e_\sigma)\right)^2\right|\end{array}\right)$$

ii. By considering $\lambda = 2$, , the GBCFWNHDDM would decrease to BCF weighted normalized Euclidean-Hausdroff distance measure (BCFWNEHDDM)

$$d_{BCFS}^{13}(\Re_1,\Re_2)=\sqrt{\sum_{\sigma=1}^{n}\omega_\sigma \max\left(\begin{array}{c}\left|\left(\mathfrak{F}_{\Re_1}^{RP}(e_\sigma)\right)^2-\left(\mathfrak{F}_{\Re_2}^{RP}(e_\sigma)\right)^2\right|,\left|\left(\mathfrak{F}_{\Re_1}^{IP}(e_\sigma)\right)^2-\left(\mathfrak{F}_{\Re_2}^{IP}(e_\sigma)\right)^2\right|\\,\left|\left(\mathfrak{F}_{\Re_1}^{RN}(e_\sigma)\right)^2-\left(\mathfrak{F}_{\Re_2}^{RN}(e_\sigma)\right)^2\right|,\left|\left(\mathfrak{F}_{\Re_1}^{IN}(e_\sigma)\right)^2-\left(\mathfrak{F}_{\Re_2}^{IN}(e_\sigma)\right)^2\right|\end{array}\right)}$$

Following, we combine GBCFWNHDM and GBCFWNHDDM to inaugurate the GHBCFWNDM. The GHBCFWNDM would give the benefits of both GBCFWNHDM and GBCFWNHDDM simultaneously.

Definition 4.6 Let two BCFSs $\Re_1 = \left\{e_\sigma, \left(\mathfrak{F}_{\Re_1}^{P}(e_\sigma), \mathfrak{F}_{\Re_1}^{N}(e_\sigma)\right) | e_\sigma \in \mathbb{E}, \sigma = 1, 2, .., n\right\} = \left\{e_\sigma, \left(\mathfrak{F}_{\Re_1}^{RP}(e_\sigma) + \iota\mathfrak{F}_{\Re_1}^{IP}(e_\sigma), \mathfrak{F}_{\Re_1}^{RN}(e_\sigma) + \iota\mathfrak{F}_{\Re_1}^{IN}(e_\sigma)\right) | e_\sigma \in \mathbb{E},\ \sigma = 1, 2, .., n\right\}$ and $\Re_2 = \left\{e, \left(\mathfrak{F}_{\Re_2}^{P}(e_\sigma), \mathfrak{F}_{\Re_2}^{N}(e_\sigma)\right) | e \in \mathbb{E}, \sigma = 1, 2, .., n\right\} = \left\{e_\sigma, \left(\mathfrak{F}_{\Re_2}^{RP}(e_\sigma) + \iota\mathfrak{F}_{\Re_2}^{IP}(e_\sigma), \mathfrak{F}_{\Re_2}^{RN}(e_\sigma) + \iota\mathfrak{F}_{\Re_2}^{IN}(e_\sigma)\right) | e_\sigma \in \mathbb{E},\quad \sigma = 1, 2, .., n\right\}$ over $\mathbb{E}$. Then the GHBCFWNDM between $\Re_1$ and $\Re_2$ is inaugurated as.

$$d_{BCFS}^{14}(\Re_1,\Re_2)$$
$$=\left(\frac{1}{2}\sum_{\sigma=1}^{n}\omega_\sigma\left(\begin{array}{c}\frac{1}{4}\left(\begin{array}{c}\left|\left(\mathfrak{F}_{\Re_1}^{RP}(e_\sigma)\right)^2-\left(\mathfrak{F}_{\Re_2}^{RP}(e_\sigma)\right)^2\right|+\left|\left(\mathfrak{F}_{\Re_1}^{IP}(e_\sigma)\right)^2-\left(\mathfrak{F}_{\Re_2}^{IP}(e_\sigma)\right)^2\right|\\+\left|\left(\mathfrak{F}_{\Re_1}^{RN}(e_\sigma)\right)^2-\left(\mathfrak{F}_{\Re_2}^{RN}(e_\sigma)\right)^2\right|+\left|\left(\mathfrak{F}_{\Re_1}^{IN}(e_\sigma)\right)^2-\left(\mathfrak{F}_{\Re_2}^{IN}(e_\sigma)\right)^2\right|\end{array}\right)\\+\max\left(\left|\left(\mathfrak{F}_{\Re_1}^{RP}(e_\sigma)\right)^2-\left(\mathfrak{F}_{\Re_2}^{RP}(e_\sigma)\right)^2\right|,\left|\left(\mathfrak{F}_{\Re_1}^{IP}(e_\sigma)\right)^2-\left(\mathfrak{F}_{\Re_2}^{IP}(e_\sigma)\right)^2\right|\right.\\\left.\left|\left(\mathfrak{F}_{\Re_1}^{RN}(e_\sigma)\right)^2-\left(\mathfrak{F}_{\Re_2}^{RN}(e_\sigma)\right)^2\right|,\left|\left(\mathfrak{F}_{\Re_1}^{IN}(e_\sigma)\right)^2-\left(\mathfrak{F}_{\Re_2}^{IN}(e_\sigma)\right)^2\right|\right)\end{array}\right)\right)^{\frac{1}{\lambda}},\lambda\ge 1$$

Observed that $\omega = \left(\omega_{1,}\omega_2, \ldots, \omega_y\right)^T$ is a weight vector and $0 \leq \omega_\sigma \leq 1$ for all σ with $\sum_{\sigma=1}^{y} \omega_\sigma = 1$.

Theorem 4.6 The GHBCFWNDM satisfies the underneath properties.

1. $0 \leq d_{BCFS}^{14}(\Re_1, \Re_2) \leq 1$
2. $d_{BCFS}^{14}(\Re_1, \Re_2) = 0 \Leftrightarrow \Re_1 = \Re_2$
3. $d_{BCFS}^{14}(\Re_1, \Re_2) = d_{BCFS}^{14}(\Re_2, \Re_1)$
4. If $\Re_1 \subset \Re_2 \subset \Re_3$, then $d_{BCFS}^{14}(\Re_1, \Re_2) \leq d_{BCFS}^{14}(\Re_1, \Re_3)$ and $d_{BCFS}^{14}(\Re_2, \Re_3) \leq d_{BCFS}^{14}(\Re_1, \Re_3)$

Proof: Omitted.

4.3 Continuous Distance Measures for BCFSs

In above developed distance measures the universal set $\mathbb{E}$ and the related weights were discrete. Now for situations where, the universal $\mathbb{E}$ and weight are continuous, we interpret the below distance measures. In this chapter the weight of $e \in \mathbb{E} = [\mathrm{m}, \mathrm{n}]$ is $\omega(e)$, $\omega(e) \in [0, 1]$ and $\int_m^n \omega(e)de$.

Definition 4.7 Let two BCFSs $\Re_1$ and $\Re_2$ over $\mathbb{E} = [\mathrm{m}, \mathrm{n}]$. Then the GCBCFWNHDM between $\Re_1$ and $\Re_2$ is inaugurated as.

$$d_{BCFS}^{15}(\Re_1, \Re_2) = \left(\frac{1}{4}\int_m^n \omega(e)\left(\begin{array}{c}\left|\left(\mathfrak{F}_{\Re_1}^{RP}(e_\sigma)\right)^2 - \left(\mathfrak{F}_{\Re_2}^{RP}(e_\sigma)\right)^2\right| + \left|\left(\mathfrak{F}_{\Re_1}^{IP}(e_\sigma)\right)^2 - \left(\mathfrak{F}_{\Re_2}^{IP}(e_\sigma)\right)^2\right| \\ + \left|\left(\mathfrak{F}_{\Re_1}^{RN}(e_\sigma)\right)^2 - \left(\mathfrak{F}_{\Re_2}^{RN}(e_\sigma)\right)^2\right| + \left|\left(\mathfrak{F}_{\Re_1}^{IN}(e_\sigma)\right)^2 - \left(\mathfrak{F}_{\Re_2}^{IN}(e_\sigma)\right)^2\right|\end{array}\right)de\right)^{\frac{1}{\lambda}}, \lambda \geq 1$$

By putting $\omega(e) = \frac{1}{m-n}$, for any $e \in [m, n]$, the GCBCFWNHDM will convert to generalized continuous bipolar complex fuzzy normalized hamming distance measures (GCBCFNHDM).

$$d_{BCFS}^{16}(\Re_1, \Re_2) = \left(\frac{1}{4(m-n)}\int_m^n \left(\begin{array}{c}\left|\left(\mathfrak{F}_{\Re_1}^{RP}(e_\sigma)\right)^2 - \left(\mathfrak{F}_{\Re_2}^{RP}(e_\sigma)\right)^2\right| + \left|\left(\mathfrak{F}_{\Re_1}^{IP}(e_\sigma)\right)^2 - \left(\mathfrak{F}_{\Re_2}^{IP}(e_\sigma)\right)^2\right| \\ + \left|\left(\mathfrak{F}_{\Re_1}^{RN}(e_\sigma)\right)^2 - \left(\mathfrak{F}_{\Re_2}^{RN}(e_\sigma)\right)^2\right| + \left|\left(\mathfrak{F}_{\Re_1}^{IN}(e_\sigma)\right)^2 - \left(\mathfrak{F}_{\Re_2}^{IN}(e_\sigma)\right)^2\right|\end{array}\right)de\right)^{\frac{1}{\lambda}}, \lambda \geq 1$$

Particular Cases

i. By considering $\lambda = 1$, the GCBCFWNHDM would convert to continuous BCFWNHDM (CBCFWNHDM)

$$d_{BCFS}^{17}(\Re_1, \Re_2) = \frac{1}{4}\int_m^n \omega(e)\left(\begin{array}{c}\left|\left(\mathfrak{F}_{\Re_1}^{RP}(e_\sigma)\right)^2 - \left(\mathfrak{F}_{\Re_2}^{RP}(e_\sigma)\right)^2\right| + \left|\left(\mathfrak{F}_{\Re_1}^{IP}(e_\sigma)\right)^2 - \left(\mathfrak{F}_{\Re_2}^{IP}(e_\sigma)\right)^2\right| \\ + \left|\left(\mathfrak{F}_{\Re_1}^{RN}(e_\sigma)\right)^2 - \left(\mathfrak{F}_{\Re_2}^{RN}(e_\sigma)\right)^2\right| + \left|\left(\mathfrak{F}_{\Re_1}^{IN}(e_\sigma)\right)^2 - \left(\mathfrak{F}_{\Re_2}^{IN}(e_\sigma)\right)^2\right|\end{array}\right)de$$

By putting $\omega(e) = \frac{1}{m-n}$, for any $e \in [m, n]$, the CBCFWNHDM will convert to continuous BCFNHDM (CBCFNHDM)

$$d_{BCFS}^{18}(\Re_1, \Re_2) = \frac{1}{4(m-n)}\int_m^n \left(\begin{array}{c}\left|\left(\mathfrak{F}_{\Re_1}^{RP}(e_\sigma)\right)^2 - \left(\mathfrak{F}_{\Re_2}^{RP}(e_\sigma)\right)^2\right| + \left|\left(\mathfrak{F}_{\Re_1}^{IP}(e_\sigma)\right)^2 - \left(\mathfrak{F}_{\Re_2}^{IP}(e_\sigma)\right)^2\right| \\ + \left|\left(\mathfrak{F}_{\Re_1}^{RN}(e_\sigma)\right)^2 - \left(\mathfrak{F}_{\Re_2}^{RN}(e_\sigma)\right)^2\right| + \left|\left(\mathfrak{F}_{\Re_1}^{IN}(e_\sigma)\right)^2 - \left(\mathfrak{F}_{\Re_2}^{IN}(e_\sigma)\right)^2\right|\end{array}\right)de$$

ii. By considering $\lambda = 2$, the GCBCFWNHDM would convert to continuous BCFWNEDM (CBCFWNEDM)

$$d_{BCFS}^{19}(\Re_1, \Re_2) = \left(\frac{1}{4}\int_m^n \omega(e)\left(\begin{array}{c}\left|\left(\mathfrak{F}_{\Re_1}^{RP}(e_\sigma)\right)^2 - \left(\mathfrak{F}_{\Re_2}^{RP}(e_\sigma)\right)^2\right| + \left|\left(\mathfrak{F}_{\Re_1}^{IP}(e_\sigma)\right)^2 - \left(\mathfrak{F}_{\Re_2}^{IP}(e_\sigma)\right)^2\right| \\ + \left|\left(\mathfrak{F}_{\Re_1}^{RN}(e_\sigma)\right)^2 - \left(\mathfrak{F}_{\Re_2}^{RN}(e_\sigma)\right)^2\right| + \left|\left(\mathfrak{F}_{\Re_1}^{IN}(e_\sigma)\right)^2 - \left(\mathfrak{F}_{\Re_2}^{IN}(e_\sigma)\right)^2\right|\end{array}\right)de\right)^{\frac{1}{2}}$$

By putting $\omega(e) = \frac{1}{m-n}$, for any $e \in [m, n]$, the CBCFWNEDM will convert to continuous BCFNEDM (CBCFNEDM)

$$d_{BCFS}^{20}(\Re_1, \Re_2) = \left(\frac{1}{4(m-n)}\int_m^n \left(\begin{array}{c}\left|\left(\mathfrak{F}_{\Re_1}^{RP}(e_\sigma)\right)^2 - \left(\mathfrak{F}_{\Re_2}^{RP}(e_\sigma)\right)^2\right| + \left|\left(\mathfrak{F}_{\Re_1}^{IP}(e_\sigma)\right)^2 - \left(\mathfrak{F}_{\Re_2}^{IP}(e_\sigma)\right)^2\right| \\ + \left|\left(\mathfrak{F}_{\Re_1}^{RN}(e_\sigma)\right)^2 - \left(\mathfrak{F}_{\Re_2}^{RN}(e_\sigma)\right)^2\right| + \left|\left(\mathfrak{F}_{\Re_1}^{IN}(e_\sigma)\right)^2 - \left(\mathfrak{F}_{\Re_2}^{IN}(e_\sigma)\right)^2\right|\end{array}\right)de\right)^{\frac{1}{2}}$$

Definition 4.8 Let two BCFSs $\Re_1$ and $\Re_2$ over $\mathbb{E} = [\mathrm{m}, \mathrm{n}]$. Then the GCBCFWNHDDM between $\Re_1$ and $\Re_2$ is inaugurated as.

$$d_{BCFS}^{21}(\Re_1, \Re_2) = \left(\int_m^n \omega(e)\max\left(\begin{array}{c}\left|\left(\mathfrak{F}_{\Re_1}^{RP}(e_\sigma)\right)^2 - \left(\mathfrak{F}_{\Re_2}^{RP}(e_\sigma)\right)^2\right|, \left|\left(\mathfrak{F}_{\Re_1}^{IP}(e_\sigma)\right)^2 - \left(\mathfrak{F}_{\Re_2}^{IP}(e_\sigma)\right)^2\right|, \\ \left|\left(\mathfrak{F}_{\Re_1}^{RN}(e_\sigma)\right)^2 - \left(\mathfrak{F}_{\Re_2}^{RN}(e_\sigma)\right)^2\right|, \left|\left(\mathfrak{F}_{\Re_1}^{IN}(e_\sigma)\right)^2 - \left(\mathfrak{F}_{\Re_2}^{IN}(e_\sigma)\right)^2\right|\end{array}\right)de\right)^{\frac{1}{\lambda}}, \lambda \geq 1$$

By putting $\omega(e) = \frac{1}{m-n}$, for any $e \in [m, n]$, the GCBCFWNHDDM will convert to generalized continuous bipolar complex fuzzy normalized Hausdorff distance measures (GCBCFNHDDM).

$$d_{BCFS}^{22}(\Re_1, \Re_2) = \left(\frac{1}{(m-n)}\int_m^n \max\left(\begin{array}{c}\left|\left(\mathfrak{F}_{\Re_1}^{RP}(e_\sigma)\right)^2 - \left(\mathfrak{F}_{\Re_2}^{RP}(e_\sigma)\right)^2\right|, \left|\left(\mathfrak{F}_{\Re_1}^{IP}(e_\sigma)\right)^2 - \left(\mathfrak{F}_{\Re_2}^{IP}(e_\sigma)\right)^2\right|, \\ \left|\left(\mathfrak{F}_{\Re_1}^{RN}(e_\sigma)\right)^2 - \left(\mathfrak{F}_{\Re_2}^{RN}(e_\sigma)\right)^2\right|, \left|\left(\mathfrak{F}_{\Re_1}^{IN}(e_\sigma)\right)^2 - \left(\mathfrak{F}_{\Re_2}^{IN}(e_\sigma)\right)^2\right|\end{array}\right)de\right)^{\frac{1}{\lambda}}, \lambda \geq 1$$

Particular cases

i. By considering $\lambda = 1$, the GCBCFWNHDDM would convert to continuous BCFWNHHDDM (CBCFWNHHDDM)

$$d_{BCFS}^{23}(\Re_1, \Re_2) = \int_m^n \omega(e)\max\left(\begin{array}{c}\left|\left(\mathfrak{F}_{\Re_1}^{RP}(e_\sigma)\right)^2 - \left(\mathfrak{F}_{\Re_2}^{RP}(e_\sigma)\right)^2\right|, \left|\left(\mathfrak{F}_{\Re_1}^{IP}(e_\sigma)\right)^2 - \left(\mathfrak{F}_{\Re_2}^{IP}(e_\sigma)\right)^2\right|, \\ \left|\left(\mathfrak{F}_{\Re_1}^{RN}(e_\sigma)\right)^2 - \left(\mathfrak{F}_{\Re_2}^{RN}(e_\sigma)\right)^2\right|, \left|\left(\mathfrak{F}_{\Re_1}^{IN}(e_\sigma)\right)^2 - \left(\mathfrak{F}_{\Re_2}^{IN}(e_\sigma)\right)^2\right|\end{array}\right)de$$

By putting $\omega(e) = \frac{1}{m-n}$, for any $e \in [m, n]$, the CBCFWNHHDDM will convert to continuous BCFNHHDDM (CBCFNHHDDM)

$$d_{BCFS}^{24}(\Re_1,\Re_2)=\frac{1}{(m-n)}\int_m^n \max\left(\begin{array}{l}\left|\left(\mathfrak{F}_{\Re_1}^{RP}(e_\sigma)\right)^2-\left(\mathfrak{F}_{\Re_2}^{RP}(e_\sigma)\right)^2\right|,\left|\left(\mathfrak{F}_{\Re_1}^{IP}(e_\sigma)\right)^2-\left(\mathfrak{F}_{\Re_2}^{IP}(e_\sigma)\right)^2\right|,\\ \left|\left(\mathfrak{F}_{\Re_1}^{RN}(e_\sigma)\right)^2-\left(\mathfrak{F}_{\Re_2}^{RN}(e_\sigma)\right)^2\right|,\left|\left(\mathfrak{F}_{\Re_1}^{IN}(e_\sigma)\right)^2-\left(\mathfrak{F}_{\Re_2}^{IN}(e_\sigma)\right)^2\right|\end{array}\right)de$$

ii. By considering $\lambda = 2$, the GCBCFWNHDDM would convert to continuous BCFWNEHDDM (CBCFWNEHDDM)

$$d_{BCFS}^{25}(\Re_1,\Re_2)=\left(\int_m^n \omega(e)\max\left(\begin{array}{l}\left|\left(\mathfrak{F}_{\Re_1}^{RP}(e_\sigma)\right)^2-\left(\mathfrak{F}_{\Re_2}^{RP}(e_\sigma)\right)^2\right|,\left|\left(\mathfrak{F}_{\Re_1}^{IP}(e_\sigma)\right)^2-\left(\mathfrak{F}_{\Re_2}^{IP}(e_\sigma)\right)^2\right|,\\ \left|\left(\mathfrak{F}_{\Re_1}^{RN}(e_\sigma)\right)^2-\left(\mathfrak{F}_{\Re_2}^{RN}(e_\sigma)\right)^2\right|,\left|\left(\mathfrak{F}_{\Re_1}^{IN}(e_\sigma)\right)^2-\left(\mathfrak{F}_{\Re_2}^{IN}(e_\sigma)\right)^2\right|\end{array}\right)de\right)^{\frac{1}{2}}$$

By putting $\omega(e) = \frac{1}{m-n}$, for any $e \in [m, n]$, the CBCFWNEHDDM will convert to continuous CBCFNEHDDM (CBCFNEHDDM)

$$d_{BCFS}^{26}(\Re_1,\Re_2)=\left(\frac{1}{m-n}\int_m^n \max\left(\begin{array}{l}\left|\left(\mathfrak{F}_{\Re_1}^{RP}(e_\sigma)\right)^2-\left(\mathfrak{F}_{\Re_2}^{RP}(e_\sigma)\right)^2\right|,\left|\left(\mathfrak{F}_{\Re_1}^{IP}(e_\sigma)\right)^2-\left(\mathfrak{F}_{\Re_2}^{IP}(e_\sigma)\right)^2\right|,\\ \left|\left(\mathfrak{F}_{\Re_1}^{RN}(e_\sigma)\right)^2-\left(\mathfrak{F}_{\Re_2}^{RN}(e_\sigma)\right)^2\right|,\left|\left(\mathfrak{F}_{\Re_1}^{IN}(e_\sigma)\right)^2-\left(\mathfrak{F}_{\Re_2}^{IN}(e_\sigma)\right)^2\right|\end{array}\right)de\right)^{\frac{1}{2}}$$

Definition 4.9 Let two BCFSs $\Re_1$ and $\Re_2$ over $\mathbb{E} = [\mathrm{m}, \mathrm{n}]$. Then the GCHBCFWNDM between $\Re_1$ and $\Re_2$ is inaugurated as.

$$d_{BCFS}^{27}(\Re_1,\Re_2)=\left(\frac{1}{2}\int_m^n \omega(e)\left(\begin{array}{l}\frac{1}{4}\left(\begin{array}{l}\left|\left(\mathfrak{F}_{\Re_1}^{RP}(e_\sigma)\right)^2-\left(\mathfrak{F}_{\Re_2}^{RP}(e_\sigma)\right)^2\right|+\left|\left(\mathfrak{F}_{\Re_1}^{IP}(e_\sigma)\right)^2-\left(\mathfrak{F}_{\Re_2}^{IP}(e_\sigma)\right)^2\right|\\ +\left|\left(\mathfrak{F}_{\Re_1}^{RN}(e_\sigma)\right)^2-\left(\mathfrak{F}_{\Re_2}^{RN}(e_\sigma)\right)^2\right|+\left|\left(\mathfrak{F}_{\Re_1}^{IN}(e_\sigma)\right)^2-\left(\mathfrak{F}_{\Re_2}^{IN}(e_\sigma)\right)^2\right|\end{array}\right)\\ +\max\left(\begin{array}{l}\left|\left(\mathfrak{F}_{\Re_1}^{RP}(e_\sigma)\right)^2-\left(\mathfrak{F}_{\Re_2}^{RP}(e_\sigma)\right)^2\right|,\left|\left(\mathfrak{F}_{\Re_1}^{IP}(e_\sigma)\right)^2-\left(\mathfrak{F}_{\Re_2}^{IP}(e_\sigma)\right)^2\right|\\ ,\left|\left(\mathfrak{F}_{\Re_1}^{RN}(e_\sigma)\right)^2-\left(\mathfrak{F}_{\Re_2}^{RN}(e_\sigma)\right)^2\right|,\left|\left(\mathfrak{F}_{\Re_1}^{IN}(e_\sigma)\right)^2-\left(\mathfrak{F}_{\Re_2}^{IN}(e_\sigma)\right)^2\right|\end{array}\right)\end{array}\right)de\right)^{\frac{1}{\lambda}},\lambda\geq 1$$

By putting $\omega(e) = \frac{1}{m-n}$, for any $e \in [m, n]$, the GCHBCFWNDM will convert to generalized continuous hybrid BCF normalized distance measure (GCHBCFNDM)

$$d_{BCFS}^{28}(\Re_1,\Re_2)=\left(\frac{1}{2(m-n)}\int_m^n\left(\begin{array}{l}\frac{1}{4}\left(\begin{array}{l}\left|\left(\mathfrak{F}_{\Re_1}^{RP}(e_\sigma)\right)^2-\left(\mathfrak{F}_{\Re_2}^{RP}(e_\sigma)\right)^2\right|+\left|\left(\mathfrak{F}_{\Re_1}^{IP}(e_\sigma)\right)^2-\left(\mathfrak{F}_{\Re_2}^{IP}(e_\sigma)\right)^2\right|\\ +\left|\left(\mathfrak{F}_{\Re_1}^{RN}(e_\sigma)\right)^2-\left(\mathfrak{F}_{\Re_2}^{RN}(e_\sigma)\right)^2\right|+\left|\left(\mathfrak{F}_{\Re_1}^{IN}(e_\sigma)\right)^2-\left(\mathfrak{F}_{\Re_2}^{IN}(e_\sigma)\right)^2\right|\end{array}\right)\\ +\max\left(\begin{array}{l}\left|\left(\mathfrak{F}_{\Re_1}^{RP}(e_\sigma)\right)^2-\left(\mathfrak{F}_{\Re_2}^{RP}(e_\sigma)\right)^2\right|,\left|\left(\mathfrak{F}_{\Re_1}^{IP}(e_\sigma)\right)^2-\left(\mathfrak{F}_{\Re_2}^{IP}(e_\sigma)\right)^2\right|\\ \left|,\left(\mathfrak{F}_{\Re_1}^{RN}(e_\sigma)\right)^2-\left(\mathfrak{F}_{\Re_2}^{RN}(e_\sigma)\right)^2\right|,\left|\left(\mathfrak{F}_{\Re_1}^{IN}(e_\sigma)\right)^2-\left(\mathfrak{F}_{\Re_2}^{IN}(e_\sigma)\right)^2\right|\end{array}\right)\end{array}\right)de\right)^{\frac{1}{\lambda}},\lambda\geq 1$$

4.4 VIKOR Approach Under BCF Information

In order to get a compromise solution relying on a specific measure, the VIKOR approach emphases on ranking and selecting among a gathering of alternatives (options) with competing criteria. The compromise ranking can be exhibited by comparing how closely each alternative resembles the ideal alternative, presuming that each alternative is assessed in accordance with each criterion function. The compromise solution is a workable solution that comes the closest to the ideal, and in this context, "compromise" refers to an agreement made by mutual concessions. The compromise solution was deduced by Zeleny [17] and Yu [18]. The $\mathbb{L}_{\mathrm{p}}$ metric, which is employed as an aggregating function in a compromise programming technique for emerging multi-criteria measure for compromise ranking is interpreted as underneath

$$\mathbb{L}_{\mathrm{p},\sigma} = \left[\sum_{\varrho=1}^{\mathrm{m}} \omega_{\varrho} \frac{\mathrm{d}\left(\mathfrak{R}_{\varrho}^{+}, \mathfrak{R}_{\sigma\varrho}\right)}{\mathrm{d}\left(\mathfrak{R}_{\varrho}^{+}, \mathfrak{R}_{\varrho}^{-}\right)}\right]^{\frac{1}{\mathrm{p}}} \tag{4.1}$$

Observed that $\mathbb{L}_{\infty,\sigma}$ signifies the minimum individual guilt of the rival and $\mathbb{L}_{\mathrm{p},\sigma}$ signifies the maximum group utility. The VIKOR technique has been applied in several sectors recently, either alone or in combination with other MCDM methods and extended with other system theories, due to its potential advantages in compromise solution-based ranking. The technique of BCF-VIKOR is the modification of VIKOR that identifies the compromise solution in awkward and complicated DM dilemmas.

Let a gathering of n alternatives (options) $\mathfrak{R}_{\mathbb{A}} = \{\mathfrak{R}_{\mathbb{A}-1}, \mathfrak{R}_{\mathbb{A}-2}, \ldots, \mathfrak{R}_{\mathbb{A}-\mathrm{n}}\}$ along with a gathering of m criteria $\mathfrak{A}_{\mathbb{C}} = \{\mathfrak{A}_{\mathbb{C}\approx-1}, \mathfrak{A}_{\mathbb{C}\approx-2}, \ldots, \mathfrak{A}_{\mathbb{C}\approx-m}\}$. Also let $\omega = (\omega_{1,}\omega_2, \ldots, \omega_m)^T$ be a weight vector for the attributes and $0 \leq \omega_{\varrho} \leq 1$ for all ϱ with $\sum_{\varrho=1}^{m} \omega_{\varrho} = 1$ established by the decision-analyst or expert by his/her own choice. The assessment values of each alternative examined by decision-analyst or expert by relying on the considered criteria are in the model of BCFS. The following steps make up the BCF-VIKOR methodology.

Step 1: The evaluated values of decision-analyst or expert would develop a BCF decision-matrix.

$$\mathfrak{D} = \left[\mathfrak{R}_{\sigma\varrho}\right]_{\mathrm{n}\times\mathrm{m}} = \begin{bmatrix} \mathfrak{R}_{11} & \mathfrak{R}_{12} & \cdots & \mathfrak{R}_{1\mathrm{m}} \\ \mathfrak{R}_{21} & \mathfrak{R}_{22} & \cdots & \mathfrak{R}_{2\mathrm{m}} \\ \vdots & \vdots & \ddots & \vdots \\ \mathfrak{R}_{\mathrm{n}1} & \mathfrak{R}_{\mathrm{n}2} & \cdots & \mathfrak{R}_{\mathrm{nm}} \end{bmatrix}$$

$$= \begin{bmatrix} \begin{pmatrix} \mathfrak{F}_{\mathfrak{R}_{11}}^{\mathrm{RP}} + \iota\mathfrak{F}_{\mathfrak{R}_{11}}^{\mathrm{IP}}, \\ \mathfrak{F}_{\mathfrak{R}_{11}}^{\mathrm{RN}} + \iota\mathfrak{F}_{\mathfrak{R}_{11}}^{\mathrm{IN}} \end{pmatrix} & \begin{pmatrix} \mathfrak{F}_{\mathfrak{R}_{12}}^{\mathrm{RP}} + \iota\mathfrak{F}_{\mathfrak{R}_{12}}^{\mathrm{IP}}, \\ \mathfrak{F}_{\mathfrak{R}_{12}}^{\mathrm{RN}} + \iota\mathfrak{F}_{\mathfrak{R}_{12}}^{\mathrm{IN}} \end{pmatrix} & \cdots & \begin{pmatrix} \mathfrak{F}_{\mathfrak{R}_{1\mathrm{m}}}^{\mathrm{RP}} + \iota\mathfrak{F}_{\mathfrak{R}_{1\mathrm{m}}}^{\mathrm{IP}}, \\ \mathfrak{F}_{\mathfrak{R}_{1\mathrm{m}}}^{\mathrm{RN}} + \iota\mathfrak{F}_{\mathfrak{R}_{1\mathrm{m}}}^{\mathrm{IN}} \end{pmatrix} \\ \begin{pmatrix} \mathfrak{F}_{\mathfrak{R}_{21}}^{\mathrm{RP}} + \iota\mathfrak{F}_{\mathfrak{R}_{21}}^{\mathrm{IP}}, \\ \mathfrak{F}_{\mathfrak{R}_{21}}^{\mathrm{RN}} + \iota\mathfrak{F}_{\mathfrak{R}_{21}}^{\mathrm{IN}} \end{pmatrix} & \begin{pmatrix} \mathfrak{F}_{\mathfrak{R}_{21}}^{\mathrm{RP}} + \iota\mathfrak{F}_{\mathfrak{R}_{21}}^{\mathrm{IP}}, \\ \mathfrak{F}_{\mathfrak{R}_{21}}^{\mathrm{RN}} + \iota\mathfrak{F}_{\mathfrak{R}_{21}}^{\mathrm{IN}} \end{pmatrix} & \cdots & \begin{pmatrix} \mathfrak{F}_{\mathfrak{R}_{2\mathrm{m}}}^{\mathrm{RP}} + \iota\mathfrak{F}_{\mathfrak{R}_{2\mathrm{m}}}^{\mathrm{IP}}, \\ \mathfrak{F}_{\mathfrak{R}_{2\mathrm{m}}}^{\mathrm{RN}} + \iota\mathfrak{F}_{\mathfrak{R}_{2\mathrm{m}}}^{\mathrm{IN}} \end{pmatrix} \\ \vdots & \vdots & \ddots & \vdots \\ \begin{pmatrix} \mathfrak{F}_{\mathfrak{R}_{\mathrm{n}1}}^{\mathrm{RP}} + \iota\mathfrak{F}_{\mathfrak{R}_{\mathrm{n}1}}^{\mathrm{IP}}, \\ \mathfrak{F}_{\mathfrak{R}_{\mathrm{n}1}}^{\mathrm{RN}} + \iota\mathfrak{F}_{\mathfrak{R}_{\mathrm{n}1}}^{\mathrm{IN}} \end{pmatrix} & \begin{pmatrix} \mathfrak{F}_{\mathfrak{R}_{\mathrm{n}2}}^{\mathrm{RP}} + \iota\mathfrak{F}_{\mathfrak{R}_{\mathrm{n}2}}^{\mathrm{IP}}, \\ \mathfrak{F}_{\mathfrak{R}_{\mathrm{n}2}}^{\mathrm{RN}} + \iota\mathfrak{F}_{\mathfrak{R}_{\mathrm{n}2}}^{\mathrm{IN}} \end{pmatrix} & \cdots & \begin{pmatrix} \mathfrak{F}_{\mathfrak{R}_{\mathrm{nm}}}^{\mathrm{RP}} + \iota\mathfrak{F}_{\mathfrak{R}_{\mathrm{nm}}}^{\mathrm{IP}}, \\ \mathfrak{F}_{\mathfrak{R}_{\mathrm{nm}}}^{\mathrm{RN}} + \iota\mathfrak{F}_{\mathfrak{R}_{\mathrm{nm}}}^{\mathrm{IN}} \end{pmatrix} \end{bmatrix}$$

Observed that $\mathfrak{R}_{\sigma\varrho} = \left(\mathfrak{F}^{\mathrm{P}}_{\mathfrak{R}_{\sigma\varrho}}, \mathfrak{F}^{\mathrm{N}}_{\mathfrak{R}_{\sigma\varrho}}\right) = \left(\mathfrak{F}^{\mathrm{RP}}_{\mathfrak{R}_{\sigma\varrho}} + \iota\mathfrak{F}^{\mathrm{IP}}_{\mathfrak{R}_{\sigma\varrho}}, \mathfrak{F}^{\mathrm{RN}}_{\mathfrak{R}_{\sigma\varrho}} + \iota\mathfrak{F}^{\mathrm{IN}}_{\mathfrak{R}_{\sigma\varrho}}\right)$, $\mathfrak{F}^{\mathrm{P}}_{\mathfrak{R}_{\sigma\varrho}}$ implies positive belonging grade and $\mathfrak{F}^{\mathrm{N}}_{\mathfrak{R}_{\sigma\varrho}}$ implies negative belonging grade.

Step 2: Achieve the finest value $\mathfrak{R}^{+}_{\varrho}$ and poorest value $\mathfrak{R}^{+}_{\varrho}$ of every criterion $\mathfrak{A}_{\mathbb{C}\approx-\varrho}$ by underneath formulas

$$\mathfrak{R}^{+}_{\varrho} = \begin{cases} \left(\max \mathfrak{F}^{\mathrm{RP}}_{\mathfrak{R}_{\sigma\varrho}} + \iota \max \mathfrak{F}^{\mathrm{IP}}_{\mathfrak{R}_{\sigma\varrho}}, \max \mathfrak{F}^{\mathrm{RN}}_{\mathfrak{R}_{\sigma\varrho}} + \iota \max \mathfrak{F}^{\mathrm{IN}}_{\mathfrak{R}_{\sigma\varrho}}\right) \mathit{forbenefitcriteria} \\ \left(\min \mathfrak{F}^{\mathrm{RP}}_{\mathfrak{R}_{\sigma\varrho}} + \iota \min \mathfrak{F}^{\mathrm{IP}}_{\mathfrak{R}_{\sigma\varrho}}, \min \mathfrak{F}^{\mathrm{RN}}_{\mathfrak{R}_{\sigma\varrho}} + \iota \min \mathfrak{F}^{\mathrm{IN}}_{\mathfrak{R}_{\sigma\varrho}}\right) \mathit{forcostcriteria} \end{cases}$$

$$\mathfrak{R}^{-}_{\varrho} = \begin{cases} \left(\min \mathfrak{F}^{\mathrm{RP}}_{\mathfrak{R}_{\sigma\varrho}} + \iota \min \mathfrak{F}^{\mathrm{IP}}_{\mathfrak{R}_{\sigma\varrho}}, \min \mathfrak{F}^{\mathrm{RN}}_{\mathfrak{R}_{\sigma\varrho}} + \iota \min \mathfrak{F}^{\mathrm{IN}}_{\mathfrak{R}_{\sigma\varrho}}\right) \mathit{forbenefitcriteria} \\ \left(\max \mathfrak{F}^{\mathrm{RP}}_{\mathfrak{R}_{\sigma\varrho}} + \iota \max \mathfrak{F}^{\mathrm{IP}}_{\mathfrak{R}_{\sigma\varrho}}, \max \mathfrak{F}^{\mathrm{RN}}_{\mathfrak{R}_{\sigma\varrho}} + \iota \max \mathfrak{F}^{\mathrm{IN}}_{\mathfrak{R}_{\sigma\varrho}}\right) \mathit{forcostcriteria} \end{cases}$$

where $\sigma = 1, 2, .., n$.

Step 3: For developing ranking measures in VIKOR technique, $\mathfrak{S}_{\sigma}$ (as $\mathbb{L}_{\mathrm{p},\sigma}$ in Eq. (4.1)) and $\mathfrak{T}_{\sigma}$ (as $\mathbb{L}_{\infty,\sigma}$ in Eq.(4.1)) would be employed. To interpret $\mathfrak{S}_{\sigma}$ and $\mathfrak{T}_{\sigma}$, we have the underneath formulas

$$\mathfrak{S}_{\sigma} = \sum_{\varrho=1}^{\mathrm{m}} \omega_{\varrho} \frac{\mathrm{d}\left(\mathfrak{R}^{+}_{\varrho}, \mathfrak{R}_{\sigma\varrho}\right)}{\mathrm{d}\left(\mathfrak{R}^{+}_{\varrho}, \mathfrak{R}^{-}_{\varrho}\right)}$$

And

$$\mathfrak{T}_{\sigma} = \max_{\varrho} \left[\omega_{\varrho} \frac{\mathrm{d}\left(\mathfrak{R}^{+}_{\varrho}, \mathfrak{R}_{\sigma\varrho}\right)}{\mathrm{d}\left(\mathfrak{R}^{+}_{\varrho}, \mathfrak{R}^{-}_{\varrho}\right)} \right]$$

Observe that d can be any distance measure in the above inaugurated measures.

Step 4: Deduce the grade of relative closeness of every alternative in terms of $\mathfrak{S}_{\sigma}$ and $\mathfrak{T}_{\sigma}$ as underneath

$$\mathcal{Q}_{\sigma} = \frac{\mathfrak{h}(\mathfrak{S}_{\sigma} - \mathfrak{S}^{+})}{(\mathfrak{S}^{-} - \mathfrak{S}^{+})} + \frac{(1 - \mathfrak{h})(\mathfrak{T}_{\sigma} - \mathfrak{T}^{+})}{(\mathfrak{T}^{-} - \mathfrak{T}^{+})}$$

Observed that

$$\mathfrak{S}^{+} = \min_{\sigma} \mathfrak{S}_{\sigma}; \quad \mathfrak{S}^{-} = \max_{\sigma} \mathfrak{S}_{\sigma}$$

$$\mathfrak{T}^{+} = \min_{\sigma} \mathfrak{T}_{\sigma}; \quad \mathfrak{T}^{-} = \max_{\sigma} \mathfrak{T}_{\sigma}$$

and $\mathfrak{h}$ and $(1 - \mathfrak{h})$ are signified as coefficients which identify the weightage of $\mathfrak{S}_{\sigma}$ and $\mathfrak{T}_{\sigma}$. The compromise solution can be chosen by majority ($\mathfrak{h} > 0.5$), consent ($\mathfrak{h} = 0.5$), or refusal ($\mathfrak{h} < 0.5$).

Step 5: For ordering the alternatives, organized the values of$\mathfrak{S}_{\sigma}$, $\mathfrak{T}_{\sigma}$ and $\mathcal{Q}_{\sigma}$ in ascending order. The finest alternative would be the one with the minimum of $\mathcal{Q}$ and that holds the underneath conditions.

CD1: (Acceptable advantage): If

$$\mathcal{Q}(\mathfrak{R}_{\mathbb{A}-2}) - \mathcal{Q}(\mathfrak{R}_{\mathbb{A}-1}) \geq \frac{1}{\mathrm{n}-1}$$

Observed that $\mathfrak{R}_{\mathbb{A}-1}$ and $\mathfrak{R}_{\mathbb{A}-2}$ are the top two alternatives in $\mathcal{Q}_{\sigma}$ column.

CD2: (Acceptable stability): In both $\mathfrak{S}_{\sigma}$ and $\mathfrak{T}_{\sigma}$ columns the alternative $\mathfrak{R}_{\mathbb{A}-1}$ should be ranked first.

If conditions CD1 and CD2 holds at a time, then the alternative $\mathfrak{R}_{\mathbb{A}-1}$ would be the finest alternative. If any of both conditions not hold at a time, then we will go for compromise solution as.

i. In case of the failure of condition CD2, the set $(\mathfrak{R}_{\mathbb{A}-1}, \mathfrak{R}_{\mathbb{A}-2})$ would be the compromise solution.
ii. In case of the failure of condition CD1, the gathering $(\mathfrak{R}_{\mathbb{A}-1}, \mathfrak{R}_{\mathbb{A}-2}, \ldots, \mathfrak{R}_{\mathbb{A}-O})$ would be the compromise solution, where O would be achieved by the underneath relation

$$\mathcal{Q}(\mathfrak{R}_{\mathbb{A}-O}) - \mathcal{Q}(\mathfrak{R}_{\mathbb{A}-1}) < \frac{1}{\mathrm{n}-1}$$

And O signifies the highest-ranking order of alternatives in column $\mathcal{Q}_{\sigma}$.

4.5 Application

Automated trading systems, commonly referred to as algorithmic trading systems or simply algo-trading, are computer programs that automatically carry out trades in financial markets using predetermined rules and statistical models. These systems are made to assess enormous volumes of data, such as market conditions, price changes, and other pertinent variables, and then decide which trades to make based on established strategies or algorithms. Automated trading systems aim to streamline the trading process and capitalize on market opportunities quickly and effectively by removing the need for manual intervention. The value of automated trading systems is found in their capacity to get beyond constraints imposed by human trading. Human decision-making can frequently be influenced by emotions like fear and greed, which can result in impulsive or unreasonable trading decisions. On the other side, automated systems only follow predetermined rules and algorithms, which removes emotional biases from the trading equation. This can aid in reducing irrational trading decisions and enhancing general trading discipline. Another essential area where automated trading systems excel is speed. They are able to recognize and take advantage of market opportunities far more quickly than human traders since these computers can examine and analyze enormous volumes of data in real-time. They can complete deals in milliseconds or even microseconds, which is practically unachievable for human traders. Profitability and efficiency can be increased by having the capacity to take advantage of transient market situations and minor price differences.

The benefit of back testing is another benefit of automated trading systems. Traders can test their algorithms using historical data to evaluate their success before deploying them in the real market. This enables traders to tweak parameters, fine-tune their strategies, and assess risk and return measures. Trading professionals can make wise selections by simulating trading situations based on historical data. This gives them vital insights into the strength and success of their methods. Another area where automated trading systems are essential is risk management. These systems can be configured to incorporate risk management procedures, such stop-loss orders or position sizing guidelines, which can help to prevent losses and safeguard capital. Automated trading systems can assist in maintaining trading discipline and reducing the effects of unfavorable market conditions by consistently executing established risk management rules. Additionally, automated trading systems give traders access to markets around-the-clock. Even when the trader is not actively monitoring the markets, they can scan for trading opportunities, monitor numerous markets at once, and place trades. Increased market coverage and the opportunity to exploit opportunities that may arise outside of typical trading hours are made possible by this round-the-clock accessibility. To sum up, automated trading systems have several benefits for investors and traders. They remove psychological biases, run quickly, permit rigorous back testing, improve risk management, and offer continuous market monitoring. These systems seek to maximize trade efficiency, execution speed, and even total profitability by utilizing technology and automation.

In the underneath numerical example, we would prioritize the strategies of automated trading system and determine the finest strategy by employing the initiated distance measures of BCFS and technique of BCF-VIKOR.

4.5.1 Numerical Example

Let's consider an Alpha trading corporation, a fictitious organization that specializes in automated trading systems. For their new automated trading system, Alpha trading aims to pick the optimal strategy. For this they initially shortlisted 5 strategies of automated trading system, which is briefly described as: $\Re_{\mathbb{A}-1}$: **Trend Following**: Trend following strategy entails spotting and trading in line with the current market trend. It establishes the trend and places trades in accordance with it using technical indicators, such as moving averages. $\Re_{\mathbb{A}-2}$: **Mean Reversion**: This strategy is predicated on the idea that over time, prices will return to their mean or average. It searches for instances where prices have significantly deviated from the mean and enters positions in the hope that they will turn around. $\Re_{\mathbb{A}-3}$: **News-Based Trading**: Reacting to breaking news and the publication of economic data is part of this strategy.

Automated systems can be set up to quickly evaluate news feeds and place trades in accordance with predetermined criteria, such as certain keywords or substantial shifts in sentiment. $\mathfrak{R}_{\mathbb{A}-4}$: **Arbitrage**: Through the use of price differences across several markets or financial instruments, arbitrage strategy aims to make money. For the purpose of profiting from the momentary mispricing, the system detects price discrepancies and performs trades concurrently. $\mathfrak{R}_{\mathbb{A}-5}$: **Breakout**: Exploiting big price changes or breakouts from established trading ranges is the goal of breakout strategies. The strategy places trades in the direction of the breakout when prices cross important lines of support or resistance. For the assessment of these strategies the Alpha trading hire an expert who assesses these strategies in the light of 4 various attributes that is $\mathfrak{A}_{\mathbb{C}\approx-1}$: Speed and efficiency, $\mathfrak{A}_{\mathbb{C}\approx-2}$: Risk management, $\mathfrak{A}_{\mathbb{C}\approx-3}$: Reduced human error, $\mathfrak{A}_{\mathbb{C}\approx-4}$: Continuous operation. To rank these strategies, we have the underneath step.

Step 1: The evaluation values of the strategies are in the model of BCFS and demonstrated in Table 4.1.

Step 2: As all criterion are benefit sorts so by utilizing Eq. (4.1), we achieved the finest value $\mathfrak{R}^+_{\varrho}$ and poorest value $\mathfrak{R}^+_{\varrho}$ of every criterion $\mathfrak{A}_{\mathbb{C}\approx-\varrho}$

$$\mathfrak{R}^+_1 = (0.761 + \iota 0.643, -0.134 - \iota 0.237);\ \mathfrak{R}^+_2 = (0.872 + \iota 0.912, -0.187 - \iota 0.135);$$

$$\mathfrak{R}^+_3 = (0.735 + \iota 0.863, -0.247 - \iota 0.196);\ \mathfrak{R}^+_4 = (0.975 + \iota 0.923, -0.153 - \iota 0.345)$$

And

$$\mathfrak{R}^-_1 = (0.176 + \iota 0.262, -0.385 - \iota 0.945);\ \mathfrak{R}^-_2 = (0.242 + \iota 0.123, -0.571 - \iota 0.656),$$

$$\mathfrak{R}^-_3 = (0.198 + \iota 0.231, -0.856 - \iota 0.974);\ \mathfrak{R}^-_4 = (0.398 + \iota 0.461, -0.475 - \iota 0.945)$$

Table 4.1 The evaluation values of the strategies

	$\mathfrak{A}_{\mathbb{C}\approx-1}$	$\mathfrak{A}_{\mathbb{C}\approx-2}$	$\mathfrak{A}_{\mathbb{C}\approx-3}$	$\mathfrak{A}_{\mathbb{C}\approx-4}$
$\mathfrak{R}_{\mathbb{A}-1}$	$\begin{pmatrix}0.485 + \iota 0.576,\\ -0.385 - \iota 0.237\end{pmatrix}$	$\begin{pmatrix}0.242 + \iota 0.445,\\ -0.571 - \iota 0.634\end{pmatrix}$	$\begin{pmatrix}0.735 + \iota 0.463,\\ -0.856 - \iota 0.573\end{pmatrix}$	$\begin{pmatrix}0.843 + \iota 0.461,\\ -0.475 - \iota 0.945\end{pmatrix}$
$\mathfrak{R}_{\mathbb{A}-2}$	$\begin{pmatrix}0.745 + \iota 0.295,\\ -0.134 - \iota 0.945\end{pmatrix}$	$\begin{pmatrix}0453. + \iota 0.123,\\ -0.865 - \iota 0.534\end{pmatrix}$	$\begin{pmatrix}0.642 + \iota 0.551,\\ -0.446 - \iota 0.362\end{pmatrix}$	$\begin{pmatrix}0.657 + \iota 0.462,\\ -0.266 - \iota 0.762\end{pmatrix}$
$\mathfrak{R}_{\mathbb{A}-3}$	$\begin{pmatrix}0.176 + \iota 0.643,\\ -0.262 - \iota 0.24\end{pmatrix}$	$\begin{pmatrix}0.872 + \iota 0.912,\\ -0.563 - \iota 0.135\end{pmatrix}$	$\begin{pmatrix}0.637 + \iota 0.231,\\ -0.247 - \iota 0.196\end{pmatrix}$	$\begin{pmatrix}0.398 + \iota 0.761,\\ -0.254 - \iota 0.655\end{pmatrix}$
$\mathfrak{R}_{\mathbb{A}-4}$	$\begin{pmatrix}0.761 + \iota 0.263,\\ -0.322 - \iota 0.625\end{pmatrix}$	$\begin{pmatrix}0.455 + \iota 0.572,\\ -0.563 - \iota 0.656\end{pmatrix}$	$\begin{pmatrix}0.275 + \iota 0.498,\\ -0.285 - \iota 0.974\end{pmatrix}$	$\begin{pmatrix}0.664 + \iota 0.764,\\ -0.153 - \iota 0.463\end{pmatrix}$
$\mathfrak{R}_{\mathbb{A}-5}$	$\begin{pmatrix}0.452 + \iota 0.262,\\ -0.137 - \iota 0.863\end{pmatrix}$	$\begin{pmatrix}0.273 + \iota 0.783,\\ -0.187 - \iota 0.254\end{pmatrix}$	$\begin{pmatrix}0.198 + \iota 0.863,\\ -0.275 - \iota 0.486\end{pmatrix}$	$\begin{pmatrix}0.975 + \iota 0.923,\\ -0.235 - \iota 0.345\end{pmatrix}$

Step 3: The values of $\mathfrak{S}_\sigma$ and $\mathfrak{T}_\sigma, \sigma = 1, 2, .., 5$ are as follows with the help of the weight $(0.24, 0.32, 0.26, 0.18)$ which is given by the expert because of the various importance of the attributes.

$$\mathfrak{S}_1 = 0.846;\ \mathfrak{S}_2 = 0.864;\ \mathfrak{S}_3 = 0.65;\ \mathfrak{S}_4 = 0.806;\ \mathfrak{S}_5 = 0.668$$

$$\mathfrak{T}_1 = 0.31;\ \mathfrak{T}_3 = 0.326;\ \mathfrak{T}_3 = 0.174;\ \mathfrak{T}_4 = 0.27;\ \ \mathfrak{T}_5 = 0.241$$

Here we utilized GBCFNHDM for finding the distance with $\lambda = 3$.

Step 4: The deduced grade of relative closeness of every alternative in terms of $\mathfrak{S}_\sigma$ and $\mathfrak{T}_\sigma$ are

$$\mathcal{Q}_1 = 0.905;\ \mathcal{Q}_2 = 1;\ \mathcal{Q}_3 = 0;\ \mathcal{Q}_4 = 0.682;\ \mathcal{Q}_5 = 0.263$$

where $\mathfrak{h} = 0.5$.

Step 5: The values of $\mathfrak{S}_\sigma$, $\mathfrak{T}_\sigma$ and $\mathcal{Q}_\sigma$ are interpreted together in Table 4.2 and their ranking is interpreted in Table 4.3.

The top two alternatives are $\mathfrak{R}_{\mathbb{A}-2}$ and $\mathfrak{R}_{\mathbb{A}-1}$ respectively in row of $\mathcal{Q}_\sigma$ and $\mathcal{Q}(\mathfrak{R}_{\mathbb{A}-2}) - \mathcal{Q}(\mathfrak{R}_{\mathbb{A}-1}) = 0.263 - 0 = 0.263 > \frac{1}{5-1} = 0.25$. Hence, CD1 is satisfied. Further, from Table (4.2) and (4.3), it is obvious that $\mathfrak{R}_{\mathbb{A}-4}$ is first in $\mathfrak{S}_\sigma$ and $\mathfrak{T}_\sigma$ rows. Thus, CD2 is also satisfied. $\mathfrak{R}_{\mathbb{A}-4}$that is arbitrage is the optimal strategy of automated trading system. The Alpha trading cooperation would like to get this arbitrage strategy.

Table 4.2 The values of $\mathfrak{S}_\sigma$, $\mathfrak{T}_\sigma$ and $\mathcal{Q}_\sigma$

	$\mathfrak{R}_{\mathbb{A}-1}$	$\mathfrak{R}_{\mathbb{A}-2}$	$\mathfrak{R}_{\mathbb{A}-3}$	$\mathfrak{R}_{\mathbb{A}-4}$	$\mathfrak{R}_{\mathbb{A}-5}$
$\mathfrak{S}_\sigma$	0.846	0.864	*Ist*	0.806	0.668
$\mathfrak{T}_\sigma$	0.31	0.326	*Ist*	0.27	0.241
$\mathcal{Q}_\sigma$	0.905	1	0	0.682	0.263

Table 4.3 The ranking of strategies

	$\mathfrak{R}_{\mathbb{A}-1}$	$\mathfrak{R}_{\mathbb{A}-2}$	$\mathfrak{R}_{\mathbb{A}-3}$	$\mathfrak{R}_{\mathbb{A}-4}$	$\mathfrak{R}_{\mathbb{A}-5}$
$\mathfrak{S}_\sigma$	*4th*	*5th*	*Ist*	*3rd*	*2nd*
$\mathfrak{T}_\sigma$	*4th*	*5th*	*Ist*	*3rd*	*2nd*
$\mathcal{Q}_\sigma$	*4th*	*5th*	*Ist*	*3rd*	*2nd*

4.6 Comparative Analysis

Let's compare the fuzzy VIKOR approach created by Wang et al. [19] and the bipolar fuzzy VIKOR technique established by Alsolame and Alshehri [20] with the initiated BCF-VIKOR approach. Because the fuzzy structure only has a belonging grade and is incapable of handling a second dimension or negative aspects, the fuzzy VIKOR technique is only able to deal with fuzzy information. It cannot handle any generalization of the FS theory. The bipolar fuzzy VIKOR approach, which is an extension of fuzzy VIKOR technique, only deals with information that comprises both positive and negative aspects of an element. However, the disadvantage of the bipolar fuzzy VIKOR technique is that it is unable to handle information with a second dimension. Both bipolar fuzzy VIKOR and fuzzy VIKOR are extensions of the interpreted BCF-VIKOR technique. The BCF-VIKOR technique has the capacity to manage second dimension as well as both positive and negative characteristics. The BCF-VIKOR approach can reduce to fuzzy, bipolar fuzzy, and complex fuzzy information to handle data in the structure of FS, BFS and CFS.

The initiated distance measures for BCFS can also be reduces to the model of FS, BFS and CFS as follows.

- By neglecting the unreal part in the both positive and negative belonging grades, the initiated distance measures would convert in the theory of BFS.
- By neglecting the negative belonging grade, the initiated distance measures would convert in the theory of CFS.
- By neglecting the unreal part in the positive belonging grade and neglecting a negative belonging grade, the initiated distance measures would convert in the theory of FS.

Further, form the literature, we know that in the model of BFS and CFS, there are not such distance measures so the initiated work is also useful for bipolar and complex fuzzy information.

Chapter 5
Similarity Measures of Bipolar Complex Fuzzy Sets and Their Application to Pattern Recognition and Medical Diagnosis

The degree of similarity between two objects can be measured mathematically. This is accomplished by using similarity metrics, also referred to as similarity measures. In several fields, such as pattern recognition, machine learning, data mining, and information retrieval, they are crucial. Comparing or relating two entities based on their shared characteristics is the aim of similarity measures. These measures of similarity provide a way to contrast and compare objects, supporting effective analysis and decision-making in a variety of applications. In order to compare and assess how similar patterns or objects are, similarity measures are crucial. These measures assess how similar or distinct the data points are in order to find patterns and make informed decisions. The cosine similarity, Manhattan distance, Euclidean distance, Hamming, Hausdorff, Jaccard similarity, and exponential similarity are examples of common similarity measures.

The discovery and interpretation of recurrent structures or regularities within data, information, or sensory input constitutes pattern recognition, a fundamental cognitive activity. It is the capacity of the human brain and computer systems to recognize meaningful patterns in a wide range of formats, including pictures, sounds, numbers, and text. This mechanism is important in many disciplines, including neuroscience, machine learning, and artificial intelligence. In order to extract pertinent characteristics from the input, classify or categorize it, and then make predictions or judgments based on the detected patterns, pattern recognition uses algorithms and procedures. This may be used in many different applications, including financial forecasting, speech recognition, and facial recognition. Using pattern recognition can help us better grasp complicated data and make sense of the world around us. For tasks like picture and speech recognition, similarity measures are essential because they quantify the similarity of patterns, allowing the detection of significant relationships and the classification of items based on their similarity.

A medical diagnosis is a critical step in the healthcare process when a healthcare expert, usually a doctor, use a variety of techniques to determine the primary reason of a patient's symptoms or health condition. This includes a thorough assessment of the

U. ur Rehman and T. Mahmood, *Bipolar Complex Fuzzy Sets*, Studies in Fuzziness and Soft Computing 439, https://doi.org/10.1007/978-3-032-07388-4_5

patient's medical history, a physical examination, and frequently, further diagnostic testing such blood tests, imaging investigations, or biopsies. This procedure aims to precisely identify the particular illness, dysfunction, or condition that the patient is experiencing. The development of an effective treatment plan depends on this information thus it is crucial. Large-scale medical expertise, clinical expertise, and availability of cutting-edge medical technology all serve as guides for the diagnosis process. It necessitates that the medical professional take into account a wide variety of variables, such as symptoms, risk factors, family history, and the findings of diagnostic testing. Making an accurate diagnosis is essential to ensure that patients get the best, most individualized care possible for their unique medical condition. In order to accurately diagnose diseases, similarity metrics in medicine are essential since they assess how closely patient data resembles known patterns. By comparing symptoms objectively, they improve accuracy and allow for prompt and individualized treatment choices.

In Sect. 5.1. of this chapter, we inaugurate the similarity measures (SMs) for BCFS based on the generalized BCF normalized Hamming distance measure, generalized BCF normalized Hausdorff distance measure, and generalized hybrid BCF normalized distance measure and on their special cases. In Sects. 5.2 and 5.3. we investigate the weighted and continuous form of the SMs initiated in Sect. 5.1. In Sect. 5.4. of this chapter, we interpret exponential and weighted exponential SMs for BCFS. Section 5.5. contains the applications of the inaugurated SMs that is pattern recognition and medical diagnosis. In Sect. 5.6. we demonstrate the comparative analysis.

5.1 Similarity Measures Based on the Developed Distance Measures

We will inaugurate the SMs from the distance measure investigated in Chap. 4 such as generalized BCF normalized Hamming SM (GBCFNHSM), generalized BCF normalized Hausdorff SM (GBCFNHDSM), generalized hybrid BCF normalized SM (GHBCFNSM) and will discuss their special cases.

The underneath Theorem demonstrate the relation between distance measures and similarity measures.

Theorem 5.1 Assume that $d(\Re_1, \Re_2)$ is a distance measure between two BCFS $\Re_1$ and $\Re_2$, then $\mathcal{S}(\Re_1, \Re_2) = 1 - d(\Re_1, \Re_2)$ is a SM of BCFS.

Proof For proving this, we have to prove the 4 properties of SM.

1. As we have that $0 \leq d(\Re_1, \Re_2) \leq 1$, then obviously $0 \leq \mathcal{S}(\Re_1, \Re_2) \leq 1$.
2. Assume that $\mathcal{S}(\Re_1, \Re_2) = 1$ and to show that $\Re_1 = \Re_2$. As $\mathcal{S}(\Re_1, \Re_2) = 1$,this implies that $d(\Re_1, \Re_2) = 0$, then by the property of distance measure we have that $\Re_1 = \Re_2$.

Conversely, assume that $\Re_1 = \Re_2$, then to prove that $\mathcal{S}(\Re_1, \Re_2) = 1$. As $\Re_1 = \Re_2$, then $d(\Re_1, \Re_2) = 0$ and this implies that $\mathcal{S}(\Re_1, \Re_2) = 1$.

3. As $d(\Re_1, \Re_2) = d(\Re_2, \Re_1)$, thus $\mathcal{S}(\Re_1, \Re_2) = \mathcal{S}(\Re_2, \Re_1)$.
4. Let $\Re_1 \subset \Re_2 \subset \Re_3$, then we have that $d(\Re_1, \Re_2) \leq d(\Re_1, \Re_3)$ and $d(\Re_2, \Re_3) \leq d(\Re_1, \Re_3)$. Now as

$$\mathcal{S}(\Re_1, \Re_2) = 1 - d(\Re_1, \Re_2) \geq 1 - d(\Re_1, \Re_3) = \mathcal{S}(\Re_1, \Re_3) \Rightarrow \mathcal{S}(\Re_1, \Re_2) \geq \mathcal{S}(\Re_1, \Re_3)$$

And

$$\mathcal{S}(\Re_2, \Re_3) = 1 - d(\Re_2, \Re_3) \geq 1 - d(\Re_1, \Re_3) = \mathcal{S}(\Re_1, \Re_3) \Rightarrow \mathcal{S}(\Re_2, \Re_3) \geq \mathcal{S}(\Re_1, \Re_3)$$

Next, we would employ the above Theorem to inaugurate the SMs from the distance measures investigated in Chap. 4.

Definition 5.1 Let two BCFSs $\Re_1 = \left\{e_\sigma, \left(\mathfrak{F}^P_{\Re_1}(e_\sigma), \mathfrak{F}^N_{\Re_1}(e_\sigma)\right) | e_\sigma \in \mathbb{E}, \sigma = 1, 2, .., n\right\} = \left\{e_\sigma, \left(\mathfrak{F}^{RP}_{\Re_1}(e_\sigma) + \iota\mathfrak{F}^{IP}_{\Re_1}(e_\sigma), \mathfrak{F}^{RN}_{\Re_1}(e_\sigma) + \iota\mathfrak{F}^{IN}_{\Re_1}(e_\sigma)\right) | e_\sigma \in \mathbb{E},\ \sigma = 1, 2, .., n\right\}$ and $\Re_2 = \left\{e, \left(\mathfrak{F}^P_{\Re_2}(e_\sigma), \mathfrak{F}^N_{\Re_2}(e_\sigma)\right) | e \in \mathbb{E},\ \sigma = 1, 2, .., n\right\} = \left\{e_\sigma, \left(\mathfrak{F}^{RP}_{\Re_2}(e_\sigma) + \iota\mathfrak{F}^{IP}_{\Re_2}(e_\sigma), \mathfrak{F}^{RN}_{\Re_2}(e_\sigma) + \iota\mathfrak{F}^{IN}_{\Re_2}(e_\sigma)\right) | e_\sigma \in \mathbb{E},\ \sigma = 1, 2, .., n\right\}$ over $\mathbb{E}$. Then the GBCFNHSM between $\Re_1$ and $\Re_2$ is inaugurated as.

$$\mathcal{S}^1_{BCFS}(\Re_1, \Re_2) = 1 - \left(\frac{1}{4n}\sum_{\sigma=1}^{n}\left(\begin{array}{c} \left|\left(\mathfrak{F}^{RP}_{\Re_1}(e_\sigma)\right)^2 - \left(\mathfrak{F}^{RP}_{\Re_2}(e_\sigma)\right)^2\right| + \left|\left(\mathfrak{F}^{IP}_{\Re_1}(e_\sigma)\right)^2 - \left(\mathfrak{F}^{IP}_{\Re_2}(e_\sigma)\right)^2\right| \\ + \left|\left(\mathfrak{F}^{RN}_{\Re_1}(e_\sigma)\right)^2 - \left(\mathfrak{F}^{RN}_{\Re_2}(e_\sigma)\right)^2\right| + \left|\left(\mathfrak{F}^{IN}_{\Re_1}(e_\sigma)\right)^2 - \left(\mathfrak{F}^{IN}_{\Re_2}(e_\sigma)\right)^2\right| \end{array}\right)\right)^{\frac{1}{\lambda}}, \lambda \geq 1$$

The underneath shows that GBCFNHSM holds the necessary conditions of SM.

Theorem 5.2 The GBCFNHSM satisfies the below properties.

1. $0 \leq \mathcal{S}^1_{BCFS}(\Re_1, \Re_2) \leq 1$
2. $\mathcal{S}^1_{BCFS}(\Re_1, \Re_2) = 1 \Leftrightarrow \Re_1 = \Re_2$
3. $\mathcal{S}^1_{BCFS}(\Re_1, \Re_2) = \mathcal{S}^1_{BCFS}(\Re_2, \Re_1)$
4. If $\Re_1 \subset \Re_2 \subset \Re_3$, then $\mathcal{S}^1_{BCFS}(\Re_1, \Re_2) \geq \mathcal{S}^1_{BCFS}(\Re_1, \Re_3)$ and $\mathcal{S}^1_{BCFS}(\Re_2, \Re_3) \geq \mathcal{S}^1_{BCFS}(\Re_1, \Re_3)$

Proof

1. We know that

$$\left(\frac{1}{4n}\sum_{\sigma=1}^{n}\left(\begin{array}{c}\left|\left(\mathfrak{F}^{RP}_{\Re_1}(e_\sigma)\right)^2-\left(\mathfrak{F}^{RP}_{\Re_2}(e_\sigma)\right)^2\right|+\left|\left(\mathfrak{F}^{IP}_{\Re_1}(e_\sigma)\right)^2-\left(\mathfrak{F}^{IP}_{\Re_2}(e_\sigma)\right)^2\right|\\+\left|\left(\mathfrak{F}^{RN}_{\Re_1}(e_\sigma)\right)^2-\left(\mathfrak{F}^{RN}_{\Re_2}(e_\sigma)\right)^2\right|+\left|\left(\mathfrak{F}^{IN}_{\Re_1}(e_\sigma)\right)^2-\left(\mathfrak{F}^{IN}_{\Re_2}(e_\sigma)\right)^2\right|\end{array}\right)\right)^{\frac{1}{\lambda}} \in [0,1]$$

$$\Rightarrow 0 \le \left(\frac{1}{4n}\sum_{\sigma=1}^{n}\left(\begin{array}{c}\left|\left(\mathfrak{F}^{RP}_{\Re_1}(e_\sigma)\right)^2-\left(\mathfrak{F}^{RP}_{\Re_2}(e_\sigma)\right)^2\right|+\left|\left(\mathfrak{F}^{IP}_{\Re_1}(e_\sigma)\right)^2-\left(\mathfrak{F}^{IP}_{\Re_2}(e_\sigma)\right)^2\right|\\+\left|\left(\mathfrak{F}^{RN}_{\Re_1}(e_\sigma)\right)^2-\left(\mathfrak{F}^{RN}_{\Re_2}(e_\sigma)\right)^2\right|+\left|\left(\mathfrak{F}^{IN}_{\Re_1}(e_\sigma)\right)^2-\left(\mathfrak{F}^{IN}_{\Re_2}(e_\sigma)\right)^2\right|\end{array}\right)\right)^{\frac{1}{\lambda}} \le 1$$

Then

$$0 \ge -\left(\frac{1}{4n}\sum_{\sigma=1}^{n}\left(\begin{array}{c}\left|\left(\mathfrak{F}^{RP}_{\Re_1}(e_\sigma)\right)^2-\left(\mathfrak{F}^{RP}_{\Re_2}(e_\sigma)\right)^2\right|+\left|\left(\mathfrak{F}^{IP}_{\Re_1}(e_\sigma)\right)^2-\left(\mathfrak{F}^{IP}_{\Re_2}(e_\sigma)\right)^2\right|\\+\left|\left(\mathfrak{F}^{RN}_{\Re_1}(e_\sigma)\right)^2-\left(\mathfrak{F}^{RN}_{\Re_2}(e_\sigma)\right)^2\right|+\left|\left(\mathfrak{F}^{IN}_{\Re_1}(e_\sigma)\right)^2-\left(\mathfrak{F}^{IN}_{\Re_2}(e_\sigma)\right)^2\right|\end{array}\right)\right)^{\frac{1}{\lambda}} \ge -1$$

$$\Rightarrow 1 \ge 1-\left(\frac{1}{4n}\sum_{\sigma=1}^{n}\left(\begin{array}{c}\left|\left(\mathfrak{F}^{RP}_{\Re_1}(e_\sigma)\right)^2-\left(\mathfrak{F}^{RP}_{\Re_2}(e_\sigma)\right)^2\right|+\left|\left(\mathfrak{F}^{IP}_{\Re_1}(e_\sigma)\right)^2-\left(\mathfrak{F}^{IP}_{\Re_2}(e_\sigma)\right)^2\right|\\+\left|\left(\mathfrak{F}^{RN}_{\Re_1}(e_\sigma)\right)^2-\left(\mathfrak{F}^{RN}_{\Re_2}(e_\sigma)\right)^2\right|+\left|\left(\mathfrak{F}^{IN}_{\Re_1}(e_\sigma)\right)^2-\left(\mathfrak{F}^{IN}_{\Re_2}(e_\sigma)\right)^2\right|\end{array}\right)\right)^{\frac{1}{\lambda}} \ge 0$$

$$\Rightarrow 0 \le 1-\left(\frac{1}{4n}\sum_{\sigma=1}^{n}\left(\begin{array}{c}\left|\left(\mathfrak{F}^{RP}_{\Re_1}(e_\sigma)\right)^2-\left(\mathfrak{F}^{RP}_{\Re_2}(e_\sigma)\right)^2\right|+\left|\left(\mathfrak{F}^{IP}_{\Re_1}(e_\sigma)\right)^2-\left(\mathfrak{F}^{IP}_{\Re_2}(e_\sigma)\right)^2\right|\\+\left|\left(\mathfrak{F}^{RN}_{\Re_1}(e_\sigma)\right)^2-\left(\mathfrak{F}^{RN}_{\Re_2}(e_\sigma)\right)^2\right|+\left|\left(\mathfrak{F}^{IN}_{\Re_1}(e_\sigma)\right)^2-\left(\mathfrak{F}^{IN}_{\Re_2}(e_\sigma)\right)^2\right|\end{array}\right)\right)^{\frac{1}{\lambda}} \le 1$$

$$\Rightarrow 0 \le \mathcal{S}^1_{BCFS}(\Re_1, \Re_2) \le 1$$

2. Let $\mathcal{S}^1_{BCFS}(\Re_1, \Re_2) = 1$. Then to reveal that $\Re_1 = \Re_2$. As $\mathcal{S}^1_{BCFS}(\Re_1, \Re_2) = 1$, then

$$1-\left(\frac{1}{4n}\sum_{\sigma=1}^{n}\left(\begin{array}{c}\left|\left(\mathfrak{F}_{\Re_1}^{RP}(e_\sigma)\right)^2-\left(\mathfrak{F}_{\Re_2}^{RP}(e_\sigma)\right)^2\right|+\left|\left(\mathfrak{F}_{\Re_1}^{IP}(e_\sigma)\right)^2-\left(\mathfrak{F}_{\Re_2}^{IP}(e_\sigma)\right)^2\right|\\+\left|\left(\mathfrak{F}_{\Re_1}^{RN}(e_\sigma)\right)^2-\left(\mathfrak{F}_{\Re_2}^{RN}(e_\sigma)\right)^2\right|+\left|\left(\mathfrak{F}_{\Re_1}^{IN}(e_\sigma)\right)^2-\left(\mathfrak{F}_{\Re_2}^{IN}(e_\sigma)\right)^2\right|\end{array}\right)\right)^{\frac{1}{\lambda}}$$
$$=1$$

$$\frac{1}{4n}\sum_{\sigma=1}^{n}\left(\begin{array}{c}\left|\left(\mathfrak{F}_{\Re_1}^{RP}(e_\sigma)\right)^2-\left(\mathfrak{F}_{\Re_2}^{RP}(e_\sigma)\right)^2\right|+\left|\left(\mathfrak{F}_{\Re_1}^{IP}(e_\sigma)\right)^2-\left(\mathfrak{F}_{\Re_2}^{IP}(e_\sigma)\right)^2\right|\\+\left|\left(\mathfrak{F}_{\Re_1}^{RN}(e_\sigma)\right)^2-\left(\mathfrak{F}_{\Re_2}^{RN}(e_\sigma)\right)^2\right|+\left|\left(\mathfrak{F}_{\Re_1}^{IN}(e_\sigma)\right)^2-\left(\mathfrak{F}_{\Re_2}^{IN}(e_\sigma)\right)^2\right|\end{array}\right)=0$$

This implies that $d_{BCFS}^1(\Re_1,\Re_2)=0$ and hence $\Re_1=\Re_2$.

Conversely, Let $\Re_1=\Re_2$. To reveal that $\mathcal{S}_{BCFS}^1(\Re_1,\Re_2)=1$. As $\Re_1=\Re_2$, then by from Theorem (4.1), we have that $d_{BCFS}^1(\Re_1,\Re_2)=0$, then,

$$\Rightarrow\left(\frac{1}{4n}\sum_{\sigma=1}^{n}\left(\begin{array}{c}\left|\left(\mathfrak{F}_{\Re_1}^{RP}(e_\sigma)\right)^2-\left(\mathfrak{F}_{\Re_2}^{RP}(e_\sigma)\right)^2\right|+\left|\left(\mathfrak{F}_{\Re_1}^{IP}(e_\sigma)\right)^2-\left(\mathfrak{F}_{\Re_2}^{IP}(e_\sigma)\right)^2\right|\\+\left|\left(\mathfrak{F}_{\Re_1}^{RN}(e_\sigma)\right)^2-\left(\mathfrak{F}_{\Re_2}^{RN}(e_\sigma)\right)^2\right|+\left|\left(\mathfrak{F}_{\Re_1}^{RN}(e_\sigma)\right)^2-\left(\mathfrak{F}_{\Re_2}^{RN}(e_\sigma)\right)^2\right|\end{array}\right)\right)^{\frac{1}{\lambda}}$$
$$\leq\left(\frac{1}{4n}\sum_{\sigma=1}^{n}\left(\begin{array}{c}\left|\left(\mathfrak{F}_{\Re_1}^{RP}(e_\sigma)\right)^2-\left(\mathfrak{F}_{\Re_3}^{RP}(e_\sigma)\right)^2\right|+\left|\left(\mathfrak{F}_{\Re_1}^{IP}(e_\sigma)\right)^2-\left(\mathfrak{F}_{\Re_3}^{IP}(e_\sigma)\right)^2\right|\\+\left|\left(\mathfrak{F}_{\Re_1}^{RN}(e_\sigma)\right)^2-\left(\mathfrak{F}_{\Re_3}^{RN}(e_\sigma)\right)^2\right|+\left|\left(\mathfrak{F}_{\Re_1}^{RN}(e_\sigma)\right)^2-\left(\mathfrak{F}_{\Re_3}^{RN}(e_\sigma)\right)^2\right|\end{array}\right)\right)^{\frac{1}{\lambda}}$$

This implies that $\mathcal{S}_{BCFS}^1(\Re_1,\Re_2)=1$

3. By employing Theorem (4.1), we get that $\mathcal{S}_{BCFS}^1(\Re_1,\Re_2)=\mathcal{S}_{BCFS}^1(\Re_2,\Re_1)$.
4. Let $\Re_1\subset\Re_2\subset\Re_3$. Then from Theorem (4.1) we get that

$$d_{BCFS}^1(\Re_2,\Re_3)\leq d_{BCFS}^1(\Re_1,\Re_3)$$

$$\Rightarrow\left(\frac{1}{4n}\sum_{\sigma=1}^{n}\left(\begin{array}{c}\left|\left(\mathfrak{F}_{\Re_1}^{RP}(e_\sigma)\right)^2-\left(\mathfrak{F}_{\Re_2}^{RP}(e_\sigma)\right)^2\right|+\left|\left(\mathfrak{F}_{\Re_1}^{IP}(e_\sigma)\right)^2-\left(\mathfrak{F}_{\Re_2}^{IP}(e_\sigma)\right)^2\right|\\+\left|\left(\mathfrak{F}_{\Re_1}^{RN}(e_\sigma)\right)^2-\left(\mathfrak{F}_{\Re_2}^{RN}(e_\sigma)\right)^2\right|+\left|\left(\mathfrak{F}_{\Re_1}^{RN}(e_\sigma)\right)^2-\left(\mathfrak{F}_{\Re_2}^{RN}(e_\sigma)\right)^2\right|\end{array}\right)\right)^{\frac{1}{\lambda}}$$
$$\leq\left(\frac{1}{4n}\sum_{\sigma=1}^{n}\left(\begin{array}{c}\left|\left(\mathfrak{F}_{\Re_1}^{RP}(e_\sigma)\right)^2-\left(\mathfrak{F}_{\Re_3}^{RP}(e_\sigma)\right)^2\right|+\left|\left(\mathfrak{F}_{\Re_1}^{IP}(e_\sigma)\right)^2-\left(\mathfrak{F}_{\Re_3}^{IP}(e_\sigma)\right)^2\right|\\+\left|\left(\mathfrak{F}_{\Re_1}^{RN}(e_\sigma)\right)^2-\left(\mathfrak{F}_{\Re_3}^{RN}(e_\sigma)\right)^2\right|+\left|\left(\mathfrak{F}_{\Re_1}^{RN}(e_\sigma)\right)^2-\left(\mathfrak{F}_{\Re_3}^{RN}(e_\sigma)\right)^2\right|\end{array}\right)\right)^{\frac{1}{\lambda}}$$

$$\Rightarrow-\left(\frac{1}{4n}\sum_{\sigma=1}^{n}\left(\begin{array}{c}\left|\left(\mathfrak{F}_{\Re_1}^{RP}(e_\sigma)\right)^2-\left(\mathfrak{F}_{\Re_2}^{RP}(e_\sigma)\right)^2\right|+\left|\left(\mathfrak{F}_{\Re_1}^{IP}(e_\sigma)\right)^2-\left(\mathfrak{F}_{\Re_2}^{IP}(e_\sigma)\right)^2\right|\\+\left|\left(\mathfrak{F}_{\Re_1}^{RN}(e_\sigma)\right)^2-\left(\mathfrak{F}_{\Re_2}^{RN}(e_\sigma)\right)^2\right|+\left|\left(\mathfrak{F}_{\Re_1}^{RN}(e_\sigma)\right)^2-\left(\mathfrak{F}_{\Re_2}^{RN}(e_\sigma)\right)^2\right|\end{array}\right)\right)^{\frac{1}{\lambda}}$$

$$\geq -\left(\frac{1}{4n}\sum_{\sigma=1}^{n}\left(\begin{array}{c}\left|\left(\mathfrak{F}_{\mathfrak{R}_1}^{RP}(e_\sigma)\right)^2-\left(\mathfrak{F}_{\mathfrak{R}_3}^{RP}(e_\sigma)\right)^2\right|+\left|\left(\mathfrak{F}_{\mathfrak{R}_1}^{IP}(e_\sigma)\right)^2-\left(\mathfrak{F}_{\mathfrak{R}_3}^{IP}(e_\sigma)\right)^2\right|\\+\left|\left(\mathfrak{F}_{\mathfrak{R}_1}^{RN}(e_\sigma)\right)^2-\left(\mathfrak{F}_{\mathfrak{R}_3}^{RN}(e_\sigma)\right)^2\right|+\left|\left(\mathfrak{F}_{\mathfrak{R}_1}^{RN}(e_\sigma)\right)^2-\left(\mathfrak{F}_{\mathfrak{R}_3}^{RN}(e_\sigma)\right)^2\right|\end{array}\right)\right)^{\frac{1}{\lambda}}$$

$$\Rightarrow 1$$

$$-\left(\frac{1}{4n}\sum_{\sigma=1}^{n}\left(\begin{array}{c}\left|\left(\mathfrak{F}_{\mathfrak{R}_1}^{RP}(e_\sigma)\right)^2-\left(\mathfrak{F}_{\mathfrak{R}_2}^{RP}(e_\sigma)\right)^2\right|+\left|\left(\mathfrak{F}_{\mathfrak{R}_1}^{IP}(e_\sigma)\right)^2-\left(\mathfrak{F}_{\mathfrak{R}_2}^{IP}(e_\sigma)\right)^2\right|\\+\left|\left(\mathfrak{F}_{\mathfrak{R}_1}^{RN}(e_\sigma)\right)^2-\left(\mathfrak{F}_{\mathfrak{R}_2}^{RN}(e_\sigma)\right)^2\right|+\left|\left(\mathfrak{F}_{\mathfrak{R}_1}^{RN}(e_\sigma)\right)^2-\left(\mathfrak{F}_{\mathfrak{R}_2}^{RN}(e_\sigma)\right)^2\right|\end{array}\right)\right)^{\frac{1}{\lambda}}$$

$$\geq 1$$

$$-\left(\frac{1}{4n}\sum_{\sigma=1}^{n}\left(\begin{array}{c}\left|\left(\mathfrak{F}_{\mathfrak{R}_1}^{RP}(e_\sigma)\right)^2-\left(\mathfrak{F}_{\mathfrak{R}_3}^{RP}(e_\sigma)\right)^2\right|+\left|\left(\mathfrak{F}_{\mathfrak{R}_1}^{IP}(e_\sigma)\right)^2-\left(\mathfrak{F}_{\mathfrak{R}_3}^{IP}(e_\sigma)\right)^2\right|\\+\left|\left(\mathfrak{F}_{\mathfrak{R}_1}^{RN}(e_\sigma)\right)^2-\left(\mathfrak{F}_{\mathfrak{R}_3}^{RN}(e_\sigma)\right)^2\right|+\left|\left(\mathfrak{F}_{\mathfrak{R}_1}^{RN}(e_\sigma)\right)^2-\left(\mathfrak{F}_{\mathfrak{R}_3}^{RN}(e_\sigma)\right)^2\right|\end{array}\right)\right)^{\frac{1}{\lambda}}$$

Consequently,

$$\mathcal{S}_{BCFS}^{1}(\mathfrak{R}_1,\mathfrak{R}_2)\geq\mathcal{S}_{BCFS}^{1}(\mathfrak{R}_1,\mathfrak{R}_3)$$

By the same procedure, $\mathcal{S}_{BCFS}^{1}(\mathfrak{R}_2,\mathfrak{R}_3)\geq\mathcal{S}_{BCFS}^{1}(\mathfrak{R}_1,\mathfrak{R}_3)$can be proved. Especially,

i. By considering $\lambda = 1$, the GBCFNHSM would decrease to BCF normalized Hamming SM (BCFNHSM)

$$\mathcal{S}_{BCFS}^{2}(\mathfrak{R}_1,\mathfrak{R}_2)=$$

$$1-\frac{1}{4n}\sum_{\sigma=1}^{n}\left(\begin{array}{c}\left|\left(\mathfrak{F}_{\mathfrak{R}_1}^{RP}(e_\sigma)\right)^2-\left(\mathfrak{F}_{\mathfrak{R}_2}^{RP}(e_\sigma)\right)^2\right|+\left|\left(\mathfrak{F}_{\mathfrak{R}_1}^{IP}(e_\sigma)\right)^2-\left(\mathfrak{F}_{\mathfrak{R}_2}^{IP}(e_\sigma)\right)^2\right|\\+\left|\left(\mathfrak{F}_{\mathfrak{R}_1}^{RN}(e_\sigma)\right)^2-\left(\mathfrak{F}_{\mathfrak{R}_2}^{RN}(e_\sigma)\right)^2\right|+\left|\left(\mathfrak{F}_{\mathfrak{R}_1}^{IN}(e_\sigma)\right)^2-\left(\mathfrak{F}_{\mathfrak{R}_2}^{IN}(e_\sigma)\right)^2\right|\end{array}\right)$$

ii. By considering $\lambda = 2$, the GBCFNHSM would decrease to BCF normalized Euclidean SM (BCFNESM)

$$\mathcal{S}_{BCFS}^{3}(\mathfrak{R}_1,\mathfrak{R}_2)=$$

$$1-\sqrt{\frac{1}{4n}\sum_{\sigma=1}^{n}\left(\begin{array}{c}\left|\left(\mathfrak{F}_{\mathfrak{R}_1}^{RP}(e_\sigma)\right)^2-\left(\mathfrak{F}_{\mathfrak{R}_2}^{RP}(e_\sigma)\right)^2\right|+\left|\left(\mathfrak{F}_{\mathfrak{R}_1}^{IP}(e_\sigma)\right)^2-\left(\mathfrak{F}_{\mathfrak{R}_2}^{IP}(e_\sigma)\right)^2\right|\\+\left|\left(\mathfrak{F}_{\mathfrak{R}_1}^{RN}(e_\sigma)\right)^2-\left(\mathfrak{F}_{\mathfrak{R}_2}^{RN}(e_\sigma)\right)^2\right|+\left|\left(\mathfrak{F}_{\mathfrak{R}_1}^{IN}(e_\sigma)\right)^2-\left(\mathfrak{F}_{\mathfrak{R}_2}^{IN}(e_\sigma)\right)^2\right|\end{array}\right)}$$

In next definition, GBCFNHDSM between two BCFSs are defined.

Definition 5.2 Let two BCFSs $\mathfrak{R}_1=\left\{e_\sigma,\left(\mathfrak{F}_{\mathfrak{R}_1}^{P}(e_\sigma),\mathfrak{F}_{\mathfrak{R}_1}^{N}(e_\sigma)\right)|e_\sigma\in\mathbb{E},\sigma=1,2,..,n\right\}=\left\{e_\sigma,\left(\mathfrak{F}_{\mathfrak{R}_1}^{RP}(e_\sigma)+\iota\mathfrak{F}_{\mathfrak{R}_1}^{IP}(e_\sigma),\mathfrak{F}_{\mathfrak{R}_1}^{RN}(e_\sigma)+\iota\mathfrak{F}_{\mathfrak{R}_1}^{IN}(e_\sigma)\right)|e_\sigma\in\mathbb{E},\ \sigma=1,2,..,n\right\}$ and

$\Re_2 = \{e, (\mathfrak{F}^{P}_{\Re_2}(e_\sigma), \mathfrak{F}^{N}_{\Re_2}(e_\sigma)) | e \in \mathbb{E},\ \sigma = 1, 2, .., n\} = \{e_\sigma, (\mathfrak{F}^{RP}_{\Re_2}(e_\sigma) + \iota\mathfrak{F}^{IP}_{\Re_2}(e_\sigma), \mathfrak{F}^{RN}_{\Re_2}(e_\sigma) + \iota\mathfrak{F}^{IN}_{\Re_2}(e_\sigma)) | e_\sigma \in \mathbb{E},\ \sigma = 1, 2, .., n\}$ over $\mathbb{E}$. Then the GBCFNHDSM between $\Re_1$ and $\Re_2$ is inaugurated as.

$$\mathcal{S}^4_{BCFS}(\Re_1, \Re_2) = 1 - \left(\frac{1}{n} \sum_{\sigma=1}^{n} \max \begin{pmatrix} \left| \left(\mathfrak{F}^{RP}_{\Re_1}(e_\sigma)\right)^2 - \left(\mathfrak{F}^{RP}_{\Re_2}(e_\sigma)\right)^2 \right|, \left| \left(\mathfrak{F}^{IP}_{\Re_1}(e_\sigma)\right)^2 - \left(\mathfrak{F}^{IP}_{\Re_2}(e_\sigma)\right)^2 \right| \\ \left| \left(\mathfrak{F}^{RN}_{\Re_1}(e_\sigma)\right)^2 - \left(\mathfrak{F}^{RN}_{\Re_2}(e_\sigma)\right)^2 \right|, \left| \left(\mathfrak{F}^{IN}_{\Re_1}(e_\sigma)\right)^2 - \left(\mathfrak{F}^{IN}_{\Re_2}(e_\sigma)\right)^2 \right| \end{pmatrix}^{\frac{1}{\lambda}} \right), \lambda \geq 1$$

Theorem 5.3 The GBCFNHDSM satisfies the underneath properties.

1. $0 \leq \mathcal{S}^4_{BCFS}(\Re_1, \Re_2) \leq 1$
2. $\mathcal{S}^4_{BCFS}(\Re_1, \Re_2) = 1 \Leftrightarrow \Re_1 = \Re_2$
3. $\mathcal{S}^4_{BCFS}(\Re_1, \Re_2) = \mathcal{S}^4_{BCFS}(\Re_2, \Re_1)$
4. If $\Re_1 \subset \Re_2 \subset \Re_3$, then $\mathcal{S}^4_{BCFS}(\Re_1, \Re_2) \geq \mathcal{S}^4_{BCFS}(\Re_1, \Re_3)$ and $\mathcal{S}^4_{BCFS}(\Re_2, \Re_3) \geq \mathcal{S}^4_{BCFS}(\Re_1, \Re_3)$

Proof

1. We know that

$$\left(\frac{1}{n} \sum_{\sigma=1}^{n} \max \begin{pmatrix} \left| \left(\mathfrak{F}^{RP}_{\Re_1}(e_\sigma)\right)^2 - \left(\mathfrak{F}^{RP}_{\Re_2}(e_\sigma)\right)^2 \right|, \left| \left(\mathfrak{F}^{IP}_{\Re_1}(e_\sigma)\right)^2 - \left(\mathfrak{F}^{IP}_{\Re_2}(e_\sigma)\right)^2 \right| \\ \left| \left(\mathfrak{F}^{RN}_{\Re_1}(e_\sigma)\right)^2 - \left(\mathfrak{F}^{RN}_{\Re_2}(e_\sigma)\right)^2 \right|, \left| \left(\mathfrak{F}^{IN}_{\Re_1}(e_\sigma)\right)^2 - \left(\mathfrak{F}^{IN}_{\Re_2}(e_\sigma)\right)^2 \right| \end{pmatrix}^{\frac{1}{\lambda}} \right) \in [0, 1]$$

$$\Rightarrow 0 \leq \left(\frac{1}{n} \sum_{\sigma=1}^{n} \max \begin{pmatrix} \left| \left(\mathfrak{F}^{RP}_{\Re_1}(e_\sigma)\right)^2 - \left(\mathfrak{F}^{RP}_{\Re_2}(e_\sigma)\right)^2 \right|, \left| \left(\mathfrak{F}^{IP}_{\Re_1}(e_\sigma)\right)^2 - \left(\mathfrak{F}^{IP}_{\Re_2}(e_\sigma)\right)^2 \right| \\ , \left| \left(\mathfrak{F}^{RN}_{\Re_1}(e_\sigma)\right)^2 - \left(\mathfrak{F}^{RN}_{\Re_2}(e_\sigma)\right)^2 \right|, \left| \left(\mathfrak{F}^{IN}_{\Re_1}(e_\sigma)\right)^2 - \left(\mathfrak{F}^{IN}_{\Re_2}(e_\sigma)\right)^2 \right| \end{pmatrix}^{\frac{1}{\lambda}} \right) \leq 1$$

Then

$$0 \geq - \left(\frac{1}{n} \sum_{\sigma=1}^{n} \max \begin{pmatrix} \left| \left(\mathfrak{I}^{RP}_{\Re_1}(e_\sigma)\right)^2 - \left(\mathfrak{I}^{RP}_{\Re_2}(e_\sigma)\right)^2 \right|, \left| \left(\mathfrak{I}^{IP}_{\Re_1}(e_\sigma)\right)^2 - \left(\mathfrak{I}^{IP}_{\Re_2}(e_\sigma)\right)^2 \right| \\ , \left| \left(\mathfrak{I}^{RN}_{\Re_1}(e_\sigma)\right)^2 - \left(\mathfrak{I}^{RN}_{\Re_2}(e_\sigma)\right)^2 \right|, \left| \left(\mathfrak{I}^{IN}_{\Re_1}(e_\sigma)\right)^2 - \left(\mathfrak{I}^{IN}_{\Re_2}(e_\sigma)\right)^2 \right| \end{pmatrix}^{\frac{1}{\lambda}} \right) \geq -1$$

$$\Rightarrow 1 \geq 1-$$

$$\left(\frac{1}{n}\sum_{\sigma=1}^{n}\max\left(\begin{array}{c}\left|\left(\mathfrak{I}_{\mathfrak{R}_1}^{RP}(e_\sigma)\right)^2-\left(\mathfrak{I}_{\mathfrak{R}_2}^{RP}(e_\sigma)\right)^2\right|,\left|\left(\mathfrak{I}_{\mathfrak{R}_1}^{IP}(e_\sigma)\right)^2-\left(\mathfrak{I}_{\mathfrak{R}_2}^{IP}(e_\sigma)\right)^2\right|\\\left|\left(\mathfrak{I}_{\mathfrak{R}_1}^{RN}(e_\sigma)\right)^2-\left(\mathfrak{I}_{\mathfrak{R}_2}^{RN}(e_\sigma)\right)^2\right|,\left|\left(\mathfrak{I}_{\mathfrak{R}_1}^{IN}(e_\sigma)\right)^2-\left(\mathfrak{I}_{\mathfrak{R}_2}^{IN}(e_\sigma)\right)^2\right|\end{array}\right)^{\frac{1}{\lambda}}\right)\geq 0$$

$$\Rightarrow 0\leq 1-\left(\frac{1}{n}\sum_{\sigma=1}^{n}\max\left(\begin{array}{c}\left|\left(\mathfrak{I}_{\mathfrak{R}_1}^{RP}(e_\sigma)\right)^2-\left(\mathfrak{I}_{\mathfrak{R}_2}^{RP}(e_\sigma)\right)^2\right|,\left|\left(\mathfrak{I}_{\mathfrak{R}_1}^{IP}(e_\sigma)\right)^2-\left(\mathfrak{I}_{\mathfrak{R}_2}^{IP}(e_\sigma)\right)^2\right|\\,\left|\left(\mathfrak{I}_{\mathfrak{R}_1}^{RN}(e_\sigma)\right)^2-\left(\mathfrak{I}_{\mathfrak{R}_2}^{RN}(e_\sigma)\right)^2\right|,\left|\left(\mathfrak{I}_{\mathfrak{R}_1}^{IN}(e_\sigma)\right)^2-\left(\mathfrak{I}_{\mathfrak{R}_2}^{IN}(e_\sigma)\right)^2\right|\end{array}\right)^{\frac{1}{\lambda}}\right)\leq 1$$

$$\Rightarrow 0\leq \mathcal{S}_{BCFS}^{4}(\mathfrak{R}_1,\mathfrak{R}_2)\leq 1$$

2. Let $\mathcal{S}_{BCFS}^{4}(\mathfrak{R}_1,\mathfrak{R}_2)=1$. Then to reveal that $\mathfrak{R}_1=\mathfrak{R}_2$. As $\mathcal{S}_{BCFS}^{4}(\mathfrak{R}_1,\mathfrak{R}_2)=1$, then

$$1-\left(\frac{1}{n}\sum_{\sigma=1}^{n}\max\left(\begin{array}{c}\left|\left(\mathfrak{F}_{\mathfrak{R}_1}^{RP}(e_\sigma)\right)^2-\left(\mathfrak{F}_{\mathfrak{R}_2}^{RP}(e_\sigma)\right)^2\right|,\left|\left(\mathfrak{F}_{\mathfrak{R}_1}^{IP}(e_\sigma)\right)^2-\left(\mathfrak{F}_{\mathfrak{R}_2}^{IP}(e_\sigma)\right)^2\right|\\\left|\left(\mathfrak{F}_{\mathfrak{R}_1}^{RN}(e_\sigma)\right)^2-\left(\mathfrak{F}_{\mathfrak{R}_2}^{RN}(e_\sigma)\right)^2\right|,\left|\left(\mathfrak{F}_{\mathfrak{R}_1}^{IN}(e_\sigma)\right)^2-\left(\mathfrak{F}_{\mathfrak{R}_2}^{IN}(e_\sigma)\right)^2\right|\end{array}\right)\right)^{\frac{1}{\lambda}}=1$$

$$\left(\frac{1}{n}\sum_{\sigma=1}^{n}\max\left(\begin{array}{c}\left|\left(\mathfrak{F}_{\mathfrak{R}_1}^{RP}(e_\sigma)\right)^2-\left(\mathfrak{F}_{\mathfrak{R}_2}^{RP}(e_\sigma)\right)^2\right|,\left|\left(\mathfrak{F}_{\mathfrak{R}_1}^{IP}(e_\sigma)\right)^2-\left(\mathfrak{F}_{\mathfrak{R}_2}^{IP}(e_\sigma)\right)^2\right|\\,\left|\left(\mathfrak{F}_{\mathfrak{R}_1}^{RN}(e_\sigma)\right)^2-\left(\mathfrak{F}_{\mathfrak{R}_2}^{RN}(e_\sigma)\right)^2\right|,\left|\left(\mathfrak{F}_{\mathfrak{R}_1}^{IN}(e_\sigma)\right)^2-\left(\mathfrak{F}_{\mathfrak{R}_2}^{IN}(e_\sigma)\right)^2\right|\end{array}\right)\right)^{\frac{1}{\lambda}}=0$$

This implies that $d_{BCFS}^{4}(\mathfrak{R}_1,\mathfrak{R}_2)=0$ and hence $\mathfrak{R}_1=\mathfrak{R}_2$.

Conversely, Let $\mathfrak{R}_1=\mathfrak{R}_2$. To reveal that $\mathcal{S}_{BCFS}^{1}(\mathfrak{R}_1,\mathfrak{R}_2)=1$. As $\mathfrak{R}_1=\mathfrak{R}_2$, then by from Theorem (4.2), we have that $d_{BCFS}^{4}(\mathfrak{R}_1,\mathfrak{R}_2)=0$, then,

$$\left(\frac{1}{n}\sum_{\sigma=1}^{n}\max\left(\begin{array}{c}\left|\left(\mathfrak{F}_{\mathfrak{R}_1}^{RP}(e_\sigma)\right)^2-\left(\mathfrak{F}_{\mathfrak{R}_2}^{RP}(e_\sigma)\right)^2\right|,\left|\left(\mathfrak{F}_{\mathfrak{R}_1}^{IP}(e_\sigma)\right)^2-\left(\mathfrak{F}_{\mathfrak{R}_2}^{IP}(e_\sigma)\right)^2\right|\\,\left|\left(\mathfrak{F}_{\mathfrak{R}_1}^{RN}(e_\sigma)\right)^2-\left(\mathfrak{F}_{\mathfrak{R}_2}^{RN}(e_\sigma)\right)^2\right|,\left|\left(\mathfrak{F}_{\mathfrak{R}_1}^{IN}(e_\sigma)\right)^2-\left(\mathfrak{F}_{\mathfrak{R}_2}^{IN}(e_\sigma)\right)^2\right|\end{array}\right)\right)^{\frac{1}{\lambda}}=0$$

$$\Rightarrow 1-\left(\frac{1}{n}\sum_{\sigma=1}^{n}\max\left(\begin{array}{c}\left|\left(\mathfrak{F}_{\mathfrak{R}_1}^{RP}(e_\sigma)\right)^2-\left(\mathfrak{F}_{\mathfrak{R}_2}^{RP}(e_\sigma)\right)^2\right|,\left|\left(\mathfrak{F}_{\mathfrak{R}_1}^{IP}(e_\sigma)\right)^2-\left(\mathfrak{F}_{\mathfrak{R}_2}^{IP}(e_\sigma)\right)^2\right|\\,\left|\left(\mathfrak{F}_{\mathfrak{R}_1}^{RN}(e_\sigma)\right)^2-\left(\mathfrak{F}_{\mathfrak{R}_2}^{RN}(e_\sigma)\right)^2\right|,\left|\left(\mathfrak{F}_{\mathfrak{R}_1}^{IN}(e_\sigma)\right)^2-\left(\mathfrak{F}_{\mathfrak{R}_2}^{IN}(e_\sigma)\right)^2\right|\end{array}\right)\right)^{\frac{1}{\lambda}}=1$$

This implies that $\mathcal{S}^4_{BCFS}(\Re_1, \Re_2) = 1$.

3. By employing Theorem (4.2), we get that $\mathcal{S}^4_{BCFS}(\Re_1, \Re_2) = \mathcal{S}^4_{BCFS}(\Re_2, \Re_1)$.
4. Let $\Re_1 \subset \Re_2 \subset \Re_3$. Then from Theorem (4.2) we get that

$$d^4_{BCFS}(\Re_2, \Re_3) \leq d^4_{BCFS}(\Re_1, \Re_3)$$

$$\Rightarrow \left(\frac{1}{n}\sum_{\sigma=1}^{n}\max\left(\begin{array}{c}\left|\left(\mathfrak{F}^{RP}_{\Re_1}(e_\sigma)\right)^2 - \left(\mathfrak{F}^{RP}_{\Re_2}(e_\sigma)\right)^2\right|, \left|\left(\mathfrak{F}^{IP}_{\Re_1}(e_\sigma)\right)^2 - \left(\mathfrak{F}^{IP}_{\Re_2}(e_\sigma)\right)^2\right| \\ , \left|\left(\mathfrak{F}^{RN}_{\Re_1}(e_\sigma)\right)^2 - \left(\mathfrak{F}^{RN}_{\Re_2}(e_\sigma)\right)^2\right|, \left|\left(\mathfrak{F}^{IN}_{\Re_1}(e_\sigma)\right)^2 - \left(\mathfrak{F}^{IN}_{\Re_2}(e_\sigma)\right)^2\right|\end{array}\right)\right)^{\frac{1}{\lambda}}$$

$$\leq \left(\frac{1}{n}\sum_{\sigma=1}^{n}\max\left(\begin{array}{c}\left|\left(\mathfrak{F}^{RP}_{\Re_1}(e_\sigma)\right)^2 - \left(\mathfrak{F}^{RP}_{\Re_3}(e_\sigma)\right)^2\right|, \left|\left(\mathfrak{F}^{IP}_{\Re_1}(e_\sigma)\right)^2 - \left(\mathfrak{F}^{IP}_{\Re_3}(e_\sigma)\right)^2\right| \\ , \left|\left(\mathfrak{F}^{RN}_{\Re_1}(e_\sigma)\right)^2 - \left(\mathfrak{F}^{RN}_{\Re_3}(e_\sigma)\right)^2\right|, \left|\left(\mathfrak{F}^{RN}_{\Re_1}(e_\sigma)\right)^2 - \left(\mathfrak{F}^{RN}_{\Re_3}(e_\sigma)\right)^2\right|\end{array}\right)\right)^{\frac{1}{\lambda}}$$

$$\Rightarrow -\left(\frac{1}{n}\sum_{\sigma=1}^{n}\max\left(\begin{array}{c}\left|\left(\mathfrak{F}^{RP}_{\Re_1}(e_\sigma)\right)^2 - \left(\mathfrak{F}^{RP}_{\Re_2}(e_\sigma)\right)^2\right|, \left|\left(\mathfrak{F}^{IP}_{\Re_1}(e_\sigma)\right)^2 - \left(\mathfrak{F}^{IP}_{\Re_2}(e_\sigma)\right)^2\right| \\ \left|\left(\mathfrak{F}^{RN}_{\Re_1}(e_\sigma)\right)^2 - \left(\mathfrak{F}^{RN}_{\Re_2}(e_\sigma)\right)^2\right|, \left|\left(\mathfrak{F}^{IN}_{\Re_1}(e_\sigma)\right)^2 - \left(\mathfrak{F}^{IN}_{\Re_2}(e_\sigma)\right)^2\right|\end{array}\right)\right)^{\frac{1}{\lambda}}$$

$$\geq -\left(\frac{1}{n}\sum_{\sigma=1}^{n}\max\left(\begin{array}{c}\left|\left(\mathfrak{F}^{RP}_{\Re_1}(e_\sigma)\right)^2 - \left(\mathfrak{F}^{RP}_{\Re_3}(e_\sigma)\right)^2\right|, \left|\left(\mathfrak{F}^{IP}_{\Re_1}(e_\sigma)\right)^2 - \left(\mathfrak{F}^{IP}_{\Re_3}(e_\sigma)\right)^2\right| \\ \left|\left(\mathfrak{F}^{RN}_{\Re_1}(e_\sigma)\right)^2 - \left(\mathfrak{F}^{RN}_{\Re_3}(e_\sigma)\right)^2\right|, \left|\left(\mathfrak{F}^{RN}_{\Re_1}(e_\sigma)\right)^2 - \left(\mathfrak{F}^{RN}_{\Re_3}(e_\sigma)\right)^2\right|\end{array}\right)\right)^{\frac{1}{\lambda}}$$

$$\Rightarrow 1-\left(\frac{1}{n}\sum_{\sigma=1}^{n}\max\left(\begin{array}{c}\left|\left(\mathfrak{F}^{RP}_{\Re_1}(e_\sigma)\right)^2 - \left(\mathfrak{F}^{RP}_{\Re_2}(e_\sigma)\right)^2\right|, \left|\left(\mathfrak{F}^{IP}_{\Re_1}(e_\sigma)\right)^2 - \left(\mathfrak{F}^{IP}_{\Re_2}(e_\sigma)\right)^2\right| \\ , \left|\left(\mathfrak{F}^{RN}_{\Re_1}(e_\sigma)\right)^2 - \left(\mathfrak{F}^{RN}_{\Re_2}(e_\sigma)\right)^2\right|, \left|\left(\mathfrak{F}^{IN}_{\Re_1}(e_\sigma)\right)^2 - \left(\mathfrak{F}^{IN}_{\Re_2}(e_\sigma)\right)^2\right|\end{array}\right)\right)^{\frac{1}{\lambda}}$$

$$\geq 1-\left(\frac{1}{n}\sum_{\sigma=1}^{n}\max\left(\begin{array}{c}\left|\left(\mathfrak{F}^{RP}_{\Re_1}(e_\sigma)\right)^2 - \left(\mathfrak{F}^{RP}_{\Re_3}(e_\sigma)\right)^2\right|, \left|\left(\mathfrak{F}^{IP}_{\Re_1}(e_\sigma)\right)^2 - \left(\mathfrak{F}^{IP}_{\Re_3}(e_\sigma)\right)^2\right| \\ \left|\left(\mathfrak{F}^{RN}_{\Re_1}(e_\sigma)\right)^2 - \left(\mathfrak{F}^{RN}_{\Re_3}(e_\sigma)\right)^2\right|, \left|\left(\mathfrak{F}^{RN}_{\Re_1}(e_\sigma)\right)^2 - \left(\mathfrak{F}^{RN}_{\Re_3}(e_\sigma)\right)^2\right|\end{array}\right)\right)^{\frac{1}{\lambda}}$$

Consequently,

$$\mathcal{S}^4_{BCFS}(\Re_1, \Re_2) \geq \mathcal{S}^4_{BCFS}(\Re_1, \Re_3)$$

By the same procedure, $\mathcal{S}^1_{BCFS}(\Re_2, \Re_3) \geq \mathcal{S}^1_{BCFS}(\Re_1, \Re_3)$ can be proved. Especially,

i. By considering $\lambda = 1$, the GBCFNHDSM would decrease to BCF normalized Hamming-Hausdroff SM (BCFNHHDSM)

$$\mathcal{S}^5_{BCFS}(\Re_1, \Re_2) = 1 - \frac{1}{n}\sum_{\sigma=1}^{n}\max\left(\begin{array}{c}\left|\left(\mathfrak{F}^{RP}_{\Re_1}(e_\sigma)\right)^2 - \left(\mathfrak{F}^{RP}_{\Re_2}(e_\sigma)\right)^2\right|, \left|\left(\mathfrak{F}^{IP}_{\Re_1}(e_\sigma)\right)^2 - \left(\mathfrak{F}^{IP}_{\Re_2}(e_\sigma)\right)^2\right| \\ , \left|\left(\mathfrak{F}^{RN}_{\Re_1}(e_\sigma)\right)^2 - \left(\mathfrak{F}^{RN}_{\Re_2}(e_\sigma)\right)^2\right|, \left|\left(\mathfrak{F}^{IN}_{\Re_1}(e_\sigma)\right)^2 - \left(\mathfrak{F}^{IN}_{\Re_2}(e_\sigma)\right)^2\right|\end{array}\right)$$

ii. By considering $\lambda = 2$, the GBCFNHDSM would decrease to BCF normalized Euclidean-Hausdroff SM (BCFNEHDDM)

$$\mathcal{S}^{6}_{BCFS}(\mathfrak{R}_1, \mathfrak{R}_2) = 1 - \sqrt{\frac{1}{n}\sum_{\sigma=1}^{n} \max\left(\begin{array}{l} \left|\left(\mathfrak{F}^{RP}_{\mathfrak{R}_1}(e_\sigma)\right)^2 - \left(\mathfrak{F}^{RP}_{\mathfrak{R}_2}(e_\sigma)\right)^2\right|, \left|\left(\mathfrak{F}^{IP}_{\mathfrak{R}_1}(e_\sigma)\right)^2 - \left(\mathfrak{F}^{IP}_{\mathfrak{R}_2}(e_\sigma)\right)^2\right| \\ , \left|\left(\mathfrak{F}^{RN}_{\mathfrak{R}_1}(e_\sigma)\right)^2 - \left(\mathfrak{F}^{RN}_{\mathfrak{R}_2}(e_\sigma)\right)^2\right|, \left|\left(\mathfrak{F}^{IN}_{\mathfrak{R}_1}(e_\sigma)\right)^2 - \left(\mathfrak{F}^{IN}_{\mathfrak{R}_2}(e_\sigma)\right)^2\right| \end{array}\right)}$$

Definition 5.3 Let two BCFSs $\mathfrak{R}_1 = \left\{e_\sigma, \left(\mathfrak{F}^{P}_{\mathfrak{R}_1}(e_\sigma), \mathfrak{F}^{N}_{\mathfrak{R}_1}(e_\sigma)\right) | e_\sigma \in \mathbb{E}, \sigma = 1, 2, .., n\right\} = \left\{e_\sigma, \left(\mathfrak{F}^{RP}_{\mathfrak{R}_1}(e_\sigma) + \iota\mathfrak{F}^{IP}_{\mathfrak{R}_1}(e_\sigma), \mathfrak{F}^{RN}_{\mathfrak{R}_1}(e_\sigma) + \iota\mathfrak{F}^{IN}_{\mathfrak{R}_1}(e_\sigma)\right) | e_\sigma \in \mathbb{E}, \sigma = 1, 2, \ldots, n\right\}$ and $\mathfrak{R}_2 = \left\{e, \left(\mathfrak{F}^{P}_{\mathfrak{R}_2}(e_\sigma), \mathfrak{F}^{N}_{\mathfrak{R}_2}(e_\sigma)\right) | e_\sigma \in \mathbb{E}, \sigma = 1, 2, .., n\right\} = \left\{e_\sigma, \left(\mathfrak{F}^{RP}_{\mathfrak{R}_2}(e_\sigma) + \iota\mathfrak{F}^{IP}_{\mathfrak{R}_2}(e_\sigma), \mathfrak{F}^{RN}_{\mathfrak{R}_2}(e_\sigma) + \iota\mathfrak{F}^{IN}_{\mathfrak{R}_2}(e_\sigma)\right) | e_\sigma \in \mathbb{E}, \sigma = 1, 2, \ldots, n\right\}$. over $\mathbb{E}$. Then the GHBCFNSM between $\mathfrak{R}_1$ and $\mathfrak{R}_2$ is inaugurated as.

$$\mathcal{S}^{7}_{BCFS}(\mathfrak{R}_1, \mathfrak{R}_2) = 1 - \left(\frac{1}{2n}\sum_{\sigma=1}^{n}\left(\begin{array}{l} \frac{1}{4}\left(\begin{array}{l} \left|\left(\mathfrak{F}^{RP}_{\mathfrak{R}_1}(e_\sigma)\right)^2 - \left(\mathfrak{F}^{RP}_{\mathfrak{R}_2}(e_\sigma)\right)^2\right| + \left|\left(\mathfrak{F}^{IP}_{\mathfrak{R}_1}(e_\sigma)\right)^2 - \left(\mathfrak{F}^{IP}_{\mathfrak{R}_2}(e_\sigma)\right)^2\right| \\ + \left|\left(\mathfrak{F}^{RN}_{\mathfrak{R}_1}(e_\sigma)\right)^2 - \left(\mathfrak{F}^{RN}_{\mathfrak{R}_2}(e_\sigma)\right)^2\right| + \left|\left(\mathfrak{F}^{IN}_{\mathfrak{R}_1}(e_\sigma)\right)^2 - \left(\mathfrak{F}^{IN}_{\mathfrak{R}_2}(e_\sigma)\right)^2\right| \end{array}\right) \\ + \max\left(\begin{array}{l} \left|\left(\mathfrak{F}^{RP}_{\mathfrak{R}_1}(e_\sigma)\right)^2 - \left(\mathfrak{F}^{RP}_{\mathfrak{R}_2}(e_\sigma)\right)^2\right|, \left|\left(\mathfrak{F}^{IP}_{\mathfrak{R}_1}(e_\sigma)\right)^2 - \left(\mathfrak{F}^{IP}_{\mathfrak{R}_2}(e_\sigma)\right)^2\right| \\ , \left|\left(\mathfrak{F}^{RN}_{\mathfrak{R}_1}(e_\sigma)\right)^2 - \left(\mathfrak{F}^{RN}_{\mathfrak{R}_2}(e_\sigma)\right)^2\right|, \left|\left(\mathfrak{F}^{IN}_{\mathfrak{R}_1}(e_\sigma)\right)^2 - \left(\mathfrak{F}^{IN}_{\mathfrak{R}_2}(e_\sigma)\right)^2\right| \end{array}\right) \end{array}\right)\right)^{\frac{1}{\lambda}},$$

$$\lambda \geq 1$$

Theorem 5.4 The GHBCFNSM satisfies the underneath properties.

1. $0 \leq \mathcal{S}^{7}_{BCFS}(\mathfrak{R}_1, \mathfrak{R}_2) \leq 1$
2. $\mathcal{S}^{7}_{BCFS}(\mathfrak{R}_1, \mathfrak{R}_2) = 1 \Leftrightarrow \mathfrak{R}_1 = \mathfrak{R}_2$
3. $\mathcal{S}^{7}_{BCFS}(\mathfrak{R}_1, \mathfrak{R}_2) = \mathcal{S}^{7}_{BCFS}(\mathfrak{R}_2, \mathfrak{R}_1)$
4. If $\mathfrak{R}_1 \subset \mathfrak{R}_2 \subset \mathfrak{R}_3$, then $\mathcal{S}^{7}_{BCFS}(\mathfrak{R}_1, \mathfrak{R}_2) \geq \mathcal{S}^{7}_{BCFS}(\mathfrak{R}_1, \mathfrak{R}_3)$ and $\mathcal{S}^{7}_{BCFS}(\mathfrak{R}_2, \mathfrak{R}_3) \geq \mathcal{S}^{7}_{BCFS}(\mathfrak{R}_1, \mathfrak{R}_3)$

Proof Correspondingly, Theorem (5.2).

5.2 Weighted Similarity Measures for BCFSs

In a few genuine-life dilemmas, the factors typically have varying degrees of significance, necessitating the assignment of various weights to them. In other cases, a decision analyst may choose to prioritize certain elements above others based on personal preferences. Decision analysts need weighted SMs in this circumstance. Therefore, in this section, we construct generalized BCF weighted normalized Hamming SM (GBCFWNHSM), generalized BCF weighted normalized Hausdorff SM (GBCFWNHDSM), generalized hybrid BCF weighted normalized SM (GHBCFWNSM) and will discuss their special cases.

Definition 5.4 Let two BCFSs $\Re_1 = \left\{e_\sigma, \left(\mathfrak{F}^P_{\Re_1}(e_\sigma), \mathfrak{F}^N_{\Re_1}(e_\sigma)\right)|e_\sigma \in \mathbb{E}, \sigma = 1, 2, .., n\right\} = \left\{e_\sigma, \left(\mathfrak{F}^{RP}_{\Re_1}(e_\sigma) + \iota\mathfrak{F}^{IP}_{\Re_1}(e_\sigma), \mathfrak{F}^{RN}_{\Re_1}(e_\sigma) + \iota\mathfrak{F}^{IN}_{\Re_1}(e_\sigma)\right)|e_\sigma \in \mathbb{E}, \sigma = 1, 2, .., n\right\}$ and $\Re_2 = \left\{e, \left(\mathfrak{F}^P_{\Re_2}(e_\sigma), \mathfrak{F}^N_{\Re_2}(e_\sigma)\right)|e \in \mathbb{E}, \sigma = 1, 2, .., n\right\} = \left\{e_\sigma, \left(\mathfrak{F}^{RP}_{\Re_2}(e_\sigma) + \iota\mathfrak{F}^{IP}_{\Re_2}(e_\sigma), \mathfrak{F}^{RN}_{\Re_2}(e_\sigma) + \iota\mathfrak{F}^{IN}_{\Re_2}(e_\sigma)\right)|e_\sigma \in \mathbb{E}, \sigma = 1, 2, .., n\right\}$ over $\mathbb{E}$. Then the GBCFWNHSM between $\Re_1$ and $\Re_2$ is inaugurated as.

$$
\begin{aligned}
&\mathcal{S}^8_{BCFS}(\Re_1, \Re_2) = 1 \\
&- \left(\frac{1}{4}\sum_{\sigma=1}^{n} \overset{\circ}{\underset{\sigma}{w}} \left(\begin{array}{c} \left|\left(\mathfrak{F}^{RP}_{\Re_1}(e_\sigma)\right)^2 - \left(\mathfrak{F}^{RP}_{\Re_2}(e_\sigma)\right)^2\right| + \left|\left(\mathfrak{F}^{IP}_{\Re_1}(e_\sigma)\right)^2 - \left(\mathfrak{F}^{IP}_{\Re_2}(e_\sigma)\right)^2\right| \\ + \left|\left(\mathfrak{F}^{RN}_{\Re_1}(e_\sigma)\right)^2 - \left(\mathfrak{F}^{RN}_{\Re_2}(e_\sigma)\right)^2\right| + \left|\left(\mathfrak{F}^{IN}_{\Re_1}(e_\sigma)\right)^2 - \left(\mathfrak{F}^{IN}_{\Re_2}(e_\sigma)\right)^2\right| \end{array} \right)\right)^{\frac{1}{\lambda}}, \\
&\ddot{u} \geq 1
\end{aligned}
$$

Observed that $\overset{\circ}{w} = \left(\overset{\circ}{w}_1, \overset{\circ}{w}_2, \ldots, \overset{\circ}{w}_y\right)^T$ is a weight vector and $0 \leq \overset{\circ}{\underset{\sigma}{w}} \leq 1$ for all σ with $\sum_{\sigma=1}^{y} \overset{\circ}{\underset{\sigma}{w}} = 1$

Theorem 5.5 The GBCFWNHDM satisfies the underneath properties.

1. $0 \leq \mathcal{S}^8_{BCFS}(\Re_1, \Re_2) \leq 1$
2. $\mathcal{S}^8_{BCFS}(\Re_1, \Re_2) = 1 \Leftrightarrow \Re_1 = \Re_2$
3. $\mathcal{S}^8_{BCFS}(\Re_1, \Re_2) = \mathcal{S}^8_{BCFS}(\Re_2, \Re_1)$
4. If $\Re_1 \subset \Re_2 \subset \Re_3$, then $\mathcal{S}^8_{BCFS}(\Re_1, \Re_2) \geq \mathcal{S}^8_{BCFS}(\Re_1, \Re_3)$ and $\mathcal{S}^8_{BCFS}(\Re_2, \Re_3) \geq \mathcal{S}^8_{BCFS}(\Re_1, \Re_3)$

Proof Omitted.

Especially,

i. By considering $\lambda = 1$, the GBCFWNHSM would transform to BCF weighted normalized Hamming SM (BCFWNHSM)

$$
\begin{aligned}
&\mathcal{S}^9_{BCFS}(\Re_1, \Re_2) = 1 \\
&- \frac{1}{4}\sum_{\sigma=1}^{n} \overset{\circ}{\underset{\sigma}{w}} \left(\begin{array}{c} \left|\left(\mathfrak{F}^{RP}_{\Re_1}(e_\sigma)\right)^2 - \left(\mathfrak{F}^{RP}_{\Re_2}(e_\sigma)\right)^2\right| + \left|\left(\mathfrak{F}^{IP}_{\Re_1}(e_\sigma)\right)^2 - \left(\mathfrak{F}^{IP}_{\Re_2}(e_\sigma)\right)^2\right| \\ + \left|\left(\mathfrak{F}^{RN}_{\Re_1}(e_\sigma)\right)^2 - \left(\mathfrak{F}^{RN}_{\Re_2}(e_\sigma)\right)^2\right| + \left|\left(\mathfrak{F}^{IN}_{\Re_1}(e_\sigma)\right)^2 - \left(\mathfrak{F}^{IN}_{\Re_2}(e_\sigma)\right)^2\right| \end{array} \right)
\end{aligned}
$$

ii. By considering $\lambda = 2$, the GBCFWNHSM would transform to BCF weighted normalized Euclidean SM (BCFWNESM)

$$\mathcal{S}^{10}_{BCFS}(\mathfrak{R}_1, \mathfrak{R}_2) = 1 - \sqrt{\frac{1}{4}\sum_{\sigma=1}^{n} \mathop{\mathbf{w}}_{\sigma}^{\circ} \begin{pmatrix} \left|\left(\mathfrak{F}^{RP}_{\mathfrak{R}_1}(e_\sigma)\right)^2 - \left(\mathfrak{F}^{RP}_{\mathfrak{R}_2}(e_\sigma)\right)^2\right| + \left|\left(\mathfrak{F}^{IP}_{\mathfrak{R}_1}(e_\sigma)\right)^2 - \left(\mathfrak{F}^{IP}_{\mathfrak{R}_2}(e_\sigma)\right)^2\right| \\ + \left|\left(\mathfrak{F}^{RN}_{\mathfrak{R}_1}(e_\sigma)\right)^2 - \left(\mathfrak{F}^{RN}_{\mathfrak{R}_2}(e_\sigma)\right)^2\right| + \left|\left(\mathfrak{F}^{IN}_{\mathfrak{R}_1}(e_\sigma)\right)^2 - \left(\mathfrak{F}^{IN}_{\mathfrak{R}_2}(e_\sigma)\right)^2\right| \end{pmatrix}}$$

Definition 5.5 Let two BCFSs $\mathfrak{R}_1 = \left\{e_\sigma, \left(\mathfrak{F}^{P}_{\mathfrak{R}_1}(e_\sigma), \mathfrak{F}^{N}_{\mathfrak{R}_1}(e_\sigma)\right) | e_\sigma \in \mathbb{E}, \sigma = 1, 2, .., n\right\} = \left\{e_\sigma, \left(\mathfrak{F}^{RP}_{\mathfrak{R}_1}(e_\sigma) + \iota\mathfrak{F}^{IP}_{\mathfrak{R}_1}(e_\sigma), \mathfrak{F}^{RN}_{\mathfrak{R}_1}(e_\sigma) + \iota\mathfrak{F}^{IN}_{\mathfrak{R}_1}(e_\sigma)\right) | e_\sigma \in \mathbb{E}, \sigma = 1, 2, .., n\right\}$ and $\mathfrak{R}_2 = \left\{e, \left(\mathfrak{F}^{P}_{\mathfrak{R}_2}(e_\sigma), \mathfrak{F}^{N}_{\mathfrak{R}_2}(e_\sigma)\right) | e \in \mathbb{E}, \sigma = 1, 2, .., n\right\} = \left\{e_\sigma, \left(\mathfrak{F}^{RP}_{\mathfrak{R}_2}(e_\sigma) + \iota\mathfrak{F}^{IP}_{\mathfrak{R}_2}(e_\sigma), \mathfrak{F}^{RN}_{\mathfrak{R}_2}(e_\sigma) + \iota\mathfrak{F}^{IN}_{\mathfrak{R}_2}(e_\sigma)\right) | e_\sigma \in \mathbb{E}, \sigma = 1, 2, .., n\right\}$ over $\mathbb{E}$. Then the GBCFWNHDSM between $\mathfrak{R}_1$ and $\mathfrak{R}_2$ is inaugurated as.

$$\mathcal{S}^{11}_{BCFS}(\mathfrak{R}_1, \mathfrak{R}_2) = 1 - \left(\sum_{\sigma=1}^{n} \mathop{w}_{\sigma}^{\circ} \max \begin{pmatrix} \left|\left(\mathfrak{F}^{RP}_{\mathfrak{R}_1}(e_\sigma)\right)^2 - \left(\mathfrak{F}^{RP}_{\mathfrak{R}_2}(e_\sigma)\right)^2\right|, \left|\left(\mathfrak{F}^{IP}_{\mathfrak{R}_1}(e_\sigma)\right)^2 - \left(\mathfrak{F}^{IP}_{\mathfrak{R}_2}(e_\sigma)\right)^2\right| \\ , \left|\left(\mathfrak{F}^{RN}_{\mathfrak{R}_1}(e_\sigma)\right)^2 - \left(\mathfrak{F}^{RN}_{\mathfrak{R}_2}(e_\sigma)\right)^2\right|, \left|\left(\mathfrak{F}^{IN}_{\mathfrak{R}_1}(e_\sigma)\right)^2 - \left(\mathfrak{F}^{IN}_{\mathfrak{R}_2}(e_\sigma)\right)^2\right| \end{pmatrix}\right)^{\frac{1}{\lambda}}, \quad \lambda \geq 1$$

Observed that $\overset{\circ}{\mathrm{w}} = \left(\overset{\circ}{\mathrm{w}}_1, \overset{\circ}{\mathrm{w}}_2, \ldots, \overset{\circ}{\mathrm{w}}_y\right)^T$ is a weight vector and $0 \leq \mathop{\mathrm{w}}_{\sigma}^{\circ} \leq 1$ for all σ with $\sum_{\sigma=1}^{y} \mathop{\mathrm{w}}_{\sigma}^{\circ} = 1$

Theorem 5.6 The GBCFWNHDSM satisfies the underneath properties.

1. $0 \leq \mathcal{S}^{11}_{BCFS}(\mathfrak{R}_1, \mathfrak{R}_2) \leq 1$
2. $\mathcal{S}^{11}_{BCFS}(\mathfrak{R}_1, \mathfrak{R}_2) = 1 \Leftrightarrow \mathfrak{R}_1 = \mathfrak{R}_2$
3. $\mathcal{S}^{11}_{BCFS}(\mathfrak{R}_1, \mathfrak{R}_2) = \mathcal{S}^{11}_{BCFS}(\mathfrak{R}_2, \mathfrak{R}_1)$
4. If $\mathfrak{R}_1 \subset \mathfrak{R}_2 \subset \mathfrak{R}_3$, then $\mathcal{S}^{11}_{BCFS}(\mathfrak{R}_1, \mathfrak{R}_2) \geq \mathcal{S}^{11}_{BCFS}(\mathfrak{R}_1, \mathfrak{R}_3)$ and $\mathcal{S}^{11}_{BCFS}(\mathfrak{R}_2, \mathfrak{R}_3) \geq \mathcal{S}^{11}_{BCFS}(\mathfrak{R}_1, \mathfrak{R}_3)$

Proof Omitted.

Especially,

i. By considering $\lambda = 1$, the GBCFWNHDSM would transform to BCF weighted normalized Hamming-Hausdorff SM (BCFWNHHDSM)

$$\mathcal{S}^{12}_{BCFS}(\mathfrak{R}_1, \mathfrak{R}_2) = 1 - \sum_{\sigma=1}^{n} \mathop{\mathrm{w}}_{\sigma}^{\circ} \max \begin{pmatrix} \left|\left(\mathfrak{F}^{RP}_{\mathfrak{R}_1}(e_\sigma)\right)^2 - \left(\mathfrak{F}^{RP}_{\mathfrak{R}_2}(e_\sigma)\right)^2\right|, \left|\left(\mathfrak{F}^{IP}_{\mathfrak{R}_1}(e_\sigma)\right)^2 - \left(\mathfrak{F}^{IP}_{\mathfrak{R}_2}(e_\sigma)\right)^2\right| \\ , \left|\left(\mathfrak{F}^{RN}_{\mathfrak{R}_1}(e_\sigma)\right)^2 - \left(\mathfrak{F}^{RN}_{\mathfrak{R}_2}(e_\sigma)\right)^2\right|, \left|\left(\mathfrak{F}^{IN}_{\mathfrak{R}_1}(e_\sigma)\right)^2 - \left(\mathfrak{F}^{IN}_{\mathfrak{R}_2}(e_\sigma)\right)^2\right| \end{pmatrix}$$

ii. By considering $\lambda = 2$, the GBCFWNHDSM would reduce to BCF weighted normalized Euclidean-Hausdorff SM (BCFWNEHDSM)

$$\mathcal{S}^{13}_{BCFS}(\Re_1, \Re_2) = 1 - \sqrt{\sum_{\sigma=1}^{n} \overset{\circ}{w}_{\sigma} \max\left(\begin{array}{c} \left|\left(\mathfrak{F}^{RP}_{\Re_1}(e_\sigma)\right)^2 - \left(\mathfrak{F}^{RP}_{\Re_2}(e_\sigma)\right)^2\right|, \left|\left(\mathfrak{F}^{IP}_{\Re_1}(e_\sigma)\right)^2 - \left(\mathfrak{F}^{IP}_{\Re_2}(e_\sigma)\right)^2\right| \\ , \left|\left(\mathfrak{F}^{RN}_{\Re_1}(e_\sigma)\right)^2 - \left(\mathfrak{F}^{RN}_{\Re_2}(e_\sigma)\right)^2\right|, \left|\left(\mathfrak{F}^{IN}_{\Re_1}(e_\sigma)\right)^2 - \left(\mathfrak{F}^{IN}_{\Re_2}(e_\sigma)\right)^2\right| \end{array} \right)}$$

Definition 5.6 Let two BCFSs $\Re_1 = \left\{e_\sigma, \left(\mathfrak{F}^{P}_{\Re_1}(e_\sigma), \mathfrak{F}^{N}_{\Re_1}(e_\sigma)\right) | e_\sigma \in \mathbb{E}, \sigma = 1, 2, .., n\right\} = \left\{e_\sigma, \left(\mathfrak{F}^{RP}_{\Re_1}(e_\sigma) + \iota\mathfrak{F}^{IP}_{\Re_1}(e_\sigma), \mathfrak{F}^{RN}_{\Re_1}(e_\sigma) + \iota\mathfrak{F}^{IN}_{\Re_1}(e_\sigma)\right) | e_\sigma \in \mathbb{E}, \sigma = 1, 2, .., n\right\}$ and $\Re_2 = \left\{e, \left(\mathfrak{F}^{P}_{\Re_2}(e_\sigma), \mathfrak{F}^{N}_{\Re_2}(e_\sigma)\right) | e \in \mathbb{E}, \sigma = 1, 2, .., n\right\} = \left\{e_\sigma, \left(\mathfrak{F}^{RP}_{\Re_2}(e_\sigma) + \iota\mathfrak{F}^{IP}_{\Re_2}(e_\sigma), \mathfrak{F}^{RN}_{\Re_2}(e_\sigma) + \iota\mathfrak{F}^{IN}_{\Re_2}(e_\sigma)\right) | e_\sigma \in \mathbb{E}, \sigma = 1, 2, .., n\right\}$ over $\mathbb{E}$. Then the GHBCFWNSM between $\Re_1$ and $\Re_2$ is inaugurated as.

$$\mathcal{S}^{14}_{BCFS}(\Re_1, \Re_2) = 1 - \left(\frac{1}{2} \sum_{\sigma=1}^{n} \overset{\circ}{w}_{\sigma} \left(\begin{array}{c} \frac{1}{4}\left(\begin{array}{c} \left|\left(\mathfrak{F}^{RP}_{\Re_1}(e_\sigma)\right)^2 - \left(\mathfrak{F}^{RP}_{\Re_2}(e_\sigma)\right)^2\right| + \left|\left(\mathfrak{F}^{IP}_{\Re_1}(e_\sigma)\right)^2 - \left(\mathfrak{F}^{IP}_{\Re_2}(e_\sigma)\right)^2\right| \\ + \left|\left(\mathfrak{F}^{RN}_{\Re_1}(e_\sigma)\right)^2 - \left(\mathfrak{F}^{RN}_{\Re_2}(e_\sigma)\right)^2\right| + \left|\left(\mathfrak{F}^{IN}_{\Re_1}(e_\sigma)\right)^2 - \left(\mathfrak{F}^{IN}_{\Re_2}(e_\sigma)\right)^2\right| \end{array} \right) \\ + \max\left(\begin{array}{c} \left|\left(\mathfrak{F}^{RP}_{\Re_1}(e_\sigma)\right)^2 - \left(\mathfrak{F}^{RP}_{\Re_2}(e_\sigma)\right)^2\right|, \left|\left(\mathfrak{F}^{IP}_{\Re_1}(e_\sigma)\right)^2 - \left(\mathfrak{F}^{IP}_{\Re_2}(e_\sigma)\right)^2\right| \\ , \left|\left(\mathfrak{F}^{RN}_{\Re_1}(e_\sigma)\right)^2 - \left(\mathfrak{F}^{RN}_{\Re_2}(e_\sigma)\right)^2\right|, \left|\left(\mathfrak{F}^{IN}_{\Re_1}(e_\sigma)\right)^2 - \left(\mathfrak{F}^{IN}_{\Re_2}(e_\sigma)\right)^2\right| \end{array} \right) \end{array} \right) \right)^{\frac{1}{\lambda}},$$

$$\lambda \geq 1$$

Observed that $\overset{\circ}{\mathrm{w}} = \left(\overset{\circ}{\mathrm{w}}_1, \overset{\circ}{\mathrm{w}}_2, \ldots, \overset{\circ}{\mathrm{w}}_y\right)^T$ is a weight vector and $0 \leq \overset{\circ}{\mathrm{w}}_\sigma \leq 1$ for all σ with $\sum_{\sigma=1}^{y} \overset{\circ}{\mathrm{w}}_\sigma = 1$

$\mathcal{S}^{14}_{BCFS}(\Re_1, \Re_2)$ also holds the necessary conditions of the SM.

5.3 Continuous Similarity Measures for BCFSs

The universal set $\mathbb{E}$ and the associated weights were discrete in the previously described SMs. Now we interpret the SMs below for scenarios where the universal $\mathbb{E}$ and weight are continuous. For this section, the weight of $e \in \mathbb{E} = [m, n]$ is $\overset{\circ}{\mathrm{w}}(e)$, $\overset{\circ}{\mathrm{w}}(e) \in [0, 1]$ and $\int_m^n \overset{\circ}{\mathrm{w}}(e)de$.

In the next definition, the generalized continuous BCF weighted normalized Hamming SM (GCBCFWNHSM) is interpreted.

Definition 5.7 Let two BCFSs $\Re_1$ and $\Re_2$ over $\mathbb{E} = [m, n]$. Then the GCBCFWNHSM between $\Re_1$ and $\Re_2$ is inaugurated as.

$$\mathcal{S}^{15}_{BCFS}(\Re_1, \Re_2) = 1 - \left(\frac{1}{4}\int_m^n \overset{\circ}{w}(e)\left(\begin{array}{c}\left|\left(\mathfrak{F}^{RP}_{\Re_1}(e_\sigma)\right)^2 - \left(\mathfrak{F}^{RP}_{\Re_2}(e_\sigma)\right)^2\right| + \left|\left(\mathfrak{F}^{IP}_{\Re_1}(e_\sigma)\right)^2 - \left(\mathfrak{F}^{IP}_{\Re_2}(e_\sigma)\right)^2\right| \\ + \left|\left(\mathfrak{F}^{RN}_{\Re_1}(e_\sigma)\right)^2 - \left(\mathfrak{F}^{RN}_{\Re_2}(e_\sigma)\right)^2\right| + \left|\left(\mathfrak{F}^{IN}_{\Re_1}(e_\sigma)\right)^2 - \left(\mathfrak{F}^{IN}_{\Re_2}(e_\sigma)\right)^2\right|\end{array}\right)de\right)^{\frac{1}{\lambda}},$$

$\lambda \geq 1.$

By putting $\overset{\circ}{w}(e) = \frac{1}{m-n}$, for any $e \in [m, n]$, the GCBCFWNHSM will convert to generalized continuous bipolar complex fuzzy normalized hamming SM (GCBCFNHSM).

$$\mathcal{S}^{16}_{BCFS}(\Re_1, \Re_2) = 1 - \left(\frac{1}{4(m-n)}\int_m^n \left(\begin{array}{c}\left|\left(\mathfrak{F}^{RP}_{\Re_1}(e_\sigma)\right)^2 - \left(\mathfrak{F}^{RP}_{\Re_2}(e_\sigma)\right)^2\right| + \left|\left(\mathfrak{F}^{IP}_{\Re_1}(e_\sigma)\right)^2 - \left(\mathfrak{F}^{IP}_{\Re_2}(e_\sigma)\right)^2\right| \\ + \left|\left(\mathfrak{F}^{RN}_{\Re_1}(e_\sigma)\right)^2 - \left(\mathfrak{F}^{RN}_{\Re_2}(e_\sigma)\right)^2\right| + \left|\left(\mathfrak{F}^{IN}_{\Re_1}(e_\sigma)\right)^2 - \left(\mathfrak{F}^{IN}_{\Re_2}(e_\sigma)\right)^2\right|\end{array}\right)de\right)^{\frac{1}{\lambda}},$$

$\ddot{u} \geq 1.$

Especially.

i. By considering $\lambda = 1$, the GCBCFWNHSM would convert to continuous BCFWNHSM (CBCFWNHSM)

$$\mathcal{S}^{17}_{BCFS}(\Re_1, \Re_2) = 1 - \frac{1}{4}\int_m^n \overset{\circ}{w}(e)\left(\begin{array}{c}\left|\left(\mathfrak{F}^{RP}_{\Re_1}(e_\sigma)\right)^2 - \left(\mathfrak{F}^{RP}_{\Re_2}(e_\sigma)\right)^2\right| + \left|\left(\mathfrak{F}^{IP}_{\Re_1}(e_\sigma)\right)^2 - \left(\mathfrak{F}^{IP}_{\Re_2}(e_\sigma)\right)^2\right| \\ + \left|\left(\mathfrak{F}^{RN}_{\Re_1}(e_\sigma)\right)^2 - \left(\mathfrak{F}^{RN}_{\Re_2}(e_\sigma)\right)^2\right| + \left|\left(\mathfrak{F}^{IN}_{\Re_1}(e_\sigma)\right)^2 - \left(\mathfrak{F}^{IN}_{\Re_2}(e_\sigma)\right)^2\right|\end{array}\right)de$$

By putting $\overset{\circ}{w}(e) = \frac{1}{m-n'}$, for any $e \in [m, n]$, the CBCFWNHSM will convert to continuous BCFNHSM (CBCFNHSM)

$$\mathcal{S}^{18}_{BCFS}(\Re_1, \Re_2) = 1 - \frac{1}{4(m-n)}\int_m^n \left(\begin{array}{c}\left|\left(\mathfrak{F}^{RP}_{\Re_1}(e_\sigma)\right)^2 - \left(\mathfrak{F}^{RP}_{\Re_2}(e_\sigma)\right)^2\right| + \left|\left(\mathfrak{F}^{IP}_{\Re_1}(e_\sigma)\right)^2 - \left(\mathfrak{F}^{IP}_{\Re_2}(e_\sigma)\right)^2\right| \\ + \left|\left(\mathfrak{F}^{RN}_{\Re_1}(e_\sigma)\right)^2 - \left(\mathfrak{F}^{RN}_{\Re_2}(e_\sigma)\right)^2\right| + \left|\left(\mathfrak{F}^{IN}_{\Re_1}(e_\sigma)\right)^2 - \left(\mathfrak{F}^{IN}_{\Re_2}(e_\sigma)\right)^2\right|\end{array}\right)de$$

ii. By considering $\lambda = 2$, the GCBCFWNHSM would convert to continuous BCFWNESM (CBCFWNESM)

$$\mathcal{S}^{19}_{BCFS}(\Re_1, \Re_2) = 1$$

$$-\left(\frac{1}{4}\int_m^n \overset{\circ}{\mathrm{w}}(e)\begin{pmatrix}\left|\left(\mathfrak{F}^{RP}_{\mathfrak{R}_1}(e_\sigma)\right)^2-\left(\mathfrak{F}^{RP}_{\mathfrak{R}_2}(e_\sigma)\right)^2\right|+\left|\left(\mathfrak{F}^{IP}_{\mathfrak{R}_1}(e_\sigma)\right)^2-\left(\mathfrak{F}^{IP}_{\mathfrak{R}_2}(e_\sigma)\right)^2\right|\\+\left|\left(\mathfrak{F}^{RN}_{\mathfrak{R}_1}(e_\sigma)\right)^2-\left(\mathfrak{F}^{RN}_{\mathfrak{R}_2}(e_\sigma)\right)^2\right|+\left|\left(\mathfrak{F}^{IN}_{\mathfrak{R}_1}(e_\sigma)\right)^2-\left(\mathfrak{F}^{IN}_{\mathfrak{R}_2}(e_\sigma)\right)^2\right|\end{pmatrix}de\right)^{\frac{1}{2}}$$

By putting $\overset{\circ}{\mathrm{W}}(e)=\frac{1}{m-n'}$, for any $e\in[m,n]$, the CBCFWNESM will convert to continuous BCFNESM (CBCFNESM)

$$\mathcal{S}^{20}_{BCFS}(\mathfrak{R}_1,\mathfrak{R}_2)=1$$
$$-\left(\frac{1}{4(m-n)}\int_m^n \begin{pmatrix}\left|\left(\mathfrak{F}^{RP}_{\mathfrak{R}_1}(e_\sigma)\right)^2-\left(\mathfrak{F}^{RP}_{\mathfrak{R}_2}(e_\sigma)\right)^2\right|+\left|\left(\mathfrak{F}^{IP}_{\mathfrak{R}_1}(e_\sigma)\right)^2-\left(\mathfrak{F}^{IP}_{\mathfrak{R}_2}(e_\sigma)\right)^2\right|\\+\left|\left(\mathfrak{F}^{RN}_{\mathfrak{R}_1}(e_\sigma)\right)^2-\left(\mathfrak{F}^{RN}_{\mathfrak{R}_2}(e_\sigma)\right)^2\right|+\left|\left(\mathfrak{F}^{IN}_{\mathfrak{R}_1}(e_\sigma)\right)^2-\left(\mathfrak{F}^{IN}_{\mathfrak{R}_2}(e_\sigma)\right)^2\right|\end{pmatrix}de\right)^{\frac{1}{2}}$$

In the next definition, the generalized continuous BCF weighted normalized Hausdorff SM (GCBCFWNHDSM) is interpreted.

Definition 5.8 Let two BCFSs $\mathfrak{R}_1$ and $\mathfrak{R}_2$ over $\mathbb{E}=[m,n]$. Then the GCBCFWNHDSM between $\mathfrak{R}_1$ and $\mathfrak{R}_2$ is inaugurated as.

$$\mathcal{S}^{21}_{BCFS}(\mathfrak{R}_1,\mathfrak{R}_2)=1$$
$$-\left(\int_m^n \overset{\circ}{\mathrm{w}}(e)\max\begin{pmatrix}\left|\left(\mathfrak{F}^{RP}_{\mathfrak{R}_1}(e_\sigma)\right)^2-\left(\mathfrak{F}^{RP}_{\mathfrak{R}_2}(e_\sigma)\right)^2\right|,\left|\left(\mathfrak{F}^{IP}_{\mathfrak{R}_1}(e_\sigma)\right)^2-\left(\mathfrak{F}^{IP}_{\mathfrak{R}_2}(e_\sigma)\right)^2\right|,\\\left|\left(\mathfrak{F}^{RN}_{\mathfrak{R}_1}(e_\sigma)\right)^2-\left(\mathfrak{F}^{RN}_{\mathfrak{R}_2}(e_\sigma)\right)^2\right|,\left|\left(\mathfrak{F}^{IN}_{\mathfrak{R}_1}(e_\sigma)\right)^2-\left(\mathfrak{F}^{IN}_{\mathfrak{R}_2}(e_\sigma)\right)^2\right|\end{pmatrix}de\right)^{\frac{1}{\lambda}},$$
$$\lambda\geq 1.$$

By putting $\overset{\circ}{\mathrm{w}}(e)=\frac{1}{m-n'}$, for any $e\in[m,n]$, the GCBCFWNHDSM will convert to generalized continuous bipolar complex fuzzy normalized Hausdorff SM (GCBCFNHDSM).

$$\mathcal{S}^{22}_{BCFS}(\mathfrak{R}_1,\mathfrak{R}_2)=1$$
$$-\left(\frac{1}{(m-n)}\int_m^n \max\begin{pmatrix}\left|\left(\mathfrak{F}^{RP}_{\mathfrak{R}_1}(e_\sigma)\right)^2-\left(\mathfrak{F}^{RP}_{\mathfrak{R}_2}(e_\sigma)\right)^2\right|,\left|\left(\mathfrak{F}^{IP}_{\mathfrak{R}_1}(e_\sigma)\right)^2-\left(\mathfrak{F}^{IP}_{\mathfrak{R}_2}(e_\sigma)\right)^2\right|,\\\left|\left(\mathfrak{F}^{RN}_{\mathfrak{R}_1}(e_\sigma)\right)^2-\left(\mathfrak{F}^{RN}_{\mathfrak{R}_2}(e_\sigma)\right)^2\right|,\left|\left(\mathfrak{F}^{IN}_{\mathfrak{R}_1}(e_\sigma)\right)^2-\left(\mathfrak{F}^{IN}_{\mathfrak{R}_2}(e_\sigma)\right)^2\right|\end{pmatrix}de\right)^{\frac{1}{\lambda}},$$
$$\lambda\geq 1.$$

Especially,

i. By considering $\lambda=1.$, the GCBCFWNHDSM would convert to continuous BCFWNHHDSM (CBCFWNHHDSM)

$$\mathcal{S}^{23}_{BCFS}(\mathfrak{R}_1,\mathfrak{R}_2)=1$$

$$-\int_m^n \overset{\circ}{w}(e)\max\begin{pmatrix}\left|\left(\Im_{\Re_1}^{RP}(e_\sigma)\right)^2-\left(\Im_{\Re_2}^{RP}(e_\sigma)\right)^2\right|,\left|\left(\Im_{\Re_1}^{IP}(e_\sigma)\right)^2-\left(\Im_{\Re_2}^{IP}(e_\sigma)\right)^2\right|,\\ \left|\left(\Im_{\Re_1}^{RN}(e_\sigma)\right)^2-\left(\Im_{\Re_2}^{RN}(e_\sigma)\right)^2\right|,\left|\left(\Im_{\Re_1}^{IN}(e_\sigma)\right)^2-\left(\Im_{\Re_2}^{IN}(e_\sigma)\right)^2\right|\end{pmatrix}de$$

By putting $\overset{\circ}{w}(e)=\frac{1}{m-n'}$, for any $e\in[m,n]$, the CBCFWNHHDSM will convert to continuous BCFNHHDSM (CBCFNHHDSM)

$$\mathcal{S}_{BCFS}^{24}(\Re_1,\Re_2)=1-\frac{1}{(m-n)}\int_m^n\max\begin{pmatrix}\left|\left(\mathfrak{F}_{\Re_1}^{RP}(e_\sigma)\right)^2-\left(\mathfrak{F}_{\Re_2}^{RP}(e_\sigma)\right)^2\right|,\left|\left(\mathfrak{F}_{\Re_1}^{IP}(e_\sigma)\right)^2-\left(\mathfrak{F}_{\Re_2}^{IP}(e_\sigma)\right)^2\right|,\\ \left|\left(\mathfrak{F}_{\Re_1}^{RN}(e_\sigma)\right)^2-\left(\mathfrak{F}_{\Re_2}^{RN}(e_\sigma)\right)^2\right|,\left|\left(\mathfrak{F}_{\Re_1}^{IN}(e_\sigma)\right)^2-\left(\mathfrak{F}_{\Re_2}^{IN}(e_\sigma)\right)^2\right|\end{pmatrix}de$$

ii. By considering $\lambda=2$, the GCBCFWNHDSM would convert to continuous BCFWNEHDSM (CBCFWNEHDSM)

$$\mathcal{S}_{BCFS}^{25}(\Re_1,\Re_2)=1-\left(\int_m^n\overset{\circ}{w}(e)\max\begin{pmatrix}\left|\left(\mathfrak{F}_{\Re_1}^{RP}(e_\sigma)\right)^2-\left(\mathfrak{F}_{\Re_2}^{RP}(e_\sigma)\right)^2\right|,\left|\left(\mathfrak{F}_{\Re_1}^{IP}(e_\sigma)\right)^2-\left(\mathfrak{F}_{\Re_2}^{IP}(e_\sigma)\right)^2\right|,\\ \left|\left(\mathfrak{F}_{\Re_1}^{RN}(e_\sigma)\right)^2-\left(\mathfrak{F}_{\Re_2}^{RN}(e_\sigma)\right)^2\right|,\left|\left(\mathfrak{F}_{\Re_1}^{IN}(e_\sigma)\right)^2-\left(\mathfrak{F}_{\Re_2}^{IN}(e_\sigma)\right)^2\right|\end{pmatrix}de\right)^{\frac{1}{2}}$$

By putting $\overset{\circ}{w}(e)=\frac{1}{m-n'}$, for any $e\in[m,n]$, the CBCFWNEHDSM will convert to continuous CBCFNEHDSSM (CBCFNEHDSM)

$$\mathcal{S}_{BCFS}^{26}(\Re_1,\Re_2)=1-\left(\frac{1}{m-n}\int_m^n\max\begin{pmatrix}\left|\left(\mathfrak{F}_{\Re_1}^{RP}(e_\sigma)\right)^2-\left(\mathfrak{F}_{\Re_2}^{RP}(e_\sigma)\right)^2\right|,\left|\left(\mathfrak{F}_{\Re_1}^{IP}(e_\sigma)\right)^2-\left(\mathfrak{F}_{\Re_2}^{IP}(e_\sigma)\right)^2\right|,\\ \left|\left(\mathfrak{F}_{\Re_1}^{RN}(e_\sigma)\right)^2-\left(\mathfrak{F}_{\Re_2}^{RN}(e_\sigma)\right)^2\right|,\left|\left(\mathfrak{F}_{\Re_1}^{IN}(e_\sigma)\right)^2-\left(\mathfrak{F}_{\Re_2}^{IN}(e_\sigma)\right)^2\right|\end{pmatrix}de\right)^{\frac{1}{2}}$$

Below the generalized continuous hybrid BCF weighted normalized SM (GCHBCFWNSM) is demonstrated.

Definition 5.9 Let two BCFSs $\Re_1$ and $\Re_2$ over $\mathbb{E}=[m,n]$. Then the GCHBCFWNSM between $\Re_1$ and $\Re_2$ is inaugurated as.

$$\delta_{BCFS}^{27}(\Re_1,\Re_2)=1-\left(\frac{1}{2}\int_m^n\overset{\circ}{w}(e)\begin{pmatrix}\frac{1}{4}\begin{pmatrix}\left|\left(\mathfrak{F}_{\Re_1}^{RP}(e_\sigma)\right)^2-\left(\mathfrak{F}_{\Re_2}^{RP}(e_\sigma)\right)^2\right|+\left|\left(\mathfrak{F}_{\Re_1}^{IP}(e_\sigma)\right)^2-\left(\mathfrak{F}_{\Re_2}^{IP}(e_\sigma)\right)^2\right|\\ +\left|\left(\mathfrak{F}_{\Re_1}^{RN}(e_\sigma)\right)^2-\left(\mathfrak{F}_{\Re_2}^{RN}(e_\sigma)\right)^2\right|+\left|\left(\mathfrak{F}_{\Re_1}^{IN}(e_\sigma)\right)^2-\left(\mathfrak{F}_{\Re_2}^{IN}(e_\sigma)\right)^2\right|\end{pmatrix}\\ +\max\begin{pmatrix}\left|\left(\mathfrak{F}_{\Re_1}^{RP}(e_\sigma)\right)^2-\left(\mathfrak{F}_{\Re_2}^{RP}(e_\sigma)\right)^2\right|,\left|\left(\mathfrak{F}_{\Re_1}^{IP}(e_\sigma)\right)^2-\left(\mathfrak{F}_{\Re_2}^{IP}(e_\sigma)\right)^2\right|\\ \left|\left(\mathfrak{F}_{\Re_1}^{RN}(e_\sigma)\right)^2-\left(\mathfrak{F}_{\Re_2}^{RN}(e_\sigma)\right)^2\right|,\left|\left(\mathfrak{F}_{\Re_1}^{IN}(e_\sigma)\right)^2-\left(\mathfrak{F}_{\Re_2}^{IN}(e_\sigma)\right)^2\right|\end{pmatrix}\end{pmatrix}de\right)^{\frac{1}{\lambda}},$$

$\lambda\geq 1$

By putting $\overset{\circ}{w}(e) = \frac{1}{m-n'}$, for any $e \in [m, n]$, the GCHBCFWNSM will convert to generalized continuous hybrid BCF normalized SM (GCHBCFNSM)

$$S^{28}_{BCFS}(\Re_1, \Re_2) = 1 - \left(\frac{1}{2(m-n)} \int_m^n \left(\frac{1}{4} \left(\begin{array}{c} \left| \left(\mathfrak{F}^{RP}_{\Re_1}(e_\sigma)\right)^2 - \left(\mathfrak{F}^{RP}_{\Re_2}(e_\sigma)\right)^2 \right| + \left| \left(\mathfrak{F}^{IP}_{\Re_1}(e_\sigma)\right)^2 - \left(\mathfrak{F}^{IP}_{\Re_2}(e_\sigma)\right)^2 \right| \\ + \left| \left(\mathfrak{F}^{RN}_{\Re_1}(e_\sigma)\right)^2 - \left(\mathfrak{F}^{RN}_{\Re_2}(e_\sigma)\right)^2 \right| + \left| \left(\mathfrak{F}^{IN}_{\Re_1}(e_\sigma)\right)^2 - \left(\mathfrak{F}^{IN}_{\Re_2}(e_\sigma)\right)^2 \right| \end{array} \right) + \max \left(\begin{array}{c} \left| \left(\mathfrak{F}^{RP}_{\Re_1}(e_\sigma)\right)^2 - \left(\mathfrak{F}^{RP}_{\Re_2}(e_\sigma)\right)^2 \right|, \left| \left(\mathfrak{F}^{IP}_{\Re_1}(e_\sigma)\right)^2 - \left(\mathfrak{F}^{IP}_{\Re_2}(e_\sigma)\right)^2 \right| \\ \left| \left(\mathfrak{F}^{RN}_{\Re_1}(e_\sigma)\right)^2 - \left(\mathfrak{F}^{RN}_{\Re_2}(e_\sigma)\right)^2 \right|, \left| \left(\mathfrak{F}^{IN}_{\Re_1}(e_\sigma)\right)^2 - \left(\mathfrak{F}^{IN}_{\Re_2}(e_\sigma)\right)^2 \right| \end{array} \right) \right) de \right)^{\frac{1}{\lambda}},$$

$\lambda \geq 1$

5.4 Exponential Similarity Measures for BCFSs

In 1987, Shepard [40] investigated a law which stated that the perceived similarities and distance are linked to an exponential function. Following are the formula of SM relying on the exponential function

$$\mathcal{S}(\Re_1, \Re_2) = e^{-d(\Re_1, \Re_2)}$$

In this part of the chapter, we inaugurate novel SMs between two BCFS $\Re_1$ and $\Re_2$ by employing exponential function that based on Hamming and Housdroff distance. Observed that $\overset{\circ}{w} = \left(\overset{\circ}{w}_1, \overset{\circ}{w}_2, \ldots, \overset{\circ}{w}_y\right)^T$ is a weight vector and $0 \leq \underset{\sigma}{\overset{\circ}{w}} \leq 1$ for all σ with $\sum_{\sigma=1}^{y} \underset{\sigma}{\overset{\circ}{w}} = 1$ in this section.

5.4.1 *Exponential SMs Based on Hamming Distance*

Following is the definition of exponential SM (ESM) for BCFSs by employing generalized normalized Hamming distance (GNHD).

Definition 5.10 Let two BCFSs $\Re_1 = \left\{e_\sigma, \left(\mathfrak{F}^{P}_{\Re_1}(e_\sigma), \mathfrak{F}^{N}_{\Re_1}(e_\sigma)\right) | e_\sigma \in \mathbb{E}, \sigma = 1, 2, .., n\right\} = \left\{e_\sigma, \left(\mathfrak{F}^{RP}_{\Re_1}(e_\sigma) + \iota\mathfrak{F}^{IP}_{\Re_1}(e_\sigma), \mathfrak{F}^{RN}_{\Re_1}(e_\sigma) + \iota\mathfrak{F}^{IN}_{\Re_1}(e_\sigma)\right) | e_\sigma \in \mathbb{E}, \sigma = 1, 2, .., n\right\}$ and $\Re_2 = \left\{e, \left(\mathfrak{F}^{P}_{\Re_2}(e_\sigma), \mathfrak{F}^{N}_{\Re_2}(e_\sigma)\right) | e \in \mathbb{E}, \sigma = 1, 2, .., n\right\} = \left\{e_\sigma, \left(\mathfrak{F}^{RP}_{\Re_2}(e_\sigma) + \iota\mathfrak{F}^{IP}_{\Re_2}(e_\sigma), \mathfrak{F}^{RN}_{\Re_2}(e_\sigma) + \iota\mathfrak{F}^{IN}_{\Re_2}(e_\sigma)\right) | e_\sigma \in \mathbb{E}, \sigma = 1, 2, .., n\right\}$ over $\mathbb{E}$. Then the ESM between $\Re_1$ and $\Re_2$ relying on GNHD is interpreted as.

$$\mathcal{S}_{BCFS}^{29}(\Re_1, \Re_2) = e^{-\left(\frac{1}{4n}\sum_{\sigma=1}^{n}\left(\begin{array}{c}\left|\left(\mathfrak{F}_{\Re_1}^{RP}(e_\sigma)\right)^2 - \left(\mathfrak{F}_{\Re_2}^{RP}(e_\sigma)\right)^2\right| + \left|\left(\mathfrak{F}_{\Re_1}^{IP}(e_\sigma)\right)^2 - \left(\mathfrak{F}_{\Re_2}^{IP}(e_\sigma)\right)^2\right| \\ + \left|\left(\mathfrak{F}_{\Re_1}^{RN}(e_\sigma)\right)^2 - \left(\mathfrak{F}_{\Re_2}^{RN}(e_\sigma)\right)^2\right| + \left|\left(\mathfrak{F}_{\Re_1}^{IN}(e_\sigma)\right)^2 - \left(\mathfrak{F}_{\Re_2}^{IN}(e_\sigma)\right)^2\right|\end{array}\right)\right)^{\frac{1}{\lambda}}}$$

Theorem 5.7 The SM $\mathcal{S}_{BCFS}^{29}(\Re_1, \Re_2)$ satisfies the below properties.

1. $0 \leq \mathcal{S}_{BCFS}^{29}(\Re_1, \Re_2) \leq 1$
2. $\mathcal{S}_{BCFS}^{29}(\Re_1, \Re_2) = 1 \Leftrightarrow \Re_1 = \Re_2$
3. $\mathcal{S}_{BCFS}^{29}(\Re_1, \Re_2) = \mathcal{S}_{BCFS}^{29}(\Re_2, \Re_1)$
4. If $\Re_1 \subset \Re_2 \subset \Re_3$, then $\mathcal{S}_{BCFS}^{29}(\Re_1, \Re_2) \geq \mathcal{S}_{BCFS}^{29}(\Re_1, \Re_3)$ and $\mathcal{S}_{BCFS}^{29}(\Re_2, \Re_3) \geq \mathcal{S}_{BCFS}^{29}(\Re_1, \Re_3)$

Proof As we know that.

$$\left(\frac{1}{4n}\sum_{\sigma=1}^{n}\left(\begin{array}{c}\left|\left(\Im_{\Re_1}^{RP}(e_\sigma)\right)^2 - \left(\Im_{\Re_2}^{RP}(e_\sigma)\right)^2\right| + \left|\left(\Im_{\Re_1}^{IP}(e_\sigma)\right)^2 - \left(\Im_{\Re_2}^{IP}(e_\sigma)\right)^2\right| \\ + \left|\left(\Im_{\Re_1}^{RN}(e_\sigma)\right)^2 - \left(\Im_{\Re_2}^{RN}(e_\sigma)\right)^2\right| + \left|\left(\Im_{\Re_1}^{IN}(e_\sigma)\right)^2 - \left(\Im_{\Re_2}^{IN}(e_\sigma)\right)^2\right|\end{array}\right)\right)^{\frac{1}{\lambda}} \in [0, 1]$$

$$\Rightarrow 0 \leq \left(\frac{1}{4n}\sum_{\sigma=1}^{n}\left(\begin{array}{c}\left|\left(\Im_{\Re_1}^{RP}(e_\sigma)\right)^2 - \left(\Im_{\Re_2}^{RP}(e_\sigma)\right)^2\right| + \left|\left(\Im_{\Re_1}^{IP}(e_\sigma)\right)^2 - \left(\Im_{\Re_2}^{IP}(e_\sigma)\right)^2\right| \\ + \left|\left(\Im_{\Re_1}^{RN}(e_\sigma)\right)^2 - \left(\Im_{\Re_2}^{RN}(e_\sigma)\right)^2\right| + \left|\left(\Im_{\Re_1}^{IN}(e_\sigma)\right)^2 - \left(\Im_{\Re_2}^{IN}(e_\sigma)\right)^2\right|\end{array}\right)\right)^{\frac{1}{\lambda}} \leq 1$$

$$\Rightarrow -1 \leq -\left(\frac{1}{4n}\sum_{\sigma=1}^{n}\left(\begin{array}{c}\left|\left(\Im_{\Re_1}^{RP}(e_\sigma)\right)^2 - \left(\Im_{\Re_2}^{RP}(e_\sigma)\right)^2\right| + \left|\left(\Im_{\Re_1}^{IP}(e_\sigma)\right)^2 - \left(\Im_{\Re_2}^{IP}(e_\sigma)\right)^2\right| \\ + \left|\left(\Im_{\Re_1}^{RN}(e_\sigma)\right)^2 - \left(\Im_{\Re_2}^{RN}(e_\sigma)\right)^2\right| + \left|\left(\Im_{\Re_1}^{IN}(e_\sigma)\right)^2 - \left(\Im_{\Re_2}^{IN}(e_\sigma)\right)^2\right|\end{array}\right)\right)^{\frac{1}{\lambda}} \leq 0$$

Thus, here the power of e would always be negative or zero and we also have that $e^0 = 1$. Consequently,

$$0 \leq e^{-\left(\frac{1}{4n}\sum_{\sigma=1}^{n}\left(\begin{array}{c}\left|\left(\Im_{\Re_1}^{RP}(e_\sigma)\right)^2 - \left(\Im_{\Re_2}^{RP}(e_\sigma)\right)^2\right| + \left|\left(\Im_{\Re_1}^{IP}(e_\sigma)\right)^2 - \left(\Im_{\Re_2}^{IP}(e_\sigma)\right)^2\right| \\ + \left|\left(\Im_{\Re_1}^{RN}(e_\sigma)\right)^2 - \left(\Im_{\Re_2}^{RN}(e_\sigma)\right)^2\right| + \left|\left(\Im_{\Re_1}^{IN}(e_\sigma)\right)^2 - \left(\Im_{\Re_2}^{IN}(e_\sigma)\right)^2\right|\end{array}\right)\right)^{\frac{1}{\lambda}}} \leq 1$$

$$\Rightarrow 0 \leq 0 \leq \mathcal{S}_{BCFS}^{29}(\Re_1, \Re_2) \leq 1.$$

2. Consider that $\mathcal{S}_{BCFS}^{29}(\Re_1, \Re_2) = 1$. To Reveal that $\Re_1 = \Re_2$. as $\mathcal{S}_{BCFS}^{29}(\Re_1, \Re_2) = 1$, Thus

$$e^{-\left(\frac{1}{4n}\sum_{\sigma=1}^{n}\left(\begin{array}{c}\left|\left(\mathfrak{F}_{\mathfrak{R}_1}^{RP}(e_\sigma)\right)^2-\left(\mathfrak{F}_{\mathfrak{R}_2}^{RP}(e_\sigma)\right)^2\right|+\left|\left(\mathfrak{F}_{\mathfrak{R}_1}^{IP}(e_\sigma)\right)^2-\left(\mathfrak{F}_{\mathfrak{R}_2}^{IP}(e_\sigma)\right)^2\right|\\+\left|\left(\mathfrak{F}_{\mathfrak{R}_1}^{RN}(e_\sigma)\right)^2-\left(\mathfrak{F}_{\mathfrak{R}_2}^{RN}(e_\sigma)\right)^2\right|+\left|\left(\mathfrak{F}_{\mathfrak{R}_1}^{IN}(e_\sigma)\right)^2-\left(\mathfrak{F}_{\mathfrak{R}_2}^{IN}(e_\sigma)\right)^2\right|\end{array}\right)\right)^{\frac{1}{\lambda}}}=1$$

$$\Rightarrow -\left(\frac{1}{4n}\sum_{\sigma=1}^{n}\left(\begin{array}{c}\left|\left(\mathfrak{F}_{\mathfrak{R}_1}^{RP}(e_\sigma)\right)^2-\left(\mathfrak{F}_{\mathfrak{R}_2}^{RP}(e_\sigma)\right)^2\right|+\left|\left(\mathfrak{F}_{\mathfrak{R}_1}^{IP}(e_\sigma)\right)^2-\left(\mathfrak{F}_{\mathfrak{R}_2}^{IP}(e_\sigma)\right)^2\right|\\+\left|\left(\mathfrak{F}_{\mathfrak{R}_1}^{RN}(e_\sigma)\right)^2-\left(\mathfrak{F}_{\mathfrak{R}_2}^{RN}(e_\sigma)\right)^2\right|+\left|\left(\mathfrak{F}_{\mathfrak{R}_1}^{IN}(e_\sigma)\right)^2-\left(\mathfrak{F}_{\mathfrak{R}_2}^{IN}(e_\sigma)\right)^2\right|\end{array}\right)\right)^{\frac{1}{\lambda}}$$
$$=0$$

Thus,

$$\left(\frac{1}{4n}\sum_{\sigma=1}^{n}\left(\begin{array}{c}\left|\left(\mathfrak{F}_{\mathfrak{R}_1}^{RP}(e_\sigma)\right)^2-\left(\mathfrak{F}_{\mathfrak{R}_2}^{RP}(e_\sigma)\right)^2\right|+\left|\left(\mathfrak{F}_{\mathfrak{R}_1}^{IP}(e_\sigma)\right)^2-\left(\mathfrak{F}_{\mathfrak{R}_2}^{IP}(e_\sigma)\right)^2\right|\\+\left|\left(\mathfrak{F}_{\mathfrak{R}_1}^{RN}(e_\sigma)\right)^2-\left(\mathfrak{F}_{\mathfrak{R}_2}^{RN}(e_\sigma)\right)^2\right|+\left|\left(\mathfrak{F}_{\mathfrak{R}_1}^{IN}(e_\sigma)\right)^2-\left(\mathfrak{F}_{\mathfrak{R}_2}^{IN}(e_\sigma)\right)^2\right|\end{array}\right)\right)^{\frac{1}{\lambda}}=0$$

Then obviously, $\mathfrak{R}_1=\mathfrak{R}_2$.

Conversely, consider that $\mathfrak{R}_1=\mathfrak{R}_2$ and to reveal that $\mathcal{S}_{BCFS}^{29}(\mathfrak{R}_1,\mathfrak{R}_2)=1$. As $\mathfrak{R}_1=\mathfrak{R}_2$, then by GBCFNHDM we have that

$$\left(\frac{1}{4n}\sum_{\sigma=1}^{n}\left(\begin{array}{c}\left|\left(\mathfrak{F}_{\mathfrak{R}_1}^{RP}(e_\sigma)\right)^2-\left(\mathfrak{F}_{\mathfrak{R}_2}^{RP}(e_\sigma)\right)^2\right|+\left|\left(\mathfrak{F}_{\mathfrak{R}_1}^{IP}(e_\sigma)\right)^2-\left(\mathfrak{F}_{\mathfrak{R}_2}^{IP}(e_\sigma)\right)^2\right|\\+\left|\left(\mathfrak{F}_{\mathfrak{R}_1}^{RN}(e_\sigma)\right)^2-\left(\mathfrak{F}_{\mathfrak{R}_2}^{RN}(e_\sigma)\right)^2\right|+\left|\left(\mathfrak{F}_{\mathfrak{R}_1}^{IN}(e_\sigma)\right)^2-\left(\mathfrak{F}_{\mathfrak{R}_2}^{IN}(e_\sigma)\right)^2\right|\end{array}\right)\right)^{\frac{1}{\lambda}}=0$$

$$\Rightarrow -\left(\frac{1}{4n}\sum_{\sigma=1}^{n}\left(\begin{array}{c}\left|\left(\mathfrak{F}_{\mathfrak{R}_1}^{RP}(e_\sigma)\right)^2-\left(\mathfrak{F}_{\mathfrak{R}_2}^{RP}(e_\sigma)\right)^2\right|+\left|\left(\mathfrak{F}_{\mathfrak{R}_1}^{IP}(e_\sigma)\right)^2-\left(\mathfrak{F}_{\mathfrak{R}_2}^{IP}(e_\sigma)\right)^2\right|\\+\left|\left(\mathfrak{F}_{\mathfrak{R}_1}^{RN}(e_\sigma)\right)^2-\left(\mathfrak{F}_{\mathfrak{R}_2}^{RN}(e_\sigma)\right)^2\right|+\left|\left(\mathfrak{F}_{\mathfrak{R}_1}^{IN}(e_\sigma)\right)^2-\left(\mathfrak{F}_{\mathfrak{R}_2}^{IN}(e_\sigma)\right)^2\right|\end{array}\right)\right)^{\frac{1}{\lambda}}$$
$$=0$$

and

$$e^{-\left(\frac{1}{4n}\sum_{\sigma=1}^{n}\left(\begin{array}{c}\left|\left(\mathfrak{F}_{\mathfrak{R}_1}^{RP}(e_\sigma)\right)^2-\left(\mathfrak{F}_{\mathfrak{R}_2}^{RP}(e_\sigma)\right)^2\right|+\left|\left(\mathfrak{F}_{\mathfrak{R}_1}^{IP}(e_\sigma)\right)^2-\left(\mathfrak{F}_{\mathfrak{R}_2}^{IP}(e_\sigma)\right)^2\right|\\+\left|\left(\mathfrak{F}_{\mathfrak{R}_1}^{RN}(e_\sigma)\right)^2-\left(\mathfrak{F}_{\mathfrak{R}_2}^{RN}(e_\sigma)\right)^2\right|+\left|\left(\mathfrak{F}_{\mathfrak{R}_1}^{IN}(e_\sigma)\right)^2-\left(\mathfrak{F}_{\mathfrak{R}_2}^{IN}(e_\sigma)\right)^2\right|\end{array}\right)\right)^{\frac{1}{\lambda}}}=e^0$$

Consequently,

$$e^{-\left(\frac{1}{4n}\sum_{\sigma=1}^{n}\left(\begin{array}{c}\left|\left(\mathfrak{F}_{\mathfrak{R}_1}^{RP}(e_\sigma)\right)^2-\left(\mathfrak{F}_{\mathfrak{R}_2}^{RP}(e_\sigma)\right)^2\right|+\left|\left(\mathfrak{F}_{\mathfrak{R}_1}^{IP}(e_\sigma)\right)^2-\left(\mathfrak{F}_{\mathfrak{R}_2}^{IP}(e_\sigma)\right)^2\right|\\+\left|\left(\mathfrak{F}_{\mathfrak{R}_1}^{RN}(e_\sigma)\right)^2-\left(\mathfrak{F}_{\mathfrak{R}_2}^{RN}(e_\sigma)\right)^2\right|+\left|\left(\mathfrak{F}_{\mathfrak{R}_1}^{IN}(e_\sigma)\right)^2-\left(\mathfrak{F}_{\mathfrak{R}_2}^{IN}(e_\sigma)\right)^2\right|\end{array}\right)\right)^{\frac{1}{\lambda}}}=1$$

$$\Rightarrow \mathcal{S}_{BCFS}^{29}(\mathfrak{R}_1,\mathfrak{R}_2)=1.$$

3. From Theorem (4.1) We Have that

$$\left(\frac{1}{4n}\sum_{\sigma=1}^{n}\left(\begin{array}{c}\left|\left(\mathfrak{F}_{\mathfrak{R}_1}^{RP}(e_\sigma)\right)^2-\left(\mathfrak{F}_{\mathfrak{R}_2}^{RP}(e_\sigma)\right)^2\right|+\left|\left(\mathfrak{F}_{\mathfrak{R}_1}^{IP}(e_\sigma)\right)^2-\left(\mathfrak{F}_{\mathfrak{R}_2}^{IP}(e_\sigma)\right)^2\right|\\+\left|\left(\mathfrak{F}_{\mathfrak{R}_1}^{RN}(e_\sigma)\right)^2-\left(\mathfrak{F}_{\mathfrak{R}_2}^{RN}(e_\sigma)\right)^2\right|+\left|\left(\mathfrak{F}_{\mathfrak{R}_1}^{IN}(e_\sigma)\right)^2-\left(\mathfrak{F}_{\mathfrak{R}_2}^{IN}(e_\sigma)\right)^2\right|\end{array}\right)\right)^{\frac{1}{\lambda}}$$

is symmetry. Thus, $\mathcal{S}_{BCFS}^{29}(\mathfrak{R}_1,\mathfrak{R}_2)=\mathcal{S}_{BCFS}^{29}(\mathfrak{R}_2,\mathfrak{R}_1)$.

4. Let $\mathfrak{R}_1\subset\mathfrak{R}_2\subset\mathfrak{R}_3$. then from Theorem (4.1) We Get that

$$d_{BCFS}^{1}(\mathfrak{R}_2,\mathfrak{R}_3)\leq d_{BCFS}^{1}(\mathfrak{R}_1,\mathfrak{R}_3)$$

$$\Rightarrow\frac{1}{4n}\sum_{\sigma=1}^{n}\left(\begin{array}{c}\left|\left(\mathfrak{F}_{\mathfrak{R}_1}^{RP}(e_\sigma)\right)^2-\left(\mathfrak{F}_{\mathfrak{R}_2}^{RP}(e_\sigma)\right)^2\right|+\left|\left(\mathfrak{F}_{\mathfrak{R}_1}^{IP}(e_\sigma)\right)^2-\left(\mathfrak{F}_{\mathfrak{R}_2}^{IP}(e_\sigma)\right)^2\right|\\+\left|\left(\mathfrak{F}_{\mathfrak{R}_1}^{RN}(e_\sigma)\right)^2-\left(\mathfrak{F}_{\mathfrak{R}_2}^{RN}(e_\sigma)\right)^2\right|+\left|\left(\mathfrak{F}_{\mathfrak{R}_1}^{RN}(e_\sigma)\right)^2-\left(\mathfrak{F}_{\mathfrak{R}_2}^{RN}(e_\sigma)\right)^2\right|\end{array}\right)\Bigg)^{\frac{1}{\lambda}}$$

$$\leq\left(\frac{1}{4n}\sum_{\sigma=1}^{n}\left(\begin{array}{c}\left|\left(\mathfrak{F}_{\mathfrak{R}_1}^{RP}(e_\sigma)\right)^2-\left(\mathfrak{F}_{\mathfrak{R}_3}^{RP}(e_\sigma)\right)^2\right|+\left|\left(\mathfrak{F}_{\mathfrak{R}_1}^{IP}(e_\sigma)\right)^2-\left(\mathfrak{F}_{\mathfrak{R}_3}^{IP}(e_\sigma)\right)^2\right|\\+\left|\left(\mathfrak{F}_{\mathfrak{R}_1}^{RN}(e_\sigma)\right)^2-\left(\mathfrak{F}_{\mathfrak{R}_3}^{RN}(e_\sigma)\right)^2\right|+\left|\left(\mathfrak{F}_{\mathfrak{R}_1}^{RN}(e_\sigma)\right)^2-\left(\mathfrak{F}_{\mathfrak{R}_3}^{RN}(e_\sigma)\right)^2\right|\end{array}\right)\right)^{\frac{1}{\lambda}}$$

Thus,

$$-\left(\frac{1}{4n}\sum_{\sigma=1}^{n}\left(\begin{array}{c}\left|\left(\mathfrak{F}_{\mathfrak{R}_1}^{RP}(e_\sigma)\right)^2-\left(\mathfrak{F}_{\mathfrak{R}_2}^{RP}(e_\sigma)\right)^2\right|+\left|\left(\mathfrak{F}_{\mathfrak{R}_1}^{IP}(e_\sigma)\right)^2-\left(\mathfrak{F}_{\mathfrak{R}_2}^{IP}(e_\sigma)\right)^2\right|\\+\left|\left(\mathfrak{F}_{\mathfrak{R}_1}^{RN}(e_\sigma)\right)^2-\left(\mathfrak{F}_{\mathfrak{R}_2}^{RN}(e_\sigma)\right)^2\right|+\left|\left(\mathfrak{F}_{\mathfrak{R}_1}^{RN}(e_\sigma)\right)^2-\left(\mathfrak{F}_{\mathfrak{R}_2}^{RN}(e_\sigma)\right)^2\right|\end{array}\right)\right)^{\frac{1}{\lambda}}$$

$$\geq-\left(\frac{1}{4n}\sum_{\sigma=1}^{n}\left(\begin{array}{c}\left|\left(\mathfrak{F}_{\mathfrak{R}_1}^{RP}(e_\sigma)\right)^2-\left(\mathfrak{F}_{\mathfrak{R}_3}^{RP}(e_\sigma)\right)^2\right|+\left|\left(\mathfrak{F}_{\mathfrak{R}_1}^{IP}(e_\sigma)\right)^2-\left(\mathfrak{F}_{\mathfrak{R}_3}^{IP}(e_\sigma)\right)^2\right|\\+\left|\left(\mathfrak{F}_{\mathfrak{R}_1}^{RN}(e_\sigma)\right)^2-\left(\mathfrak{F}_{\mathfrak{R}_3}^{RN}(e_\sigma)\right)^2\right|+\left|\left(\mathfrak{F}_{\mathfrak{R}_1}^{RN}(e_\sigma)\right)^2-\left(\mathfrak{F}_{\mathfrak{R}_3}^{RN}(e_\sigma)\right)^2\right|\end{array}\right)\right)^{\frac{1}{\lambda}}$$

And

$$e^{-\left(\frac{1}{4n}\sum_{\sigma=1}^{n}\left(\begin{array}{c}\left|\left(\mathfrak{F}_{\mathfrak{R}_1}^{RP}(e_\sigma)\right)^2-\left(\mathfrak{F}_{\mathfrak{R}_2}^{RP}(e_\sigma)\right)^2\right|+\left|\left(\mathfrak{F}_{\mathfrak{R}_1}^{IP}(e_\sigma)\right)^2-\left(\mathfrak{F}_{\mathfrak{R}_2}^{IP}(e_\sigma)\right)^2\right|\\+\left|\left(\mathfrak{F}_{\mathfrak{R}_1}^{RN}(e_\sigma)\right)^2-\left(\mathfrak{F}_{\mathfrak{R}_2}^{RN}(e_\sigma)\right)^2\right|+\left|\left(\mathfrak{F}_{\mathfrak{R}_1}^{RN}(e_\sigma)\right)^2-\left(\mathfrak{F}_{\mathfrak{R}_2}^{RN}(e_\sigma)\right)^2\right|\end{array}\right)\right)^{\frac{1}{\lambda}}}$$

$$\geq e^{-\left(\frac{1}{4n}\sum_{\sigma=1}^{n}\left(\begin{array}{c}\left|\left(\mathfrak{F}_{\mathfrak{R}_1}^{RP}(e_\sigma)\right)^2-\left(\mathfrak{F}_{\mathfrak{R}_3}^{RP}(e_\sigma)\right)^2\right|+\left|\left(\mathfrak{F}_{\mathfrak{R}_1}^{IP}(e_\sigma)\right)^2-\left(\mathfrak{F}_{\mathfrak{R}_3}^{IP}(e_\sigma)\right)^2\right|\\+\left|\left(\mathfrak{F}_{\mathfrak{R}_1}^{RN}(e_\sigma)\right)^2-\left(\mathfrak{F}_{\mathfrak{R}_3}^{RN}(e_\sigma)\right)^2\right|+\left|\left(\mathfrak{F}_{\mathfrak{R}_1}^{RN}(e_\sigma)\right)^2-\left(\mathfrak{F}_{\mathfrak{R}_3}^{RN}(e_\sigma)\right)^2\right|\end{array}\right)\right)^{\frac{1}{\lambda}}}$$

This implies that $\mathcal{S}_{BCFS}^{29}(\mathfrak{R}_1,\mathfrak{R}_2)\geq\mathcal{S}_{BCFS}^{29}(\mathfrak{R}_1,\mathfrak{R}_3)$. Likewise, we can reveal that $\mathcal{S}_{BCFS}^{29}(\mathfrak{R}_2,\mathfrak{R}_3)\geq\mathcal{S}_{BCFS}^{29}(\mathfrak{R}_1,\mathfrak{R}_3)$.

Special Cases of $\mathcal{S}_{BCFS}^{29}(\mathfrak{R}_1,\mathfrak{R}_2)$.

The SM $\mathcal{S}^{29}_{BCFS}(\Re_1, \Re_2)$ has underneath special cases.

1. By assuming the value of $\lambda = 1$, we achieve the ESM relying on normalized Hamming distance, demonstrated as

$$\mathcal{S}^{30}_{BCFS}(\Re_1, \Re_2) = e^{-\left(\frac{1}{4n}\Sigma_{\sigma=1}^{n}\left(\begin{array}{l}\left|\left(\mathfrak{F}^{RP}_{\Re_1}(e_\sigma)\right)^2-\left(\mathfrak{F}^{RP}_{\Re_2}(e_\sigma)\right)^2\right|+\left|\left(\mathfrak{F}^{IP}_{\Re_1}(e_\sigma)\right)^2-\left(\mathfrak{F}^{IP}_{\Re_2}(e_\sigma)\right)^2\right|\\+\left|\left(\mathfrak{F}^{RN}_{\Re_1}(e_\sigma)\right)^2-\left(\mathfrak{F}^{RN}_{\Re_2}(e_\sigma)\right)^2\right|+\left|\left(\mathfrak{F}^{IN}_{\Re_1}(e_\sigma)\right)^2-\left(\mathfrak{F}^{IN}_{\Re_2}(e_\sigma)\right)^2\right|\end{array}\right)\right)}$$

2. By assuming the value of $\lambda = \frac{1}{2}$, we achieve the ESM relying on normalized Euclidean distance, demonstrated as

$$\mathcal{S}^{31}_{BCFS}(\Re_1, \Re_2) = e^{-\sqrt{\frac{1}{4n}\Sigma_{\sigma=1}^{n}\left(\begin{array}{l}\left|\left(\mathfrak{F}^{RP}_{\Re_1}(e_\sigma)\right)^2-\left(\mathfrak{F}^{RP}_{\Re_2}(e_\sigma)\right)^2\right|+\left|\left(\mathfrak{F}^{IP}_{\Re_1}(e_\sigma)\right)^2-\left(\mathfrak{F}^{IP}_{\Re_2}(e_\sigma)\right)^2\right|\\+\left|\left(\mathfrak{F}^{RN}_{\Re_1}(e_\sigma)\right)^2-\left(\mathfrak{F}^{RN}_{\Re_2}(e_\sigma)\right)^2\right|+\left|\left(\mathfrak{F}^{IN}_{\Re_1}(e_\sigma)\right)^2-\left(\mathfrak{F}^{IN}_{\Re_2}(e_\sigma)\right)^2\right|\end{array}\right)}}$$

Following is the definition of weighted ESM for BCFSs by employing GNHD.

Definition 5.11 Let two BCFSs $\Re_1 = \left\{e_\sigma, \left(\mathfrak{F}^{P}_{\Re_1}(e_\sigma), \mathfrak{F}^{N}_{\Re_1}(e_\sigma)\right)|e_\sigma \in \mathbb{E}, \sigma = 1, 2, .., n\right\} = \left\{e_\sigma, \left(\mathfrak{F}^{RP}_{\Re_1}(e_\sigma) + \iota\mathfrak{F}^{IP}_{\Re_1}(e_\sigma), \mathfrak{F}^{RN}_{\Re_1}(e_\sigma) + \iota\mathfrak{F}^{IN}_{\Re_1}(e_\sigma)\right)|e_\sigma \in \mathbb{E}, \sigma = 1, 2, .., n\right\}$ and $\Re_2 = \left\{e, \left(\mathfrak{F}^{P}_{\Re_2}(e_\sigma), \mathfrak{F}^{N}_{\Re_2}(e_\sigma)\right)|e \in \mathbb{E}, \sigma = 1, 2, .., n\right\} = \left\{e_\sigma, \left(\mathfrak{F}^{RP}_{\Re_2}(e_\sigma) + \iota\mathfrak{F}^{IP}_{\Re_2}(e_\sigma), \mathfrak{F}^{RN}_{\Re_2}(e_\sigma) + \iota\mathfrak{F}^{IN}_{\Re_2}(e_\sigma)\right)|e_\sigma \in \mathbb{E}, \sigma = 1, 2, .., n\right\}$ over $\mathbb{E}$. Then the weighted ESM between $\Re_1$ and $\Re_2$ relying on GNHD is interpreted as.

$$\mathcal{S}^{32}_{BCFS}(\Re_1, \Re_2) = e^{-\left(\sum_{\sigma=1}^{n}\mathrm{w}^{\circ}_{\sigma}\left(\begin{array}{l}\left|\left(\mathfrak{F}^{RP}_{\Re_1}(e_\sigma)\right)^2-\left(\mathfrak{F}^{RP}_{\Re_2}(e_\sigma)\right)^2\right|+\left|\left(\mathfrak{F}^{IP}_{\Re_1}(e_\sigma)\right)^2-\left(\mathfrak{F}^{IP}_{\Re_2}(e_\sigma)\right)^2\right|\\+\left|\left(\mathfrak{F}^{RN}_{\Re_1}(e_\sigma)\right)^2-\left(\mathfrak{F}^{RN}_{\Re_2}(e_\sigma)\right)^2\right|+\left|\left(\mathfrak{F}^{IN}_{\Re_1}(e_\sigma)\right)^2-\left(\mathfrak{F}^{IN}_{\Re_2}(e_\sigma)\right)^2\right|\end{array}\right)\right)^{\frac{1}{\lambda}}}$$

Theorem 5.8 The SM $\mathcal{S}^{32}_{BCFS}(\Re_1, \Re_2)$ satisfies the below properties.

1. $0 \leq \mathcal{S}^{32}_{BCFS}(\Re_1, \Re_2) \leq 1$
2. $\mathcal{S}^{29}_{BCFS}\mathcal{S}^{32}_{BCFS}(\Re_1, \Re_2) = 1 \Leftrightarrow \Re_1 = \Re_2$
3. $\mathcal{S}^{32}_{BCFS}(\Re_1, \Re_2) = \mathcal{S}^{32}_{BCFS}(\Re_2, \Re_1)$
4. If $\Re_1 \subset \Re_2 \subset \Re_3$, then $\mathcal{S}^{32}_{BCFS}(\Re_1, \Re_2) \geq \mathcal{S}^{32}_{BCFS}(\Re_1, \Re_3)$ and $\mathcal{S}^{32}_{BCFS}(\Re_2, \Re_3) \geq \mathcal{S}^{32}_{BCFS}(\Re_1, \Re_3)$

Proof Omitted.

On similar pattern, we can discuss of special cases of weighted ESM relying on GNHD.

5.4.2 Exponential SMs Based on Hausdorff Distance

In the underneath definition, we inaugurate ESM based on generalized normalized Hausdorff distance (GNHDD).

Definition 5.12 Let two BCFSs $\Re_1 = \left\{e_\sigma, \left(\mathfrak{F}^P_{\Re_1}(e_\sigma), \mathfrak{F}^N_{\Re_1}(e_\sigma)\right)|e_\sigma \in \mathbb{E}, \sigma = 1, 2, .., n\right\} = \left\{e_\sigma, \left(\mathfrak{F}^{RP}_{\Re_1}(e_\sigma) + \iota\mathfrak{F}^{IP}_{\Re_1}(e_\sigma), \mathfrak{F}^{RN}_{\Re_1}(e_\sigma) + \iota\mathfrak{F}^{IN}_{\Re_1}(e_\sigma)\right)|e_\sigma \in \mathbb{E}, \sigma = 1, 2, .., n\right\}$ and $\Re_2 = \left\{e, \left(\mathfrak{F}^P_{\Re_2}(e_\sigma), \mathfrak{F}^N_{\Re_2}(e_\sigma)\right)|e \in \mathbb{E}, \sigma = 1, 2, .., n\right\} = \left\{e_\sigma, \left(\mathfrak{F}^{RP}_{\Re_2}(e_\sigma) + \iota\mathfrak{F}^{IP}_{\Re_2}(e_\sigma), \mathfrak{F}^{RN}_{\Re_2}(e_\sigma) + \iota\mathfrak{F}^{IN}_{\Re_2}(e_\sigma)\right)|e_\sigma \in \mathbb{E}, \sigma = 1, 2, .., n\right\}$ over $\mathbb{E}$. Then the ESM between $\Re_1$ and $\Re_2$ relying on GNHDD is interpreted as.

$$\mathcal{S}^{33}_{BCFS}(\Re_1, \Re_2) = e^{-\left(\frac{1}{n}\sum_{\sigma=1}^{n}\max\left(\begin{array}{c}\left|\left(\mathfrak{F}^{RP}_{\Re_1}(e_\sigma)\right)^2 - \left(\mathfrak{F}^{RP}_{\Re_2}(e_\sigma)\right)^2\right|, \left|\left(\mathfrak{F}^{IP}_{\Re_1}(e_\sigma)\right)^2 - \left(\mathfrak{F}^{IP}_{\Re_2}(e_\sigma)\right)^2\right| \\ , \left|\left(\mathfrak{F}^{RN}_{\Re_1}(e_\sigma)\right)^2 - \left(\mathfrak{F}^{RN}_{\Re_2}(e_\sigma)\right)^2\right|, \left|\left(\mathfrak{F}^{IN}_{\Re_1}(e_\sigma)\right)^2 - \left(\mathfrak{F}^{IN}_{\Re_2}(e_\sigma)\right)^2\right|\end{array}\right)\right)^{\frac{1}{\lambda}}}$$

Theorem 5.9 The SM $\mathcal{S}^{33}_{BCFS}(\Re_1, \Re_2)$ satisfies the below properties.

1. $0 \le \mathcal{S}^{33}_{BCFS}(\Re_1, \Re_2) \le 1$
2. $\mathcal{S}^{33}_{BCFS}(\Re_1, \Re_2) = 1 \Leftrightarrow \Re_1 = \Re_2$
3. $\mathcal{S}^{33}_{BCFS}(\Re_1, \Re_2) = \mathcal{S}^{33}_{BCFS}(\Re_2, \Re_1)$
4. If $\Re_1 \subset \Re_2 \subset \Re_3$, then $\mathcal{S}^{33}_{BCFS}(\Re_1, \Re_2) \ge \mathcal{S}^{33}_{BCFS}(\Re_1, \Re_3)$ and $\mathcal{S}^{33}_{BCFS}(\Re_2, \Re_3) \ge \mathcal{S}^{33}_{BCFS}(\Re_1, \Re_3)$

Proof As we know that.

$$\left(\frac{1}{n}\sum_{\sigma=1}^{n}\max\left(\begin{array}{c}\left|\left(\mathfrak{F}^{RP}_{\Re_1}(e_\sigma)\right)^2 - \left(\mathfrak{F}^{RP}_{\Re_2}(e_\sigma)\right)^2\right|, \left|\left(\mathfrak{F}^{IP}_{\Re_1}(e_\sigma)\right)^2 - \left(\mathfrak{F}^{IP}_{\Re_2}(e_\sigma)\right)^2\right| \\ , \left|\left(\mathfrak{F}^{RN}_{\Re_1}(e_\sigma)\right)^2 - \left(\mathfrak{F}^{RN}_{\Re_2}(e_\sigma)\right)^2\right|, \left|\left(\mathfrak{F}^{IN}_{\Re_1}(e_\sigma)\right)^2 - \left(\mathfrak{F}^{IN}_{\Re_2}(e_\sigma)\right)^2\right|\end{array}\right)\right)^{\frac{1}{\lambda}} \in [0, 1]$$

$$\Rightarrow 0 \le \left(\frac{1}{n}\sum_{\sigma=1}^{n}\max\left(\begin{array}{c}\left|\left(\mathfrak{F}^{RP}_{\Re_1}(e_\sigma)\right)^2 - \left(\mathfrak{F}^{RP}_{\Re_2}(e_\sigma)\right)^2\right|, \left|\left(\mathfrak{F}^{IP}_{\Re_1}(e_\sigma)\right)^2 - \left(\mathfrak{F}^{IP}_{\Re_2}(e_\sigma)\right)^2\right| \\ , \left|\left(\mathfrak{F}^{RN}_{\Re_1}(e_\sigma)\right)^2 - \left(\mathfrak{F}^{RN}_{\Re_2}(e_\sigma)\right)^2\right|, \left|\left(\mathfrak{F}^{IN}_{\Re_1}(e_\sigma)\right)^2 - \left(\mathfrak{F}^{IN}_{\Re_2}(e_\sigma)\right)^2\right|\end{array}\right)\right)^{\frac{1}{\lambda}} \le 1$$

$$\Rightarrow -1 \le -\left(\frac{1}{n}\sum_{\sigma=1}^{n}\max\left(\begin{array}{c}\left|\left(\mathfrak{F}^{RP}_{\Re_1}(e_\sigma)\right)^2 - \left(\mathfrak{F}^{RP}_{\Re_2}(e_\sigma)\right)^2\right|, \left|\left(\mathfrak{F}^{IP}_{\Re_1}(e_\sigma)\right)^2 - \left(\mathfrak{F}^{IP}_{\Re_2}(e_\sigma)\right)^2\right| \\ , \left|\left(\mathfrak{F}^{RN}_{\Re_1}(e_\sigma)\right)^2 - \left(\mathfrak{F}^{RN}_{\Re_2}(e_\sigma)\right)^2\right|, \left|\left(\mathfrak{F}^{IN}_{\Re_1}(e_\sigma)\right)^2 - \left(\mathfrak{F}^{IN}_{\Re_2}(e_\sigma)\right)^2\right|\end{array}\right)\right)^{\frac{1}{\lambda}}$$

$$\leq 0$$

Thus, here the power of e would always be negative or zero and we also have that $e^0 = 1$. Consequently,

$$0 \leq e^{-\left(\frac{1}{n}\sum_{\sigma=1}^{n}\max\left(\begin{array}{l}\left|\left(\mathfrak{F}^{RP}_{\mathfrak{R}_1}(e_\sigma)\right)^2 - \left(\mathfrak{F}^{RP}_{\mathfrak{R}_2}(e_\sigma)\right)^2\right|, \left|\left(\mathfrak{F}^{IP}_{\mathfrak{R}_1}(e_\sigma)\right)^2 - \left(\mathfrak{F}^{IP}_{\mathfrak{R}_2}(e_\sigma)\right)^2\right| \\ , \left|\left(\mathfrak{F}^{RN}_{\mathfrak{R}_1}(e_\sigma)\right)^2 - \left(\mathfrak{F}^{RN}_{\mathfrak{R}_2}(e_\sigma)\right)^2\right|, \left|\left(\mathfrak{F}^{IN}_{\mathfrak{R}_1}(e_\sigma)\right)^2 - \left(\mathfrak{F}^{IN}_{\mathfrak{R}_2}(e_\sigma)\right)^2\right|\end{array}\right)\right)^{\frac{1}{\lambda}}} \leq 1$$

$$\Rightarrow 0 \leq \mathcal{S}^{33}_{BCFS}(\mathfrak{R}_1, \mathfrak{R}_2) \leq 1.$$

2. Consider that $\mathcal{S}^{33}_{BCFS}(\mathfrak{R}_1, \mathfrak{R}_2) = 1$. To reveal that $\mathfrak{R}_1 = \mathfrak{R}_2$. As $\mathcal{S}^{33}_{BCFS}(\mathfrak{R}_1, \mathfrak{R}_2) = 1$, thus

$$e^{-\left(\frac{1}{n}\sum_{\sigma=1}^{n}\max\left(\begin{array}{l}\left|\left(\mathfrak{F}^{RP}_{\mathfrak{R}_1}(e_\sigma)\right)^2 - \left(\mathfrak{F}^{RP}_{\mathfrak{R}_2}(e_\sigma)\right)^2\right|, \left|\left(\mathfrak{F}^{IP}_{\mathfrak{R}_1}(e_\sigma)\right)^2 - \left(\mathfrak{F}^{IP}_{\mathfrak{R}_2}(e_\sigma)\right)^2\right| \\ , \left|\left(\mathfrak{F}^{RN}_{\mathfrak{R}_1}(e_\sigma)\right)^2 - \left(\mathfrak{F}^{RN}_{\mathfrak{R}_2}(e_\sigma)\right)^2\right|, \left|\left(\mathfrak{F}^{IN}_{\mathfrak{R}_1}(e_\sigma)\right)^2 - \left(\mathfrak{F}^{IN}_{\mathfrak{R}_2}(e_\sigma)\right)^2\right|\end{array}\right)\right)^{\frac{1}{\lambda}}} = 1$$

$$\Rightarrow$$

$$-\left(\frac{1}{n}\sum_{\sigma=1}^{n}\max\left(\begin{array}{l}\left|\left(\mathfrak{F}^{RP}_{\mathfrak{R}_1}(e_\sigma)\right)^2 - \left(\mathfrak{F}^{RP}_{\mathfrak{R}_2}(e_\sigma)\right)^2\right|, \left|\left(\mathfrak{F}^{IP}_{\mathfrak{R}_1}(e_\sigma)\right)^2 - \left(\mathfrak{F}^{IP}_{\mathfrak{R}_2}(e_\sigma)\right)^2\right| \\ , \left|\left(\mathfrak{F}^{RN}_{\mathfrak{R}_1}(e_\sigma)\right)^2 - \left(\mathfrak{F}^{RN}_{\mathfrak{R}_2}(e_\sigma)\right)^2\right|, \left|\left(\mathfrak{F}^{IN}_{\mathfrak{R}_1}(e_\sigma)\right)^2 - \left(\mathfrak{F}^{IN}_{\mathfrak{R}_2}(e_\sigma)\right)^2\right|\end{array}\right)\right)^{\frac{1}{\lambda}} = 0$$

Thus,

$$\left(\frac{1}{n}\sum_{\sigma=1}^{n}\max\left(\begin{array}{l}\left|\left(\mathfrak{F}^{RP}_{\mathfrak{R}_1}(e_\sigma)\right)^2 - \left(\mathfrak{F}^{RP}_{\mathfrak{R}_2}(e_\sigma)\right)^2\right|, \left|\left(\mathfrak{F}^{IP}_{\mathfrak{R}_1}(e_\sigma)\right)^2 - \left(\mathfrak{F}^{IP}_{\mathfrak{R}_2}(e_\sigma)\right)^2\right| \\ , \left|\left(\mathfrak{F}^{RN}_{\mathfrak{R}_1}(e_\sigma)\right)^2 - \left(\mathfrak{F}^{RN}_{\mathfrak{R}_2}(e_\sigma)\right)^2\right|, \left|\left(\mathfrak{F}^{IN}_{\mathfrak{R}_1}(e_\sigma)\right)^2 - \left(\mathfrak{F}^{IN}_{\mathfrak{R}_2}(e_\sigma)\right)^2\right|\end{array}\right)\right)^{\frac{1}{\lambda}} = 0$$

Then obviously, $\mathfrak{R}_1 = \mathfrak{R}_2$.

Conversely, consider that $\mathfrak{R}_1 = \mathfrak{R}_2$ and to reveal that $\mathcal{S}^{33}_{BCFS}(\mathfrak{R}_1, \mathfrak{R}_2) = 1$. As $\mathfrak{R}_1 = \mathfrak{R}_2$, then by GBCFNHDM we have that

$$\left(\frac{1}{n}\sum_{\sigma=1}^{n}\max\left(\begin{array}{l}\left|\left(\mathfrak{F}^{RP}_{\mathfrak{R}_1}(e_\sigma)\right)^2 - \left(\mathfrak{F}^{RP}_{\mathfrak{R}_2}(e_\sigma)\right)^2\right|, \left|\left(\mathfrak{F}^{IP}_{\mathfrak{R}_1}(e_\sigma)\right)^2 - \left(\mathfrak{F}^{IP}_{\mathfrak{R}_2}(e_\sigma)\right)^2\right| \\ , \left|\left(\mathfrak{F}^{RN}_{\mathfrak{R}_1}(e_\sigma)\right)^2 - \left(\mathfrak{F}^{RN}_{\mathfrak{R}_2}(e_\sigma)\right)^2\right|, \left|\left(\mathfrak{F}^{IN}_{\mathfrak{R}_1}(e_\sigma)\right)^2 - \left(\mathfrak{F}^{IN}_{\mathfrak{R}_2}(e_\sigma)\right)^2\right|\end{array}\right)\right)^{\frac{1}{\lambda}} = 0$$

$$\Rightarrow -\left(\frac{1}{n}\sum_{\sigma=1}^{n}\max\left(\begin{array}{l}\left|\left(\mathfrak{F}^{RP}_{\mathfrak{R}_1}(e_\sigma)\right)^2 - \left(\mathfrak{F}^{RP}_{\mathfrak{R}_2}(e_\sigma)\right)^2\right|, \left|\left(\mathfrak{F}^{IP}_{\mathfrak{R}_1}(e_\sigma)\right)^2 - \left(\mathfrak{F}^{IP}_{\mathfrak{R}_2}(e_\sigma)\right)^2\right| \\ , \left|\left(\mathfrak{F}^{RN}_{\mathfrak{R}_1}(e_\sigma)\right)^2 - \left(\mathfrak{F}^{RN}_{\mathfrak{R}_2}(e_\sigma)\right)^2\right|, \left|\left(\mathfrak{F}^{IN}_{\mathfrak{R}_1}(e_\sigma)\right)^2 - \left(\mathfrak{F}^{IN}_{\mathfrak{R}_2}(e_\sigma)\right)^2\right|\end{array}\right)\right)^{\frac{1}{\lambda}}$$

$= 0$

and

$$e^{-\left(\frac{1}{n}\sum_{\sigma=1}^{n}\max\left(\begin{array}{c}\left|\left(\mathfrak{F}_{\mathfrak{R}_1}^{RP}(e_\sigma)\right)^2-\left(\mathfrak{F}_{\mathfrak{R}_2}^{RP}(e_\sigma)\right)^2\right|,\left|\left(\mathfrak{F}_{\mathfrak{R}_1}^{IP}(e_\sigma)\right)^2-\left(\mathfrak{F}_{\mathfrak{R}_2}^{IP}(e_\sigma)\right)^2\right|\\,\left|\left(\mathfrak{F}_{\mathfrak{R}_1}^{RN}(e_\sigma)\right)^2-\left(\mathfrak{F}_{\mathfrak{R}_2}^{RN}(e_\sigma)\right)^2\right|,\left|\left(\mathfrak{F}_{\mathfrak{R}_1}^{IN}(e_\sigma)\right)^2-\left(\mathfrak{F}_{\mathfrak{R}_2}^{IN}(e_\sigma)\right)^2\right|\end{array}\right)\right)^{\frac{1}{\lambda}}} = e^0$$

Consequently,

$$e^{-\left(\frac{1}{n}\sum_{\sigma=1}^{n}\max\left(\begin{array}{c}\left|\left(\mathfrak{F}_{\mathfrak{R}_1}^{RP}(e_\sigma)\right)^2-\left(\mathfrak{F}_{\mathfrak{R}_2}^{RP}(e_\sigma)\right)^2\right|,\left|\left(\mathfrak{F}_{\mathfrak{R}_1}^{IP}(e_\sigma)\right)^2-\left(\mathfrak{F}_{\mathfrak{R}_2}^{IP}(e_\sigma)\right)^2\right|\\,\left|\left(\mathfrak{F}_{\mathfrak{R}_1}^{RN}(e_\sigma)\right)^2-\left(\mathfrak{F}_{\mathfrak{R}_2}^{RN}(e_\sigma)\right)^2\right|,\left|\left(\mathfrak{F}_{\mathfrak{R}_1}^{IN}(e_\sigma)\right)^2-\left(\mathfrak{F}_{\mathfrak{R}_2}^{IN}(e_\sigma)\right)^2\right|\end{array}\right)\right)^{\frac{1}{\lambda}}} = 1$$

$$\Rightarrow \mathcal{S}_{BCFS}^{33}(\mathfrak{R}_1, \mathfrak{R}_2) = 1.$$

3. From Theorem (4.2), we have that

$$\frac{1}{n}\sum_{\sigma=1}^{n}\max\left(\begin{array}{c}\left|\left(\mathfrak{F}_{\mathfrak{R}_1}^{RP}(e_\sigma)\right)^2-\left(\mathfrak{F}_{\mathfrak{R}_2}^{RP}(e_\sigma)\right)^2\right|,\left|\left(\mathfrak{F}_{\mathfrak{R}_1}^{IP}(e_\sigma)\right)^2-\left(\mathfrak{F}_{\mathfrak{R}_2}^{IP}(e_\sigma)\right)^2\right|\\,\left|\left(\mathfrak{F}_{\mathfrak{R}_1}^{RN}(e_\sigma)\right)^2-\left(\mathfrak{F}_{\mathfrak{R}_2}^{RN}(e_\sigma)\right)^2\right|,\left|\left(\mathfrak{F}_{\mathfrak{R}_1}^{IN}(e_\sigma)\right)^2-\left(\mathfrak{F}_{\mathfrak{R}_2}^{IN}(e_\sigma)\right)^2\right|\end{array}\right)\Bigg)^{\frac{1}{\lambda}}$$

$$= \left(\frac{1}{n}\sum_{\sigma=1}^{n}\max\left(\begin{array}{c}\left|\left(\mathfrak{F}_{\mathfrak{R}_2}^{RP}(e_\sigma)\right)^2-\left(\mathfrak{F}_{\mathfrak{R}_2}^{RP}(e_\sigma)\right)^2\right|,\left|\left(\mathfrak{F}_{\mathfrak{R}_2}^{IP}(e_\sigma)\right)^2-\left(\mathfrak{F}_{\mathfrak{R}_1}^{IP}(e_\sigma)\right)^2\right|\\,\left|\left(\mathfrak{F}_{\mathfrak{R}_2}^{RN}(e_\sigma)\right)^2-\left(\mathfrak{F}_{\mathfrak{R}_2}^{RN}(e_\sigma)\right)^2\right|,\left|\left(\mathfrak{F}_{2}^{IN}(e_\sigma)\right)^2-\left(\mathfrak{F}_{\mathfrak{R}_1}^{IN}(e_\sigma)\right)^2\right|\end{array}\right)\right)^{\frac{1}{\lambda}}$$

Thus, $\mathcal{S}_{BCFS}^{33}(\mathfrak{R}_1, \mathfrak{R}_2) = \mathcal{S}_{BCFS}^{33}(\mathfrak{R}_2, \mathfrak{R}_1)$.

4. Let $\mathfrak{R}_1 \subset \mathfrak{R}_2 \subset \mathfrak{R}_3$. Then from Theorem (4.2) we get that

$$d_{BCFS}^{4}(\mathfrak{R}_2, \mathfrak{R}_3) \le d_{BCFS}^{4}(\mathfrak{R}_1, \mathfrak{R}_3)$$

$$\Rightarrow \left(\frac{1}{n}\sum_{\sigma=1}^{n}\max\left(\begin{array}{c}\left|\left(\mathfrak{F}_{\mathfrak{R}_1}^{RP}(e_\sigma)\right)^2-\left(\mathfrak{F}_{\mathfrak{R}_2}^{RP}(e_\sigma)\right)^2\right|,\left|\left(\mathfrak{F}_{\mathfrak{R}_1}^{IP}(e_\sigma)\right)^2-\left(\mathfrak{F}_{\mathfrak{R}_2}^{IP}(e_\sigma)\right)^2\right|,\\\left|\left(\mathfrak{F}_{\mathfrak{R}_1}^{RN}(e_\sigma)\right)^2-\left(\mathfrak{F}_{\mathfrak{R}_2}^{RN}(e_\sigma)\right)^2\right|,\left|\left(\mathfrak{F}_{\mathfrak{R}_1}^{RN}(e_\sigma)\right)^2-\left(\mathfrak{F}_{\mathfrak{R}_2}^{RN}(e_\sigma)\right)^2\right|\end{array}\right)\right)^{\frac{1}{\lambda}}$$

$$\le \left(\frac{1}{n}\sum_{\sigma=1}^{n}\max\left(\begin{array}{c}\left|\left(\mathfrak{F}_{\mathfrak{R}_1}^{RP}(e_\sigma)\right)^2-\left(\mathfrak{F}_{\mathfrak{R}_3}^{RP}(e_\sigma)\right)^2\right|,\left|\left(\mathfrak{F}_{\mathfrak{R}_1}^{IP}(e_\sigma)\right)^2-\left(\mathfrak{F}_{\mathfrak{R}_3}^{IP}(e_\sigma)\right)^2\right|\\,\left|\left(\mathfrak{F}_{\mathfrak{R}_1}^{RN}(e_\sigma)\right)^2-\left(\mathfrak{F}_{\mathfrak{R}_3}^{RN}(e_\sigma)\right)^2\right|,\left|\left(\mathfrak{F}_{\mathfrak{R}_1}^{RN}(e_\sigma)\right)^2-\left(\mathfrak{F}_{\mathfrak{R}_3}^{RN}(e_\sigma)\right)^2\right|\end{array}\right)\right)^{\frac{1}{\lambda}}$$

Thus,

$$-\left(\frac{1}{n}\sum_{\sigma=1}^{n}\max\begin{pmatrix}\left|\left(\mathfrak{F}_{\Re_1}^{RP}(e_\sigma)\right)^2-\left(\mathfrak{F}_{\Re_2}^{RP}(e_\sigma)\right)^2\right|,\left|\left(\mathfrak{F}_{\Re_1}^{IP}(e_\sigma)\right)^2-\left(\mathfrak{F}_{\Re_2}^{IP}(e_\sigma)\right)^2\right|,\\\left|\left(\mathfrak{F}_{\Re_1}^{RN}(e_\sigma)\right)^2-\left(\mathfrak{F}_{\Re_2}^{RN}(e_\sigma)\right)^2\right|,\left|\left(\mathfrak{F}_{\Re_1}^{RN}(e_\sigma)\right)^2-\left(\mathfrak{F}_{\Re_2}^{RN}(e_\sigma)\right)^2\right|\end{pmatrix}\right)^{\frac{1}{\lambda}}$$

$$\geq-\left(\frac{1}{n}\sum_{\sigma=1}^{n}\max\begin{pmatrix}\left|\left(\mathfrak{F}_{\Re_1}^{RP}(e_\sigma)\right)^2-\left(\mathfrak{F}_{\Re_3}^{RP}(e_\sigma)\right)^2\right|,\left|\left(\mathfrak{F}_{\Re_1}^{IP}(e_\sigma)\right)^2-\left(\mathfrak{F}_{\Re_3}^{IP}(e_\sigma)\right)^2\right|\\,\left|\left(\mathfrak{F}_{\Re_1}^{RN}(e_\sigma)\right)^2-\left(\mathfrak{F}_{\Re_3}^{RN}(e_\sigma)\right)^2\right|,\left|\left(\mathfrak{F}_{\Re_1}^{RN}(e_\sigma)\right)^2-\left(\mathfrak{F}_{\Re_3}^{RN}(e_\sigma)\right)^2\right|\end{pmatrix}\right)^{\frac{1}{\lambda}}$$

And

$$e^{-\left(\frac{1}{n}\sum_{\sigma=1}^{n}\max\begin{pmatrix}\left|\left(\mathfrak{F}_{\Re_1}^{RP}(e_\sigma)\right)^2-\left(\mathfrak{F}_{\Re_2}^{RP}(e_\sigma)\right)^2\right|,\left|\left(\mathfrak{F}_{\Re_1}^{IP}(e_\sigma)\right)^2-\left(\mathfrak{F}_{\Re_2}^{IP}(e_\sigma)\right)^2\right|,\\\left|\left(\mathfrak{F}_{\Re_1}^{RN}(e_\sigma)\right)^2-\left(\mathfrak{F}_{\Re_2}^{RN}(e_\sigma)\right)^2\right|,\left|\left(\mathfrak{F}_{\mathcal{R}_1}^{RN}(e_\sigma)\right)^2-\left(\mathfrak{F}_{\mathcal{R}_2}^{RN}(e_\sigma)\right)^2\right|\end{pmatrix}\right)^{\frac{1}{\lambda}}}$$

$$\geq e^{-\left(\frac{1}{n}\sum_{\sigma=1}^{n}\max\begin{pmatrix}\left|\left(\mathfrak{F}_{\Re_1}^{RP}(e_\sigma)\right)^2-\left(\mathfrak{F}_{\Re_3}^{RP}(e_\sigma)\right)^2\right|,\left|\left(\mathfrak{F}_{\Re_1}^{IP}(e_\sigma)\right)^2-\left(\mathfrak{F}_{\Re_3}^{IP}(e_\sigma)\right)^2\right|\\\left|\left(\mathfrak{F}_{\Re_1}^{RN}(e_\sigma)\right)^2-\left(\mathfrak{F}_{\Re_3}^{RN}(e_\sigma)\right)^2\right|,\left|\left(\mathfrak{F}_{\Re_1}^{RN}(e_\sigma)\right)^2-\left(\mathfrak{F}_{\Re_3}^{RN}(e_\sigma)\right)^2\right|\end{pmatrix}\right)^{\frac{1}{\lambda}}}$$

This implies that $\mathcal{S}_{BCFS}^{33}(\Re_1,\Re_2)\geq\mathcal{S}_{BCFS}^{33}(\Re_1,\Re_3)$. Likewise, we can reveal that $\mathcal{S}_{BCFS}^{33}(\Re_2,\Re_3)\geq\mathcal{S}_{BCFS}^{33}(\Re_1,\Re_3)$.

Special Cases $\mathcal{S}_{BCFS}^{33}(\Re_1,\Re_2)$.

The SM $\mathcal{S}_{BCFS}^{33}(\Re_1,\Re_2)$ has underneath special cases.

1. By assuming the value of $\lambda=1$, we achieve the ESM relying on normalized Hamming-Hausdroff distance, demonstrated as

$$\mathcal{S}_{BCFS}^{34}(\Re_1,\Re_2)=$$
$$e^{-\left(\frac{1}{n}\sum_{\sigma=1}^{n}\max\begin{pmatrix}\left|\left(\mathfrak{F}_{\Re_1}^{RP}(e_\sigma)\right)^2-\left(\mathfrak{F}_{\Re_2}^{RP}(e_\sigma)\right)^2\right|,\left|\left(\mathfrak{F}_{\Re_1}^{IP}(e_\sigma)\right)^2-\left(\mathfrak{F}_{\Re_2}^{IP}(e_\sigma)\right)^2\right|\\,\left|\left(\mathfrak{F}_{\Re_1}^{RN}(e_\sigma)\right)^2-\left(\mathfrak{F}_{\Re_2}^{RN}(e_\sigma)\right)^2\right|,\left|\left(\mathfrak{F}_{\Re_1}^{IN}(e_\sigma)\right)^2-\left(\mathfrak{F}_{\Re_2}^{IN}(e_\sigma)\right)^2\right|\end{pmatrix}\right)}$$

2. By assuming the value of $\lambda=\frac{1}{2}$, we achieve the ESM relying on normalized Euclidean-Hausdroff distance, demonstrated as

$$\mathcal{S}_{BCFS}^{35}(\Re_1,\Re_2)=$$
$$e^{-\sqrt{\frac{1}{n}\sum_{\sigma=1}^{n}\max\begin{pmatrix}\left|\left(\mathfrak{F}_{\Re_1}^{RP}(e_\sigma)\right)^2-\left(\mathfrak{F}_{\Re_2}^{RP}(e_\sigma)\right)^2\right|,\left|\left(\mathfrak{F}_{\Re_1}^{IP}(e_\sigma)\right)^2-\left(\mathfrak{F}_{\Re_2}^{IP}(e_\sigma)\right)^2\right|\\,\left|\left(\mathfrak{F}_{\Re_1}^{RN}(e_\sigma)\right)^2-\left(\mathfrak{F}_{\Re_2}^{RN}(e_\sigma)\right)^2\right|,\left|\left(\mathfrak{F}_{\Re_1}^{IN}(e_\sigma)\right)^2-\left(\mathfrak{F}_{\Re_2}^{IN}(e_\sigma)\right)^2\right|\end{pmatrix}}}$$

In the underneath definition, we inaugurate weighted ESM based on GNHDD.

Definition 5.13 Let two BCFSs $\mathfrak{R}_1 = \left\{e_\sigma, \left(\mathfrak{F}^P_{\mathfrak{R}_1}(e_\sigma), \mathfrak{F}^N_{\mathfrak{R}_1}(e_\sigma)\right)|e_\sigma \in \mathbb{E}, \sigma = 1, 2, .., n\right\} = \left\{e_\sigma, \left(\mathfrak{F}^{RP}_{\mathfrak{R}_1}(e_\sigma) + \iota\mathfrak{F}^{IP}_{\mathfrak{R}_1}(e_\sigma), \mathfrak{F}^{RN}_{\mathfrak{R}_1}(e_\sigma) + \iota\mathfrak{F}^{IN}_{\mathfrak{R}_1}(e_\sigma)\right)|e_\sigma \in \mathbb{E}, \sigma = 1, 2, .., n\right\}$ and $\mathfrak{R}_2 = \left\{e, \left(\mathfrak{F}^P_{\mathfrak{R}_2}(e_\sigma), \mathfrak{F}^N_{\mathfrak{R}_2}(e_\sigma)\right)|e \in \mathbb{E}, \sigma = 1, 2, .., n\right\} = \left\{e_\sigma, \left(\mathfrak{F}^{RP}_{\mathfrak{R}_2}(e_\sigma) + \iota\mathfrak{F}^{IP}_{\mathfrak{R}_2}(e_\sigma), \mathfrak{F}^{RN}_{\mathfrak{R}_2}(e_\sigma) + \iota\mathfrak{F}^{IN}_{\mathfrak{R}_2}(e_\sigma)\right)|e_\sigma \in \mathbb{E}, \sigma = 1, 2, .., n\right\}$ over $\mathbb{E}$. Then the weighted ESM between $\mathfrak{R}_1$ and $\mathfrak{R}_2$ relying on generalized normalized Hausdroff distance (GNHDD) is interpreted as.

$$\mathcal{S}^{36}_{BCFS}(\mathfrak{R}_1, \mathfrak{R}_2) = e^{-\left(\sum_{\sigma=1}^{n} \overset{\circ}{\mathrm{w}}_\sigma \max\left(\begin{array}{l}\left|\left(\mathfrak{F}^{RP}_{\mathfrak{R}_1}(e_\sigma)\right)^2 - \left(\mathfrak{F}^{RP}_{\mathfrak{R}_2}(e_\sigma)\right)^2\right|, \left|\left(\mathfrak{F}^{IP}_{\mathfrak{R}_1}(e_\sigma)\right)^2 - \left(\mathfrak{F}^{IP}_{\mathfrak{R}_2}(e_\sigma)\right)^2\right| \\ , \left|\left(\mathfrak{F}^{RN}_{\mathfrak{R}_1}(e_\sigma)\right)^2 - \left(\mathfrak{F}^{RN}_{\mathfrak{R}_2}(e_\sigma)\right)^2\right|, \left|\left(\mathfrak{F}^{IN}_{\mathfrak{R}_1}(e_\sigma)\right)^2 - \left(\mathfrak{F}^{IN}_{\mathfrak{R}_2}(e_\sigma)\right)^2\right|\end{array}\right)\right)^{\frac{1}{\lambda}}}$$

Theorem 5.10 The SM $\mathcal{S}^{36}_{BCFS}(\mathfrak{R}_1, \mathfrak{R}_2)$ satisfies the below properties.

1. $0 \leq \mathcal{S}^{36}_{BCFS}(\mathfrak{R}_1, \mathfrak{R}_2) \leq 1$
2. $\mathcal{S}^{36}_{BCFS}(\mathfrak{R}_1, \mathfrak{R}_2) = 1 \Leftrightarrow \mathfrak{R}_1 = \mathfrak{R}_2$
3. $\mathcal{S}^{36}_{BCFS}(\mathfrak{R}_1, \mathfrak{R}_2) = \mathcal{S}^{36}_{BCFS}(\mathfrak{R}_2, \mathfrak{R}_1)$
4. If $\mathfrak{R}_1 \subset \mathfrak{R}_2 \subset \mathfrak{R}_3$, then $\mathcal{S}^{36}_{BCFS}(\mathfrak{R}_1, \mathfrak{R}_2) \geq \mathcal{S}^{36}_{BCFS}(\mathfrak{R}_1, \mathfrak{R}_3)$ and $\mathcal{S}^{36}_{BCFS}(\mathfrak{R}_2, \mathfrak{R}_3) \geq \mathcal{S}^{36}_{BCFS}(\mathfrak{R}_1, \mathfrak{R}_3)$

Proof Omitted.

On similar pattern, we can discuss of special cases of weighted ESM relying on GNHDD.

5.5 Applications of the Inaugurated Similarity Measures

Here, we inspect two genuine life applications of the inaugurated SMs that is pattern recognition and medical diagnosis.

5.5.1 Pattern Recognition

A cognitive process called pattern recognition includes spotting and categorizing patterns or regularities in data or sensory input. It describes the capacity to identify and comprehend recurrent structures, relationships, or patterns within a given set of data. Pattern recognition can take place in a variety of contexts, including linguistic, auditory, and visual. For instance, in visual pattern recognition, the brain analyses visual input and spots recurring traits or patterns to identify objects or forms. Similar to visual pattern recognition, aural pattern recognition uses sound patterns to help the brain identify well-known melodies or spoken phrases. In addition to perception, pattern recognition is essential for cognition and DM. Pattern recognition in disciplines like computer science and artificial intelligence entails the creation of algorithms and models to automate the process of locating and classifying patterns in huge datasets. Pattern recognition can be done using a variety of methods, such as statistical techniques, machine learning methods, and neural networks. These techniques seek to identify patterns in data, learn from them, and generalize them to solve classification or prediction problems.

By offering a quantitative evaluation of the similarity between patterns or objects, fuzzy SMs play a crucial role in pattern recognition. The objective of pattern recognition is to recognize and classify patterns or objects according to their inborn traits or properties. BCF SMs make it possible to compare patterns in a more flexible and nuanced way, especially when patterns may only be partially similar or unknown. In uncertain pattern recognition tasks, BCF SMs aid the process of DM. BCF SMs establish a foundation for DM in circumstances when patterns only exhibit partial similarity or when the data is ambiguous by quantifying the degree of similarity between patterns. Consequently, underneath, we handle a pattern recognition dilemma by employing few of the inaugurated SMs.

Example 5.1 In order to characterize and identify different features of building materials, pattern recognition is essential for analysis and comprehension. Take into account a situation where there are four known materials that is $\mathfrak{R}_{\mathbb{A}-1}$: Concrete, $\mathfrak{R}_{\mathbb{A}-2}$:Stone, $\mathfrak{R}_{\mathbb{A}-3}$:Brick, and $\mathfrak{R}_{\mathbb{A}-4}$: Cement, and one unknown material $\mathfrak{R}_{\mathbb{A}}$ that needs to be identified using pattern recognition. The information of these materials based on their 4 various features that is e_1, e_2, e_3 and e_4 are interpreted as below in the structure of BCFS.

$$\mathfrak{R}_{\mathbb{A}-1} = \left\{ \begin{array}{l} e_1, \begin{pmatrix} 0.935 + \iota 0.235, \\ -0.45 - \iota 0.342 \end{pmatrix}, e_2, \begin{pmatrix} 0.563 + \iota 0.634, \\ -0.733 - \iota 0.575 \end{pmatrix}, \\ e_3, \begin{pmatrix} 0.346 + \iota 0.163, \\ -0.233 - \iota 0.774 \end{pmatrix}, e_4, \begin{pmatrix} 0.204 + \iota 0.320, \\ -0.156 - \iota 0.732 \end{pmatrix} \end{array} \right\}$$

$$\mathfrak{R}_{\mathbb{A}-2} = \left\{ \begin{array}{l} e_1, \begin{pmatrix} 0.323 + \iota 0.653, \\ -0.702 - \iota 0.476 \end{pmatrix}, e_2, \begin{pmatrix} 0.282 + \iota 0.845, \\ -0.294 - \iota 0.235 \end{pmatrix}, \\ e_3, \begin{pmatrix} 0.752 + \iota 0.235, \\ -0.923 - \iota 0129 \end{pmatrix}, e_4, \begin{pmatrix} 0.203 + \iota 0.973, \\ -0.837 - \iota 0.375 \end{pmatrix} \end{array} \right\}$$

$$\mathfrak{R}_{\mathbb{A}-3} = \left\{ \begin{array}{l} e_1, \begin{pmatrix} 0.275 + \iota 0.375, \\ -0.375 - \iota 0.754 \end{pmatrix}, e_2, \begin{pmatrix} 0552. + \iota 0.245, \\ -0.468 - \iota 0.135 \end{pmatrix}, \\ e_3, \begin{pmatrix} 0.745 + \iota 0.942, \\ -0.852 - \iota 0.22 \end{pmatrix}, e_4, \begin{pmatrix} 0.545 + \iota 0.265, \\ -0.372 - \iota 0.265 \end{pmatrix} \end{array} \right\}$$

$$\Re_{\mathbb{A}-4} = \left\{ \begin{array}{l} e_1, \begin{pmatrix} 0.746 + \iota 0.386, \\ -0.123 - \iota 0.225 \end{pmatrix}, e_2, \begin{pmatrix} 0.234 + \iota 0.645, \\ -0.364 - \iota 0.807 \end{pmatrix}, \\ e_3, \begin{pmatrix} 0.375 + \iota 0.257, \\ -0.435 - \iota 0.649 \end{pmatrix}, e_4, \begin{pmatrix} 0.283 + \iota 0.821, \\ -0.435 - \iota 0.325 \end{pmatrix} \end{array} \right\}$$

And

$$\Re_{\mathbb{A}} = \left\{ \begin{array}{l} e_1, \begin{pmatrix} 0.857 + \iota 0.873, \\ -0.675 - \iota 0.445 \end{pmatrix}, e_2, \begin{pmatrix} 0.234 + \iota 0.712, \\ -0.846 - \iota 0.135 \end{pmatrix}, \\ e_3, \begin{pmatrix} 0.375 + \iota 0.957, \\ -0.256 - \iota 0.345 \end{pmatrix}, e_4, \begin{pmatrix} 0.373 + \iota 0.462, \\ -0.754 - \iota 0.356 \end{pmatrix} \end{array} \right\}$$

To identified that the unknown material, we employ a few of the existing SMs to get the degree of similarity between the unknown and each known material. The required outcomes are revealed in Table 5.1.

By employing the SMs $\mathcal{S}^1_{BCFS}$ and $\mathcal{S}^{29}_{BCFS}$ and the SMs $\mathcal{S}^4_{BCFS}$ and $\mathcal{S}^{33}_{BCFS}$, the unknown building material $\Re_{\mathbb{A}}$, is said to resemble the building material, $\Re_{\mathbb{A}-1}$, which is concrete, and the building material, $\Re_{\mathbb{A}-3}$, which is brick. The other SMs

Table 5.1 The required outcomes of Example 5.1. after employing a few initiated SMs with $\lambda = 3$

SMs	Degree of similarity	Ranking
$\mathcal{S}^1_{BCFS}$	$\mathcal{S}^1_{BCFS}(\Re_{\mathbb{A}}, \Re_{\mathbb{A}-1}) = 0.3399$, $\mathcal{S}^1_{BCFS}(\Re_{\mathbb{A}}, \Re_{\mathbb{A}-2}) = 0.3177$, $\mathcal{S}^1_{BCFS}(\Re_{\mathbb{A}}, \Re_{\mathbb{A}-3}) = 0.316$, $\mathcal{S}^1_{BCFS}(\Re_{\mathbb{A}}, \Re_{\mathbb{A}-4}) = 0.3268$	$\Re_{\mathbb{A}-1} > \Re_{\mathbb{A}-4} > \Re_{\mathbb{A}-2} > \Re_{\mathbb{A}-3}$
$\mathcal{S}^4_{BCFS}$	$\mathcal{S}^4_{BCFS}(\Re_{\mathbb{A}}, \Re_{\mathbb{A}-1}) = 0.1504$, $\mathcal{S}^4_{BCFS}(\Re_{\mathbb{A}}, \Re_{\mathbb{A}-2}) = 0.1065$, $\mathcal{S}^4_{BCFS}(\Re_{\mathbb{A}}, \Re_{\mathbb{A}-3}) = 0.175$, $\mathcal{S}^4_{BCFS}(\Re_{\mathbb{A}}, \Re_{\mathbb{A}-4}) = 0.1386$	$\Re_{\mathbb{A}-3} > \Re_{\mathbb{A}-1} > \Re_{\mathbb{A}-4} > \Re_{\mathbb{A}-2}$
$\mathcal{S}^{29}_{BCFS}$	$\mathcal{S}^{29}_{BCFS}(\Re_{\mathbb{A}}, \Re_{\mathbb{A}-1}) = 0.5168$, $\mathcal{S}^{29}_{BCFS}(\Re_{\mathbb{A}}, \Re_{\mathbb{A}-2}) = 0.5055$, $\mathcal{S}^{29}_{BCFS}(\Re_{\mathbb{A}}, \Re_{\mathbb{A}-3}) = 0.5046$, $\mathcal{S}^{29}_{BCFS}(\Re_{\mathbb{A}}, \Re_{\mathbb{A}-4}) = 0.5101$	$\Re_{\mathbb{A}-1} > \Re_{\mathbb{A}-4} > \Re_{\mathbb{A}-2} > \Re_{\mathbb{A}-3}$
$\mathcal{S}^{33}_{BCFS}$	$\mathcal{S}^{33}_{BCFS}(\Re_{\mathbb{A}}, \Re_{\mathbb{A}-1}) = 0.4276$, $\mathcal{S}^{33}_{BCFS}(\Re_{\mathbb{A}}, \Re_{\mathbb{A}-2}) = 0.4092$, $\mathcal{S}^{33}_{BCFS}(\Re_{\mathbb{A}}, \Re_{\mathbb{A}-3}) = 0.4382$, $\mathcal{S}^{33}_{BCFS}(\Re_{\mathbb{A}}, \Re_{\mathbb{A}-4}) = 0.4226$	$\Re_{\mathbb{A}-3} > \Re_{\mathbb{A}-1} > \Re_{\mathbb{A}-4} > \Re_{\mathbb{A}-2}$

Table 5.2 The required result of Example 5.1 after employing a few initiated weighted SMs with $\lambda = 3$

SMs	Degree of similarity	Ranking
$\mathcal{S}^{8}_{BCFS}$	$\mathcal{S}^{8}_{BCFS}(\Re_{\mathbb{A}}, \Re_{\mathbb{A}-1}) = 0.3446$, $\mathcal{S}^{8}_{BCFS}(\Re_{\mathbb{A}}, \Re_{\mathbb{A}-2}) = 0.3279$, $\mathcal{S}^{8}_{BCFS}(\Re_{\mathbb{A}}, \Re_{\mathbb{A}-3}) = 0.2857$, $\mathcal{S}^{8}_{BCFS}(\Re_{\mathbb{A}}, \Re_{\mathbb{A}-4}) = 0.3141$	$\Re_{\mathbb{A}-1} > \Re_{\mathbb{A}-2} > \Re_{\mathbb{A}-4} > \Re_{\mathbb{A}-3}$
$\mathcal{S}^{11}_{BCFS}$	$\mathcal{S}^{11}_{BCFS}(\Re_{\mathbb{A}}, \Re_{\mathbb{A}-1}) = 0.1525$, $\mathcal{S}^{11}_{BCFS}(\Re_{\mathbb{A}}, \Re_{\mathbb{A}-2}) = 0.1179$, $\mathcal{S}^{11}_{BCFS}(\Re_{\mathbb{A}}, \Re_{\mathbb{A}-3}) = 0.1629$, $\mathcal{S}^{11}_{BCFS}(\Re_{\mathbb{A}}, \Re_{\mathbb{A}-4}) = 0.1332$	$\Re_{\mathbb{A}-3} > \Re_{\mathbb{A}-1} > \Re_{\mathbb{A}-4} > \Re_{\mathbb{A}-2}$
$\mathcal{S}^{32}_{BCFS}$	$\mathcal{S}^{32}_{BCFS}(\Re_{\mathbb{A}}, \Re_{\mathbb{A}-1}) = 0.5192$, $\mathcal{S}^{32}_{BCFS}(\Re_{\mathbb{A}}, \Re_{\mathbb{A}-2}) = 0.5106$, $\mathcal{S}^{32}_{BCFS}(\Re_{\mathbb{A}}, \Re_{\mathbb{A}-3}) = 0.4895$, $\mathcal{S}^{32}_{BCFS}(\Re_{\mathbb{A}}, \Re_{\mathbb{A}-4}) = 0.5036$	$\Re_{\mathbb{A}-1} > \Re_{\mathbb{A}-2} > \Re_{\mathbb{A}-4} > \Re_{\mathbb{A}-3}$
$\mathcal{S}^{36}_{BCFS}$	$\mathcal{S}^{36}_{BCFS}(\Re_{\mathbb{A}}, \Re_{\mathbb{A}-1}) = 0.4285$, $\mathcal{S}^{36}_{BCFS}(\Re_{\mathbb{A}}, \Re_{\mathbb{A}-2}) = 0.4139$, $\mathcal{S}^{36}_{BCFS}(\Re_{\mathbb{A}}, \Re_{\mathbb{A}-3}) = 0.4328$, $\mathcal{S}^{36}_{BCFS}(\Re_{\mathbb{A}}, \Re_{\mathbb{A}-4}) = 0.4203$	$\Re_{\mathbb{A}-3} > \Re_{\mathbb{A}-1} > \Re_{\mathbb{A}-4} > \Re_{\mathbb{A}-2}$

that have been created can also be used to assess how similar the unknown and known materials are to one another.

In a few genuine-world scenarios, the elements commonly have varying degrees of significance, necessitating the allocation of various weights to them. To keep this in mind, we assign weight $(0.4, 0.3, 0.2, 0.1)$ and will employ the weighted SMs. The required result is interpreted in Table 5.2.

By employing the SMs $\mathcal{S}^{8}_{BCFS}$ and $\mathcal{S}^{32}_{BCFS}$ and the SMs $\mathcal{S}^{11}_{BCFS}$ and $\mathcal{S}^{36}_{BCFS}$, the unknown building material $\Re_{\mathbb{A}}$, is said to resemble the building material, $\Re_{\mathbb{A}-1}$, which is concrete, and the building material, $\Re_{\mathbb{A}-3}$, which is brick. The other weighted SMs that have been created can also be used to assess how similar the unknown and known materials are to one another in the situation where weight is involved.

5.5.2 Medical Diagnosis

Finding the nature and cause of a person's symptoms, signs, and medical condition is the process of making a medical diagnosis. It includes reviewing the patient's medical history, performing a physical exam, and, as appropriate, requesting a number of diagnostic tests or procedures. Finding the underlying illness or condition that is causing a patient's symptoms is the aim of a medical diagnostic. Healthcare professionals, such as doctors, often make medical diagnoses by interpreting the data acquired during the diagnostic procedure using their clinical knowledge and expertise. In order to aid in their diagnosis, they may also study medical guidelines, reference materials, or technology tools. The correct treatment strategy and patient care techniques are guided by the medical diagnosis once it has been determined. It enables medical professionals to choose the best actions, therapies, or drugs to treat the identified illness. It's crucial to remember that medical diagnoses must be made by licensed healthcare experts and should not be based primarily on personal beliefs or internet research. It is advised that you seek medical advice from a professional for an accurate assessment and diagnosis if you are experiencing any particular medical issues or symptoms.

By allowing for the evaluation of similarity or dissimilarity between medical instances or patterns, BCF SMs play a key role in medical diagnosis. BCF SMs are particularly helpful in medical diagnosis when working with ambiguous or inaccurate data, which is frequent in healthcare settings. DM that helps medical professionals make precise diagnoses can incorporate BCF SMs. These systems can rank potential diagnoses according to how closely they match the patient's symptoms and medical history using patient data, medical knowledge bases, and fuzzy similarity measurements. Consequently, underneath, we handle a medical diagnosis dilemma by employing few of the inaugurated SMs.

Example 5.2 Patient X, a 45-year-old person, went to the doctor after experiencing a number of unsettling symptoms that were interfering with their daily activities. With their extensive knowledge and state-of-the-art diagnostic equipment, the diligent team of medical professionals undertook a thorough medical investigation to ascertain the underlying cause of these distressing symptoms. The medical team's potential diagnoses were reduced to four different diseases: $\mathfrak{R}_{\mathbb{D}-1}$: Rheumatoid Arthritis, $\mathfrak{R}_{\mathbb{D}-2}$: Hypothyroidism, $\mathfrak{R}_{\mathbb{D}-3}$: Lyme Disease, and $\mathfrak{R}_{\mathbb{D}-4}$: Fibromyalgia. after carefully reviewing Patient X's medical history, performing exhaustive physical examinations, and carefully discussing the nature and progression of the symptoms that is e_1: Fever, e_2: Pain, e_3: tiredness, and e_4: Headache. The medical team provide the data of the diseases and patient $\mathfrak{R}_{\mathbb{D}}$ in the model of BCFS as follows.

$$\mathfrak{R}_{\mathbb{D}-1} = \left\{ \begin{array}{l} e_1, \begin{pmatrix} 0.573 + \iota 0.467, \\ -0.345 - \iota 0.434 \end{pmatrix}, e_2, \begin{pmatrix} 0.125 + \iota 0.473, \\ -0.237 - \iota 0.973 \end{pmatrix}, \\ e_3, \begin{pmatrix} 0.632 + \iota 0.745, \\ -0.534 - \iota 0.23 \end{pmatrix}, e_4, \begin{pmatrix} 0.423 + \iota 0.653, \\ -0.644 - \iota 0.355 \end{pmatrix} \end{array} \right\}$$

$$\mathfrak{R}_{\mathbb{D}-2} = \left\{ \begin{array}{l} e_1, \begin{pmatrix} 0.675 + \iota 0.849, \\ -0.337 - \iota 0.875 \end{pmatrix}, e_2, \begin{pmatrix} 0.246 + \iota 0.84, \\ -0.227 - \iota 0.345 \end{pmatrix}, \\ e_3, \begin{pmatrix} 0.745 + \iota 0.75, \\ -0.845 - \iota 0.245 \end{pmatrix}, e_4, \begin{pmatrix} 0.573 + \iota 0.485, \\ -0.346 - \iota 0.534 \end{pmatrix} \end{array} \right\}$$

$$\mathfrak{R}_{\mathbb{D}-3} = \left\{ \begin{array}{l} e_1, \begin{pmatrix} 0.453 + \iota 0.262, \\ -0.967 - \iota 0.346 \end{pmatrix}, e_2, \begin{pmatrix} 0.185 + \iota 0.346, \\ -0.124 - \iota 0.85 \end{pmatrix}, \\ e_3, \begin{pmatrix} 0.634 + \iota 0.734, \\ -0.213 - \iota 0.745 \end{pmatrix}, e_4, \begin{pmatrix} 0.635 + \iota 0.4534, \\ -0.634 - \iota 0.424 \end{pmatrix} \end{array} \right\}$$

$$\mathfrak{R}_{\mathbb{D}-4} = \left\{ \begin{array}{l} \mathrm{e}_1, \begin{pmatrix} 0.245 + 0.467 \\ -0917. - \iota 0.574 \end{pmatrix}, \mathrm{e}_2, \begin{pmatrix} 0.476 + \iota 0.956, \\ -0.536 - \iota 0.83 \end{pmatrix}, \\ \mathrm{e}_3, \begin{pmatrix} 0.785 + \iota 0.563, \\ -0.22 - \iota 0.852 \end{pmatrix}, \mathrm{e}_4, \begin{pmatrix} 0.524 + \iota 0.734, \\ -0.424 - \iota 0.958 \end{pmatrix} \end{array} \right\}$$

and

$$\mathfrak{R}_{\mathbb{D}} = \left\{ \begin{array}{l} e_1, \begin{pmatrix} 0.775 + \iota 0.684, \\ -0.556 - \iota 0.353 \end{pmatrix}, e_2, \begin{pmatrix} 0.574 + \iota 0.485, \\ -0.345 - \iota 0.245 \end{pmatrix}, \\ e_3, \begin{pmatrix} 0.575 + \iota 0.675, \\ -0.686 - \iota 0.34 \end{pmatrix}, e_4, \begin{pmatrix} 0.756 + \iota 0.345, \\ -0.245 - \iota 0.898 \end{pmatrix} \end{array} \right\}$$

To diagnosis the diseases of the patient $\mathfrak{R}_{\mathbb{D}}$, we employ a few of the existing SMs to get the degree of similarity between the patient symptoms and the diseases symptoms. The required outcomes are revealed in Table 5.2.

According to the Table 5.3, patient, $\mathfrak{R}_{\mathbb{D}}$ has the disease $\mathfrak{R}_{\mathbb{D}-1}$ that is Rheumatoid Arthritis by using the SMs $\mathcal{S}^{1}_{BCFS}$ and $\mathcal{S}^{29}_{BCFS}$ and has the disease $\mathfrak{R}_{\mathbb{D}-3}$ that is Lyme Disease by utilizing the SMs $\mathcal{S}^{4}_{BCFS}$ and $\mathcal{S}^{33}_{BCFS}$. Similarly, one can employ the other initiated SMs to determine the disease of the patient.

In a few medical cases, there are need of the weight, thus, here we assume the weight (0.1, 0.2, 0.3, 0.4) and then, we will employ the weighted SMs. The required result is interpreted in Table 5.4.

According to the Table 5.4, patient, $\mathfrak{R}_{\mathbb{D}}$ has the disease $\mathfrak{R}_{\mathbb{D}-1}$ that is "Rheumatoid Arthritis" by using the SMs $\mathcal{S}^{8}_{BCFS}$ and $\mathcal{S}^{32}_{BCFS}$ and has the disease $\mathfrak{R}_{\mathbb{D}-3}$ that is "Lyme Disease" by utilizing the SMs $\mathcal{S}^{11}_{BCFS}$ and $\mathcal{S}^{36}_{BCFS}$. Similarly, one can employ the other initiated weighted SMs to determine the disease of the patient.

Table 5.3 The required result of Example 5.2 after employing a few initiated SMs with $\lambda = 3$

SMs	Degree of similarity	Ranking
$\mathcal{S}^{1}_{BCFS}$	$\mathcal{S}^{1}_{BCFS}(\Re_{\mathbb{D}}, \Re_{\mathbb{D}-1}) = 0.3878$, $\mathcal{S}^{1}_{BCFS}(\Re_{\mathbb{D}}, \Re_{\mathbb{D}-2}) = 0.3418$, $\mathcal{S}^{1}_{BCFS}(\Re_{\mathbb{D}}, \Re_{\mathbb{D}-3}) = 0.2991$, $\mathcal{S}^{1}_{BCFS}(\Re_{\mathbb{D}}, \Re_{\mathbb{D}-4}) = 0.3596$	$\Re_{\mathbb{D}-1} > \Re_{\mathbb{D}-4} > \Re_{\mathbb{D}-2} > \Re_{\mathbb{D}-3}$
$\mathcal{S}^{4}_{BCFS}$	$\mathcal{S}^{4}_{BCFS}(\Re_{\mathbb{D}}, \Re_{\mathbb{D}-1}) = 0.223$, $\mathcal{S}^{4}_{BCFS}(\Re_{\mathbb{D}}, \Re_{\mathbb{D}-2}) = 0.174$, $\mathcal{S}^{4}_{BCFS}(\Re_{\mathbb{D}}, \Re_{\mathbb{D}-3}) = 0.1746$, $\mathcal{S}^{4}_{BCFS}(\Re_{\mathbb{D}}, \Re_{\mathbb{D}-4}) = 0.203$	$\Re_{\mathbb{D}-3} > \Re_{\mathbb{D}-1} > \Re_{\mathbb{D}-4} > \Re_{\mathbb{D}-2}$
$\mathcal{S}^{29}_{BCFS}$	$\mathcal{S}^{29}_{BCFS}(\Re_{\mathbb{D}}, \Re_{\mathbb{D}-1}) = 0.5422$, $\mathcal{S}^{29}_{BCFS}(\Re_{\mathbb{D}}, \Re_{\mathbb{D}-2}) = 0.5178$, $\mathcal{S}^{29}_{BCFS}(\Re_{\mathbb{D}}, \Re_{\mathbb{D}-3}) = 0.4961$, $\mathcal{S}^{29}_{BCFS}(\Re_{\mathbb{D}}, \Re_{\mathbb{D}-4}) = 0.5271$	$\Re_{\mathbb{D}-1} > \Re_{\mathbb{D}-4} > \Re_{\mathbb{D}-2} > \Re_{\mathbb{D}-3}$
$\mathcal{S}^{33}_{BCFS}$	$\mathcal{S}^{33}_{BCFS}(\Re_{\mathbb{D}}, \Re_{\mathbb{D}-1}) = 0.4598$, $\mathcal{S}^{33}_{BCFS}(\Re_{\mathbb{D}}, \Re_{\mathbb{D}-2}) = 0.4381$, $\mathcal{S}^{33}_{BCFS}(\Re_{\mathbb{D}}, \Re_{\mathbb{D}-3}) = 0.4381$, $\mathcal{S}^{33}_{BCFS}(\Re_{\mathbb{D}}, \Re_{\mathbb{D}-4}) = 0.4507$	$\Re_{\mathbb{D}-3} > \Re_{\mathbb{D}-1} > \Re_{\mathbb{D}-4} > \Re_{\mathbb{D}-2}$

5.6 Comparative Analysis

Comparing two things helps us see familiar things in fresh ways, which is why comparison is so essential. Although "common sense" would have us believe that two items are the same, a close comparison reveals that they are actually very different. Even while common sense may claim that two items are completely incompatible, careful comparison reveals their affinities. Thus, we compare a few of the existing SMs with our original SMs in this section.

Let us consider the SMs initiated by Rajeshwari et al. [21], Patrascu [22] and the initiated SMs and apply these SMs to deal with Example 5.1. The required result is revealed in Table 5.5.

This result demonstrated that the prevailing SMs initiated by Rajeshwari et al. [21] and Patrascu [22] are not giving any sort of value and failed to cope with Example 5.1. It is because the SMs initiated by Rajeshwari et al. [21] and Patrascu [22] are in the model of BFS and can't overcome with 2nd dimension and the information of Example 5.1 is two-dimensional. Thus, merely, the inaugurated SMs are giving us the required result. By employing the SMs $\mathcal{S}^{1}_{BCFS}$ and $\mathcal{S}^{29}_{BCFS}$ and the SMs $\mathcal{S}^{4}_{BCFS}$ and $\mathcal{S}^{33}_{BCFS}$, the unknown building material $\Re_{\mathbb{A}}$, is said to resemble the building material, $\Re_{\mathbb{A}-1}$, which is concrete, and the building material, $\Re_{\mathbb{A}-3}$, which is brick.

Table 5.4 The required result of Example 5.2 after employing a few initiated weighted SMs with $\lambda = 3$

SMs	Degree of similarity	Ranking
$\mathcal{S}^{8}_{BCFS}$	$\mathcal{S}^{8}_{BCFS}(\Re_{\mathbb{D}}, \Re_{\mathbb{D}-1}) = 0.4005$, $\mathcal{S}^{8}_{BCFS}(\Re_{\mathbb{D}}, \Re_{\mathbb{D}-2}) = 0.3577$, $\mathcal{S}^{8}_{BCFS}(\Re_{\mathbb{D}}, \Re_{\mathbb{D}-3}) = 0.3152$, $\mathcal{S}^{8}_{BCFS}(\Re_{\mathbb{D}}, \Re_{\mathbb{D}-4}) = 0.3397$	$\Re_{\mathbb{D}-1} > \Re_{\mathbb{D}-2} > \Re_{\mathbb{D}-4} > \Re_{\mathbb{D}-3}$
$\mathcal{S}^{11}_{BCFS}$	$\mathcal{S}^{11}_{BCFS}(\Re_{\mathbb{D}}, \Re_{\mathbb{D}-1}) = 0.2396$, $\mathcal{S}^{11}_{BCFS}(\Re_{\mathbb{D}}, \Re_{\mathbb{D}-2}) = 0.188$, $\mathcal{S}^{11}_{BCFS}(\Re_{\mathbb{D}}, \Re_{\mathbb{D}-3}) = 0.1895$, $\mathcal{S}^{11}_{BCFS}(\Re_{\mathbb{D}}, \Re_{\mathbb{D}-4}) = 0.1895$	$\Re_{\mathbb{D}-3} > \Re_{\mathbb{D}-1} > \Re_{\mathbb{D}-4} > \Re_{\mathbb{D}-2}$
$\mathcal{S}^{32}_{BCFS}$	$\mathcal{S}^{32}_{BCFS}(\Re_{\mathbb{D}}, \Re_{\mathbb{D}-1}) = 0.5491$, $\mathcal{S}^{32}_{BCFS}(\Re_{\mathbb{D}}, \Re_{\mathbb{D}-2}) = 0.5261$, $\mathcal{S}^{32}_{BCFS}(\Re_{\mathbb{D}}, \Re_{\mathbb{D}-3}) = 0.5042$, $\mathcal{S}^{32}_{BCFS}(\Re_{\mathbb{D}}, \Re_{\mathbb{D}-4}) = 0.5167$	$\Re_{\mathbb{D}-1} > \Re_{\mathbb{D}-2} > \Re_{\mathbb{D}-4} > \Re_{\mathbb{D}-3}$
$\mathcal{S}^{36}_{BCFS}$	$\mathcal{S}^{36}_{BCFS}(\Re_{\mathbb{D}}, \Re_{\mathbb{D}-1}) = 0.4675$, $\mathcal{S}^{36}_{BCFS}(\Re_{\mathbb{D}}, \Re_{\mathbb{D}-2}) = 0.444$, $\mathcal{S}^{36}_{BCFS}(\Re_{\mathbb{D}}, \Re_{\mathbb{D}-3}) = 0.4428$, $\mathcal{S}^{36}_{BCFS}(\Re_{\mathbb{D}}, \Re_{\mathbb{D}-4}) = 0.4447$	$\Re_{\mathbb{D}-3} > \Re_{\mathbb{D}-1} > \Re_{\mathbb{D}-4} > \Re_{\mathbb{D}-2}$

Now let us construct the same Example 5.1 of pattern recognition in the model of BFS. Let there are four known materials that is $\Re_{\mathbb{EA}-1}$: concrete, $\Re_{\mathbb{EA}-2}$:stone, $\Re_{\mathbb{EA}-3}$:brick, and $\Re_{\mathbb{EA}-4}$: cement, and one unknown material $\Re_{\mathbb{EA}}$ that needs to be identified using pattern recognition. The information of these materials interpreted as below in the structure of BFS.

$$\Re_{\mathbb{BA}-1} = \left\{ \begin{array}{l} e_1, (0.935, -0.45),\ e_2, (0.563 - 0.733),\ e_3, (0.346, \ -0.233), \\ e_4, (0.204, \ -0.156) \end{array} \right\}$$

$$\Re_{\mathbb{BA}-2} = \left\{ \begin{array}{l} e_1, (0.323, -0.702),\ e_2, (0.282, \ -0.294),\ e_3, (0.752, \ -0.923), \\ e_4, (0.203, \ -0.837) \end{array} \right\}$$

$$\Re_{\mathbb{BA}-3} = \left\{ \begin{array}{l} e_1, (0.275, -0.375),\ e_2, (0.552, \ -0.468),\ e_3, (0.745, \ -0.852), \\ e_4, (0.545, \ -0.372) \end{array} \right\}$$

$$\Re_{\mathbb{BA}-3} = \left\{ \begin{array}{l} e_1, (0.746, -0.123),\ e_2, (0.234, \ -0.364),\ e_3, (0.375, \ -0.435), \\ e_4, (0.283, \ -0.435) \end{array} \right\}$$

and

Table 5.5 The result of comparison of the a few initiated SMs with a few prevailing SMs

SMs	Degree of similarity	Ranking
Rajeshwari et al. [21]	$\times \rightarrow \times \rightarrow \times$	$\times \rightarrow \times \rightarrow \times$
Rajeshwari et al. [21]	$\times \rightarrow \times \rightarrow \times$	$\times \rightarrow \times \rightarrow \times$
Patrascu [22]	$\times \rightarrow \times \rightarrow \times$	$\times \rightarrow \times \rightarrow \times$
Patrascu [22]	$\times \rightarrow \times \rightarrow \times$	$\times \rightarrow \times \rightarrow \times$
Patrascu [22]	$\times \rightarrow \times \rightarrow \times$	$\times \rightarrow \times \rightarrow \times$
Inaugurated SM	$S^1_{BCFS}(\Re_{\mathbb{A}}, \Re_{\mathbb{A}-1}) = 0.3399$, $S^1_{BCFS}(\Re_{\mathbb{A}}, \Re_{\mathbb{A}-2}) = 0.3177$, $S^1_{BCFS}(\Re_{\mathbb{A}}, \Re_{\mathbb{A}-3}) = 0.316$, $S^1_{BCFS}(\Re_{\mathbb{A}}, \Re_{\mathbb{A}-4}) = 0.3268$	$\Re_{\mathbb{A}-1} > \Re_{\mathbb{A}-4} > \Re_{\mathbb{A}-2} > \Re_{\mathbb{A}-3}$
Inaugurated SM	$S^4_{BCFS}(\Re_{\mathbb{A}}, \Re_{\mathbb{A}-1}) = 0.1504$, $S^4_{BCFS}(\Re_{\mathbb{A}}, \Re_{\mathbb{A}-2}) = 0.1065$, $S^4_{BCFS}(\Re_{\mathbb{A}}, \Re_{\mathbb{A}-3}) = 0.175$, $S^4_{BCFS}(\Re_{\mathbb{A}}, \Re_{\mathbb{A}-4}) = 0.1386$	$\Re_{\mathbb{A}-3} > \Re_{\mathbb{A}-1} > \Re_{\mathbb{A}-4} > \Re_{\mathbb{A}-2}$
Inaugurated SM	$S^{29}_{BCFS}(\Re_{\mathbb{A}}, \Re_{\mathbb{A}-1}) = 0.5168$, $S^{29}_{BCFS}(\Re_{\mathbb{A}}, \Re_{\mathbb{A}-2}) = 0.5055$, $S^{29}_{BCFS}(\Re_{\mathbb{A}}, \Re_{\mathbb{A}-3}) = 0.5046$, $S^{29}_{BCFS}(\Re_{\mathbb{A}}, \Re_{\mathbb{A}-4}) = 0.5101$	$\Re_{\mathbb{A}-1} > \Re_{\mathbb{A}-4} > \Re_{\mathbb{A}-2} > \Re_{\mathbb{A}-3}$
Inaugurated SM	$S^{33}_{BCFS}(\Re_{\mathbb{A}}, \Re_{\mathbb{A}-1}) = 0.4276$, $S^{33}_{BCFS}(\Re_{\mathbb{A}}, \Re_{\mathbb{A}-2}) = 0.4092$, $S^{33}_{BCFS}(\Re_{\mathbb{A}}, \Re_{\mathbb{A}-3}) = 0.4382$, $S^{33}_{BCFS}(\Re_{\mathbb{A}}, \Re_{\mathbb{A}-4}) = 0.4226$	$\Re_{\mathbb{A}-3} > \Re_{\mathbb{A}-1} > \Re_{\mathbb{A}-4} > \Re_{\mathbb{A}-2}$

$$\Re_{\mathbb{BA}} = \left\{ \begin{array}{l} e_1, (0.857, -0.675),\ e_2, (0.234, \quad -0.846),\ e_3, (0.375, \quad -0.256), \\ e_4, (0.373, \quad -0.754) \end{array} \right\}$$

To identified that the unknown material $\Re_{\mathbb{BA}}$, we employ the SMs diagnosed by Rajeshwari et al. [21], Patrascu [22] and the initiated SMs. The required result is verified in Table 5.6.

The outcomes of Table 7.6 demonstrated that the prevailing SMs investigated by Rajeshwari et al. [21], Patrascu [22] and the diagnosed SMs are giving that the unknown $\Re_{\mathbb{BA}}$ has the resembles with known building material $\Re_{\mathbb{BA}-1}$ "Concrete". Further, the if weight is involved in any situation, then the SMs can replace by weighted SMs. This demonstrated that the diagnosed SMs can cope with bipolar fuzzy information. Similarly, the developed SMs of this chapter can handle the information in the model of FS and CFS.

Table 5.6 The required result for the pattern recognition in the model of BFS. (for initiated SMs $\lambda = 3$)

SMs	Degree of similarity	Ranking
Rajeshwari et al. [21]	$\mathcal{S}_{NHD}(\Re_{\mathbb{BA}}, \Re_{\mathbb{BA}-1}) = 0.609$ $\mathcal{S}_{NHD}(\Re_{\mathbb{BA}}, \Re_{\mathbb{BA}-2}) = 0.283$ $\mathcal{S}_{NHD}(\Re_{\mathbb{BA}}, \Re_{\mathbb{BA}-3}) = 0.226$ $\mathcal{S}_{NHD}(\Re_{\mathbb{BA}}, \Re_{\mathbb{BA}-4}) = 0.567$	$\Re_{\mathbb{BA}-1} > \Re_{\mathbb{BA}-4} > \Re_{\mathbb{BA}-2} > \Re_{\mathbb{BA}-3}$
Rajeshwari et al. [21]	$\mathcal{S}_{NED}(\Re_{\mathbb{BA}}, \Re_{\mathbb{BA}-1}) = 0.624$ $\mathcal{S}_{NED}(\Re_{\mathbb{BA}}, \Re_{\mathbb{BA}-2}) = 0.432$ $\mathcal{S}_{NED}(\Re_{\mathbb{BA}}, \Re_{\mathbb{BA}-3}) = 0.421$ $\mathcal{S}_{NED}(\Re_{\mathbb{BA}}, \Re_{\mathbb{BA}-4}) = 0.584$	$\Re_{\mathbb{BA}-1} > \Re_{\mathbb{BA}-4} > \Re_{\mathbb{BA}-2} > \Re_{\mathbb{BA}-3}$
Patrascu [22]	$\mathcal{S}_{PH}(\Re_{\mathbb{BA}}, \Re_{\mathbb{BA}-1}) = 0.827$ $\mathcal{S}_{PH}(\Re_{\mathbb{BA}}, \Re_{\mathbb{BA}-2}) = 0.715$ $\mathcal{S}_{PH}(\Re_{\mathbb{BA}}, \Re_{\mathbb{BA}-3}) = 0.687$ $\mathcal{S}_{PH}(\Re_{\mathbb{BA}}, \Re_{\mathbb{BA}-4}) = 0.804$	$\Re_{\mathbb{BA}-1} > \Re_{\mathbb{BA}-4} > \Re_{\mathbb{BA}-2} > \Re_{\mathbb{BA}-3}$
Patrascu [22]	$\mathcal{S}_{PE}(\Re_{\mathbb{BA}}, \Re_{\mathbb{BA}-1}) = 0.95$ $\mathcal{S}_{PE}(\Re_{\mathbb{BA}}, \Re_{\mathbb{BA}-2}) = 0.901$ $\mathcal{S}_{PE}(\Re_{\mathbb{BA}}, \Re_{\mathbb{BA}-3}) = 0.894$ $\mathcal{S}_{PE}(\Re_{\mathbb{BA}}, \Re_{\mathbb{BA}-4}) = 0.942$	$\Re_{\mathbb{BA}-1} > \Re_{\mathbb{BA}-4} > \Re_{\mathbb{BA}-2} > \Re_{\mathbb{BA}-3}$
Patrascu [22]	$\mathcal{S}_{PP}(\Re_{\mathbb{BA}}, \Re_{\mathbb{BA}-1}) = 0.778$ $\mathcal{S}_{PP}(\Re_{\mathbb{BA}}, \Re_{\mathbb{BA}-2}) = 0.637$ $\mathcal{S}_{PP}(\Re_{\mathbb{BA}}, \Re_{\mathbb{BA}-3}) = 0.606$ $\mathcal{S}_{PP}(\Re_{\mathbb{BA}}, \Re_{\mathbb{BA}-4}) = 0.752$	$\Re_{\mathbb{BA}-1} > \Re_{\mathbb{BA}-4} > \Re_{\mathbb{BA}-2} > \Re_{\mathbb{BA}-3}$
Inaugurated SM	$\mathcal{S}^{1}_{BCFS}(\Re_{\mathbb{BA}}, \Re_{\mathbb{BA}-1}) = 0.545,$ $\mathcal{S}^{1}_{BCFS}(\Re_{\mathbb{BA}}, \Re_{\mathbb{BA}-2}) = 0.443,$ $\mathcal{S}^{1}_{BCFS}(\Re_{\mathbb{BA}}, \Re_{\mathbb{BA}-3}) = 0.404,$ $\mathcal{S}^{1}_{BCFS}(\Re_{\mathbb{BA}}, \Re_{\mathbb{BA}-4}) = 0.521$	$\Re_{\mathbb{BA}-1} > \Re_{\mathbb{BA}-4} > \Re_{\mathbb{BA}-2} > \Re_{\mathbb{BA}-3}$
Inaugurated SM	$\mathcal{S}^{4}_{BCFS}(\Re_{\mathbb{BA}}, \Re_{\mathbb{BA}-1}) = 0.354,$ $\mathcal{S}^{4}_{BCFS}(\Re_{\mathbb{BA}}, \Re_{\mathbb{BA}-2}) = 0.183,$ $\mathcal{S}^{4}_{BCFS}(\Re_{\mathbb{BA}}, \Re_{\mathbb{BA}-3}) = 0.175,$ $\mathcal{S}^{4}_{BCFS}(\Re_{\mathbb{BA}}, \Re_{\mathbb{BA}-4}) = 0.275$	$\Re_{\mathbb{BA}-1} > \Re_{\mathbb{BA}-4} > \Re_{\mathbb{BA}-2} > \Re_{\mathbb{BA}-3}$
Inaugurated SM	$\mathcal{S}^{29}_{BCFS}(\Re_{\mathbb{BA}}, \Re_{\mathbb{BA}-1}) = 0.634,$ $\mathcal{S}^{29}_{BCFS}(\Re_{\mathbb{BA}}, \Re_{\mathbb{BA}-2}) = 0.573,$ $\mathcal{S}^{29}_{BCFS}(\Re_{\mathbb{BA}}, \Re_{\mathbb{BA}-3}) = 0.551,$ $\mathcal{S}^{29}_{BCFS}(\Re_{\mathbb{BA}}, \Re_{\mathbb{BA}-4}) = 0.619$	$\Re_{\mathbb{BA}-1} > \Re_{\mathbb{BA}-4} > \Re_{\mathbb{BA}-2} > \Re_{\mathbb{BA}-3}$
Inaugurated SM	$\mathcal{S}^{33}_{BCFS}(\Re_{\mathbb{BA}}, \Re_{\mathbb{BA}-1}) = 0.524,$ $\mathcal{S}^{33}_{BCFS}(\Re_{\mathbb{BA}}, \Re_{\mathbb{BA}-2}) = 0.442,$ $\mathcal{S}^{33}_{BCFS}(\Re_{\mathbb{BA}}, \Re_{\mathbb{BA}-3}) = 0.438,$ $\mathcal{S}^{33}_{BCFS}(\Re_{\mathbb{BA}}, \Re_{\mathbb{BA}-4}) = 0.484$	$\Re_{\mathbb{BA}-1} > \Re_{\mathbb{BA}-4} > \Re_{\mathbb{BA}-2} > \Re_{\mathbb{BA}-3}$

Chapter 6
TOPSIS Technique Based on Correlation Coefficient and Entropy Measures in the Environment of Bipolar Complex Fuzzy Set

A statistical measure that expresses the magnitude and direction of the association between two variables is the correlation coefficient. How closely the variables move together is usefully revealed by the correlation coefficient. If the correlation coefficient is positive, it shows that if one variable grows, the other variable tends to increase as well. In contrast, a negative correlation coefficient denotes a tendency for the other variable to drop as the first increases. If we investigate, for instance, the link between temperature and ice cream sales, we can discover a positive correlation coefficient, indicating that ice cream sales typically grow as temperatures rise. In a variety of disciplines, including economics, social sciences, and finance, the correlation coefficient is frequently utilized. In order to make predictions or conclusions based on the links observed, it aids researchers and analysts in understanding the interdependence of variables. However, it is essential to evaluate correlation coefficients in light of the precise variables under study and to take into account additional variables that might have an impact on the relationship. The correlation coefficient offers a numerical assessment of the intensity and axis of the association between two variables. It is a helpful tool for evaluating relationships, spotting trends, and forecasting behavior. Correlation does not necessarily imply causality hence thorough analysis is required to get significant conclusions from correlation coefficients.

Entropy assesses the uncertainty or randomness in a set of data or a probability distribution and is a basic concept in information theory and statistics. It offers a technique to measure how much information or surprise is there in a system. Entropy is frequently employed in the context of information theory to assess the typical amount of information required to encode or transmit a message from a specific source. More information is needed to represent a source with a higher entropy since it is more uncertain or unpredictable. Lower entropy, on the other hand, denotes a source that is more predictable or certain and requires less data to represent it. In information theory, statistical analysis, machine learning, physics, and economics, it is essential. We can learn more about the structure, predictability, and complexity of

U. ur Rehman and T. Mahmood, *Bipolar Complex Fuzzy Sets*, Studies in Fuzziness and Soft Computing 439, https://doi.org/10.1007/978-3-032-07388-4_6

various processes in the environment around us by comprehending and using entropy metrics.

In Sect. 6.1, we investigate correlation coefficient (CC) and weighted correlation coefficient under the environment of BCF information. In Sect. 6.2, we invent entropy measures (EM) in the model of BCF information. In Sect. 6.3, we develop an approach of TOPSIS in the setting of BCFS relying on the initiated CC and EMS. Section 6.4, contains the demonstrative examples of the BCF-TOPSIS approach to reveal the applicability of the proposed BCF-TOPSIS approach. In Sect. 6.4, we have the comparative analysis.

6.1 Correlation Coefficient Based on BCFSs

The correlation coefficient, which ranges from -1 to 1, is a measure of association among two variables. The correlation coefficient will either be 1 or -1 if the two variables are perfectly linearly related to one another. Whether the variables are positively or negatively correlated determines the sign. If there is no linear relationship between the variables, the correlation coefficient is 0. There are two distinct kinds of correlation coefficients in use. One is based on the rank relationship between variables and is known as the Pearson product moment correlation coefficient, while the other is known as the Spearman rank correlation coefficient. The correlation between two variables is more frequently measured using the Pearson product-moment correlation coefficient. Provided paired measurements $(\mathcal{X}_1, \mathcal{Y}_1), (\mathcal{X}_2, \mathcal{Y}_2), \ldots, (\mathcal{X}_n, \mathcal{Y}_n)$, as a measure of association, the Pearson product moment correlation coefficient is demonstrated by

$$\mathfrak{r}_{\mathfrak{P}} = \frac{\sum_{\sigma=1}^{n}(\mathcal{X}_\sigma - \overline{\mathcal{X}}_\sigma)(\mathcal{Y}_\sigma - \overline{\mathcal{Y}}_\sigma)}{\sum_{\sigma=1}^{n}(\mathcal{X}_\sigma - \overline{\mathcal{X}}_\sigma)^2 \sum_{\sigma=1}^{n}(\mathcal{Y}_\sigma - \overline{\mathcal{Y}}_\sigma)^2}$$

Noted that $\overline{\mathcal{X}}$ and $\overline{\mathcal{Y}}$ are the mean of $\mathcal{X}_1, \mathcal{X}_2, \ldots, \mathcal{X}_n$ and $\mathcal{Y}_1, \mathcal{Y}_2, \ldots, \mathcal{Y}_n$ respectively.

The informational energy of a BCFS $\mathfrak{R}_1 = \left\{e_\sigma, \left(\mathfrak{F}^{RP}_{\mathfrak{R}_1}(e_\sigma) + \iota\, \mathfrak{F}^{IP}_{\mathfrak{R}_1}(e_\sigma),\ \mathfrak{F}^{RN}_{\mathfrak{R}_1}(e_\sigma) + \iota\, \mathfrak{F}^{IN}_{\mathfrak{R}_1}(e_\sigma)\right) | e_\sigma \in \mathbb{E},\ \sigma = 1, 2, .., n\right\}$ is identified and demonstrated as

$$\underline{\mathrm{T}}(\mathfrak{R}_1) = \sum_{\sigma=1}^{n}\left(\left(\mathfrak{F}^{RP}_{\mathfrak{R}_1}(e_\sigma)\right)^2 + \left(\mathfrak{F}^{IP}_{\mathfrak{R}_1}(e_\sigma)\right)^2 + \left(\mathfrak{F}^{RN}_{\mathfrak{R}_1}(e_\sigma)\right)^2 + \left(\mathfrak{F}^{IN}_{\mathfrak{R}_1}(e_\sigma)\right)^2\right)$$

and the correlation between $\mathfrak{R}_1 = \left\{e_\sigma, \left(\mathfrak{F}^{RP}_{\mathfrak{R}_1}(e_\sigma) + \iota\mathfrak{F}^{IP}_{\mathfrak{R}_1}(e_\sigma), \mathfrak{F}^{RN}_{\mathfrak{R}_1}(e_\sigma) + \iota\mathfrak{F}^{IN}_{\mathfrak{R}_1}(e_\sigma)\right) | e_\sigma \in \mathbb{E}, \sigma = 1, 2, .., n\right\}$ and $\mathfrak{R}_2 = \left\{e_\sigma, \left(\mathfrak{F}^{RP}_{\mathfrak{R}_2}(e_\sigma) + \iota\mathfrak{F}^{IP}_{\mathfrak{R}_2}(e_\sigma), \mathfrak{F}^{RN}_{\mathfrak{R}_2}(e_\sigma) + \iota\mathfrak{F}^{IN}_{\mathfrak{R}_2}(e_\sigma)\right) | e_\sigma \in \mathbb{E}, \sigma = 1, 2, .., n\right\}$ is identified and demonstrated as

$$C(\mathfrak{R}_1,\ \mathfrak{R}_2) = \sum_{\sigma=1}^{n}\begin{pmatrix} \mathfrak{F}^{RP}_{\mathfrak{R}_1}(e_\sigma)\mathfrak{F}^{RP}_{\mathfrak{R}_2}(e_\sigma) + \mathfrak{F}^{IP}_{\mathfrak{R}_1}(e_\sigma)\mathfrak{F}^{IP}_{\mathfrak{R}_2}(e_\sigma) + \mathfrak{F}^{RN}_{\mathfrak{R}_1}(e_\sigma)\mathfrak{F}^{RN}_{\mathfrak{R}_2}(e_\sigma) \\ +\mathfrak{F}^{IN}_{\mathfrak{R}_1}(e_\sigma)\mathfrak{F}^{IN}_{\mathfrak{R}_2}(e_\sigma) \end{pmatrix}$$

holds the underneath axioms.

1. $C(\mathfrak{R}_1, \mathfrak{R}_2) = \underline{T}(\mathfrak{R})$
2. $C(\mathfrak{R}_1, \mathfrak{R}_2) = C(\mathfrak{R}_2, \mathfrak{R}_1)$

Underneath, we define the CC among BCFSs.

Definition 6.1 Let two BCFSs $\mathfrak{R}_1 = \left\{e_\sigma, \left(\mathfrak{F}^{P}_{\mathfrak{R}_1}(e_\sigma), \mathfrak{F}^{N}_{\mathfrak{R}_1}(e_\sigma)\right) | e_\sigma \in \mathbb{E}, \sigma = 1, 2, .., n\right\} = \left\{e_\sigma, \left(\mathfrak{F}^{RP}_{\mathfrak{R}_1}(e_\sigma) + \iota\mathfrak{F}^{IP}_{\mathfrak{R}_1}(e_\sigma), \mathfrak{F}^{RN}_{\mathfrak{R}_1}(e_\sigma) + \iota\mathfrak{F}^{IN}_{\mathfrak{R}_1}(e_\sigma)\right) | e_\sigma \in \mathbb{E}, \sigma = 1, 2, .., n\right\}$ and $\mathfrak{R}_2 = \left\{e, \left(\mathfrak{F}^{P}_{\mathfrak{R}_2}(e_\sigma), \mathfrak{F}^{N}_{\mathfrak{R}_2}(e_\sigma)\right) | e \in \mathbb{E}, \sigma = 1, 2, .., n\right\} = \left\{e_\sigma, \left(\mathfrak{F}^{RP}_{\mathfrak{R}_2}(e_\sigma) + \iota\mathfrak{F}^{IP}_{\mathfrak{R}_2}(e_\sigma), \mathfrak{F}^{RN}_{\mathfrak{R}_2}(e_\sigma) + \iota\mathfrak{F}^{IN}_{\mathfrak{R}_2}(e_\sigma)\right) | e_\sigma \in \mathbb{E}, \sigma = 1, 2, .., n\right\}$ over $\mathbb{E}$. Then CC among $\mathfrak{R}_1$ and $\mathfrak{R}_2$ is deduced as

$$K(\mathfrak{R}_1, \mathfrak{R}_2) = \frac{C(\mathfrak{R}_1, \mathfrak{R}_2)}{\left(\underline{T}(\mathfrak{R}_1)\underline{T}(\mathfrak{R}_2)\right)^{\frac{1}{2}}}$$

$$= \frac{\sum_{\sigma=1}^{n}\left(\mathfrak{F}^{RP}_{\mathfrak{R}_1}(e_\sigma)\mathfrak{F}^{RP}_{\mathfrak{R}_2}(e_\sigma) + \mathfrak{F}^{IP}_{\mathfrak{R}_1}(e_\sigma)\mathfrak{F}^{IP}_{\mathfrak{R}_2}(e_\sigma) + \mathfrak{F}^{RN}_{\mathfrak{R}_1}(e_\sigma)\mathfrak{F}^{RN}_{\mathfrak{R}_2}(e_\sigma) + \mathfrak{F}^{IN}_{\mathfrak{R}_1}(e_\sigma)\mathfrak{F}^{IN}_{\mathfrak{R}_2}(e_\sigma)\right)}{\begin{pmatrix}\left(\sum_{\sigma=1}^{n}\left(\left(\mathfrak{F}^{RP}_{\mathfrak{R}_1}(e_\sigma)\right)^2 + \left(\mathfrak{F}^{IP}_{\mathfrak{R}_1}(e_\sigma)\right)^2 + \left(\mathfrak{F}^{RN}_{\mathfrak{R}_1}(e_\sigma)\right)^2 + \left(\mathfrak{F}^{IN}_{\mathfrak{R}_1}(e_\sigma)\right)^2\right)\right) \\ \left(\sum_{\sigma=1}^{n}\left(\left(\mathfrak{F}^{RP}_{\mathfrak{R}_2}(e_\sigma)\right)^2 + \left(\mathfrak{F}^{IP}_{\mathfrak{R}_2}(e_\sigma)\right)^2 + \left(\mathfrak{F}^{RN}_{\mathfrak{R}_2}(e_\sigma)\right)^2 + \left(\mathfrak{F}^{IN}_{\mathfrak{R}_2}(e_\sigma)\right)^2\right)\right)\end{pmatrix}^{\frac{1}{2}}} \tag{6.1}$$

The CC of BCFSs holds the underneath axioms.

1. $0 \le K(\mathfrak{R}_1, \mathfrak{R}_2) \le 1$,
2. $K(\mathfrak{R}_1, \mathfrak{R}_2) = K(\mathfrak{R}_2, \mathfrak{R}_1)$,
3. $\mathfrak{R}_1 = \mathfrak{R}_2 \Rightarrow K(\mathfrak{R}_1, \mathfrak{R}_2) = 1$.

Proof

1. Employing Eq. (6.1), we obviously have that $K(\mathfrak{R}_1, \mathfrak{R}_2) \ge 0$, now we need to reveal the other part that $K(\mathfrak{R}_2, \mathfrak{R}_1) \le 1$.

$$C(\mathfrak{R}_1, \mathfrak{R}_2) = \sum_{\sigma=1}^{n}\left(\mathfrak{F}^{RP}_{\mathfrak{R}_1}(e_\sigma)\mathfrak{F}^{RP}_{\mathfrak{R}_2}(e_\sigma) + \mathfrak{F}^{IP}_{\mathfrak{R}_1}(e_\sigma)\mathfrak{F}^{IP}_{\mathfrak{R}_2}(e_\sigma) + \mathfrak{F}^{RN}_{\mathfrak{R}_1}(e_\sigma)\mathfrak{F}^{RN}_{\mathfrak{R}_2}(e_\sigma) + \mathfrak{F}^{IN}_{\mathfrak{R}_1}(e_\sigma)\mathfrak{F}^{IN}_{\mathfrak{R}_2}(e_\sigma)\right)$$

$$= \begin{pmatrix} \left(\mathfrak{F}^{RP}_{\mathfrak{R}_1}(e_1)\mathfrak{F}^{RP}_{\mathfrak{R}_2}(e_1) + \mathfrak{F}^{IP}_{\mathfrak{R}_1}(e_1)\mathfrak{F}^{IP}_{\mathfrak{R}_2}(e_1) + \mathfrak{F}^{RN}_{\mathfrak{R}_1}(e_1)\mathfrak{F}^{RN}_{\mathfrak{R}_2}(e_1) + \mathfrak{F}^{IN}_{\mathfrak{R}_1}(e_1)\mathfrak{F}^{IN}_{\mathfrak{R}_2}(e_1)\right) \\ +\left(\mathfrak{F}^{RP}_{\mathfrak{R}_1}(e_2)\mathfrak{F}^{RP}_{\mathfrak{R}_2}(e_2) + \mathfrak{F}^{IP}_{\mathfrak{R}_1}(e_2)\mathfrak{F}^{IP}_{\mathfrak{R}_2}(e_2) + \mathfrak{F}^{RN}_{\mathfrak{R}_1}(e_2)\mathfrak{F}^{RN}_{\mathfrak{R}_2}(e_2) + \mathfrak{F}^{IN}_{\mathfrak{R}_1}(e_2)\mathfrak{F}^{IN}_{\mathfrak{R}_2}(e_2)\right) \\ +\left(\mathfrak{F}^{RP}_{\mathfrak{R}_1}(e_3)\mathfrak{F}^{RP}_{\mathfrak{R}_2}(e_3) + \mathfrak{F}^{IP}_{\mathfrak{R}_1}(e_3)\mathfrak{F}^{IP}_{\mathfrak{R}_2}(e_3) + \mathfrak{F}^{RN}_{\mathfrak{R}_1}(e_3)\mathfrak{F}^{RN}_{\mathfrak{R}_2}(e_3) + \mathfrak{F}^{IN}_{\mathfrak{R}_1}(e_3)\mathfrak{F}^{IN}_{\mathfrak{R}_2}(e_3)\right) + \cdots \\ +\left(\mathfrak{F}^{RP}_{\mathfrak{R}_1}(e_n)\mathfrak{F}^{RP}_{\mathfrak{R}_2}(e_n) + \mathfrak{F}^{IP}_{\mathfrak{R}_1}(e_n)\mathfrak{F}^{IP}_{\mathfrak{R}_2}(e_n) + \mathfrak{F}^{RN}_{\mathfrak{R}_1}(e_n)\mathfrak{F}^{RN}_{\mathfrak{R}_2}(e_n) + \mathfrak{F}^{IN}_{\mathfrak{R}_1}(e_n)\mathfrak{F}^{IN}_{\mathfrak{R}_2}(e_n)\right) \end{pmatrix}$$

By utilizing Cauchy-Schwarz inequality

$$(x_1z_1 + x_2z_2 + \ldots + x_nz_n)^2 \le \left(x_1^2 + x_2^2 + \ldots + x_n^2\right)\left(z_1^2 + z_2^2 + \ldots + z_n^2\right),$$

then

$$(\mathrm{C}(\Re_1,\Re_2))^2 \leq \left(\begin{pmatrix} \left(\mathfrak{F}^{RP}_{\Re_1}(e_1)\right)^2+\left(\mathfrak{F}^{IP}_{\Re_1}(e_1)\right)^2+\left(\mathfrak{F}^{RN}_{\Re_1}(e_1)\right)^2+\left(\mathfrak{F}^{IN}_{\Re_1}(e_1)\right)^2 \\ +\left(\mathfrak{F}^{RP}_{\Re_1}(e_2)\right)^2+\left(\mathfrak{F}^{IP}_{\Re_1}(e_2)\right)^2+\left(\mathfrak{F}^{RN}_{\Re_1}(e_2)\right)^2+\left(\mathfrak{F}^{IN}_{\Re_1}(e_2)\right)^2 \\ +\left(\mathfrak{F}^{RP}_{\Re_1}(e_3)\right)^2+\left(\mathfrak{F}^{IP}_{\Re_1}(e_3)\right)^2+\left(\mathfrak{F}^{RN}_{\Re_1}(e_3)\right)^2+\left(\mathfrak{F}^{IN}_{\Re_1}(e_3)\right)^2+\cdots \\ +\left(\mathfrak{F}^{RP}_{\Re_1}(e_n)\right)^2+\left(\mathfrak{F}^{IP}_{\Re_1}(e_n)\right)^2+\left(\mathfrak{F}^{RN}_{\Re_1}(e_n)\right)^2+\left(\mathfrak{F}^{IN}_{\Re_1}(e_n)\right)^2 \end{pmatrix}\times \\ \begin{pmatrix} \left(\mathfrak{F}^{RP}_{\Re_2}(e_1)\right)^2+\left(\mathfrak{F}^{IP}_{\Re_2}(e_1)\right)^2+\left(\mathfrak{F}^{RN}_{\Re_2}(e_1)\right)^2+\left(\mathfrak{F}^{IN}_{\Re_2}(e_1)\right)^2 \\ +\left(\mathfrak{F}^{RP}_{\Re_2}(e_2)\right)^2+\left(\mathfrak{F}^{IP}_{\Re_2}(e_2)\right)^2+\left(\mathfrak{F}^{RN}_{\Re_2}(e_2)\right)^2+\left(\mathfrak{F}^{IN}_{\Re_2}(e_2)\right)^2 \\ +\left(\mathfrak{F}^{RP}_{\Re_2}(e_3)\right)^2+\left(\mathfrak{F}^{IP}_{\Re_2}(e_3)\right)^2+\left(\mathfrak{F}^{RN}_{\Re_2}(e_3)\right)^2+\left(\mathfrak{F}^{IN}_{\Re_2}(e_3)\right)^2+\cdots \\ +\left(\mathfrak{F}^{RP}_{\Re_2}(e_n)\right)^2+\left(\mathfrak{F}^{IP}_{\Re_2}(e_n)\right)^2+\left(\mathfrak{F}^{RN}_{\Re_2}(e_n)\right)^2+\left(\mathfrak{F}^{IN}_{\Re_2}(e_n)\right)^2 \end{pmatrix}\right)$$

$$= \left(\begin{array}{c} \left(\sum_{\sigma=1}^{n}\left(\left(\mathfrak{F}^{RP}_{\Re_1}(e_\sigma)\right)^2+\left(\mathfrak{F}^{IP}_{\Re_1}(e_\sigma)\right)^2+\left(\mathfrak{F}^{RN}_{\Re_1}(e_\sigma)\right)^2+\left(\mathfrak{F}^{IN}_{\Re_1}(e_\sigma)\right)^2\right)\right)\times \\ \left(\sum_{\sigma=1}^{n}\left(\left(\mathfrak{F}^{RP}_{\Re_2}(e_\sigma)\right)^2+\left(\mathfrak{F}^{IP}_{\Re_2}(e_\sigma)\right)^2+\left(\mathfrak{F}^{RN}_{\Re_2}(e_\sigma)\right)^2+\left(\mathfrak{F}^{IN}_{\Re_2}(e_\sigma)\right)^2\right)\right) \end{array}\right)$$

$$= \underline{\mathrm{T}}(\Re_1)\underline{\mathrm{T}}(\Re_2)$$

$$K(\Re_1,\Re_2) = \frac{\mathrm{C}(\Re_1,\Re_2)}{\left(\underline{\mathrm{T}}(\Re_1)\underline{\mathrm{T}}(\Re_2)\right)^{\frac{1}{2}}} \leq 1$$

Hence, $0 \leq K(\Re_1,\Re_2) \leq 1$.

2. Consider equation (5), we have

$$K(\Re_1,\Re_2) = \frac{\sum_{\sigma=1}^{n}\left(\mathfrak{F}^{RP}_{\Re_1}(e_\sigma)\mathfrak{F}^{RP}_{\Re_2}(e_\sigma)+\mathfrak{F}^{IP}_{\Re_1}(e_\sigma)\mathfrak{F}^{IP}_{\Re_2}(e_\sigma)+\mathfrak{F}^{RN}_{\Re_1}(e_\sigma)\mathfrak{F}^{RN}_{\Re_2}(e_\sigma)+\mathfrak{F}^{IN}_{\Re_1}(e_\sigma)\mathfrak{F}^{IN}_{\Re_2}(e_\sigma)\right)}{\left(\begin{array}{c}\left(\sum_{\sigma=1}^{n}\left(\left(\mathfrak{F}^{RP}_{\Re_1}(e_\sigma)\right)^2+\left(\mathfrak{F}^{IP}_{\Re_1}(e_\sigma)\right)^2+\left(\mathfrak{F}^{RN}_{\Re_1}(e_\sigma)\right)^2+\left(\mathfrak{F}^{IN}_{\Re_1}(e_\sigma)\right)^2\right)\right) \\ \left(\sum_{\sigma=1}^{n}\left(\left(\mathfrak{F}^{RP}_{\Re_2}(e_\sigma)\right)^2+\left(\mathfrak{F}^{IP}_{\Re_2}(e_\sigma)\right)^2+\left(\mathfrak{F}^{RN}_{\Re_2}(e_\sigma)\right)^2+\left(\mathfrak{F}^{IN}_{\Re_2}(e_\sigma)\right)^2\right)\right)\end{array}\right)^{\frac{1}{2}}}$$

$$= \frac{\sum_{\sigma=1}^{n}\left(\mathfrak{F}^{RP}_{\Re_2}(e_\sigma)\mathfrak{F}^{RP}_{\Re_1}(e_\sigma)+\mathfrak{F}^{IP}_{\Re_2}(e_\sigma)\mathfrak{F}^{IP}_{\Re_1}(e_\sigma)+\mathfrak{F}^{RN}_{\Re_2}(e_\sigma)\mathfrak{F}^{RN}_{\Re_1}(e_\sigma)+\mathfrak{F}^{IN}_{\Re_2}(e_\sigma)\mathfrak{F}^{IN}_{\Re_1}(e_\sigma)\right)}{\left(\begin{array}{c}\left(\sum_{\sigma=1}^{n}\left(\left(\mathfrak{F}^{RP}_{\Re_2}(e_\sigma)\right)^2+\left(\mathfrak{F}^{IP}_{\Re_2}(e_\sigma)\right)^2+\left(\mathfrak{F}^{RN}_{\Re_2}(e_\sigma)\right)^2+\left(\mathfrak{F}^{IN}_{\Re_2}(e_\sigma)\right)^2\right)\right) \\ \left(\sum_{\sigma=1}^{n}\left(\left(\mathfrak{F}^{RP}_{\Re_1}(e_\sigma)\right)^2+\left(\mathfrak{F}^{IP}_{\Re_1}(e_\sigma)\right)^2+\left(\mathfrak{F}^{RN}_{\Re_1}(e_\sigma)\right)^2+\left(\mathfrak{F}^{IN}_{\Re_1}(e_\sigma)\right)^2\right)\right)\end{array}\right)^{\frac{1}{2}}}$$

$$= K(\Re_2,\Re_1)$$

3. Consider Eq. (6.1), and let $\Re_1 = \Re_2$, i.e. $\mathfrak{F}^{RP}_{\Re_1}(e_\sigma) = \mathfrak{F}^{RP}_{\Re_2}(e_\sigma)$, $\mathfrak{F}^{IP}_{\Re_1}(e_\sigma) = \mathfrak{F}^{IP}_{\Re_2}(e_\sigma)$, $\mathfrak{F}^{RN}_{\Re_1}(e_\sigma) = \mathfrak{F}^{RN}_{\Re_2}(e_\sigma)$, and $\mathfrak{F}^{IN}_{\Re_1}(e_\sigma) = \mathfrak{F}^{IN}_{\Re_2}(e_\sigma)$, then

$$K(\Re_1,\Re_2) = \frac{\sum_{\sigma=1}^{n}\left(\mathfrak{F}^{RP}_{\Re_1}(e_\sigma)\mathfrak{F}^{RP}_{\Re_2}(e_\sigma)+\mathfrak{F}^{IP}_{\Re_1}(e_\sigma)\mathfrak{F}^{IP}_{\Re_2}(e_\sigma)+\mathfrak{F}^{RN}_{\Re_1}(e_\sigma)\mathfrak{F}^{RN}_{\Re_2}(e_\sigma)+\mathfrak{F}^{IN}_{\Re_1}(e_\sigma)\mathfrak{F}^{IN}_{\Re_2}(e_\sigma)\right)}{\left(\begin{array}{c}\left(\sum_{\sigma=1}^{n}\left(\left(\mathfrak{F}^{RP}_{\Re_1}(e_\sigma)\right)^2+\left(\mathfrak{F}^{IP}_{\Re_1}(e_\sigma)\right)^2+\left(\mathfrak{F}^{RN}_{\Re_1}(e_\sigma)\right)^2+\left(\mathfrak{F}^{IN}_{\Re_1}(e_\sigma)\right)^2\right)\right) \\ \left(\sum_{\sigma=1}^{n}\left(\left(\mathfrak{F}^{RP}_{\Re_2}(e_\sigma)\right)^2+\left(\mathfrak{F}^{IP}_{\Re_2}(e_\sigma)\right)^2+\left(\mathfrak{F}^{RN}_{\Re_2}(e_\sigma)\right)^2+\left(\mathfrak{F}^{IN}_{\Re_2}(e_\sigma)\right)^2\right)\right)\end{array}\right)^{\frac{1}{2}}} = 1$$

This completed the proof.

Remark 6.1 when we suppose that the imaginary part of Eq. (6.1) is zero, then Eq. (6.1) will be transformed to BFS.

There are various circumstances, where, the weight is required. In view of this, next, we would deduce weighted CC for BCFSs.

Definition 6.2 Let two BCFSs $\Re_1 = \left\{e_\sigma, \left(\mathfrak{F}^P_{\Re_1}(e_\sigma), \mathfrak{F}^N_{\Re_1}(e_\sigma)\right) | e_\sigma \in \mathbb{E}, \sigma = 1, 2, .., n\right\} = \left\{e_\sigma, \left(\mathfrak{F}^{RP}_{\Re_1}(e_\sigma) + \iota\mathfrak{F}^{IP}_{\Re_1}(e_\sigma), \mathfrak{F}^{RN}_{\Re_1}(e_\sigma) + \iota\mathfrak{F}^{IN}_{\Re_1}(e_\sigma)\right) | e_\sigma \in \mathbb{E}, \quad \sigma = 1, 2, .., n\right\}$ and $\Re_2 = \left\{e_\sigma, \left(\mathfrak{F}^P_{\Re_1}(e_\sigma), \mathfrak{F}^N_{\Re_1}(e_\sigma)\right) | e_\sigma \in \mathbb{E}, \sigma = 1, 2, .., n\right\} = \left\{e_\sigma, \left(\mathfrak{F}^{RP}_{\Re_1}(e_\sigma) + \iota\mathfrak{F}^{IP}_{\Re_1}(e_\sigma), \mathfrak{F}^{RN}_{\Re_1}(e_\sigma) + \iota\mathfrak{F}^{IN}_{\Re_1}(e_\sigma)\right) | e_\sigma \in \mathbb{E}, \quad \sigma = 1, 2, .., n\right\}$ over $\mathbb{E}$. Then weighted CC among $\Re_1$ and $\Re_2$ is deduced as

$$K_\omega(\Re_1, \Re_2) = \frac{\mathrm{C}_\omega(\Re_1, \Re_2)}{\left(\underline{\mathrm{T}}_\omega(\Re_1)\underline{\mathrm{T}}_\omega(\Re_2)\right)^{\frac{1}{2}}}$$

$$= \frac{\sum_{\sigma=1}^{n} \omega\left(\mathfrak{F}^{RP}_{\Re_1}(e_\sigma)\mathfrak{F}^{RP}_{\Re_2}(e_\sigma) + \mathfrak{F}^{IP}_{\Re_1}(e_\sigma)\mathfrak{F}^{IP}_{\Re_2}(e_\sigma) + \mathfrak{F}^{RN}_{\Re_1}(e_\sigma)\mathfrak{F}^{RN}_{\Re_2}(e_\sigma) + \mathfrak{F}^{IN}_{\Re_1}(e_\sigma)\mathfrak{F}^{IN}_{\Re_2}(e_\sigma)\right)}{\left(\begin{array}{c}\left(\sum_{\sigma=1}^{n} \omega\left(\left(\mathfrak{F}^{RP}_{\Re_1}(e_\sigma)\right)^2 + \left(\mathfrak{F}^{IP}_{\Re_1}(e_\sigma)\right)^2 + \left(\mathfrak{F}^{RN}_{\Re_1}(e_\sigma)\right)^2 + \left(\mathfrak{F}^{IN}_{\Re_1}(e_\sigma)\right)^2\right)\right) \\ \left(\sum_{\sigma=1}^{n} \omega\left(\left(\mathfrak{F}^{RP}_{\Re_2}(e_\sigma)\right)^2 + \left(\mathfrak{F}^{IP}_{\Re_2}(e_\sigma)\right)^2 + \left(\mathfrak{F}^{RN}_{\Re_2}(e_\sigma)\right)^2 + \left(\mathfrak{F}^{IN}_{\Re_2}(e_\sigma)\right)^2\right)\right)\end{array}\right)^{\frac{1}{2}}} \tag{6.2}$$

where,

$$\mathrm{C}(\Re_1, \Re_2) = \sum_{\sigma=1}^{n} \omega\left(\mathfrak{F}^{RP}_{\Re_1}(e_\sigma)\mathfrak{F}^{RP}_{\Re_2}(e_\sigma) + \mathfrak{F}^{IP}_{\Re_1}(e_\sigma)\mathfrak{F}^{IP}_{\Re_2}(e_\sigma) + \mathfrak{F}^{RN}_{\Re_1}(e_\sigma)\mathfrak{F}^{RN}_{\Re_2}(e_\sigma) + \mathfrak{F}^{IN}_{\Re_1}(e_\sigma)\mathfrak{F}^{IN}_{\Re_2}(e_\sigma)\right)$$

$$\underline{\mathrm{T}}(\Re) = \sum_{\sigma=1}^{n} \omega\left(\left(\mathfrak{F}^{RP}_{\Re_1}(e_\sigma)\right)^2 + \left(\mathfrak{F}^{IP}_{\Re_1}(e_\sigma)\right)^2 + \left(\mathfrak{F}^{RN}_{\Re_1}(e_\sigma)\right)^2 + \left(\mathfrak{F}^{IN}_{\Re_1}(e_\sigma)\right)^2\right)$$

and

$$\underline{\mathrm{T}}(\Re_2) = \sum_{\sigma=1}^{n} \omega\left(\left(\mathfrak{F}^{RP}_{\Re_2}(e_\sigma)\right)^2 + \left(\mathfrak{F}^{IP}_{\Re_2}(e_\sigma)\right)^2 + \left(\mathfrak{F}^{RN}_{\Re_2}(e_\sigma)\right)^2 + \left(\mathfrak{F}^{IN}_{\Re_2}(e_\sigma)\right)^2\right)$$

interpret the correlation and information energy for BCFSs along with conditions that $\mathrm{C}_\omega(\Re_1, \Re_2) = \underline{\mathrm{T}}_\omega(\Re)$ and $\mathrm{C}_\omega(\Re_1, \Re_2) = \mathrm{C}_\omega(\Re_2, \Re_1)$. The weighted CC of BCFSs holds the underneath axioms.

1. $0 \leq K_\omega(\Re_1, \Re_2) \leq 1$,
2. $K_\omega(\Re_1, \Re_2) = K_\omega(\Re_2, \Re_1)$,
3. $\Re_1 = \Re_2 \Rightarrow K_\omega(\Re_1, \Re_2) = 1$.

Proof

1. From Eq. (6.2) it is obvious that $K_\omega(\Re_1, \Re_2) \geq 0$, so it remains only to prove that $K_\omega(\Re_1, \Re_2) \leq$. Next $\mathrm{C}_{\omega_\sigma}(\Re_1, \Re_2) = \sum_{\sigma=1}^{n} \omega_\sigma\left(\mathfrak{F}^{RP}_{\Re_1}(e_\sigma)\mathfrak{F}^{RP}_{\Re_2}(e_\sigma) + \mathfrak{F}^{IP}_{\Re_1}(e_\sigma)\mathfrak{F}^{IP}_{\Re_2}(e_\sigma) + \mathfrak{F}^{RN}_{\Re_1}(e_\sigma)\mathfrak{F}^{RN}_{\Re_2}(e_\sigma) + \mathfrak{F}^{IN}_{\Re_1}(e_\sigma)\mathfrak{F}^{IN}_{\Re_2}(e_\sigma)\right)$
$$= \left(\begin{array}{c}\omega_1\left(\mathfrak{F}^{RP}_{\Re_1}(e_1)\mathfrak{F}^{RP}_{\Re_2}(e_1) + \mathfrak{F}^{IP}_{\Re_1}(e_1)\mathfrak{F}^{IP}_{\Re_2}(e_1) + \mathfrak{F}^{RN}_{\Re_1}(e_1)\mathfrak{F}^{RN}_{\Re_2}(e_1) + \mathfrak{F}^{IN}_{\Re_1}(e_1)\mathfrak{F}^{IN}_{\Re_2}(e_1)\right) \\ +\omega_2\left(\mathfrak{F}^{RP}_{\Re_1}(e_2)\mathfrak{F}^{RP}_{\Re_2}(e_2) + \mathfrak{F}^{IP}_{\Re_1}(e_2)\mathfrak{F}^{IP}_{\Re_2}(e_2) + \mathfrak{F}^{RN}_{\Re_1}(e_2)\mathfrak{F}^{RN}_{\Re_2}(e_2) + \mathfrak{F}^{IN}_{\Re_1}(e_2)\mathfrak{F}^{IN}_{\Re_2}(e_2)\right) \\ +\omega_3\left(\mathfrak{F}^{RP}_{\Re_1}(e_3)\mathfrak{F}^{RP}_{\Re_2}(e_3) + \mathfrak{F}^{IP}_{\Re_1}(e_3)\mathfrak{F}^{IP}_{\Re_2}(e_3) + \mathfrak{F}^{RN}_{\Re_1}(e_3)\mathfrak{F}^{RN}_{\Re_2}(e_3) + \mathfrak{F}^{IN}_{\Re_1}(e_3)\mathfrak{F}^{IN}_{\Re_2}(e_3)\right) + \cdots \\ +\omega_4\left(\mathfrak{F}^{RP}_{\Re_1}(e_n)\mathfrak{F}^{RP}_{\Re_2}(e_n) + \mathfrak{F}^{IP}_{\Re_1}(e_n)\mathfrak{F}^{IP}_{\Re_2}(e_n) + \mathfrak{F}^{RN}_{\Re_1}(e_n)\mathfrak{F}^{RN}_{\Re_2}(e_n) + \mathfrak{F}^{IN}_{\Re_1}(e_n)\mathfrak{F}^{IN}_{\Re_2}(e_n)\right)\end{array}\right)$$

By utilizing Cauchy-Schwarz inequality

$$(x_1z_1 + x_2z_2 + \ldots + x_nz_n)^2 \le \left(x_1^2 + x_2^2 + \ldots + x_n^2\right)\left(z_1^2 + z_2^2 + \ldots + z_n^2\right),$$

then

$$(\mathrm{C}_\omega(\Re_1, \Re_2))^2 \le \left(\begin{array}{c} \left(\begin{array}{c} \omega_1\left(\mathfrak{F}_{\Re_1}^{RP}(e_1)\right)^2 + \left(\mathfrak{F}_{\Re_1}^{IP}(e_1)\right)^2 + \left(\mathfrak{F}_{\Re_1}^{RN}(e_1)\right)^2 + \left(\mathfrak{F}_{\Re_1}^{IN}(e_1)\right)^2 \\ +\omega_2\left(\mathfrak{F}_{\Re_1}^{RP}(e_2)\right)^2 + \left(\mathfrak{F}_{\Re_1}^{IP}(e_2)\right)^2 + \left(\mathfrak{F}_{\Re_1}^{RN}(e_2)\right)^2 + \left(\mathfrak{F}_{\Re_1}^{IN}(e_2)\right)^2 \\ +\omega_3\left(\mathfrak{F}_{\Re_1}^{RP}(e_3)\right)^2 + \left(\mathfrak{F}_{\Re_1}^{IP}(e_3)\right)^2 + \left(\mathfrak{F}_{\Re_1}^{RN}(e_3)\right)^2 + \left(\mathfrak{F}_{\Re_1}^{IN}(e_3)\right)^2 + \cdots \\ +\omega_n\left(\mathfrak{F}_{\Re_1}^{RP}(e_n)\right)^2 + \left(\mathfrak{F}_{\Re_1}^{IP}(e_n)\right)^2 + \left(\mathfrak{F}_{\Re_1}^{RN}(e_n)\right)^2 + \left(\mathfrak{F}_{\Re_1}^{IN}(e_n)\right)^2 \end{array}\right) \times \\ \left(\begin{array}{c} \omega_1\left(\mathfrak{F}_{\Re_2}^{RP}(e_1)\right)^2 + \left(\mathfrak{F}_{\Re_2}^{IP}(e_1)\right)^2 + \left(\mathfrak{F}_{\Re_2}^{RN}(e_1)\right)^2 + \left(\mathfrak{F}_{\Re_2}^{IN}(e_1)\right)^2 \\ +\omega_2\left(\mathfrak{F}_{\Re_2}^{RP}(e_2)\right)^2 + \left(\mathfrak{F}_{\Re_2}^{IP}(e_2)\right)^2 + \left(\mathfrak{F}_{\Re_2}^{RN}(e_2)\right)^2 + \left(\mathfrak{F}_{\Re_2}^{IN}(e_2)\right)^2 \\ +\omega_3\left(\mathfrak{F}_{\Re_2}^{RP}(e_3)\right)^2 + \left(\mathfrak{F}_{\Re_2}^{IP}(e_3)\right)^2 + \left(\mathfrak{F}_{\Re_2}^{RN}(e_3)\right)^2 + \left(\mathfrak{F}_{\Re_2}^{IN}(e_3)\right)^2 + \cdots \\ +\omega_n\left(\mathfrak{F}_{\Re_2}^{RP}(e_n)\right)^2 + \left(\mathfrak{F}_{\Re_2}^{IP}(e_n)\right)^2 + \left(\mathfrak{F}_{\Re_2}^{RN}(e_n)\right)^2 + \left(\mathfrak{F}_{\Re_2}^{IN}(e_n)\right)^2 \end{array}\right) \end{array}\right)$$

$$= \left(\begin{array}{c} \left(\sum_{\sigma=1}^{n} \omega_\sigma\left(\left(\mathfrak{F}_{\Re_1}^{RP}(e_\sigma)\right)^2 + \left(\mathfrak{F}_{\Re_1}^{IP}(e_\sigma)\right)^2 + \left(\mathfrak{F}_{\Re_1}^{RN}(e_\sigma)\right)^2 + \left(\mathfrak{F}_{\Re_1}^{IN}(e_\sigma)\right)^2\right)\right) \times \\ \left(\sum_{\sigma=1}^{n} \omega_\sigma\left(\left(\mathfrak{F}_{\Re_2}^{RP}(e_\sigma)\right)^2 + \left(\mathfrak{F}_{\Re_2}^{IP}(e_\sigma)\right)^2 + \left(\mathfrak{F}_{\Re_2}^{RN}(e_\sigma)\right)^2 + \left(\mathfrak{F}_{\Re_2}^{IN}(e_\sigma)\right)^2\right)\right) \end{array}\right)$$

$$= \underline{\mathrm{T}}_\omega(\Re_1)\underline{\mathrm{T}}_\omega(\Re_2)$$

$$K_\omega(\Re_1, \Re_2) = \frac{\mathrm{C}_\omega(\Re_1, \Re_2)}{\left(\underline{\mathrm{T}}_\omega(\Re_1)\underline{\mathrm{T}}_\omega(\Re_2)\right)^{\frac{1}{2}}} \le 1$$

Hence, $0 \le K_\omega(\Re_1, \Re_2) \le 1$.

2. Consider Eq. (6.2), we have

$$K_\omega(\Re_1, \Re_2) = \frac{\sum_{\sigma=1}^{n} \omega_\sigma\left(\mathfrak{F}_{\Re_1}^{RP}(e_\sigma)\mathfrak{F}_{\Re_2}^{RP}(e_\sigma) + \mathfrak{F}_{\Re_1}^{IP}(e_\sigma)\mathfrak{F}_{\Re_2}^{IP}(e_\sigma) + \mathfrak{F}_{\Re_1}^{RN}(e_\sigma)\mathfrak{F}_{\Re_2}^{RN}(e_\sigma) + \mathfrak{F}_{\Re_1}^{IN}(e_\sigma)\mathfrak{F}_{\Re_2}^{IN}(e_\sigma)\right)}{\left(\begin{array}{c}\left(\sum_{\sigma=1}^{n} \omega_\sigma\left(\left(\mathfrak{F}_{\Re_1}^{RP}(e_\sigma)\right)^2 + \left(\mathfrak{F}_{\Re_1}^{IP}(e_\sigma)\right)^2 + \left(\mathfrak{F}_{\Re_1}^{RN}(e_\sigma)\right)^2 + \left(\mathfrak{F}_{\Re_1}^{IN}(e_\sigma)\right)^2\right)\right) \\ \left(\sum_{\sigma=1}^{n} \omega_\sigma\left(\left(\mathfrak{F}_{\Re_2}^{RP}(e_\sigma)\right)^2 + \left(\mathfrak{F}_{\Re_2}^{IP}(e_\sigma)\right)^2 + \left(\mathfrak{F}_{\Re_2}^{RN}(e_\sigma)\right)^2 + \left(\mathfrak{F}_{\Re_2}^{IN}(e_\sigma)\right)^2\right)\right)\end{array}\right)^{\frac{1}{2}}}$$

$$= \frac{\sum_{\sigma=1}^{n} \omega_\sigma\left(\mathfrak{F}_{\Re_2}^{RP}(e_\sigma)\mathfrak{F}_{\Re_1}^{RP}(e_\sigma) + \mathfrak{F}_{\Re_2}^{IP}(e_\sigma)\mathfrak{F}_{\Re_1}^{IP}(e_\sigma) + \mathfrak{F}_{\Re_2}^{RN}(e_\sigma)\mathfrak{F}_{\Re_1}^{RN}(e_\sigma) + \mathfrak{F}_{\Re_2}^{IN}(e_\sigma)\mathfrak{F}_{\Re_1}^{IN}(e_\sigma)\right)}{\left(\begin{array}{c}\left(\sum_{\sigma=1}^{n} \omega_\sigma\left(\left(\mathfrak{F}_{\Re_2}^{RP}(e_\sigma)\right)^2 + \left(\mathfrak{F}_{\Re_2}^{IP}(e_\sigma)\right)^2 + \left(\mathfrak{F}_{\Re_2}^{RN}(e_\sigma)\right)^2 + \left(\mathfrak{F}_{\Re_2}^{IN}(e_\sigma)\right)^2\right)\right) \\ \left(\sum_{\sigma=1}^{n} \omega_\sigma\left(\left(\mathfrak{F}_{\Re_1}^{RP}(e_\sigma)\right)^2 + \left(\mathfrak{F}_{\Re_1}^{IP}(e_\sigma)\right)^2 + \left(\mathfrak{F}_{\Re_1}^{RN}(e_\sigma)\right)^2 + \left(\mathfrak{F}_{\Re_1}^{IN}(e_\sigma)\right)^2\right)\right)\end{array}\right)^{\frac{1}{2}}}$$

$$= K_\omega(\Re_2, \Re_1)$$

4. Consider Eq. (6.2), and let $\Re_1 = \Re_2$, i.e. $\mathfrak{F}_{\Re_1}^{RP}(e_\sigma) = \mathfrak{F}_{\Re_2}^{RP}(e_\sigma)$, $\mathfrak{F}_{\Re_1}^{IP}(e_\sigma) = \mathfrak{F}_{\Re_2}^{IP}(e_\sigma)$, $\mathfrak{F}_{\Re_1}^{RN}(e_\sigma) = \mathfrak{F}_{\Re_2}^{RN}(e_\sigma)$, and $\mathfrak{F}_{\Re_1}^{IN}(e_\sigma) = \mathfrak{F}_{\Re_2}^{IN}(e_\sigma)$, then

$$K_\omega(\Re_1, \Re_2) = \frac{\sum_{\sigma=1}^{n} \omega_\sigma\left(\mathfrak{F}_{\Re_1}^{RP}(e_\sigma)\mathfrak{F}_{\Re_2}^{RP}(e_\sigma) + \mathfrak{F}_{\Re_1}^{IP}(e_\sigma)\mathfrak{F}_{\Re_2}^{IP}(e_\sigma) + \mathfrak{F}_{\Re_1}^{RN}(e_\sigma)\mathfrak{F}_{\Re_2}^{RN}(e_\sigma) + \mathfrak{F}_{\Re_1}^{IN}(e_\sigma)\mathfrak{F}_{\Re_2}^{IN}(e_\sigma)\right)}{\left(\begin{array}{c}\left(\sum_{\sigma=1}^{n} \omega_\sigma\left(\left(\mathfrak{F}_{\Re_1}^{RP}(e_\sigma)\right)^2 + \left(\mathfrak{F}_{\Re_1}^{IP}(e_\sigma)\right)^2 + \left(\mathfrak{F}_{\Re_1}^{RN}(e_\sigma)\right)^2 + \left(\mathfrak{F}_{\Re_1}^{IN}(e_\sigma)\right)^2\right)\right) \\ \left(\sum_{\sigma=1}^{n} \omega_\sigma\left(\left(\mathfrak{F}_{\Re_2}^{RP}(e_\sigma)\right)^2 + \left(\mathfrak{F}_{\Re_2}^{IP}(e_\sigma)\right)^2 + \left(\mathfrak{F}_{\Re_2}^{RN}(e_\sigma)\right)^2 + \left(\mathfrak{F}_{\Re_2}^{IN}(e_\sigma)\right)^2\right)\right)\end{array}\right)^{\frac{1}{2}}} = 1$$

This completed the proof.

Remark 6.2 when we suppose that the imaginary part of Eq. (6.2) is zero, then Eq. (6.2) will be transformed to BFS.

In the prevailing literature, there is not research on the entropy measures (EMs) under the setting of BCFS. Thus, in the next Section, we would discuss the EMs for BCFS.

6.2 The Entropy Measures for BCFSs

The BCFS considers perfectly the individual comments of the decision-maker. It might have vague, or incomplete information. Consequently, there are information measures to evaluate the vagueness and imprecision of the BCF information. Entropy is one of them. In this subsection, we introduce two EMs for BCFSs.

Definition 6.3 Let two BCFSs $\mathfrak{R}_1 = \left\{e_\sigma, \left(\mathfrak{F}^{P}_{\mathfrak{R}_1}(e_\sigma), \mathfrak{F}^{N}_{\mathfrak{R}_1}(e_\sigma)\right) | e_\sigma \in \mathbb{E},\ \sigma = 1, 2, .., n\right\} = \left\{e_\sigma, \left(\mathfrak{F}^{RP}_{\mathfrak{R}_1}(e_\sigma) + \iota\,\mathfrak{F}^{IP}_{\mathfrak{R}_1}(e_\sigma), \mathfrak{F}^{RN}_{\mathfrak{R}_1}(e_\sigma) + \iota\mathfrak{F}^{IN}_{\mathfrak{R}_1}(e_\sigma)\right) | e_\sigma \in \mathbb{E},\ \sigma = 1, 2, .., n\right\}$ and $\mathfrak{R}_2 = \left\{e, \left(\mathfrak{F}^{P}_{\mathfrak{R}_2}(e_\sigma), \mathfrak{F}^{N}_{\mathfrak{R}_2}(e_\sigma)\right) | e \in \mathbb{E},\ \sigma = 1, 2, .., n\right\} = \left\{e_\sigma, \left(\mathfrak{F}^{RP}_{\mathfrak{R}_2}(e_\sigma) + \iota\mathfrak{F}^{IP}_{\mathfrak{R}_2}(e_\sigma), \mathfrak{F}^{RN}_{\mathfrak{R}_2}(e_\sigma) + \iota\mathfrak{F}^{IN}_{\mathfrak{R}_2}(e_\sigma)\right) | e_\sigma \in \mathbb{E},\ \sigma = 1, 2, .., n\right\}$ over $\mathbb{E}$. Then two EMs of $\mathfrak{R}$ are deduced as

$$\mathcal{E}_1(\mathfrak{R}) = \frac{1}{n}\sum_{\sigma=1}^{n}\left\{\left\{\begin{array}{l}\sin\left(\dfrac{\left(\pi\times\left(2+\mathfrak{F}^{RP}_{\mathfrak{R}}(e_\sigma)+\mathfrak{F}^{IP}_{\mathfrak{R}}(e_\sigma)-\left|\mathfrak{F}^{RN}_{\mathfrak{R}}(e_\sigma)\right|-\left|\mathfrak{F}^{IP}_{\mathfrak{R}}(e_\sigma)\right|\right)\right)}{8}\right)+\\ \sin\left(\dfrac{\left(\pi\times\left(2-\mathfrak{F}^{RP}_{\mathfrak{R}}(e_\sigma)-\mathfrak{F}^{IP}_{\mathfrak{R}}(e_\sigma)+\left|\mathfrak{F}^{RN}_{\mathfrak{R}}(e_\sigma)\right|+\left|\mathfrak{F}^{IP}_{\mathfrak{R}}(e_\sigma)\right|\right)\right)}{8}\right)-1\end{array}\right\}\times\frac{1}{(2)^{\frac{1}{2}}-1}\right\} \tag{6.3}$$

$$\mathcal{E}_2(\mathfrak{R}) = \frac{1}{n}\sum_{\sigma=1}^{n}\left\{\left\{\begin{array}{l}\cos\left(\dfrac{\left(\pi\times\left(2+\mathfrak{F}^{RP}_{\mathfrak{R}}(e_\sigma)+\mathfrak{F}^{IP}_{\mathfrak{R}}(e_\sigma)-\left|\mathfrak{F}^{RN}_{\mathfrak{R}}(e_\sigma)\right|-\left|\mathfrak{F}^{IP}_{\mathfrak{R}}(e_\sigma)\right|\right)\right)}{8}\right)+\\ \cos\left(\dfrac{\left(\pi\times\left(2-\mathfrak{F}^{RP}_{\mathfrak{R}}(e_\sigma)-\mathfrak{F}^{IP}_{\mathfrak{R}}(e_\sigma)+\left|\mathfrak{F}^{RN}_{\mathfrak{R}}(e_\sigma)\right|+\left|\mathfrak{F}^{IP}_{\mathfrak{R}}(e_\sigma)\right|\right)\right)}{8}\right)-1\end{array}\right\}\times\frac{1}{(2)^{\frac{1}{2}}-1}\right\} \tag{6.4}$$

The above-defined EMs satisfy the following axioms:

1. If $\mathfrak{F}^{RP}_{\mathfrak{R}}(e_\sigma) = 1$, $\mathfrak{F}^{IP}_{\mathfrak{R}}(e_\sigma) = 1$ $\mathfrak{F}^{RN}_{\mathfrak{R}}(e_\sigma) = 0$, $\mathfrak{F}^{IN}_{\mathfrak{R}}(e_\sigma) = 0$, or $\mathfrak{F}^{RP}_{\mathfrak{R}}(e_\sigma) = 0$, $\mathfrak{F}^{IP}_{\mathfrak{R}}(e_\sigma) = 0$ $\mathfrak{F}^{RN}_{\mathfrak{R}}(e_\sigma) = -1$, $\mathfrak{F}^{IN}_{\mathfrak{R}}(e_\sigma) = -1$, then $\mathcal{E}_1(\mathfrak{R}) = \mathcal{E}_2(\mathfrak{R}) = 0$, for all $\sigma = 1, 2, 3, .., n$
2. If $\mathfrak{F}^{RP}_{\mathfrak{R}}(e_\sigma) = -\mathfrak{F}^{RN}_{\mathfrak{R}}(e_\sigma)$ and $\mathfrak{F}^{IP}_{\mathfrak{R}}(e_\sigma) = -\mathfrak{F}^{IN}_{\mathfrak{R}}(e_\sigma)$, then $\mathcal{E}_1(\mathfrak{R}) = \mathcal{E}_2(\mathfrak{R}) = 1$.
3. If $\mathfrak{F}^{RP}_{\mathfrak{R}_1}(e_\sigma) \leq \mathfrak{F}^{RP}_{\mathfrak{R}_2}(e_\sigma)$, $\mathfrak{F}^{IP}_{\mathfrak{R}_1}(e_\sigma) \leq \mathfrak{F}^{IP}_{\mathfrak{R}_2}(e_\sigma)$, $\mathfrak{F}^{RN}_{\mathfrak{R}_1}(e_\sigma) \leq \mathfrak{F}^{RN}_{\mathfrak{R}_2}(e_\sigma)$ and $\mathfrak{F}^{IN}_{\mathfrak{R}_1}(e_\sigma) \leq \mathfrak{F}^{IN}_{\mathfrak{R}_2}(e_\sigma)$ then $\mathcal{E}_1(\mathfrak{R}_1) \leq \mathcal{E}_1(\mathfrak{R}_2)$, and $\mathcal{E}_1(\mathfrak{R}_1) \leq \mathcal{E}_1(\mathfrak{R}_2)$.
4. $\mathcal{E}_1(\mathfrak{R}) = \mathcal{E}_1(\mathfrak{R}^c)$ and $\mathcal{E}_2(\mathfrak{R}) = \mathcal{E}_2(\mathfrak{R}^c)$.

Proof For defined two EMs we prove the above 4 axioms as follows

1. Let $\mathfrak{F}^{RP}_{\mathfrak{R}}(e_\sigma) = 1$, $\mathfrak{F}^{IP}_{\mathfrak{R}}(e_\sigma) = 1$ $\mathfrak{F}^{RN}_{\mathfrak{R}}(e_\sigma) = 0$, and $\mathfrak{F}^{IN}_{\mathfrak{R}}(e_\sigma) = 0$, for all $\sigma = 1, 2, 3, .., n$ then we have

$$\mathcal{E}_1(\Re) = \frac{1}{n}\left\{\begin{array}{c} \left\{\begin{array}{c} \sin\left(\frac{(\pi\times(2+1+1-0-0))}{8}\right)+ \\ \sin\left(\frac{(\pi\times(2-1-1+0+0))}{8}\right)-1 \end{array}\right\}\times\frac{1}{(2)^{\frac{1}{2}}-1}+ \\ \left\{\begin{array}{c} \sin\left(\frac{(\pi\times(2+1+1-0-0))}{8}\right)+ \\ \sin\left(\frac{(\pi\times(2-1-1+0+0))}{8}\right)-1 \end{array}\right\}\times\frac{1}{(2)^{\frac{1}{2}}-1} \\ +\ldots+ \\ \left\{\begin{array}{c} \sin\left(\frac{(\pi\times(2+1+1-0-0))}{8}\right)+ \\ \sin\left(\frac{(\pi\times(2-1-1+0+0))}{8}\right)-1 \end{array}\right\}\times\frac{1}{(2)^{\frac{1}{2}}-1} \end{array}\right\}$$

$$\mathcal{E}_2(\Re) = \frac{1}{n}\left\{\begin{array}{c} \left\{\begin{array}{c} \cos\left(\frac{(\pi\times(2+1+1-0-0))}{8}\right)+ \\ \cos\left(\frac{(\pi\times(2-1-1+0+0))}{8}\right)-1 \end{array}\right\}\times\frac{1}{(2)^{\frac{1}{2}}-1}+ \\ \left\{\begin{array}{c} \cos\left(\frac{(\pi\times(2+1+1-0-0))}{8}\right)+ \\ \cos\left(\frac{(\pi\times(2-1-1+0+0))}{8}\right)-1 \end{array}\right\}\times\frac{1}{(2)^{\frac{1}{2}}-1} \\ +\ldots+ \\ \left\{\begin{array}{c} \cos\left(\frac{(\pi\times(2+1+1-0-0))}{8}\right)+ \\ \cos\left(\frac{(\pi\times(2-1-1+0+0))}{8}\right)-1 \end{array}\right\}\times\frac{1}{(2)^{\frac{1}{2}}-1} \end{array}\right\}$$

$$\mathcal{E}_1(\Re) = \frac{1}{n}\left\{\begin{array}{c} \left\{\sin\left(\frac{\pi}{2}\right)+\sin(0)-1\right\}\times\frac{1}{(2)^{\frac{1}{2}}-1}+\left\{\sin\left(\frac{\pi}{2}\right)+\sin(0)-1\right\}\times\frac{1}{(2)^{\frac{1}{2}}-1}+\ldots+ \\ \left\{\sin\left(\frac{\pi}{2}\right)+\sin(0)-1\right\}\times\frac{1}{(2)^{\frac{1}{2}}-1} \end{array}\right\}$$

$$\mathcal{E}_2(\Re) = \frac{1}{n}\left\{\begin{array}{c} \left\{\cos\left(\frac{\pi}{2}\right)+\cos(0)-1\right\}\times\frac{1}{(2)^{\frac{1}{2}}-1}+\left\{\cos\left(\frac{\pi}{2}\right)+\cos(0)-1\right\}\times\frac{1}{(2)^{\frac{1}{2}}-1}+\ldots+ \\ \left\{\cos\left(\frac{\pi}{2}\right)+\cos(0)-1\right\}\times\frac{1}{(2)^{\frac{1}{2}}-1} \end{array}\right\}$$

$$\mathcal{E}_1(\Re) = 0$$
$$\mathcal{E}_2(\Re) = 0$$

Now let $\mathfrak{F}_{\Re}^{RP}(e_\sigma) = 0$, $\mathfrak{F}_{\Re}^{IP}(e_\sigma) = 0$ $\mathfrak{F}_{\Re}^{RN}(e_\sigma) = -1$, and $\mathfrak{F}_{\Re}^{IN}(e_\sigma) = -1$, for all $\sigma = 1, 2, 3, .., n$ then we have

$$\mathcal{E}_1(\Re) = \frac{1}{n}\left\{\begin{array}{c} \left\{\begin{array}{c} \sin\left(\frac{(\pi\times(2+0+0-1-1))}{8}\right)+ \\ \sin\left(\frac{(\pi\times(2-0-0+1+1))}{8}\right)-1 \end{array}\right\}\times\frac{1}{(2)^{\frac{1}{2}}-1}+ \\ \left\{\begin{array}{c} \sin\left(\frac{(\pi\times(2+0+0-1-1))}{8}\right)+ \\ \sin\left(\frac{(\pi\times(2-0-0+1+1))}{8}\right)-1 \end{array}\right\}\times\frac{1}{(2)^{\frac{1}{2}}-1} \\ +\ldots+ \\ \left\{\begin{array}{c} \sin\left(\frac{(\pi\times(2+0+0-1-1))}{8}\right)+ \\ \sin\left(\frac{(\pi\times(2-0-0+1+1))}{8}\right)-1 \end{array}\right\}\times\frac{1}{(2)^{\frac{1}{2}}-1} \end{array}\right\}$$

$$\mathcal{E}_2(\Re) = \frac{1}{n}\left\{\begin{array}{c}\left\{\begin{array}{c}\cos\left(\frac{(\pi\times(2+0+0-1-1))}{8}\right)+\\ \cos\left(\frac{(\pi\times(2-0-0+1+1))}{8}\right)-1\end{array}\right\}\times\frac{1}{(2)^{\frac{1}{2}}-1}+\\ \left\{\begin{array}{c}\cos\left(\frac{(\pi\times(2+0+0-1-1))}{8}\right)+\\ \cos\left(\frac{(\pi\times(2-0-0+1+1))}{8}\right)-1\end{array}\right\}\times\frac{1}{(2)^{\frac{1}{2}}-1}\\ +\ldots+\\ \left\{\begin{array}{c}\cos\left(\frac{(\pi\times(2+0+0-1-1))}{8}\right)+\\ \cos\left(\frac{(\pi\times(2-0-0+1+1))}{8}\right)-1\end{array}\right\}\times\frac{1}{(2)^{\frac{1}{2}}-1}\end{array}\right\}$$

$$\mathcal{E}_1(\Re) = \frac{1}{n}\left\{\begin{array}{c}\{\sin(0)+\sin(\frac{\pi}{2})-1\}\times\frac{1}{(2)^{\frac{1}{2}}-1}+\{\sin(0)+\sin(\frac{\pi}{2})-1\}\times\frac{1}{(2)^{\frac{1}{2}}-1}+\ldots+\\ \{\sin(0)+\sin(\frac{\pi}{2})-1\}\times\frac{1}{(2)^{\frac{1}{2}}-1}\end{array}\right\}$$

$$\mathcal{E}_2(\Re) = \frac{1}{n}\left\{\begin{array}{c}\{\cos(0)+\cos(\frac{\pi}{2})-1\}\times\frac{1}{(2)^{\frac{1}{2}}-1}+\{\cos(0)+\cos(\frac{\pi}{2})-1\}\times\frac{1}{(2)^{\frac{1}{2}}-1}+\ldots+\\ \{\cos(0)+\cos(\frac{\pi}{2})-1\}\times\frac{1}{(2)^{\frac{1}{2}}-1}\end{array}\right\}$$

$$\mathcal{E}_1(\Re) = 0$$
$$\mathcal{E}_2(\Re) = 0$$

2. Let $\mathfrak{F}_{\Re}^{RP}(e_\sigma) = -\mathfrak{F}_{\Re}^{RN}(e_\sigma)$ and $\mathfrak{F}_{\Re}^{IP}(e_\sigma) = -\mathfrak{F}_{\Re}^{IN}(e_\sigma)$, for all $\sigma = 1, 2, 3, .., n$ then we have

$$\mathcal{E}_1(\Re) = \frac{1}{n}\sum_{\sigma=1}^{n}\left\{\left\{\begin{array}{c}\sin\left(\frac{\left(\pi\times\left(2+\mathfrak{F}_{\Re}^{RP}(e_\sigma)+\mathfrak{F}_{\Re}^{IP}(e_\sigma)-\mathfrak{F}_{\Re}^{RP}(e_\sigma)-\mathfrak{F}_{\Re}^{IP}(e_\sigma)\right)\right)}{8}\right)+\\ \sin\left(\frac{\left(\pi\times\left(2-\mathfrak{F}_{\Re}^{RP}(e_\sigma)-\mathfrak{F}_{\Re}^{IP}(e_\sigma)+\mathfrak{F}_{\Re}^{RP}(e_\sigma)+\mathfrak{F}_{\Re}^{IP}(e_\sigma)\right)\right)}{8}\right)-1\end{array}\right\}\times\frac{1}{(2)^{\frac{1}{2}}-1}\right\}$$

$$\mathcal{E}_2(\Re) = \frac{1}{n}\sum_{\sigma=1}^{n}\left\{\left\{\begin{array}{c}\cos\left(\frac{\left(\pi\times\left(2+\mathfrak{F}_{\Re}^{RP}(e_\sigma)+\mathfrak{F}_{\Re}^{IP}(e_\sigma)-\mathfrak{F}_{\Re}^{RP}(e_\sigma)-\mathfrak{F}_{\Re}^{IP}(e_\sigma)\right)\right)}{8}\right)+\\ \cos\left(\frac{\left(\pi\times\left(2-\mathfrak{F}_{\Re}^{RP}(e_\sigma)-\mathfrak{F}_{\Re}^{IP}(e_\sigma)+\mathfrak{F}_{\Re}^{RP}(e_\sigma)+\mathfrak{F}_{\Re}^{IP}(e_\sigma)\right)\right)}{8}\right)-1\end{array}\right\}\times\frac{1}{(2)^{\frac{1}{2}}-1}\right\}$$

$$\mathcal{E}_1(\Re) = \frac{1}{n}\sum_{\sigma=1}^{n}\left\{\begin{array}{c}\left\{\begin{array}{c}\sin\left(\frac{(\pi\times(2+\mathfrak{F}_{\Re}^{RP}(e_1)+\mathfrak{F}_{\Re}^{IP}(e_1)-\mathfrak{F}_{\Re}^{RP}(e_1)-\mathfrak{F}_{\Re}^{IP}(e_1)))}{8}\right)+\\ \sin\left(\frac{(\pi\times(2-\mathfrak{F}_{\Re}^{RP}(e_1)-\mathfrak{F}_{\Re}^{IP}(e_1)+\mathfrak{F}_{\Re}^{RP}(e_1)+\mathfrak{F}_{\Re}^{IP}(e_1)))}{8}\right)-1\end{array}\right\}\times\frac{1}{(2)^{\frac{1}{2}}-1}+\\ \left\{\begin{array}{c}\sin\left(\frac{(\pi\times(2+\mathfrak{F}_{\Re}^{RP}(e_2)+\mathfrak{F}_{\Re}^{IP}(e_2)-\mathfrak{F}_{\Re}^{RP}(e_2)-\mathfrak{F}_{\Re}^{IP}(e_2)))}{8}\right)+\\ \sin\left(\frac{(\pi\times(2-\mathfrak{F}_{\Re}^{RP}(e_2)-\mathfrak{F}_{\Re}^{IP}(e_2)+\mathfrak{F}_{\Re}^{RP}(e_2)+\mathfrak{F}_{\Re}^{IP}(e_2)))}{8}\right)-1\end{array}\right\}\times\frac{1}{(2)^{\frac{1}{2}}-1}+\ldots+\\ \left\{\begin{array}{c}\sin\left(\frac{(\pi\times(2+\mathfrak{F}_{\Re}^{RP}(e_n)+\mathfrak{F}_{\Re}^{IP}(e_n)-\mathfrak{F}_{\Re}^{RP}(e_n)-\mathfrak{F}_{\Re}^{IP}(e_n)))}{8}\right)+\\ \sin\left(\frac{(\pi\times(2-\mathfrak{F}_{\Re}^{RP}(e_n)-\mathfrak{F}_{\Re}^{IP}(e_n)+\mathfrak{F}_{\Re}^{RP}(e_n)+\mathfrak{F}_{\Re}^{IP}(e_n)))}{8}\right)\end{array}\right\}\times\frac{1}{(2)^{\frac{1}{2}}-1}\end{array}\right\}$$

$$\mathcal{E}_2(\Re) = \frac{1}{n}\sum_{\sigma=1}^{n}\left\{\begin{array}{l}\left\{\begin{array}{l}\cos\left(\frac{(\pi\times(2+\mathfrak{F}_{\Re}^{RP}(e_1)+\mathfrak{F}_{\Re}^{IP}(e_1)-\mathfrak{F}_{\Re}^{RP}(e_1)-\mathfrak{F}_{\Re}^{IP}(e_1)))}{8}\right)+\\ \cos\left(\frac{(\pi\times(2-\mathfrak{F}_{\Re}^{RP}(e_1)-\mathfrak{F}_{\Re}^{IP}(e_1)+\mathfrak{F}_{\Re}^{RP}(e_1)+\mathfrak{F}_{\Re}^{IP}(e_1)))}{8}\right)-1\end{array}\right\}\times\frac{1}{(2)^{\frac{1}{2}}-1}+\\ \left\{\begin{array}{l}\cos\left(\frac{(\pi\times(2+\mathfrak{F}_{\Re}^{RP}(e_2)+\mathfrak{F}_{\Re}^{IP}(e_2)-\mathfrak{F}_{\Re}^{RP}(e_2)-\mathfrak{F}_{\Re}^{IP}(e_2)))}{8}\right)+\\ \cos\left(\frac{(\pi\times(2-\mathfrak{F}_{\Re}^{RP}(e_2)-\mathfrak{F}_{\Re}^{IP}(e_2)+\mathfrak{F}_{\Re}^{RP}(e_2)+\mathfrak{F}_{\Re}^{IP}(e_2)))}{8}\right)-1\end{array}\right\}\times\frac{1}{(2)^{\frac{1}{2}}-1}+\ldots+\\ \left\{\begin{array}{l}\cos\left(\frac{(\pi\times(2+\mathfrak{F}_{\Re}^{RP}(e_n)+\mathfrak{F}_{\Re}^{IP}(e_n)-\mathfrak{F}_{\Re}^{RP}(e_n)-\mathfrak{F}_{\Re}^{IP}(e_n)))}{8}\right)+\\ \cos\left(\frac{(\pi\times(2-\mathfrak{F}_{\Re}^{RP}(e_n)-\mathfrak{F}_{\Re}^{IP}(e_n)+\mathfrak{F}_{\Re}^{RP}(e_n)+\mathfrak{F}_{\Re}^{IP}(e_n)))}{8}\right)-1\end{array}\right\}\times\frac{1}{(2)^{\frac{1}{2}}-1}\end{array}\right\}$$

$$\mathcal{E}_1(\Re) = \frac{1}{n}\left\{\begin{array}{c}\left\{\sin\left(\frac{\pi}{4}\right)+\sin\left(\frac{\pi}{4}\right)-1\right\}\times\frac{1}{(2)^{\frac{1}{2}}-1}+\left\{\sin\left(\frac{\pi}{4}\right)+\sin\left(\frac{\pi}{4}\right)-1\right\}\times\frac{1}{(2)^{\frac{1}{2}}-1}+\ldots+\\ \left\{\sin\left(\frac{\pi}{4}\right)+\sin\left(\frac{\pi}{4}\right)-1\right\}\times\frac{1}{(2)^{\frac{1}{2}}-1}\end{array}\right\}$$

$$\mathcal{E}_2(\Re) = \frac{1}{n}\left\{\begin{array}{c}\left\{\cos\left(\frac{\pi}{4}\right)+\cos\left(\frac{\pi}{4}\right)-1\right\}\times\frac{1}{(2)^{\frac{1}{2}}-1}+\left\{\cos\left(\frac{\pi}{4}\right)+\cos\left(\frac{\pi}{4}\right)-1\right\}\times\frac{1}{(2)^{\frac{1}{2}}-1}+\ldots+\\ \left\{\cos\left(\frac{\pi}{4}\right)+\cos\left(\frac{\pi}{4}\right)-1\right\}\times\frac{1}{(2)^{\frac{1}{2}}-1}\end{array}\right\}$$

$$\mathcal{E}_1(\Re) = \frac{1}{n}\left\{\begin{array}{c}\left\{\frac{\sqrt{2}}{2}+\frac{\sqrt{2}}{2}-1\right\}\times\frac{1}{(2)^{\frac{1}{2}}-1}+\left\{\frac{\sqrt{2}}{2}+\frac{\sqrt{2}}{2}-1\right\}\times\frac{1}{(2)^{\frac{1}{2}}-1}+\ldots+\\ \left\{\frac{\sqrt{2}}{2}+\frac{\sqrt{2}}{2}-1\right\}\times\frac{1}{(2)^{\frac{1}{2}}-1}\end{array}\right\}=1$$

$$\mathcal{E}_2(\Re) = \frac{1}{n}\left\{\begin{array}{c}\left\{\frac{\sqrt{2}}{2}+\frac{\sqrt{2}}{2}-1\right\}\times\frac{1}{(2)^{\frac{1}{2}}-1}+\left\{\frac{\sqrt{2}}{2}+\frac{\sqrt{2}}{2}-1\right\}\times\frac{1}{(2)^{\frac{1}{2}}-1}+\ldots+\\ \left\{\frac{\sqrt{2}}{2}+\frac{\sqrt{2}}{2}-1\right\}\times\frac{1}{(2)^{\frac{1}{2}}-1}\end{array}\right\}=1$$

This implies that

$$\mathcal{E}_1(\Re) = 1$$
$$\mathcal{E}_2(\Re) = 1$$

3. Omitted.
4. Since we have $\Re^c = \{(e_\sigma, (1-\mathfrak{F}_{\Re}^{RP}(e)+\iota(1-\mathfrak{F}_{\Re}^{IP}(e)), (-1-\mathfrak{F}_{\Re}^{RN}(e))+\iota(-1-\mathfrak{F}_{\Re}^{IN}(e))))|e_\sigma \in \mathbb{E}\}$ then

$$\mathcal{E}_1(\Re^c) = \frac{1}{n}\sum_{\sigma=1}^{n}\left\{\left\{\begin{array}{l}\sin\left(\frac{\left(\pi\times\left(\begin{array}{c}2+(1-\mathfrak{F}_{\Re}^{RP}(e_\sigma))+(1-\mathfrak{F}_{\Re}^{IP}(e_\sigma))\\ -|-1-\mathfrak{F}_{\Re}^{RN}(e_\sigma)|-|-1-\mathfrak{F}_{\Re}^{IP}(e_\sigma)|\end{array}\right)\right)}{8}\right)+\\ \sin\left(\frac{\left(\pi\times\left(\begin{array}{c}2-(1-\mathfrak{F}_{\Re}^{RP}(e_\sigma))-(1-\mathfrak{F}_{\Re}^{IP}(e_\sigma))\\ +|-1-\mathfrak{F}_{\Re}^{RN}(e_\sigma)|+|-1-\mathfrak{F}_{\Re}^{IP}(e_\sigma)|\end{array}\right)\right)}{8}\right)-1\end{array}\right\}\times\frac{1}{(2)^{\frac{1}{2}}-1}\right\}$$

$$\mathcal{E}_2(\Re^c) = \frac{1}{n}\sum_{\sigma=1}^{n}\left\{\left\{\begin{array}{l}\cos\left(\frac{\left(\pi\times\left(\begin{array}{c}2+(1-\mathfrak{F}_{\Re}^{RP}(e_\sigma))+(1-\mathfrak{F}_{\Re}^{IP}(e_\sigma))\\ -|-1-\mathfrak{F}_{\Re}^{RN}(e_\sigma)|-|-1-\mathfrak{F}_{\Re}^{IP}(e_\sigma)|\end{array}\right)\right)}{8}\right)+\\ \cos\left(\frac{\left(\pi\times\left(\begin{array}{c}2-(1-\mathfrak{F}_{\Re}^{RP}(e_\sigma))-(1-\mathfrak{F}_{\Re}^{IP}(e_\sigma))\\ +|-1-\mathfrak{F}_{\Re}^{RN}(e_\sigma)|+|-1-\mathfrak{F}_{\Re}^{IP}(e_\sigma)|\end{array}\right)\right)}{8}\right)-1\end{array}\right\}\times\frac{1}{(2)^{\frac{1}{2}}-1}\right\}$$

$$\mathcal{E}_1\left(\mathfrak{R}^c\right)=\frac{1}{n}\sum_{\sigma=1}^{n}\left\{\left\{\begin{array}{l}\sin\left(\dfrac{\left(\pi\times\left(\begin{array}{c}2+1-\mathfrak{F}_{\mathfrak{R}}^{RP}(e_\sigma)+1-\mathfrak{F}_{\mathfrak{R}}^{IP}(e_\sigma)\\-1-\mathfrak{F}_{\mathfrak{R}}^{RN}(e_\sigma)-1-\mathfrak{F}_{\mathfrak{R}}^{IP}(e_\sigma)\end{array}\right)\right)}{8}\right)+\\ \sin\left(\dfrac{\left(\pi\times\left(\begin{array}{c}2-1+\mathfrak{F}_{\mathfrak{R}}^{RP}(e_\sigma)-1+\mathfrak{F}_{\mathfrak{R}}^{IP}(e_\sigma)\\+1+\mathfrak{F}_{\mathfrak{R}}^{RN}(e_\sigma)+1+\mathfrak{F}_{\mathfrak{R}}^{IP}(e_\sigma)\end{array}\right)\right)}{8}\right)-1\end{array}\right\}\times\frac{1}{(2)^{\frac{1}{2}}-1}\right\}$$

$$\mathcal{E}_2\left(\mathfrak{R}^c\right)=\frac{1}{n}\sum_{\sigma=1}^{n}\left\{\left\{\begin{array}{l}\cos\left(\dfrac{\left(\pi\times\left(\begin{array}{c}2+1-\mathfrak{F}_{\mathfrak{R}}^{RP}(e_\sigma)+1-\mathfrak{F}_{\mathfrak{R}}^{IP}(e_\sigma)\\-1-\mathfrak{F}_{\mathfrak{R}}^{RN}(e_\sigma)-1-\mathfrak{F}_{\mathfrak{R}}^{IP}(e_\sigma)\end{array}\right)\right)}{8}\right)+\\ \cos\left(\dfrac{\left(\pi\times\left(\begin{array}{c}2-1+\mathfrak{F}_{\mathfrak{R}}^{RP}(e_\sigma)-1+\mathfrak{F}_{\mathfrak{R}}^{IP}(e_\sigma)\\+1+\mathfrak{F}_{\mathfrak{R}}^{RN}(e_\sigma)+1+\mathfrak{F}_{\mathfrak{R}}^{IP}(e_\sigma)\end{array}\right)\right)}{8}\right)-1\end{array}\right\}\times\frac{1}{(2)^{\frac{1}{2}}-1}\right\}$$

$$\mathcal{E}_1\left(\mathfrak{R}^c\right)=\frac{1}{n}\sum_{\sigma=1}^{n}\left\{\left\{\begin{array}{l}\sin\left(\frac{\left(\pi\times\left(2-\mathfrak{F}_{\mathfrak{R}}^{RP}(e_\sigma)-\mathfrak{F}_{\mathfrak{R}}^{IP}(e_\sigma)-\mathfrak{F}_{\mathfrak{R}}^{RN}(e_\sigma)-\mathfrak{F}_{\mathfrak{R}}^{IP}(e_\sigma)\right)\right)}{8}\right)+\\ \sin\left(\frac{\left(\pi\times\left(2+\mathfrak{F}_{\mathfrak{R}}^{RP}(e_\sigma)+\mathfrak{F}_{\mathfrak{R}}^{IP}(e_\sigma)+\mathfrak{F}_{\mathfrak{R}}^{RN}(e_\sigma)+\mathfrak{F}_{\mathfrak{R}}^{IP}(e_\sigma)\right)\right)}{8}\right)-1\end{array}\right\}\times\frac{1}{(2)^{\frac{1}{2}}-1}\right\}$$

$$\mathcal{E}_2\left(\mathfrak{R}^c\right)=\frac{1}{n}\sum_{\sigma=1}^{n}\left\{\left\{\begin{array}{l}\cos\left(\frac{\left(\pi\times\left(2-\mathfrak{F}_{\mathfrak{R}}^{RP}(e_\sigma)-\mathfrak{F}_{\mathfrak{R}}^{IP}(e_\sigma)-\mathfrak{F}_{\mathfrak{R}}^{RN}(e_\sigma)-\mathfrak{F}_{\mathfrak{R}}^{IP}(e_\sigma)\right)\right)}{8}\right)+\\ \cos\left(\frac{\left(\pi\times\left(2+\mathfrak{F}_{\mathfrak{R}}^{RP}(e_\sigma)+\mathfrak{F}_{\mathfrak{R}}^{IP}(e_\sigma)+\mathfrak{F}_{\mathfrak{R}}^{RN}(e_\sigma)+\mathfrak{F}_{\mathfrak{R}}^{IP}(e_\sigma)\right)\right)}{8}\right)-1\end{array}\right\}\times\frac{1}{(2)^{\frac{1}{2}}-1}\right\}$$

As $\mathfrak{F}_{\mathfrak{R}}^{RN}(e_\sigma),\mathfrak{F}_{\mathfrak{R}}^{IN}(e_\sigma)\in[-1,0]$, thus, $-\mathfrak{F}_{\mathfrak{R}}^{RN}(e_\sigma)=\left|\mathfrak{F}_{\mathfrak{R}}^{RN}(e_\sigma)\right|$ and $-\mathfrak{F}_{\mathfrak{R}}^{IN}(e_\sigma)=\left|\mathfrak{F}_{\mathfrak{R}}^{IN}(e_\sigma)\right|$.

$$\mathcal{E}_1\left(\mathfrak{R}^c\right)=\frac{1}{n}\sum_{\sigma=1}^{n}\left\{\left\{\begin{array}{l}\sin\left(\dfrac{\left(\pi\times\left(\begin{array}{c}2+\mathfrak{F}_{\mathfrak{R}}^{RP}(e_\sigma)+\mathfrak{F}_{\mathfrak{R}}^{IP}(e_\sigma)\\-\left|\mathfrak{F}_{\mathfrak{R}}^{RN}(e_\sigma)\right|-\left|\mathfrak{F}_{\mathfrak{R}}^{IP}(e_\sigma)\right|\end{array}\right)\right)}{8}\right)+\\ \sin\left(\dfrac{\left(\pi\times\left(\begin{array}{c}2-\mathfrak{F}_{\mathfrak{R}}^{RP}(e_\sigma)-\mathfrak{F}_{\mathfrak{R}}^{IP}(e_\sigma)\\+\left|\mathfrak{F}_{\mathfrak{R}}^{RN}(e_\sigma)\right|+\left|\mathfrak{F}_{\mathfrak{R}}^{IP}(e_\sigma)\right|\end{array}\right)\right)}{8}\right)-1\end{array}\right\}\times\frac{1}{(2)^{\frac{1}{2}}-1}\right\}=\mathcal{E}_1(\mathfrak{R})$$

$$\mathcal{E}_2\left(\mathfrak{R}^c\right)=\frac{1}{n}\sum_{\sigma=1}^{n}\left\{\left\{\begin{array}{l}\cos\left(\dfrac{\left(\pi\times\left(\begin{array}{c}2+\mathfrak{F}_{\mathfrak{R}}^{RP}(e_\sigma)+\mathfrak{F}_{\mathfrak{R}}^{IP}(e_\sigma)\\-\left|\mathfrak{F}_{\mathfrak{R}}^{RN}(e_\sigma)\right|-\left|\mathfrak{F}_{\mathfrak{R}}^{IP}(e_\sigma)\right|\end{array}\right)\right)}{8}\right)+\\ \cos\left(\dfrac{\left(\pi\times\left(\begin{array}{c}2-\mathfrak{F}_{\mathfrak{R}}^{RP}(e_\sigma)-\mathfrak{F}_{\mathfrak{R}}^{IP}(e_\sigma)\\+\left|\mathfrak{F}_{\mathfrak{R}}^{RN}(e_\sigma)\right|+\left|\mathfrak{F}_{\mathfrak{R}}^{IP}(e_\sigma)\right|\end{array}\right)\right)}{8}\right)-1\end{array}\right\}\times\frac{1}{(2)^{\frac{1}{2}}-1}\right\}=\mathcal{E}_2(\mathfrak{R})$$

This implies that

$$\mathcal{E}_1\left(\mathfrak{R}^c\right)=\mathcal{E}_1(\mathfrak{R})$$
$$\mathcal{E}_2\left(\mathfrak{R}^c\right)=\mathcal{E}_2(\mathfrak{R}).$$

This completes the proof.

6.3 TOPSIS Approach Based on CC and EMs under BCF Information

A DM procedure identified as TOPSIS (Technique for Order Preference by Similarity to Ideal Solution) is employed to assess and rank a group of choices based on several criteria. Since its initial introduction by Hwang and Yoon [23] in 1981, it has grown to be a well-liked technique in the multi-criteria DM (MCDM) community. Various researchers utilized the TOPSIS approach to cope DM dilemmas [24–26]. The TOPSIS technique takes into account a number of decision-relevant alternatives as well as a number of criteria. Based on how well it meets these criteria, each alternative is evaluated. The approach takes as given that the criteria are either costs or benefits (maximization criterion) (minimization criteria). The following steps are part of the TOPSIS approach:

Step 1 (Interpret the dilemma): Clearly state the DM issue and the standards by which the alternatives will be assessed. Ascertain whether each condition is a cost or a benefit.

Step 2 (develop the decision matrix): Make a matrix with a row for each alternative and a column for each criterion. Based on how each alternative performed against each criterion, give it a score or rating.

Step 3 (Normalize the decision matrix): To remove the impact of various measurement scales, normalize the decision matrix. This stage makes sure that every criterion has a comparable scale.

Step 4 (Find the positive and negative ideal solutions): Recognize the normalized decision matrix's best and worst values for each criterion. The optimal outcome for benefit criteria is the highest possible value for each criterion, but for cost criteria the ideal outcome is the lowest possible value. The inverse of the ideal answer is the negative ideal solution.

Step 5 (Compute the distances): Compute the distance between each alternative and the positive and negative ideal solutions. For each criterion, the distance shows how close an alternative is to the positive ideal or negative ideal solutions.

Step 6 (Compute the relative closeness): Calculate the ratio of the distance from the negative ideal solution to the total distances from the positive ideal and negative ideal solutions to determine how close each alternative is to the positive ideal solution.

Step 7: (Rank the alternatives): Based on the relative closeness values, order the alternatives. The alternative that has higher degree of closeness is supposed to be the finest one.

Underneath we interpret TOPSIS approach by employing the initiated CC and EMs in the environment of BCF information and will be identified as BCF-TOPSIS. In essence, an approach of BCF-TOPSIS employing CC and EMs is significant to deal with MADM dilemmas.

Step 1 (Interpret the dilemma): Let us assume a MADM dilemma, where the m number of alternatives and n number attributes are signified by $\mathcal{A}_{\mathbb{A}} = \{\mathcal{A}_{\mathbb{A}-1}, \mathcal{A}_{\mathbb{A}-2}, \mathcal{A}_{\mathbb{A}-3}, \ldots, \mathcal{A}_{\mathbb{A}-m}\}$ and $\mathfrak{A}_{\mathbb{C}\approx} = \{\mathfrak{A}_{\mathbb{C}\approx-1}, \mathfrak{A}_{\mathbb{C}\approx-2}, \ldots, \mathfrak{A}_{\mathbb{C}\approx-n}\}$ and consider $\omega = (\omega_1, \omega_2, \omega_3, \ldots, \omega_n)^{\mathrm{T}}$ as a weight vector with properties that $\omega_\varrho \in [0, 1] \forall \varrho$ and $\sum_{\varrho=1}^{n} \omega_\varrho = 1$ demonstrated by the expert or decision analyst by keeping in view his/her preference. The assessment value of every attribute of every alternative demonstrated by the expert is presented in the form of BCFN that is $\mathfrak{R}_{\sigma\varrho} = \left(\mathfrak{F}^{P}_{\mathfrak{R}_{\sigma\varrho}}, \mathfrak{F}^{N}_{\mathfrak{R}_{\sigma\varrho}}\right) = \left(\mathfrak{F}^{RP}_{\mathfrak{R}_{\sigma\varrho}} + \iota\, \mathfrak{F}^{IP}_{\mathfrak{R}_{\sigma\varrho}}, \mathfrak{F}^{RN}_{\mathfrak{R}_{\sigma\varrho}} + \iota\, \mathfrak{F}^{IN}_{\mathfrak{R}_{\sigma\varrho}}\right)$, where $\mathfrak{F}^{RP}_{\mathfrak{R}_{\sigma\varrho}}, \mathfrak{F}^{IP}_{\mathfrak{R}_{\sigma\varrho}} \in [0, 1]$ and $\mathfrak{F}^{RN}_{\mathfrak{R}_{\sigma\varrho}}, \mathfrak{F}^{IN}_{\mathfrak{R}_{\sigma\varrho}} \in [-1, 0]$

Step 2 (develop the decision matrix): Construct a BCF decision matrix i.e., $\mathfrak{D} = \left[\mathfrak{R}_{\sigma\varrho}\right]_{m\times n} = \left(\mathfrak{F}^{P}_{\mathfrak{R}_{\sigma\varrho}}, \mathfrak{F}^{N}_{\mathfrak{R}_{\sigma\varrho}}\right)_{m\times n} = \left(\mathfrak{F}^{RP}_{\mathfrak{R}_{\sigma\varrho}} + \iota\, \mathfrak{F}^{IP}_{\mathfrak{R}_{\sigma\varrho}}, \mathfrak{F}^{RN}_{\mathfrak{R}_{\sigma\varrho}} + \iota\, \mathfrak{F}^{IN}_{\mathfrak{R}_{\sigma\varrho}}\right)_{m\times n}$

Step 3 (Normalize the decision matrix): In the MADM problems usually cost and benefits sort of information are existing. Therefore, we have the following formula to normalize the decision matrix

$$\mathfrak{R}_{\sigma\varrho} = \begin{cases} \left(\mathfrak{F}^{RP}_{\mathfrak{R}_{\sigma\varrho}} + \iota \mathfrak{F}^{RP}_{\mathfrak{R}_{\sigma\varrho}}, \mathfrak{F}^{RN}_{\mathfrak{R}_{\sigma\varrho}} + \iota \mathfrak{F}^{IN}_{\mathfrak{R}_{\sigma\varrho}}\right) & \textit{for benefit sort of attributes} \\ \left(1 - \mathfrak{F}^{RP}_{\mathfrak{R}_{\sigma\varrho}} + \iota\left(-1 - \mathfrak{F}^{IP}_{\mathfrak{R}_{\sigma\varrho}}\right), 1 - \mathfrak{F}^{RN}_{\mathfrak{R}_{\sigma\varrho}} + \iota\left(-1 - \mathfrak{F}^{IN}_{\mathfrak{R}_{\sigma\varrho}}\right)\right) & \textit{for cost sort of attributes} \end{cases}$$

Step 4 (Compute the weight): It is challenging to give the weight information of the attribute in advance due to the stress, constrained time of the DMs, and complex concerns. To address these issues, we establish attribute weights and assume that the initiated EM will assess the weight of each property as

$$\omega_\varrho = \frac{1 - \mathfrak{P}_\varrho}{n - \sum_{\sigma=1}^{n} \mathfrak{P}_\sigma} \tag{6.5}$$

where the $\mathfrak{P}_\varrho \in [0, 1]$, $\varrho = 1, 2, \ldots, n$ is interpreted as

$$\mathfrak{P}_\varrho = \frac{1}{m}\sum_{\sigma=1}^{m}\left\{\left\{\begin{array}{l} \sin\left(\frac{\left(\pi\times\left(2+\mathfrak{F}^{RP}_{\mathfrak{R}_{\sigma\varrho}}+\mathfrak{F}^{IP}_{\mathfrak{R}_{\sigma\varrho}}-\left|\mathfrak{F}^{RN}_{\mathfrak{R}_{\sigma\varrho}}\right|-\left|\mathfrak{F}^{IP}_{\mathfrak{R}_{\sigma\varrho}}\right|\right)\right)}{8}\right)+ \\ \sin\left(\frac{\left(\pi\times\left(2-\mathfrak{F}^{RP}_{\mathfrak{R}_{\sigma\varrho}}-\mathfrak{F}^{IP}_{\mathfrak{R}_{\sigma\varrho}}+\left|\mathfrak{F}^{RN}_{\mathfrak{R}_{\sigma\varrho}}\right|+\left|\mathfrak{F}^{IP}_{\mathfrak{R}_{\sigma\varrho}}\right|\right)\right)}{8}\right)-1 \end{array}\right\}\times\frac{1}{(2)^{\frac{1}{2}}-1}\right\}$$

$$\mathfrak{P}_\varrho = \frac{1}{m}\sum_{\sigma=1}^{m}\left\{\left\{\begin{array}{l} \cos\left(\frac{\left(\pi\times\left(2+\mathfrak{F}^{RP}_{\mathfrak{R}_{\sigma\varrho}}+\mathfrak{F}^{IP}_{\mathfrak{R}_{\sigma\varrho}}-\left|\mathfrak{F}^{RN}_{\mathfrak{R}_{\sigma\varrho}}\right|-\left|\mathfrak{F}^{IP}_{\mathfrak{R}_{\sigma\varrho}}\right|\right)\right)}{8}\right)+ \\ \cos\left(\frac{\left(\pi\times\left(2-\mathfrak{F}^{RP}_{\mathfrak{R}_{\sigma\varrho}}-\mathfrak{F}^{IP}_{\mathfrak{R}_{\sigma\varrho}}+\left|\mathfrak{F}^{RN}_{\mathfrak{R}_{\sigma\varrho}}\right|+\left|\mathfrak{F}^{IP}_{\mathfrak{R}_{\sigma\varrho}}\right|\right)\right)}{8}\right)-1 \end{array}\right\}\times\frac{1}{(2)^{\frac{1}{2}}-1}\right\}$$

Step 5 (Find the BCF positive and BCF negative ideal solutions): We compute the BCF positive ideal solution (BCF-PIS) $\mathfrak{R}^{+}$ and BCF negative ideal solution (BCF-NIS) $\mathfrak{R}^{-}$ by employing the underneath Equations.

$$\mathfrak{R}^{+} = \left\{ \begin{pmatrix} \mathfrak{F}_{\mathfrak{R}_1}^{+RP} + \iota\mathfrak{F}_{\mathfrak{R}_1}^{+IP}, \\ \mathfrak{F}_{\mathfrak{R}_1}^{+RN} + \iota\mathfrak{F}_{\mathfrak{R}_1}^{+IN} \end{pmatrix}, \begin{pmatrix} \mathfrak{F}_{\mathfrak{R}_2}^{+RP} + \iota\mathfrak{F}_{\mathfrak{R}_2}^{+IP}, \\ \mathfrak{F}_{\mathfrak{R}_2}^{+RN} + \iota\mathfrak{F}_{\mathfrak{R}_2}^{+IN} \end{pmatrix}, \ldots, \begin{pmatrix} \mathfrak{F}_{\mathfrak{R}_n}^{+RP} + \iota\mathfrak{F}_{\mathfrak{R}_n}^{+IP}, \\ \mathfrak{F}_{\mathfrak{R}_n}^{+RN} + \iota\mathfrak{F}_{\mathfrak{R}_n}^{+IN} \end{pmatrix} \right\}$$

$$\mathfrak{R}^{-} \left\{ \begin{pmatrix} \mathfrak{F}_{\mathfrak{R}_1}^{-RP} + \iota\mathfrak{F}_{\mathfrak{R}_1}^{-IP}, \\ \mathfrak{F}_{\mathfrak{R}_1}^{-RN} + \iota\mathfrak{F}_{\mathfrak{R}_1}^{-IN} \end{pmatrix}, \begin{pmatrix} \mathfrak{F}_{\mathfrak{R}_2}^{-RP} + \iota\mathfrak{F}_{\mathfrak{R}_2}^{-IP}, \\ \mathfrak{F}_{\mathfrak{R}_2}^{-RN} + \iota\mathfrak{F}_{\mathfrak{R}_2}^{-IN} \end{pmatrix}, \ldots, \begin{pmatrix} \mathfrak{F}_{\mathfrak{R}_n}^{+RP} - \iota\mathfrak{F}_{\mathfrak{R}_n}^{-IP}, \\ \mathfrak{F}_{\mathfrak{R}_n}^{-RN} + \iota\mathfrak{F}_{\mathfrak{R}_n}^{-IN} \end{pmatrix} \right\}$$

Noted that $\mathfrak{F}_{\mathfrak{R}_\varrho}^{+RP} = \max_{\varrho}\left\{\mathfrak{F}_{\mathfrak{R}_{\sigma\varrho}}^{RP}\right\}$, $\mathfrak{F}_{\mathfrak{R}_\varrho}^{+IP} = \max_{\varrho}\left\{\mathfrak{F}_{\mathfrak{R}_{\sigma\varrho}}^{IP}\right\}$, $\mathfrak{F}_{\mathfrak{R}_\varrho}^{+RN} = \max_{\varrho}\left\{\mathfrak{F}_{\mathfrak{R}_{\sigma\varrho}}^{RN}\right\}$, $\mathfrak{F}_{\mathfrak{R}_\varrho}^{+IN} = \max_{\varrho}\left\{\mathfrak{F}_{\mathfrak{R}_{\sigma\varrho}}^{IN}\right\}$, and $\mathfrak{F}_{\mathfrak{R}_\varrho}^{-RP} = \min_{\varrho}\left\{\mathfrak{F}_{\mathfrak{R}_{\sigma\varrho}}^{RP}\right\}$, $\mathfrak{F}_{\mathfrak{R}_\varrho}^{-IP} = \min_{\varrho}\left\{\mathfrak{F}_{\mathfrak{R}_{\sigma\varrho}}^{IP}\right\}$, $\mathfrak{F}_{\mathfrak{R}_\varrho}^{-RN} = \min_{\varrho}\left\{\mathfrak{F}_{\mathfrak{R}_{\sigma\varrho}}^{RN}\right\}$, $\mathfrak{F}_{\mathfrak{R}_\varrho}^{-IN} = \min_{\varrho}\left\{\mathfrak{F}_{\mathfrak{R}_{\sigma\varrho}}^{IN}\right\}$

Step 6: Compute the CC) We compute the CC among $\mathfrak{R}_{\sigma\varrho}(\sigma = 1, 2, \ldots, m)$ and BCF-PIS and CC among $\mathfrak{R}_{\sigma\varrho}(\sigma = 1, 2, \ldots, m)$ and BCF-NIS as follows

$$K_\sigma\left(\mathfrak{R}_{\sigma\varrho}, \mathfrak{R}^{+}\right) = \frac{\sum_{\sigma=1}^{m} \omega_\sigma \left(\mathfrak{F}_{\mathfrak{R}_{\sigma\varrho}}^{RP}\mathfrak{F}_{\mathfrak{R}_\varrho}^{+RP} + \mathfrak{F}_{\mathfrak{R}_{\sigma\varrho}}^{IP}\mathfrak{F}_{\mathfrak{R}_\varrho}^{+IP} + \mathfrak{F}_{\mathfrak{R}_{\sigma\varrho}}^{RN}\mathfrak{F}_{\mathfrak{R}_\varrho}^{+RN} + \mathfrak{F}_{\mathfrak{R}_{\sigma\varrho}}^{IN}\mathfrak{F}_{\mathfrak{R}_\varrho}^{+IN}\right)}{\begin{pmatrix} \left(\sum_{\sigma=1}^{m} \omega_\sigma \left(\left(\mathfrak{F}_{\mathfrak{R}_{\sigma\varrho}}^{RP}\right)^2 + \left(\mathfrak{F}_{\mathfrak{R}_{\sigma\varrho}}^{IP}\right)^2 + \left(\mathfrak{F}_{\mathfrak{R}_{\sigma\varrho}}^{RN}\right)^2 + \left(\mathfrak{F}_{\mathfrak{R}_{\sigma\varrho}}^{IN}\right)^2\right)\right) \\ \left(\sum_{\sigma=1}^{m} \omega_\sigma \left(\left(\mathfrak{F}_{\mathfrak{R}_\varrho}^{+RP}\right)^2 + \left(\mathfrak{F}_{\mathfrak{R}_\varrho}^{+IP}\right)^2 + \left(\mathfrak{F}_{\mathfrak{R}_\varrho}^{+RN}\right)^2 + \left(\mathfrak{F}_{\mathfrak{R}_\varrho}^{+IN}\right)^2\right)\right) \end{pmatrix}^{\frac{1}{2}}} \tag{6.6}$$

$$K_\sigma\left(\mathfrak{R}_{\sigma\varrho}, \mathfrak{R}^{-}\right) = \frac{\sum_{\sigma=1}^{m} \omega_\sigma \left(\mathfrak{F}_{\mathfrak{R}_{\sigma\varrho}}^{RP}\mathfrak{F}_{\mathfrak{R}_\varrho}^{-RP} + \mathfrak{F}_{\mathfrak{R}_{\sigma\varrho}}^{IP}\mathfrak{F}_{\mathfrak{R}_\varrho}^{-IP} + \mathfrak{F}_{\mathfrak{R}_{\sigma\varrho}}^{RN}\mathfrak{F}_{\mathfrak{R}_\varrho}^{-RN} + \mathfrak{F}_{\mathfrak{R}_{\sigma\varrho}}^{IN}\mathfrak{F}_{\mathfrak{R}_\varrho}^{-IN}\right)}{\begin{pmatrix} \left(\sum_{\sigma=1}^{m} \omega_\sigma \left(\left(\mathfrak{F}_{\mathfrak{R}_{\sigma\varrho}}^{RP}\right)^2 + \left(\mathfrak{F}_{\mathfrak{R}_{\sigma\varrho}}^{IP}\right)^2 + \left(\mathfrak{F}_{\mathfrak{R}_{\sigma\varrho}}^{RN}\right)^2 + \left(\mathfrak{F}_{\mathfrak{R}_{\sigma\varrho}}^{IN}\right)^2\right)\right) \\ \left(\sum_{\sigma=1}^{m} \omega_\sigma \left(\left(\mathfrak{F}_{\mathfrak{R}_\varrho}^{-RP}\right)^2 + \left(\mathfrak{F}_{\mathfrak{R}_\varrho}^{-IP}\right)^2 + \left(\mathfrak{F}_{\mathfrak{R}_\varrho}^{-RN}\right)^2 + \left(\mathfrak{F}_{\mathfrak{R}_\varrho}^{-IN}\right)^2\right)\right) \end{pmatrix}^{\frac{1}{2}}} \tag{6.7}$$

Step 7 (Compute the closeness): We compute the closeness of every alternative by employing the underneath Eq.

$$\mathcal{A}_{\mathbb{A}\sigma} = \frac{K_\sigma\left(\mathfrak{R}_{\sigma\varrho}, \mathfrak{R}^{+}\right)}{K_\sigma\left(\mathfrak{R}_{\sigma\varrho}, \mathfrak{R}^{+}\right) + K_\sigma\left(\mathfrak{R}_{\sigma\varrho}, \mathfrak{R}^{-}\right)} \tag{6.8}$$

Step 8 (Rank the alternatives): We compute the best alternative by ranking all the alternatives.

Step 7: Finished.

Fig (6.1) demonstrates the flow chart of the initiated BCF-TOPSIS approach.

6.4 Demonstrative Examples

In this section, we will use the established BCF-TOPSIS method for selecting and analyzing the emergency project assessment for the emergency management department and selection of the best e-learning program.

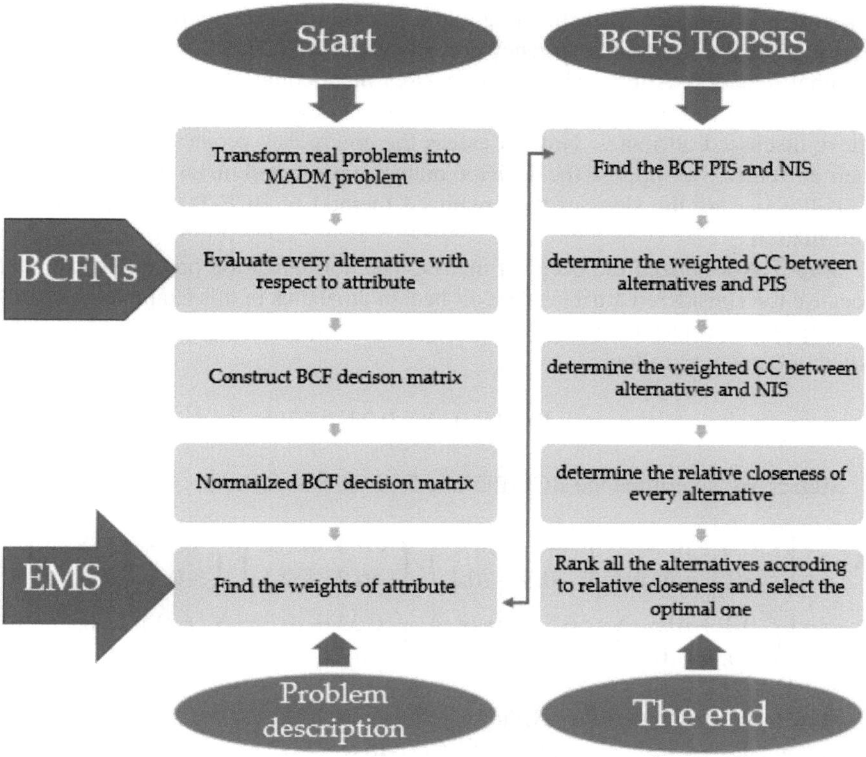

Fig. 6.1. Flowchart of the established TOPSIS method based on BCFSs

6.4.1 Emergency Project Assessment for Emergency Management Department

In this part, with the help of the following Example 6.1, we will show that how decision-makers select an emergency project for the emergency management department in the environment of BCFSs by utilizing the BCF-TOPSIS method.

Example 6.1 In the last few years, numerous natural disasters are the reason for the destruction of human life and property. Considering the key features of natural disasters, how to use realistic and effective emergency alternatives to decrease the damage caused by an unexpected disaster has become a major focus of decision science and management science research. The current situation requires coping with the emergency and lessen the damage as much as could be expected. By studying the features of emergencies and summing up the current literature, we shortlisted four important indicators as the assessment attributes of the emergency project i.e., $\mathfrak{A}_{\mathbb{C}\approx-1}$: response capability, $\mathfrak{A}_{\mathbb{C}\approx-2}$: Mitigation capability, $\mathfrak{A}_{\mathbb{C}\approx-3}$: recovery capability, and $\mathfrak{A}_{\mathbb{C}\approx-4}$: rescue capability. Following pre-choice of the projects gathered by the expert assessment committee corresponding to the essential criteria, representatives of the expert group finalize four emergency projects $\mathcal{A}_{\mathbb{A}-\sigma}(\sigma = 1, 2, 3, 4)$ to participate in the final selection. Afterward, the expert assessment committee will select the best emergency project for the emergency management department in the opinion of the above-discussed attributes. Now to assess the emergency projects with respect to their attributes, we suppose the decision matrix is presented in Table 6.1.

Following are the steps of the organized method of BCF-TOPSIS under BCF information.

Step 3: Normalized the decision matrix. The normalization process is excluded because the considered attributes are all benefit attributes in this example.

Step 4: We determine the wight vector of every attribute by utilizing Eq. (6.5), which are given as follows

$$\omega = \{0.346, 0.389, 0.24, 0.026\}$$

Step 5: We determine the BCF-PIS and BCF-NIS as follows

$$\mathfrak{R}^{+} = \left\{\begin{pmatrix} 0.6 + \iota 0.8, \\ -0.1 - \iota 0.2 \end{pmatrix}, \begin{pmatrix} 44 + \iota 0.87, \\ -0.3 - \iota 0.1 \end{pmatrix}, \begin{pmatrix} 1.0 + \iota 0.98, \\ -0.2 - \iota 0.1 \end{pmatrix}, \begin{pmatrix} 0.7 + \iota 0.88, \\ -0.2 - \iota 0.28 \end{pmatrix}\right\}$$

$$\mathfrak{R}^{-} = \left\{\begin{pmatrix} 0.1 + \iota 0.3, \\ -0.9 - \iota 0.9 \end{pmatrix}, \begin{pmatrix} 0.2 + \iota 0.18, \\ -0.8 - \iota 0.5 \end{pmatrix}, \begin{pmatrix} 0.34 + \iota 0.33, \\ -0.54 - \iota 0.76 \end{pmatrix}, \begin{pmatrix} 0.1 + \iota 0.17, \\ -0.79 - \iota 0.8 \end{pmatrix}\right\}$$

Step 6: We computed the CC among $\mathfrak{R}_{\sigma Q}(\sigma = 1, 2, \ldots, m)$ and BCF-PIS and CC among $\mathfrak{R}_{\sigma Q}(\sigma = 1, 2, \ldots, m)$ and BCF-NIS as

$K_1(\mathfrak{R}_{1Q}, \mathfrak{R}^{+}) = 0.802$, $K_2(\mathfrak{R}_{2Q}, \mathfrak{R}^{+}) = 0.876$, $K_3(\mathfrak{R}_{3Q}, \mathfrak{R}^{+}) = 0.809$, $K_4(\mathfrak{R}_{4Q}, \mathfrak{R}^{+}) = 0.78$, and $K_1(\mathfrak{R}_{1Q}, \mathfrak{R}^{-}) = 0.811$, $K_2(\mathfrak{R}_{2Q}, \mathfrak{R}^{-}) = 0.69$, $K_3(\mathfrak{R}_{3Q}, \mathfrak{R}^{-}) = 0.746$, $K_4(\mathfrak{R}_{4Q}, \mathfrak{R}^{-}) = 0.799$

Step 7: We computed the closeness of every alternative by employing Eq. (6.8),

Table 6.1 The decision matrix of example 6.1

	$\mathfrak{A}_{C\approx-1}$	$\mathfrak{A}_{C\approx-2}$	$\mathfrak{A}_{C\approx-3}$	$\mathfrak{A}_{C\approx-4}$
$\mathcal{A}_{\mathbb{A}-1}$	$\begin{pmatrix} 0.5 + \iota 0.7, \\ -0.6 - \iota 0.9 \end{pmatrix}$	$\begin{pmatrix} 0.25 + \iota 0.65, \\ -0.5 - \iota 0.5 \end{pmatrix}$	$\begin{pmatrix} 0.34 + \iota 0.98, \\ -0.2 - \iota 0.1 \end{pmatrix}$	$\begin{pmatrix} 0.15 + \iota 0.17, \\ -0.36 - \iota 0.29 \end{pmatrix}$
$\mathcal{A}_{\mathbb{A}-2}$	$\begin{pmatrix} 0.6 + \iota 0.3, \\ -0.1 - \iota 0.2 \end{pmatrix}$	$\begin{pmatrix} 0.2 + \iota 0.66, \\ -0.8 - \iota 0.5 \end{pmatrix}$	$\begin{pmatrix} 1 + \iota 0.8, \\ -0.25 - \iota 0.76 \end{pmatrix}$	$\begin{pmatrix} 0.61 + \iota 0.33, \\ -0.19 - \iota 0.28 \end{pmatrix}$
$\mathcal{A}_{\mathbb{A}-3}$	$\begin{pmatrix} 0.1 + \iota 0.8, \\ -0.7 - \iota 0.3 \end{pmatrix}$	$\begin{pmatrix} 0.44 + \iota 0.87, \\ -0.17 - \iota 0.23 \end{pmatrix}$	$\begin{pmatrix} 0.56 + \iota 0.33, \\ -0.54 - \iota 0.45 \end{pmatrix}$	$\begin{pmatrix} 0.1 + \iota 0.88, \\ -0.79 - \iota 0.36 \end{pmatrix}$
$\mathcal{A}_{\mathbb{A}-4}$	$\begin{pmatrix} 0.6 + \iota 0.6, \\ -0.9 - \iota 0.2 \end{pmatrix}$	$\begin{pmatrix} 0.28 + \iota 0.18, \\ -0.3 - \iota 0.1 \end{pmatrix}$	$\begin{pmatrix} 0.36 + \iota 0.56, \\ -0.24 - \iota 0.45 \end{pmatrix}$	$\begin{pmatrix} 0.7 + \iota 0.2, \\ -0.2 - \iota 0.8 \end{pmatrix}$

$$\mathcal{A}_{\mathbb{A}-1} = 0.497, \mathcal{A}_{\mathbb{A}-2} = 0.559, \mathcal{A}_{\mathbb{A}-3} = 0.52, \mathcal{A}_{\mathbb{A}-4} = 0.494$$

Step 8: We compute the finest optimal one by ranking all alternatives i.e.

$$\mathcal{A}_{\mathbb{A}-2} > \mathcal{A}_{\mathbb{A}-3} > \mathcal{A}_{\mathbb{A}-1} > \mathcal{A}_{\mathbb{A}-4}$$

From the ranking, we have that alternative $\mathcal{A}_{\mathbb{A}-2}$ is the best emergency project for the emergency management department.

Step 9: Finished.

6.4.2 Selection of Best Electronic Learning (e-learning) Program

In this section, we'll demonstrate how the decision-maker chooses the optimal e-learning application in the context of BCFSs by employing the initiated BCF-TOPSIS approach with the aid of the Example ().

Example 6.2 With the development of information technology, e-learning has played a crucial role in learning and teaching, which has become increasingly trendy these days all over the world due to Covid 19. The selection of the best e-learning program is a challenge for educational institutions. The appropriate e-learning relies on different features or criteria. Here we suppose that an educational expert summarized 4 essential features or criteria i.e., $\mathcal{P}_{\mathbb{C}\approx-1}$: material of e-learning, $\mathcal{P}_{\mathbb{C}\approx-2}$: responsive design, $\mathcal{P}_{\mathbb{C}\approx-2}$: reporting and data analysis, and $\mathcal{P}_{\mathbb{C}\approx-2}$: User-friendly and intuitive learner interface to select the best e-learning program. Suppose 4 e-learning programs $\mathcal{A}_{\mathbb{L}-\sigma}(\sigma = 1, 2, 3, 4)$ with respect to the above 4 criteria by utilizing the BCFS. Now to select the best e-learning program with respect to their attributes, we suppose the decision matrix is presented in Table 6.2.

Following are the steps of the organized method of BCF-TOPSIS under BCF information.

Step 3: Normalized the decision matrix. The normalization process is excluded because the considered attributes are all benefit attributes in this example.

Step 4: We determine the wight vector of every attribute by utilizing Eq. (6.5), which is interpreted as

$$W = \{0.355, 0.243, 0.166, 0.236\}$$

Step 5: We determined the BCF-PIS and BCF-NIS as follows

$$\Re^{+} = \left\{\begin{pmatrix} 1.0 + \iota 0.9, \\ -0.1 - \iota 0.1 \end{pmatrix}, \begin{pmatrix} 0.8 + \iota 0.7, \\ -0.19 - \iota 0.33 \end{pmatrix}, \begin{pmatrix} 0.7 + \iota 0.5, \\ -0.5 - \iota 0.4 \end{pmatrix}, \begin{pmatrix} 0.9 + \iota 0.78, \\ -0.1 - \iota 0.6 \end{pmatrix}\right\}$$

Table 6.2 The decision matrix of example 6.2

	$\mathcal{P}_{\mathbb{C}\approx-1}$	$\mathcal{P}_{\mathbb{C}\approx-2}$	$\mathcal{P}_{\mathbb{C}\approx-3}$	$\mathcal{P}_{\mathbb{C}\approx-4}$
$\mathcal{A}_{\mathbb{L}-1}$	$\begin{pmatrix} 0.1+\iota 0.9, \\ -0.7-\iota 0.3 \end{pmatrix}$	$\begin{pmatrix} 0.2+\iota 0.5, \\ -0.5-\iota 0.5 \end{pmatrix}$	$\begin{pmatrix} 0.25+\iota 0.45, \\ -0.8-\iota 0.9 \end{pmatrix}$	$\begin{pmatrix} 0.3+\iota 0.6, \\ -0.6-\iota 0.98 \end{pmatrix}$
$\mathcal{A}_{\mathbb{L}-2}$	$\begin{pmatrix} 0.46+\iota 0.23, \\ -0.9-\iota 0.3 \end{pmatrix}$	$\begin{pmatrix} 0.12+\iota 0.6, \\ -0.5-\iota 0.45 \end{pmatrix}$	$\begin{pmatrix} 0.4+\iota 0.2, \\ -0.5-\iota 0.8 \end{pmatrix}$	$\begin{pmatrix} 0.1+\iota 0.3, \\ -0.1-\iota 0.8 \end{pmatrix}$
$\mathcal{A}_{\mathbb{L}-3}$	$\begin{pmatrix} 0.7+\iota 0.2, \\ -0.1-\iota 0.6 \end{pmatrix}$	$\begin{pmatrix} 0.4+\iota 0.7, \\ -0.19-\iota 0.33 \end{pmatrix}$	$\begin{pmatrix} 0.66+\iota 0.43, \\ -0.64-\iota 0.55 \end{pmatrix}$	$\begin{pmatrix} 0.15+\iota 0.78, \\ -0.9-\iota 0.6 \end{pmatrix}$
$\mathcal{A}_{\mathbb{L}-4}$	$\begin{pmatrix} 1+\iota 0.1, \\ -0.2-\iota 0.1 \end{pmatrix}$	$\begin{pmatrix} 0.8+\iota 0.5, \\ -0.7-\iota 0.4 \end{pmatrix}$	$\begin{pmatrix} 0.7+\iota 0.5, \\ -0.5-\iota 0.4 \end{pmatrix}$	$\begin{pmatrix} 0.9+\iota 0.1, \\ -0.6-\iota 0.9 \end{pmatrix}$

$$\Re^- = \left\{ \begin{pmatrix} 0.1+\iota 0.1, \\ -0.6-\iota 0.9 \end{pmatrix}, \begin{pmatrix} 0.12+\iota 0.5, \\ -0.7-\iota 0.5 \end{pmatrix}, \begin{pmatrix} 0.25+\iota 0.2, \\ -0.9-\iota 0.8 \end{pmatrix}, \begin{pmatrix} 0.1+\iota 0.1, \\ -0.9-\iota 0.98 \end{pmatrix} \right\}$$

Step 6: We computed the CC among $\Re_{\sigma\varrho}(\sigma = 1, 2, \ldots, m)$ and BCF-PIS and CC among $\Re_{\sigma\varrho}(\sigma = 1, 2, \ldots, m)$ and BCF-NIS as

$K_1(\Re_{1\varrho}, \Re^+) = 0.751$, $K_2(\Re_{2\varrho}, \Re^+) = 0.708$, $K_3(\Re_{3\varrho}, \Re^+) = 0.834$, $K_4(\Re_{4\varrho}, \Re^+) = 0.872$, and $K_1(\Re_{1\varrho}, \Re^-) = 0.89$, $K_2(\Re_{2\varrho}, \Re^-) = 0.908$, $K_3(\Re_{3\varrho}, \Re^-) = 0.751$, $K_4(\Re_{4\varrho}, \Re^-) = 0.681$

Step 7: We computed the closeness of every alternative by employing Eq. (6.8)

$$\mathcal{A}_{\mathbb{L}-1} = 0.457, \mathcal{A}_{\mathbb{L}-2} = 0.438, \mathcal{A}_{\mathbb{L}-3} = 0.526, \mathcal{A}_{\mathbb{L}-4} = 0.561$$

Step 8: We compute the finest optimal one by ranking all alternatives i.e.

$$\mathcal{A}_{\mathbb{L}-4} > \mathcal{A}_{\mathbb{L}-3} > \mathcal{A}_{\mathbb{L}-1} > \mathcal{A}_{\mathbb{L}-2}$$

From the ranking, we have that alternative $\mathcal{A}_{\mathbb{L}-4}$ is the best e-learning program.

Step 9: Finished.

6.5 Comparative Analysis

To describe the validity and advantages of the BCF-TOPSIS approach based on CC and EM, we compare the interpreted theory with a few prevailing notions in this part. Through the following examples, we will show that how our initiated notion is superior to the existing notions.

6.5.1 *Voice and Speech Recognition*

Voice and speech recognition technology has become thankful to Al helpers like Amazon's Alexa, Microsoft's Cortana, Google's OK Google, and Apple's siri. In voice recognition, a spoken word is the cause of information or data. It can be utilized for speech-to-text conversion and also for automatic subtitling. Corresponding to the research which is done in the voice research field, 30–50% of users depend on it for their everyday tasks. According to the survey organized by Adobe in the United States, which shows in Fig (6.2) that these are the reasons for which people utilize voice search

To demonstrate the efficacy and advantages of the established work, we will apply voice and speech recognition using both established methods and newly developed methods in the example that follows.

Example 6.3 Suppose that four known voices that is $\mathcal{V}_\sigma (\sigma = 1, 2, 3, 4)$ and one unknown voice $\mathcal{V}$ which is need to be identified. The voice analysis expert has to identify the unknown voice based on the 4 various features of the voice that is e_1, e_2, e_3 and e_4. The information of these features of each voice are demonstrated in the model of BFS as underneath.

$$\mathcal{V}_1 = \{(e_1, (0.1, -0.8)), (e_2, (0.9, -0.5)), (e_3, (0.76, -0.3)), (e_4, (0.5, -0.5))\}$$
$$\mathcal{V}_2 = \{(e_1, (0.6, -0.7)), (e_2, (0.2, -0.8)), (e_3, (0.26, -0.1)), (e_4, (0.2, -0.9))\}$$
$$\mathcal{V}_3 = \{(e_1, (1, -0.43)), (e_2, (0.45, -0.56)), (e_3, (0.6, -0.0)), (e_4, (0.3, -0.7))\}$$

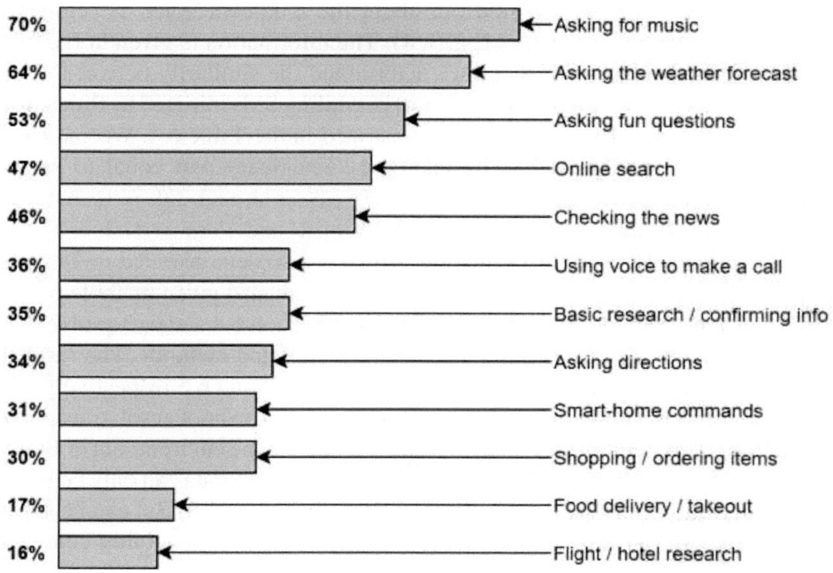

Fig. 6.2 A survey was organized by Adobe in the United States

$\mathcal{V}_4 = \{(e_1, (0.5, -0.3)), (e_2, (0.4, -0.34)), (e_3, (0.21, -0.23)), (e_4, (0.7, -0.67))\}$

and

$$\mathcal{V} = \{(e_1, (1.0, -1.0)), (e_2, (0.9, -0.2)), (e_3, (0.6, -0.12)), (e_4, (0.7, -0.9))\}$$

In the context of BCFSs, we may modify the aforementioned data by taking 0.0 $= \mathfrak{F}^{IP}_{\mathcal{V}}(e)$ and $0.0 = \mathfrak{F}^{IP}_{\mathcal{V}}(e)$, then

$$\mathcal{V}_1 = \left\{ \begin{array}{l} (e_1, (0.1 + \iota 0.0, -0.8 - \iota 0.0)), (e_2, (0.9 + \iota 0.0, -0.5 - \iota 0.0)), \\ (e_3, (0.76 + \iota 0.0, -0.3 - \iota 0.0)), (e_4, (0.5 + \iota 0.0, -0.5 - \iota 0.0)) \end{array} \right\}$$

$$\mathcal{V}_2 = \left\{ \begin{array}{l} (e_1, (0.6 + \iota 0.0, -0.7 - \iota 0.0)), (e_2, (0.2 + \iota 0.0, -0.8 - \iota 0.0)), \\ (e_3, (0.26 + \iota 0.0, -0.1 - \iota 0.0)), (e_4, (0.2 + \iota 0.0, -0.9 - \iota 0.0)) \end{array} \right\}$$

$$\mathcal{V}_3 = \left\{ \begin{array}{c} (e_1, (1 + \iota 0.0, -0.43 - \iota 0.0)), (e_2, (0.45 + \iota 0.0, -0.56 - \iota 0.0)), \\ (e_3, (0.6 + \iota 0.0, -0.0 - \iota 0.0)), (e_4, (0.3 + \iota 0.0, -0.7 - \iota 0.0)) \end{array} \right\}$$

$$\mathcal{V}_4 = \left\{ \begin{array}{c} (e_1, (0.5 + \iota 0.0, -0.3 - \iota 0.0)), (e_2, (0.4 + \iota 0.0, -0.34 - \iota 0.0)), \\ (e_3, (0.21 + \iota 0.0, -0.23 - \iota 0.0)), (e_4, (0.7 + \iota 0.0, -0.67 - \iota 0.0)) \end{array} \right\}$$

and

$$\mathcal{V} = \left\{ \begin{array}{l} (e_1, (1.0 + \iota 0.0, -1.0 - \iota 0.0)), (e_2, (0.9 + \iota 0.0, -0.2 - \iota 0.0)), \\ (e_3, (0.6 + \iota 0.0, -0.12 - \iota 0.0)), (e_4, (0.7 + \iota 0.0, -0.9 - \iota 0.0)) \end{array} \right\}$$

In Example 6.3, we have to find out about the unknown voice $\mathcal{V}$ belongs to which of the known voice $\mathcal{V}_\sigma (\sigma = 1, 2, 3, 4)$. The information is given in Example 6.3 is in an environment of BFSs so we determined the similarity between $\mathcal{V}$ and $\mathcal{V}_\sigma (\sigma = 1, 2, 3, 4)$ with the assistance of a few prevailing SMs initiated by Rajeshwari et al. [21] and Patrascu [22], which is demonstrated in the Table 6.3. We can easily convert the BFSs into the BCFSs by taking the imaginary part equal to zero in both positive and negative belonging grades i.e., $\mathfrak{F}^{IP}_{\mathcal{V}} = 0$, and $\mathfrak{F}^{IN}_{\mathcal{V}} = 0$. After the conversation, we determined the similarity between $\mathcal{V}$ and $\mathcal{V}_\sigma (\sigma = 1, 2, 3, 4)$ with help of our initiated BCF-TOPSIS method, which is also demonstrated in Table 6.3 Ranking of the initiated and some existing measures are also given in Table 6.3.

Next, consider examples (6.1) and (6.2) given in Section 6.4 and try to solve these examples with help of some existing measures and initiated methods. The resultant values of examples (6.2) and (6.3) are given in Tables 6.4 and 6.5 respectively.

Tables 6.4 and 6.5 show that the prevailing measures cannot resolve examples (6.1) and (6.2). This demonstrates that the data provided in the environment of BCFSs cannot be resolved by the prevailing measures. Such forms of data can only be solved using our innovative approach. Both types of data (BCFSs and BFSs) can be solved using our newly developed method, demonstrating its greater generality compared to previously used measures in the literature. Further, the proposed BCF-TOPSIS approach can generalize the notion of bipolar fuzzy TOPSIS, complex fuzzy TOPSIS

Table 6.3 The comparison between initiated and already established work in example 6.3

Source	SMs	Ranking
Rajeshwari et al. [21]	$\delta_1(\mathcal{V}_1, \mathcal{V}) = 0.415, \delta_1(\mathcal{V}_2, \mathcal{V}) = 0.285,$ $\delta_1(\mathcal{V}_3, \mathcal{V}) = 0.475, \delta_1(\mathcal{V}_4, \mathcal{V}) = 0.3575,$	$\mathcal{V}_3 \geq \mathcal{V}_1 \geq \mathcal{V}_4 \geq \mathcal{V}_2$
Rajeshwari et al. [21]	$\delta_2(\mathcal{V}_1, \mathcal{V}) = 0.453, \delta_2(\mathcal{V}_2, \mathcal{V}) = 0.395,$ $\delta_2(\mathcal{V}_3, \mathcal{V}) = 0.533, \delta_2(\mathcal{V}_4, \mathcal{V}) = 0.446,$	$\mathcal{V}_3 \geq \mathcal{V}_1 \geq \mathcal{V}_4 \geq \mathcal{V}_2$
Patrascu [22]	$S_{ph}(\mathcal{V}_1, \mathcal{V}) = 0.778, S_{ph}(\mathcal{V}_2, \mathcal{V}) = 0.721,$ $S_{ph}(\mathcal{V}_3, \mathcal{V}) = 0.795, S_{ph}(\mathcal{V}_4, \mathcal{V}) = 0.738,$	$\mathcal{V}_3 \geq \mathcal{V}_1 \geq \mathcal{V}_4 \geq \mathcal{V}_2$
Patrascu [22]	$S_{pe}(\mathcal{V}_1, \mathcal{V}) = 0.95, S_{pe}(\mathcal{V}_2, \mathcal{V}) = 0.89,$ $S_{pe}(\mathcal{V}_3, \mathcal{V}) = 0.933, S_{pe}(\mathcal{V}_4, \mathcal{V}) = 0.913,$	$\mathcal{V}_1 \geq \mathcal{V}_3 \geq \mathcal{V}_4 \geq \mathcal{V}_2$
Patrascu [22]	$S_{pp}(\mathcal{V}_1, \mathcal{V}) = 0.694, S_{pp}(\mathcal{V}_2, \mathcal{V}) = 0.645,$ $S_{pp}(\mathcal{V}_3, \mathcal{V}) = 0.72, S_{pp}(\mathcal{V}_4, \mathcal{V}) = 0.679,$	$\mathcal{V}_3 \geq \mathcal{V}_1 \geq \mathcal{V}_4 \geq \mathcal{V}_2$
Initiated method	$\mathcal{V}_1 = 0.498, \mathcal{V}_2 = 502,$ $\mathcal{V}_3 = 0.51, \mathcal{V}_4 = 0.536$	$\mathcal{V}_4 \geq \mathcal{V}_2 \geq \mathcal{V}_2 \geq \mathcal{V}_1$

Table 6.4 The comparison between initiated and already established work is in example 6.1

Methods	Similarity measures	Ranking
Rajeshwari et al. [21]	Unable to calculate	Unable to calculate
Rajeshwari et al. [21]	Unable to calculate	Unable to calculate
Patrascu [22]	Unable to calculate	Unable to calculate
Patrascu [22]	Unable to calculate	Unable to calculate
Patrascu [22]	Unable to calculate	Unable to calculate
Initiated method	$\mathcal{G}_1 = 0.507, \mathcal{G}_2 = 0.499,$ $\mathcal{G}_1 = 0.51, \mathcal{G}_4 = 0.512$	$\mathcal{V}_4 \geq \mathcal{V}_3 \geq \mathcal{V}_1 \geq \mathcal{V}_2$

Table 6.5 The comparison between initiated and already established work is in example 6.2

Methods	Similarity measures	Ranking
Rajeshwari et al. [21]	Unable to calculate	Unable to calculate
Rajeshwari et al. [21]	Unable to calculate	Unable to calculate
Patrascu [22]	Unable to calculate	Unable to calculate
Patrascu [22]	Unable to calculate	Unable to calculate
Patrascu [22]	Unable to calculate	Unable to calculate
Initiated method	$\mathcal{G}_1 = 0.53, \mathcal{G}_2 = 0.527,$ $\mathcal{G}_1 = 0.556, \mathcal{G}_4 = 0.533$	$\mathcal{V}_3 \geq \mathcal{V}_4 \geq \mathcal{V}_1 \geq \mathcal{V}_2$

and fuzzy TOPSIS. The proposed CC and EMs can also reduce to the model of BFS, CFS and FS.

Chapter 7
Bipolar Complex Fuzzy Hamy Aggregation Operators and Their Application in Investment Sectors

Mathematical functions or procedures known as "aggregation operators" are used to combine or summarize several values into a single outcome. These operators are essential to database management and data analysis systems. Aggregation operators enable the extraction of valuable insights and a deeper comprehension of the underlying data. A single output value is produced by aggregation operators after they work on a group of input values. They may carry out several actions, including sums, averages, determining minimum and maximum values, counting occurrences, and computing statistical measures like variance or standard deviation. Depending on the unique needs of the analysis, the input values may come from a single dataset or a collection of datasets. In many different fields, including finance, economics, business intelligence, and data mining, aggregate operators are frequently utilized. They make it possible for researchers and analysts to distill massive amounts of data into summaries, facilitating easier interpretation and decision-making. When working with complicated datasets that contain a large number of variables or observations, these operators are quite helpful.

Hamy mean and dual Hamy mean operators are the modified version of the averaging and geometric aggregation operators. They are utilized to assess the gathering of a finite number of values into a singleton set. The Hamy mean and dual Hamy mean operators are highly effective and beneficial concepts for handling difficult and problematic information. The link between aggregated arguments is effectively captured by Hamy mean and dual Hamy mean operators, which is a key advantage.

In Sect. 7.1, we investigate BCF Hamy mean, BCF weighted Hamy mean, and their related results. In Sect. 7.2., we discuss BCF dual Hamy mean and BCF weighted dual Hamy mean operators along with the associated results. In Sect. 7.3, we develop multi-attribute decision-making (MADM) by employing the initiated aggregation operators (AOs) for Sects. 7.1, and 7.2 and then investigate a case study to demonstrate the applicability of the proposed theory. In Sect. 7.4, we interpret the comparative analysis of the developed work with certain prevailing work.

U. ur Rehman and T. Mahmood, *Bipolar Complex Fuzzy Sets*, Studies in Fuzziness and Soft Computing 439, https://doi.org/10.1007/978-3-032-07388-4_7

7.1 Bipolar Complex Fuzzy Hamy Mean AOs

We began with the basic notion of the Hamy operator which is initiated by Hara [27].

The Hamy mean (HM) operator is demonstrated underneath

$$HM^{\xi}\left(\mathfrak{O}_1, \mathfrak{O}_2, \ldots, \mathfrak{O}_y\right) = \frac{\sum_{1 \le \sigma_1 < \ldots < \sigma_\xi \le y}\left(\prod_{\sigma=1}^{\xi} \mathfrak{O}_{\sigma_\varrho}\right)^{\frac{1}{\xi}}}{\mathbb{C}_y^{\xi}}$$

where, ξ is signified as a parameter and $\xi = 1, 2, \ldots, y, \sigma_1, \sigma_2, \ldots, \sigma_\xi$ are ξ integer values taken from the gathering of y integer values i.e., $\{1, 2, \ldots, y\}$ and $\mathbb{C}_y^{\xi} = \frac{y!}{\xi!(y-\xi)!}$ Identify the binomial coefficient. The HM operator holds the underneath 3 axioms

1. $HM^{\xi}\left(\mathfrak{O}_1, \mathfrak{O}_2, \ldots, \mathfrak{O}_y\right) = \mathfrak{O}$, when $\mathfrak{O}_\sigma = \mathfrak{O}, \forall \sigma$
2. $HM^{\xi}\left(\mathfrak{O}_1, \mathfrak{O}_2, \ldots, \mathfrak{O}_y\right) \le HM^{\xi}\left(\mathfrak{O}_1^*, \mathfrak{O}_2^*, \ldots, \mathfrak{O}_y^*\right)$, when $\mathfrak{O}_\sigma \le \mathfrak{O}_\sigma^*, \forall \sigma$
3. $\min\{\mathfrak{O}_\sigma\} \le HM^{\xi}\left(\mathfrak{O}_1, \mathfrak{O}_2, \ldots, \mathfrak{O}_y\right) \le \max\{\mathfrak{O}_\sigma\}$.

Underneath are the two special cases of an HM operator.

1. By taking $\xi = 1$, the HM operator reduces to the arithmetic mean operator.
2. By taking $\xi = y$, the HM operator reduces to the geometric mean operator.

The concept of the BCF Hamy mean (BCF-HM) operator is investigated in the underneath definition.

Definition 7.1 The BCF-HM operator relying on the gathering of BCFNs $\mathfrak{R}_\sigma = \left(\mathfrak{F}_{\mathfrak{R}_\sigma}^{P}, \mathfrak{F}_{\mathfrak{R}_\sigma}^{N}\right) = \left(\mathfrak{F}_{\mathfrak{R}_\sigma}^{RP} + \iota \mathfrak{F}_{\mathfrak{R}_\sigma}^{IP}, \mathfrak{F}_{\mathfrak{R}_\sigma}^{RN} + \iota \mathfrak{F}_{\mathfrak{R}_\sigma}^{IN}\right)\sigma = 1, 2, \ldots, y$, is spotted as.

$$BCF - HM^{\xi}\left(\mathfrak{R}_1, \mathfrak{R}_2, \ldots, \mathfrak{R}_y\right) = \frac{\underset{1 \le \sigma_1 < \ldots < \sigma_\xi \le y}{\oplus}\left(\underset{\varrho=1}{\overset{\xi}{\otimes}} \mathfrak{R}_{\sigma_\varrho}\right)^{\frac{1}{\xi}}}{\mathbb{C}_y^{\xi}} \tag{7.1}$$

In Eq. (7.1), ξ is utilized as a parameter, $\xi = 1, 2, \ldots, y, \sigma_1, \sigma_2, \ldots, \sigma_\xi$ are ξ integer values occupied from the gathering of y integer values that is $\{1, 2, \ldots, y\}$ and $\mathbb{C}_y^{\xi} = \frac{y!}{\xi!(y-\xi)!}$ Identify the binomial coefficient.

Underneath Theorem shows that the BCF-HM operator will again give us a BCFN.

Theorem 7.1 Let a gathering of BCFNs $\mathfrak{R}_\sigma = \left(\mathfrak{F}_{\mathfrak{R}_\sigma}^{P}, \mathfrak{F}_{\mathfrak{R}_\sigma}^{N}\right) = \left(\mathfrak{F}_{\mathfrak{R}_\sigma}^{RP} + \iota \mathfrak{F}_{\mathfrak{R}_\sigma}^{IP}, \mathfrak{F}_{\mathfrak{R}_\sigma}^{RN} + \iota \mathfrak{F}_{\mathfrak{R}_\sigma}^{IN}\right)\sigma = 1, 2, \ldots, y$. The aggregated outcome of Eq. (7.1) is a BCFN and.

$$BCF-HM^{\xi}(\mathfrak{R}_1,\mathfrak{R}_2,\ldots,\mathfrak{R}_y)=\left(\begin{array}{c} 1-\left(\prod_{1\le\sigma_1<\ldots<\sigma_\xi\le y}\left(1-\left(\prod_{\varrho=1}^{\xi}\left(\mathfrak{F}^{RP}_{\mathfrak{R}_{\sigma_\varrho}}\right)\right)^{\frac{1}{\xi}}\right)\right)^{\frac{1}{\mathbb{C}_y^{\xi}}} \\ +\iota\left(1-\left(\prod_{1\le\sigma_1<\ldots<\sigma_\xi\le y}\left(1-\left(\prod_{\varrho=1}^{\xi}\left(\mathfrak{F}^{IP}_{\mathfrak{R}_{\sigma_\varrho}}\right)\right)^{\frac{1}{\xi}}\right)\right)^{\frac{1}{\mathbb{C}_y^{\xi}}}\right) \\ -\left|(-1)^{y+1}\prod_{1\le\sigma_1<\ldots<\sigma_\xi\le y}\left(-1+\left(\prod_{\varrho=1}^{\xi}\left(1+\mathfrak{F}^{RN}_{\mathfrak{R}_{\sigma_\varrho}}\right)\right)^{\frac{1}{\xi}}\right)\right|^{\frac{1}{\mathbb{C}_y^{\xi}}} \\ +\iota\left(-\left|(-1)^{y+1}\prod_{1\le\sigma_1<\ldots<\sigma_\xi\le y}\left(-1+\left(\prod_{\varrho=1}^{\xi}\left(1+\mathfrak{F}^{IN}_{\mathfrak{R}_{\sigma_\varrho}}\right)\right)^{\frac{1}{\xi}}\right)\right|^{\frac{1}{\mathbb{C}_y^{\xi}}}\right) \end{array}\right)$$

Proof With the assistance of definition (2.9), we have.

$$\underset{\varrho=1}{\overset{\xi}{\otimes}}\mathfrak{R}_{\sigma_\varrho}=\prod_{\varrho=1}^{\xi}\left(\mathfrak{F}^{RP}_{\mathfrak{R}_{\sigma_\varrho}}\right)+\iota\prod_{\varrho=1}^{\xi}\left(\mathfrak{F}^{IP}_{\mathfrak{R}_{\sigma_\varrho}}\right),-1+\prod_{\varrho=1}^{\xi}\left(1+\mathfrak{F}^{RN}_{\mathfrak{R}_{\sigma_\varrho}}\right)+\iota\left(-1+\prod_{\varrho=1}^{\xi}\left(1+\mathfrak{F}^{RN}_{\mathfrak{R}_{\sigma_\varrho}}\right)\right)$$

Consequently,

$$\left(\underset{\varrho=1}{\overset{\xi}{\otimes}}\mathfrak{R}_{\sigma_\varrho}\right)^{\frac{1}{\xi}}=\left(\begin{array}{c}\left(\prod_{\varrho=1}^{\xi}\left(\mathfrak{F}^{RP}_{\mathfrak{R}_{\sigma_\varrho}}\right)\right)^{\frac{1}{\xi}}+\iota\left(\prod_{\varrho=1}^{\xi}\left(\mathfrak{F}^{IP}_{\mathfrak{R}_{\sigma_\varrho}}\right)\right)^{\frac{1}{\xi}}, \\ -1+\left(\prod_{\varrho=1}^{\xi}\left(1+\mathfrak{F}^{RN}_{\mathfrak{R}_{\sigma_\varrho}}\right)\right)^{\frac{1}{\xi}}+\iota\left(-1+\left(\prod_{\varrho=1}^{\xi}\left(1+\mathfrak{F}^{IN}_{\mathfrak{R}_{\sigma_\varrho}}\right)\right)^{\frac{1}{\xi}}\right)\end{array}\right)$$

Further,

$$\underset{1\le\sigma_1<\ldots<\sigma_\xi\le y}{\oplus}\left(\underset{\varrho=1}{\overset{\xi}{\otimes}}\mathfrak{R}_{\sigma_\varrho}\right)^{\frac{1}{\xi}}=\left(\begin{array}{c} 1-\prod_{1\le\sigma_1<\ldots<\sigma_\xi\le y}\left(1-\left(\prod_{\varrho=1}^{\xi}\left(\mathfrak{F}^{RP}_{\mathfrak{R}_{\sigma_\varrho}}\right)\right)^{\frac{1}{\xi}}\right) \\ +\iota\left(1-\prod_{1\le\sigma_1<\ldots<\sigma_\xi\le y}\left(1-\left(\prod_{\varrho=1}^{\xi}\left(\mathfrak{F}^{IP}_{\mathfrak{R}_{\sigma_\varrho}}\right)\right)^{\frac{1}{\xi}}\right)\right) \\ (-1)^{y+1}\prod_{1\le\sigma_1<\ldots<\sigma_\xi\le y}\left(-1+\left(\prod_{\varrho=1}^{\xi}\left(1+\mathfrak{F}^{RN}_{\mathfrak{R}_{\sigma_\varrho}}\right)\right)^{\frac{1}{\xi}}\right) \\ +\iota\left((-1)^{y+1}\prod_{1\le\sigma_1<\ldots<\sigma_\xi\le y}\left(-1+\left(\prod_{\varrho=1}^{\xi}\left(1+\mathfrak{F}^{IN}_{\mathfrak{R}_{\sigma_\varrho}}\right)\right)^{\frac{1}{\xi}}\right)\right)\end{array}\right)$$

Next,

$$\frac{\underset{1\le\sigma_1<\ldots<\sigma_\xi\le y}{\oplus}\left(\underset{\varrho=1}{\overset{\xi}{\otimes}}\mathfrak{R}_{\sigma_\varrho}\right)^{\frac{1}{\xi}}}{\mathbb{C}_y^{\xi}}$$

$$= \begin{pmatrix} 1 - \left(\prod_{1 \le \sigma_1 < \ldots < \sigma_\xi \le y} \left(1 - \left(\prod_{\varrho=1}^{\xi} \left(\mathfrak{F}^{RP}_{\mathfrak{R}_{\sigma_\varrho}} \right) \right)^{\frac{1}{\xi}} \right) \right)^{\frac{1}{C_y^\xi}} \\ +\iota \left(1 - \left(\prod_{1 \le \sigma_1 < \ldots < \sigma_\xi \le y} \left(1 - \left(\prod_{\varrho=1}^{\xi} \left(\mathfrak{F}^{IP}_{\mathfrak{R}_{\sigma_\varrho}} \right) \right)^{\frac{1}{\xi}} \right) \right)^{\frac{1}{C_y^\xi}} \right), \\ -\left| (-1)^{y+1} \prod_{1 \le \sigma_1 < \ldots < \sigma_\xi \le y} \left(-1 + \left(\prod_{\varrho=1}^{\xi} \left(1 + \mathfrak{F}^{RN}_{\mathfrak{R}_{\sigma_\varrho}} \right) \right)^{\frac{1}{\xi}} \right) \right|^{\frac{1}{C_y^\xi}} \\ +\iota \left(-\left| (-1)^{y+1} \prod_{1 \le \sigma_1 < \ldots < \sigma_\xi \le y} \left(-1 + \left(\prod_{\varrho=1}^{\xi} \left(1 + \mathfrak{F}^{IN}_{\mathfrak{R}_{\sigma_\varrho}} \right) \right)^{\frac{1}{\xi}} \right) \right|^{\frac{1}{C_y^\xi}} \right) \end{pmatrix}$$

$$= BCF - HM^{\xi}(\mathfrak{R}_1, \mathfrak{R}_2, \ldots, \mathfrak{R}_y)$$

Underneath we discuss the proposition of the BCF-HM operator.

Proposition 7.1 (Idempotency): Let a gathering of BCFNs $\mathfrak{R}_\sigma = \left(\mathfrak{F}^P_{\mathfrak{R}_\sigma}, \mathfrak{F}^N_{\mathfrak{R}_\sigma}\right) = \left(\mathfrak{F}^{RP}_{\mathfrak{R}_\sigma} + \iota\mathfrak{F}^{IP}_{\mathfrak{R}_\sigma}, \mathfrak{F}^{RN}_{\mathfrak{R}_\sigma} + \iota\mathfrak{F}^{IN}_{\mathfrak{R}_\sigma}\right)\sigma = 1, 2, \ldots, y$. If $\forall\sigma\, \mathfrak{R}_\sigma = \mathfrak{R}$, then

$$BCF - HM^{\xi}(\mathfrak{R}_1, \mathfrak{R}_2, \ldots, \mathfrak{R}_y) = \mathfrak{R}$$

Proof From Theorem 7.1, we have.

$$BCF - HM^{\xi}(\mathfrak{R}_1, \mathfrak{R}_2, \ldots, \mathfrak{R}_y) = \begin{pmatrix} 1 - \left(\prod_{1 \le \sigma_1 < \ldots < \sigma_\xi \le y} \left(1 - \left(\prod_{\varrho=1}^{\xi} \left(\mathfrak{F}^{RP}_{\mathfrak{R}_{\sigma_\varrho}} \right) \right)^{\frac{1}{\xi}} \right) \right)^{\frac{1}{C_y^\xi}} \\ +\iota \left(1 - \left(\prod_{1 \le \sigma_1 < \ldots < \sigma_\xi \le y} \left(1 - \left(\prod_{\varrho=1}^{\xi} \left(\mathfrak{F}^{IP}_{\mathfrak{R}_{\sigma_\varrho}} \right) \right)^{\frac{1}{\xi}} \right) \right)^{\frac{1}{C_y^\xi}} \right) \\ -\left| (-1)^{y+1} \prod_{1 \le \sigma_1 < \ldots < \sigma_\xi \le y} \left(-1 + \left(\prod_{\varrho=1}^{\xi} \left(1 + \mathfrak{F}^{RN}_{\mathfrak{R}_{\sigma_\varrho}} \right) \right)^{\frac{1}{\xi}} \right) \right|^{\frac{1}{C_y^\xi}} \\ +\iota \left(-\left| (-1)^{y+1} \prod_{1 \le \sigma_1 < \ldots < \sigma_\xi \le y} \left(-1 + \left(\prod_{\varrho=1}^{\xi} \left(1 + \mathfrak{F}^{IN}_{\mathfrak{R}_{\sigma_\varrho}} \right) \right)^{\frac{1}{\xi}} \right) \right|^{\frac{1}{C_y^\xi}} \right) \end{pmatrix}$$

Now as $\forall\sigma\, \mathfrak{R}_\sigma = \mathfrak{R} = \left(\mathfrak{F}^P_{\mathfrak{R}}, \mathfrak{F}^N_{\mathfrak{R}}\right) = \left(\mathfrak{F}^{RP}_{\mathfrak{R}} + \iota\mathfrak{F}^{IP}_{\mathfrak{R}}, \mathfrak{F}^{RN}_{\mathfrak{R}} + \iota\mathfrak{F}^{IN}_{\mathfrak{R}}\right)$, then

$$BCF - HM^{y}(\mathfrak{R}_1, \mathfrak{R}_2, \ldots, \mathfrak{R}_y)$$

$$=\left(\begin{array}{c} 1-\left(\prod_{1\le\sigma_1<\ldots<\sigma_\xi\le y}\left(1-\left(\prod_{\varrho=1}^{\xi}\left(\mathfrak{F}^{RP}_{\mathfrak{R}_{\sigma_\varrho}}\right)\right)^{\frac{1}{\xi}}\right)\right)^{\frac{1}{\mathbb{C}_y^{\xi}}} \\ +\iota\left(1-\left(\prod_{1\le\sigma_1<\ldots<\sigma_\xi\le y}\left(1-\left(\prod_{\varrho=1}^{\xi}\left(\mathfrak{F}^{IP}_{\mathfrak{R}_{\sigma_\varrho}}\right)\right)^{\frac{1}{\xi}}\right)\right)^{\frac{1}{\mathbb{C}_y^{\xi}}}\right) \\ -\left|(-1)^{y+1}\prod_{1\le\sigma_1<\ldots<\sigma_\xi\le y}\left(-1+\left(\prod_{\varrho=1}^{\xi}\left(1+\mathfrak{F}^{RN}_{\mathfrak{R}_{\sigma_\varrho}}\right)\right)^{\frac{1}{\xi}}\right)\right|^{\frac{1}{\mathbb{C}_y^{\xi}}} \\ +\iota\left(-\left|(-1)^{y+1}\prod_{1\le\sigma_1<\ldots<\sigma_\xi\le y}\left(-1+\left(\prod_{\varrho=1}^{\xi}\left(1+\mathfrak{F}^{IN}_{\mathfrak{R}_{\sigma_\varrho}}\right)\right)^{\frac{1}{\xi}}\right)\right|^{\frac{1}{\mathbb{C}_y^{\xi}}}\right) \end{array}\right)$$

$$=\left(\begin{array}{c} 1-\left(\left(1-\left(\left(\mathfrak{F}^{RP}_{\mathfrak{R}}\right)^{\xi}\right)^{\frac{1}{\xi}}\right)^{\mathbb{C}_y^{\xi}}\right)^{\frac{1}{\mathbb{C}_y^{\xi}}}+\iota\left(1-\left(\left(1-\left(\left(\mathfrak{F}^{IP}_{\mathfrak{R}}\right)^{\xi}\right)^{\frac{1}{\xi}}\right)^{\mathbb{C}_y^{\xi}}\right)^{\frac{1}{\mathbb{C}_y^{\xi}}}\right), \\ -\left|(-1)^{y+1}\left(-1+\left(\left(1+\mathfrak{F}^{RN}_{\mathfrak{R}}\right)^{\xi}\right)^{\frac{1}{\xi}}\right)^{\mathbb{C}_y^{\xi}}\right|^{\frac{1}{\mathbb{C}_y^{\xi}}}+\iota\left(-\left|(-1)^{y+1}\left(-1+\left(\left(1+\mathfrak{F}^{IN}_{\mathfrak{R}}\right)^{\xi}\right)^{\frac{1}{\xi}}\right)^{\mathbb{C}_y^{\xi}}\right|^{\frac{1}{\mathbb{C}_y^{\xi}}}\right) \end{array}\right)$$

$$=\left(1-\left(1-\mathfrak{F}^{RP}_{\mathfrak{R}}\right)+\iota\left(1-\left(1-\mathfrak{F}^{IP}_{\mathfrak{R}}\right)\right),-\left|-1+\left(1+\mathfrak{F}^{RN}_{\mathfrak{R}}\right)\right|+\iota\left(-\left|-1+\left(1+\mathfrak{F}^{IN}_{\mathfrak{R}}\right)\right|\right)\right)$$

As $\mathfrak{F}^{RN}_{\mathfrak{R}}, \mathfrak{F}^{IN}_{\mathfrak{R}} \in [-1, 0]$, so $\left|\mathfrak{F}^{RN}_{\mathfrak{R}}\right| = -\left(-\mathfrak{F}^{RN}_{\mathfrak{R}}\right)$ and $\left|\mathfrak{F}^{IN}_{\mathfrak{R}}\right| = -\left(-\mathfrak{F}^{IN}_{\mathfrak{R}}\right)$, then we have.

$\left(\mathfrak{F}^{RP}_{\mathfrak{R}} + \iota\mathfrak{F}^{IP}_{\mathfrak{R}}, \mathfrak{F}^{RN}_{\mathfrak{R}} + \iota\mathfrak{F}^{IN}_{\mathfrak{R}}\right) = \mathfrak{R}.$

Proposition 7.2 (Monotonicity): Let two gatherings of BCFNs $\mathfrak{R}_\sigma = \left(\mathfrak{F}^{P}_{\mathfrak{R}_\sigma}, \mathfrak{F}^{N}_{\mathfrak{R}_\sigma}\right) = \left(\mathfrak{F}^{RP}_{\mathfrak{R}_\sigma} + \iota\mathfrak{F}^{IP}_{\mathfrak{R}_\sigma}, \mathfrak{F}^{RN}_{\mathfrak{R}_\sigma} + \iota\mathfrak{F}^{IN}_{\mathfrak{R}_\sigma}\right)$ and $\mathfrak{R}'_\sigma = \left(\mathfrak{F}^{P}_{\mathfrak{R}'_\sigma}, \mathfrak{F}^{N}_{\mathfrak{R}'_\sigma}\right) = \left(\mathfrak{F}^{RP}_{\mathfrak{R}'_\sigma} + \iota\mathfrak{F}^{IP}_{\mathfrak{R}'_\sigma}, \mathfrak{F}^{RN}_{\mathfrak{R}'_\sigma} + \iota\mathfrak{F}^{IN}_{\mathfrak{R}'_\sigma}\right)\sigma = 1, 2, \ldots, y$. If $\mathfrak{F}^{RP}_{\mathfrak{R}_\sigma} \le \mathfrak{F}^{RP}_{\mathfrak{R}'_\sigma}$, $\mathfrak{F}^{IP}_{\mathfrak{R}_\sigma} \le \mathfrak{F}^{IP}_{\mathfrak{R}'_\sigma}$, $\mathfrak{F}^{RN}_{\mathfrak{R}_\sigma} \le \mathfrak{F}^{RN}_{\mathfrak{R}'_\sigma}$, and $\mathfrak{F}^{IN}_{\mathfrak{R}_\sigma} \le \mathfrak{F}^{IN}_{\mathfrak{R}'_\sigma}$, then

$$BCF-HM^{\xi}\left(\mathfrak{R}_1, \mathfrak{R}_2, \ldots, \mathfrak{R}_y\right) \le BCF-HM^{\xi}\left(\mathfrak{R}'_1, \mathfrak{R}'_2, \ldots, \mathfrak{R}'_y\right)$$

Proof As $\xi \ge 1$ and $\mathfrak{F}^{RP}_{\mathfrak{R}_\sigma} \le \mathfrak{F}^{RP}_{\mathfrak{R}'_\sigma}$, then.

$$\left(\prod_{\varrho=1}^{\xi}\left(\mathfrak{F}^{RP}_{\mathfrak{R}_{\sigma_\varrho}}\right)\right)^{\frac{1}{\xi}} \le \left(\prod_{\varrho=1}^{\xi}\left(\mathfrak{F}^{RP}_{\mathfrak{R}'_{\sigma_\varrho}}\right)\right)^{\frac{1}{\xi}} \Rightarrow 1-\left(\prod_{\varrho=1}^{\xi}\left(\mathfrak{F}^{RP}_{\mathfrak{R}_{\sigma_\varrho}}\right)\right)^{\frac{1}{\xi}} \ge 1-\left(\prod_{\varrho=1}^{\xi}\left(\mathfrak{F}^{RP}_{\mathfrak{R}'_{\sigma_\varrho}}\right)\right)^{\frac{1}{\xi}}$$

$$\Rightarrow \left(\prod_{1\le\sigma_1<\ldots<\sigma_\xi\le y}\left(1-\left(\prod_{\varrho=1}^{\xi}\left(\mathfrak{F}^{RP}_{\mathfrak{R}_{\sigma_\varrho}}\right)\right)^{\frac{1}{\xi}}\right)\right)^{\frac{1}{\mathbb{C}_y^{\xi}}} \ge \left(\prod_{1\le\sigma_1<\ldots<\sigma_\xi\le y}\left(1-\left(\prod_{\varrho=1}^{\xi}\left(\mathfrak{F}^{RP}_{\mathfrak{R}'_{\sigma_\varrho}}\right)\right)^{\frac{1}{\xi}}\right)\right)^{\frac{1}{\mathbb{C}_y^{\xi}}}$$

$$\Rightarrow 1-\left(\prod_{1\le\sigma_1<\ldots<\sigma_\xi\le y}\left(1-\left(\prod_{\varrho=1}^{\xi}\left(\mathfrak{F}^{RP}_{\mathfrak{R}_{\sigma_\varrho}}\right)\right)^{\frac{1}{\xi}}\right)\right)^{\frac{1}{\mathbb{C}_y^{\xi}}} \le 1-\left(\prod_{1\le\sigma_1<\ldots<\sigma_\xi\le y}\left(1-\left(\prod_{\varrho=1}^{\xi}\left(\mathfrak{F}^{RP}_{\mathfrak{R}'_{\sigma_\varrho}}\right)\right)^{\frac{1}{\xi}}\right)\right)^{\frac{1}{\mathbb{C}_y^{\xi}}}$$

Likewise as $\mathfrak{F}^{IP}_{\mathfrak{R}_\sigma} \leq \mathfrak{F}^{IP}_{\mathfrak{R}'_\sigma}$, then

$$1-\left(\prod_{1\leq\sigma_1<\ldots<\sigma_\xi\leq y}\left(1-\left(\prod_{\varrho=1}^{\xi}\left(\mathfrak{F}^{IP}_{\mathfrak{R}_{\sigma_\varrho}}\right)\right)^{\frac{1}{\xi}}\right)\right)^{\frac{1}{\mathbb{C}^{\xi}_{y}}} \leq 1-\left(\prod_{1\leq\sigma_1<\ldots<\sigma_\xi\leq y}\left(1-\left(\prod_{\varrho=1}^{\xi}\left(\mathfrak{F}^{IP}_{\mathfrak{R}'_{\sigma_\varrho}}\right)\right)^{\frac{1}{\xi}}\right)\right)^{\frac{1}{\mathbb{C}^{\xi}_{y}}}$$

Next, as $\mathfrak{F}^{RN}_{\mathfrak{R}_\sigma} \leq \mathfrak{F}^{RN}_{\mathfrak{R}'_\sigma}$, then

$$1+\mathfrak{F}^{RN}_{\mathfrak{R}_{\sigma_\varrho}} \leq 1+\mathfrak{F}^{RN}_{\mathfrak{R}'_{\sigma_\varrho}} \Rightarrow \left(\prod_{\varrho=1}^{\xi}\left(1+\mathfrak{F}^{RN}_{\mathfrak{R}_{\sigma_\varrho}}\right)\right)^{\frac{1}{\xi}} \leq \left(\prod_{\varrho=1}^{\xi}\left(1+\mathfrak{F}^{RN}_{\mathfrak{R}'_{\sigma_\varrho}}\right)\right)^{\frac{1}{\xi}}$$

$$\Rightarrow -1+\left(\prod_{\varrho=1}^{\xi}\left(1+\mathfrak{F}^{RN}_{\mathfrak{R}_{\sigma_\varrho}}\right)\right)^{\frac{1}{\xi}} \geq -1+\left(\prod_{\varrho=1}^{\xi}\left(1+\mathfrak{F}^{RN}_{\mathfrak{R}'_{\sigma_\varrho}}\right)\right)^{\frac{1}{\xi}}$$

$$\Rightarrow \prod_{1\leq\sigma_1<\ldots<\sigma_\xi\leq y}\left(-1+\left(\prod_{\varrho=1}^{\xi}\left(1+\mathfrak{F}^{RN}_{\mathfrak{R}_{\sigma_\varrho}}\right)\right)^{\frac{1}{\xi}}\right) \geq \prod_{1\leq\sigma_1<\ldots<\sigma_\xi\leq y}\left(-1+\left(\prod_{\varrho=1}^{\xi}\left(1+\mathfrak{F}^{RN}_{\mathfrak{R}'_{\sigma_\varrho}}\right)\right)^{\frac{1}{\xi}}\right)$$

Multiply the above Eq. with $(-1)^{y+1}$, then for y is odd we get

$$\Rightarrow (-1)^{y+1}\prod_{1\leq\sigma_1<\ldots<\sigma_\xi\leq y}\left(-1+\left(\prod_{\varrho=1}^{\xi}\left(1+\mathfrak{F}^{RN}_{\mathfrak{R}_{\sigma_\varrho}}\right)\right)^{\frac{1}{\xi}}\right) \leq (-1)^{y+1}\prod_{1\leq\sigma_1<\ldots<\sigma_\xi\leq y}\left(-1+\left(\prod_{\varrho=1}^{\xi}\left(1+\mathfrak{F}^{RN}_{\mathfrak{R}'_{\sigma_\varrho}}\right)\right)^{\frac{1}{\xi}}\right)$$

And for y even if we get

$$\Rightarrow (-1)^{y+1}\prod_{1\leq\sigma_1<\ldots<\sigma_\xi\leq y}\left(-1+\left(\prod_{\varrho=1}^{\xi}\left(1+\mathfrak{F}^{RN}_{\mathfrak{R}_{\sigma_\varrho}}\right)\right)^{\frac{1}{\xi}}\right) \geq (-1)^{y+1}\prod_{1\leq\sigma_1<\ldots<\sigma_\xi\leq y}\left(-1+\left(\prod_{\varrho=1}^{\xi}\left(1+\mathfrak{F}^{RN}_{\mathfrak{R}'_{\sigma_\varrho}}\right)\right)^{\frac{1}{\xi}}\right)$$

After taking the modulus both equations for y is odd and y is even coinciding as

$$\Rightarrow \left|(-1)^{y+1}\prod_{1\leq\sigma_1<\ldots<\sigma_\xi\leq y}\left(-1+\left(\prod_{\varrho=1}^{\xi}\left(1+\mathfrak{F}^{RN}_{\mathfrak{R}_{\sigma_\varrho}}\right)\right)^{\frac{1}{\xi}}\right)\right|^{\frac{1}{\mathbb{C}^{\xi}_{y}}}$$

$$\geq \left|(-1)^{y+1}\prod_{1\leq\sigma_1<\ldots<\sigma_\xi\leq y}\left(-1+\left(\prod_{\varrho=1}^{\xi}\left(1+\mathfrak{F}^{RN}_{\mathfrak{R}'_{\sigma_\varrho}}\right)\right)^{\frac{1}{\xi}}\right)\right|^{\frac{1}{\mathbb{C}^{\xi}_{y}}}$$

$$\Rightarrow -\left|(-1)^{y+1}\prod_{1\leq\sigma_1<\ldots<\sigma_\xi\leq y}\left(-1+\left(\prod_{\varrho=1}^{\xi}\left(1+\mathfrak{F}^{RN}_{\mathfrak{R}_{\sigma_\varrho}}\right)\right)^{\frac{1}{\xi}}\right)\right|^{\frac{1}{\mathbb{C}^{\xi}_{y}}}$$

$$\leq -\left|(-1)^{y+1}\prod_{1\leq\sigma_1<\ldots<\sigma_\xi\leq y}\left(-1+\left(\prod_{\varrho=1}^{\xi}\left(1+\mathfrak{F}^{RN}_{\mathfrak{R}'_{\sigma_\varrho}}\right)\right)^{\frac{1}{\xi}}\right)\right|^{\frac{1}{\mathbb{C}^{\xi}_{y}}}$$

Likewise, as $\mathfrak{F}^{IN}_{\mathfrak{R}_\sigma} \leq \mathfrak{F}^{IN}_{\mathfrak{R}'_\sigma}$, *then*

$$-\left|(-1)^{y+1}\prod_{1\le\sigma_1<\ldots<\sigma_\xi\le y}\left(-1+\left(\prod_{\varrho=1}^{\xi}\left(1+\mathfrak{F}^{IN}_{\mathfrak{R}_{\sigma_\varrho}}\right)\right)^{\frac{1}{\xi}}\right)\right|^{\frac{1}{\mathbb{C}_y^\xi}}$$

$$\le-\left|(-1)^{y+1}\prod_{1\le\sigma_1<\ldots<\sigma_\xi\le y}\left(-1+\left(\prod_{\varrho=1}^{\xi}\left(1+\mathfrak{F}^{IN}_{\mathfrak{R}'_{\sigma_\varrho}}\right)\right)^{\frac{1}{\xi}}\right)\right|^{\frac{1}{\mathbb{C}_y^\xi}}$$

This completed the proof.

Proposition 7.3 (boundedness): Let a gathering of BCFNs $\mathfrak{R}_\sigma = \left(\mathfrak{F}^P_{\mathfrak{R}_\sigma}, \mathfrak{F}^N_{\mathfrak{R}_\sigma}\right) = \left(\mathfrak{F}^{RP}_{\mathfrak{R}_\sigma} + \iota\mathfrak{F}^{IP}_{\mathfrak{R}_\sigma}, \mathfrak{F}^{RN}_{\mathfrak{R}_\sigma} + \iota\mathfrak{F}^{IN}_{\mathfrak{R}_\sigma}\right)\sigma = 1, 2, \ldots, y$. If $\mathfrak{R}^- = \begin{pmatrix}\min_\sigma\{\mathfrak{F}^{RP}_{\mathfrak{R}_\sigma}\} + \iota\min_\sigma\{\mathfrak{F}^{IP}_{\mathfrak{R}_\sigma}\}, \\ \min_\sigma\{\mathfrak{F}^{RN}_{\mathfrak{R}_\sigma}\} + \iota\min_\sigma\{\mathfrak{F}^{IN}_{\mathfrak{R}_\sigma}\}\end{pmatrix}$, and $\mathfrak{R}^+ = \begin{pmatrix}\max_\sigma\{\mathfrak{F}^{RP}_{\mathfrak{R}_\sigma}\} + \iota\max_\sigma\{\mathfrak{F}^{IP}_{\mathfrak{R}_\sigma}\}, \\ \max_\sigma\{\mathfrak{F}^{RN}_{\mathfrak{R}_\sigma}\} + \iota\max_\sigma\{\mathfrak{F}^{IN}_{\mathfrak{R}_\sigma}\}\end{pmatrix}$, then

$$\mathfrak{R}^- \le BCF - HM^\xi\left(\mathfrak{R}_1, \mathfrak{R}_2, \ldots, \mathfrak{R}_y\right) \le \mathfrak{R}^+$$

Proof by the assistance of Proposition 7.1 we get.

$BCF - HM^\xi\left(\mathfrak{R}^-, \mathfrak{R}^-, \ldots, \mathfrak{R}^-\right) = \mathfrak{R}^-$ and $BCF - HM^\xi\left(\mathfrak{R}^+, \mathfrak{R}^+, \ldots, \mathfrak{R}^+\right) = \mathfrak{R}^+$.

And by the assistance of axiom 2, we get

$$\mathfrak{R}^- \le BCF - HM^\xi\left(\mathfrak{R}_1, \mathfrak{R}_2, \ldots, \mathfrak{R}_y\right) \le \mathfrak{R}^+$$

By putting $\xi = 1$ and $\xi = y$, we achieve the underneath special cases.

1. Put $\xi = 1$, the BCF-HM operator would transform to BCF arithmetic average i.e.

$$BCF - HM^1\left(\mathfrak{R}_1, \mathfrak{R}_2, \ldots, \mathfrak{R}_y\right) = \begin{pmatrix} 1-\left(\prod_{1\le\sigma_1<\ldots<\sigma_\xi\le y}\left(1-\mathfrak{F}^{RP}_{\mathfrak{R}_\sigma}\right)\right)^{\frac{1}{y}} + \iota\left(1-\left(\prod_{1\le\sigma_1<\ldots<\sigma_\xi\le y}\left(1-\mathfrak{F}^{IP}_{\mathfrak{R}_\sigma}\right)\right)^{\frac{1}{y}}\right), \\ -\left|(-1)^{y+1}\prod_{1\le\sigma_1<\ldots<\sigma_\xi\le y}\left(\mathfrak{F}^{RN}_{\mathfrak{R}_\sigma}\right)\right|^{\frac{1}{y}} + \iota\left(-\left|(-1)^{y+1}\prod_{1\le\sigma_1<\ldots<\sigma_\xi\le y}\left(\mathfrak{F}^{IN}_{\mathfrak{R}_\sigma}\right)\right|^{\frac{1}{y}}\right)\end{pmatrix}$$

$$= \frac{1}{y}\bigoplus_{\sigma=1}^{y}\mathfrak{R}_\sigma$$

2. Put $\xi = y$, the BCF-HM operator would transform to BCF geometric operator. i.e.

$$BCF-HM^{y}(\mathfrak{R}_1,\mathfrak{R}_2,\ldots,\mathfrak{R}_y)=\left(\begin{array}{c}1-\left(\prod_{1\le\sigma_1<\ldots<\sigma_y\le y}\left(1-\left(\prod_{\varrho=1}^{y}\left(\mathfrak{F}^{RP}_{\mathfrak{R}_{\sigma_\varrho}}\right)\right)^{\frac{1}{y}}\right)\right)^{\frac{1}{\mathbb{C}_y^y}}\\ +\iota\left(1-\left(\prod_{1\le\sigma_1<\ldots<\sigma_y\le y}\left(1-\left(\prod_{\varrho=1}^{y}\left(\mathfrak{F}^{IP}_{\mathfrak{R}_{\sigma_\varrho}}\right)\right)^{\frac{1}{y}}\right)\right)^{\frac{1}{\mathbb{C}_y^y}}\right),\\ -\left|(-1)^{y+1}\prod_{1\le\sigma_1<\ldots<\sigma_y\le y}\left(-1+\left(\prod_{\varrho=1}^{y}\left(1+\mathfrak{F}^{RN}_{\mathfrak{R}_{\sigma_\varrho}}\right)\right)^{\frac{1}{y}}\right)\right|^{\frac{1}{\mathbb{C}_y^y}}\\ +\iota\left(-\left|(-1)^{y+1}\prod_{1\le\sigma_1<\ldots<\sigma_y\le y}\left(-1+\left(\prod_{\varrho=1}^{y}\left(1+\mathfrak{F}^{IN}_{\mathfrak{R}_{\sigma_\varrho}}\right)\right)^{\frac{1}{y}}\right)\right|^{\frac{1}{\mathbb{C}_y^y}}\right)\end{array}\right)$$

$$=\left(\begin{array}{c}\left(\prod_{\varrho=1}^{y}\left(\mathfrak{F}^{RP}_{\mathfrak{R}_{\sigma_\varrho}}\right)\right)^{\frac{1}{y}}+\iota\left(\prod_{\varrho=1}^{y}\left(\mathfrak{F}^{IP}_{\mathfrak{R}_{\sigma_\varrho}}\right)\right)^{\frac{1}{y}},\\ -1+\left(\prod_{\varrho=1}^{y}\left(1+\mathfrak{F}^{RN}_{\mathfrak{R}_{\sigma_\varrho}}\right)\right)^{\frac{1}{y}}+\iota\left(-1+\left(\prod_{\varrho=1}^{y}\left(1+\mathfrak{F}^{IN}_{\mathfrak{R}_{\sigma_\varrho}}\right)\right)^{\frac{1}{y}}\right)\end{array}\right)=\left(\bigotimes_{\sigma=1}^{y}\mathfrak{R}_\sigma\right)^{\frac{1}{y}}$$

Weighted AOs are vital because they allow us to give various grades of significance to various elements or attributes while fusing them into a single result. We can use weights to show the relative importance or significance of different components within a dataset or DM process. Thus, the next definition is presenting BCF weighted HM (BCF-WHM) operator.

Definition 7.2 The BCF-WHM operator relying on the gathering of BCFNs $\mathfrak{R}_\sigma=\left(\mathfrak{F}^{P}_{\mathfrak{R}_\sigma},\mathfrak{F}^{N}_{\mathfrak{R}_\sigma}\right)=\left(\mathfrak{F}^{RP}_{\mathfrak{R}_\sigma}+\iota\mathfrak{F}^{IP}_{\mathfrak{R}_\sigma},\mathfrak{F}^{RN}_{\mathfrak{R}_\sigma}+\iota\mathfrak{F}^{IN}_{\mathfrak{R}_\sigma}\right)\sigma=1,2,\ldots,y$, is spotted as.

$$BCF-WHM^{\xi}(\mathfrak{R}_1,\mathfrak{R}_2,\ldots,\mathfrak{R}_y)=\frac{\bigoplus_{1\le\sigma_1<\ldots<\sigma_\xi\le y}\left(\bigotimes_{\varrho=1}^{\xi}\left(\mathfrak{R}_{\sigma_\varrho}\right)^{\omega_{\sigma_\varrho}}\right)^{\frac{1}{\xi}}}{\mathbb{C}_y^{\xi}} \tag{7.2}$$

In Eq. (7.2), $\omega=\left(\omega_1,\omega_2,\ldots,\omega_y\right)^T$ is a weight vector and $0\le\omega_\sigma\le1$ for all σ with $\sum_{\sigma}^{y}\omega_\sigma=1$, ξ is utilized as a parameter, $\xi=1,2,\ldots,y,\sigma_1,\sigma_2,\ldots,\sigma_\xi$ are ξ integer values occupied from the gathering of y integer values that is $\{1,2,\ldots,y\}$ and $\mathbb{C}_y^{\xi}=\frac{y!}{\xi!(y-\xi)!}$ Identify the binomial coefficient.

Theorem 7.2 Let a gathering of BCFNs $\mathfrak{R}_\sigma=\left(\mathfrak{F}^{P}_{\mathfrak{R}_\sigma},\mathfrak{F}^{N}_{\mathfrak{R}_\sigma}\right)=\left(\mathfrak{F}^{RP}_{\mathfrak{R}_\sigma}+\iota\mathfrak{F}^{IP}_{\mathfrak{R}_\sigma},\mathfrak{F}^{RN}_{\mathfrak{R}_\sigma}+\iota\mathfrak{F}^{IN}_{\mathfrak{R}_\sigma}\right)\sigma=1,2,\ldots,y$. The aggregated outcome of Eq. (7.2) is a BCFN and.

$$BCF-WHM^{\xi}(\mathfrak{R}_1,\mathfrak{R}_2,\ldots,\mathfrak{R}_y)=\left(\begin{array}{c}1-\left(\prod_{1\leq\sigma_1<\ldots<\sigma_y\leq y}\left(1-\left(\prod_{\varrho=1}^{\xi}\left(\mathfrak{F}^{RP}_{\mathfrak{R}_{\sigma_\varrho}}\right)^{\omega\sigma_\varrho}\right)^{\frac{1}{\xi}}\right)\right)^{\frac{1}{\mathbb{C}_y^{\xi}}}\\+\iota\left(1-\left(\prod_{1\leq\sigma_1<\ldots<\sigma_\xi\leq y}\left(1-\left(\prod_{\varrho=1}^{\xi}\left(\mathfrak{F}^{IP}_{\mathfrak{R}_{\sigma_\varrho}}\right)^{\omega\sigma_\varrho}\right)^{\frac{1}{\xi}}\right)\right)^{\frac{1}{\mathbb{C}_y^{\xi}}}\right),\\-\left|(-1)^{y+1}\prod_{1\leq\sigma_1<\ldots<\sigma_\xi\leq y}\left(-1+\left(\prod_{\varrho=1}^{\xi}\left(1+\mathfrak{F}^{RN}_{\mathfrak{R}_{\sigma_\varrho}}\right)^{\omega\sigma_\varrho}\right)^{\frac{1}{\xi}}\right)\right|^{\frac{1}{\mathbb{C}_y^{\xi}}}\\+\iota\left(-\left|(-1)^{y+1}\prod_{1\leq\sigma_1<\ldots<\sigma_\xi\leq y}\left(-1+\left(\prod_{\varrho=1}^{\xi}\left(1+\mathfrak{F}^{IN}_{\mathfrak{R}_{\sigma_\varrho}}\right)^{\omega\sigma_\varrho}\right)^{\frac{1}{\xi}}\right)\right|^{\frac{1}{\mathbb{C}_{y\xi}}}\right)\end{array}\right)$$

Proof With the assistance of definition (2.9), we have.

$$\left(\mathfrak{R}_{\sigma_\varrho}\right)^{\omega\sigma_\varrho}=\left(\left(\mathfrak{F}^{RP}_{\mathfrak{R}_{\sigma_\varrho}}\right)^{\omega\sigma_\varrho}+\iota\left(\mathfrak{F}^{IP}_{\mathfrak{R}_{\sigma_\varrho}}\right)^{\omega\sigma_\varrho},-1+\left(1+\mathfrak{F}^{RN}_{\mathfrak{R}_{\sigma_\varrho}}\right)^{\omega\sigma_\varrho}+\iota\left(-1+\left(1+\mathfrak{F}^{IN}_{\mathfrak{R}_{\sigma_\varrho}}\right)^{\omega\sigma_\varrho}\right)\right)$$

$$\underset{\varrho=1}{\overset{\xi}{\otimes}}\left(\mathfrak{R}_{\sigma_\varrho}\right)^{\omega_{\sigma_\varrho}}=\left(\begin{array}{c}\prod_{\varrho=1}^{\xi}\left(\mathfrak{F}^{RP}_{\mathfrak{R}_{\sigma_\varrho}}\right)^{\omega_{\sigma_\varrho}}+\iota\prod_{\varrho=1}^{\xi}\left(\mathfrak{F}^{IP}_{\mathfrak{R}_{\sigma_\varrho}}\right)^{\omega_{\sigma_\varrho}},\\-1+\prod_{\varrho=1}^{\xi}\left(1+\mathfrak{F}^{RN}_{\mathfrak{R}_{\sigma_\varrho}}\right)^{\omega_{\sigma_\varrho}}+\iota\left(-1+\prod_{\varrho=1}^{\xi}\left(1+\mathfrak{F}^{IN}_{\mathfrak{R}_{\sigma_\varrho}}\right)^{\omega_{\sigma_\varrho}}\right)\end{array}\right)$$

Consequently,

$$\left(\underset{\varrho=1}{\overset{\xi}{\otimes}}\left(\mathfrak{R}_{\sigma_\varrho}\right)^{\omega\sigma_\varrho}\right)^{\frac{1}{\xi}}=\left(\begin{array}{c}\left(\prod_{\varrho=1}^{\xi}\left(\mathfrak{F}^{RP}_{\mathfrak{R}_{\sigma_\varrho}}\right)^{\omega\sigma_\varrho}\right)^{\frac{1}{\xi}}+\iota\left(\prod_{\varrho=1}^{\xi}\left(\mathfrak{F}^{IP}_{\mathfrak{R}_{\sigma_\varrho}}\right)^{\omega\sigma_\varrho}\right)^{\frac{1}{\xi}},\\-1+\left(\prod_{\varrho=1}^{\xi}\left(1+\mathfrak{F}^{RN}_{\mathfrak{R}_{\sigma_\varrho}}\right)^{\omega\sigma_\varrho}\right)^{\frac{1}{\xi}}+\iota\left(-1+\left(\prod_{\varrho=1}^{\xi}\left(1+\mathfrak{F}^{IN}_{\mathfrak{R}_{\sigma_\varrho}}\right)^{\omega\sigma_\varrho}\right)^{\frac{1}{\xi}}\right)\end{array}\right)$$

Further,

$$\underset{1\leq\sigma_1<\ldots<\sigma_\xi\leq y}{\oplus}\left(\underset{\varrho=1}{\overset{\xi}{\otimes}}\left(\mathfrak{R}_{\sigma_\varrho}\right)^{\omega\sigma_\varrho}\right)^{\frac{1}{\xi}}$$

$$= \left(\begin{array}{c} 1 - \prod_{1 \le \sigma_1 < \ldots < \sigma_\xi \le y} \left(1 - \left(\prod_{\varrho=1}^{\xi} \left(\mathfrak{F}^{RP}_{\mathfrak{R}_{\sigma_\varrho}} \right)^{\omega \sigma_\varrho} \right)^{\frac{1}{\xi}} \right) \\ +\iota \left(1 - \prod_{1 \le \sigma_1 < \ldots < \sigma_\xi \le y} \left(1 - \left(\prod_{\varrho=1}^{\xi} \left(\mathfrak{F}^{IP}_{\mathfrak{R}_{\sigma_\varrho}} \right)^{\omega \sigma_\varrho} \right)^{\frac{1}{\xi}} \right) \right) \\ (-1)^{y+1} \prod_{1 \le \sigma_1 < \ldots < \sigma_\xi \le y} \left(-1 + \left(\prod_{\varrho=1}^{\xi} \left(1 + \mathfrak{F}^{RN}_{\mathfrak{R}_{\sigma_\varrho}} \right)^{\omega \sigma_\varrho} \right)^{\frac{1}{\xi}} \right) \\ +\iota \left((-1)^{y+1} \prod_{1 \le \sigma_1 < \ldots < \sigma_\xi \le y} \left(-1 + \left(\prod_{\varrho=1}^{\xi} \left(1 + \mathfrak{F}^{IN}_{\mathfrak{R}_{\sigma_\varrho}} \right)^{\omega \sigma_\varrho} \right)^{\frac{1}{\xi}} \right) \right) \end{array} \right)$$

Next,

$$\frac{\bigoplus_{1 \le \sigma_1 < \ldots < \sigma_\xi \le y} \left(\bigotimes_{\varrho = 1}^{\xi} \left(\mathfrak{R}_{\sigma_\varrho} \right)^{\omega \sigma_\varrho} \right)^{\frac{1}{\xi}}}{\mathbb{C}_y^\xi}$$

$$= \left(\begin{array}{c} 1 - \left(\prod_{1 \le \sigma_1 < \ldots < \sigma_\xi \le y} \left(1 - \left(\prod_{\varrho=1}^{\xi} \left(\mathfrak{F}^{RP}_{\mathfrak{R}_{\sigma_\varrho}} \right)^{\omega \sigma_\varrho} \right)^{\frac{1}{\xi}} \right) \right)^{\frac{1}{\mathbb{C}_y^\xi}} \\ +\iota \left(1 - \left(\prod_{1 \le \sigma_1 < \ldots < \sigma_\xi \le y} \left(1 - \left(\prod_{\varrho=1}^{\xi} \left(\mathfrak{F}^{IP}_{\mathfrak{R}_{\sigma_\varrho}} \right)^{\omega \sigma_\varrho} \right)^{\frac{1}{\xi}} \right) \right)^{\frac{1}{\mathbb{C}_y^\xi}} \right), \\ - \left| (-1)^{y+1} \prod_{1 \le \sigma_1 < \ldots < \sigma_\xi \le y} \left(-1 + \left(\prod_{\varrho=1}^{\xi} \left(1 + \mathfrak{F}^{RN}_{\mathfrak{R}_{\sigma_\varrho}} \right)^{\omega \sigma_\varrho} \right)^{\frac{1}{\xi}} \right) \right|^{\frac{1}{\mathbb{C}_y^\xi}} \\ +\iota \left(- \left| (-1)^{y+1} \prod_{1 \le \sigma_1 < \ldots < \sigma_\xi \le y} \left(-1 + \left(\prod_{\varrho=1}^{\xi} \left(1 + \mathfrak{F}^{IN}_{\mathfrak{R}_{\sigma_\varrho}} \right)^{\omega \sigma_\varrho} \right)^{\frac{1}{\xi}} \right) \right|^{\frac{1}{\mathbb{C}_y^\xi}} \right) \end{array} \right)$$

$$= BCF - WHM^{\xi} \left(\mathfrak{R}_1, \mathfrak{R}_2, \ldots, \mathfrak{R}_y \right)$$

Underneath we discuss the axioms of the BCF-WHM operator.

Proposition 7.4 (Idempotency): Let a gathering of BCFNs $\mathfrak{R}_\sigma = \left(\mathfrak{F}^P_{\mathfrak{R}_\sigma}, \mathfrak{F}^N_{\mathfrak{R}_\sigma} \right) = \left(\mathfrak{F}^{RP}_{\mathfrak{R}_\sigma} + \iota \mathfrak{F}^{IP}_{\mathfrak{R}_\sigma}, \mathfrak{F}^{RN}_{\mathfrak{R}_\sigma} + \iota \mathfrak{F}^{IN}_{\mathfrak{R}_\sigma} \right) \sigma = 1, 2, \ldots, y$. If $\forall \sigma \ \mathfrak{R}_\sigma = \mathfrak{R}$, then

$$BCF - WHM^{\xi} \left(\mathfrak{R}_1, \mathfrak{R}_2, \ldots, \mathfrak{R}_y \right) = \mathfrak{R}$$

Proof **Omitted.**

Proposition 7.5 (Monotonicity): Let two gatherings of BCFNs $\mathfrak{R}_\sigma = \left(\mathfrak{F}^P_{\mathfrak{R}_\sigma}, \mathfrak{F}^N_{\mathfrak{R}_\sigma}\right) = \left(\mathfrak{F}^{RP}_{\mathfrak{R}_\sigma} + \iota\mathfrak{F}^{IP}_{\mathfrak{R}_\sigma}, \mathfrak{F}^{RN}_{\mathfrak{R}_\sigma} + \iota\mathfrak{F}^{IN}_{\mathfrak{R}_\sigma}\right)$ and $\mathfrak{R}'_\sigma = \left(\mathfrak{F}^P_{\mathfrak{R}'_\sigma}, \mathfrak{F}^N_{\mathfrak{R}'_\sigma}\right) = \left(\mathfrak{F}^{RP}_{\mathfrak{R}'_\sigma} + \iota\mathfrak{F}^{IP}_{\mathfrak{R}'_\sigma}, \mathfrak{F}^{RN}_{\mathfrak{R}'_\sigma} + \iota\mathfrak{F}^{IN}_{\mathfrak{R}'_\sigma}\right)\sigma = 1, 2, \ldots, y$. If $\mathfrak{F}^{RP}_{\mathfrak{R}_\sigma} \le \mathfrak{F}^{RP}_{\mathfrak{R}'_\sigma}$, $\mathfrak{F}^{IP}_{\mathfrak{R}_\sigma} \le \mathfrak{F}^{IP}_{\mathfrak{R}'_\sigma}$, $\mathfrak{F}^{RN}_{\mathfrak{R}_\sigma} \le \mathfrak{F}^{RN}_{\mathfrak{R}'_\sigma}$, and $\mathfrak{F}^{IN}_{\mathfrak{R}_\sigma} \le \mathfrak{F}^{IN}_{\mathfrak{R}'_\sigma}$, then

$$BCF - WHM^\xi(\mathfrak{R}_1, \mathfrak{R}_2, \ldots, \mathfrak{R}_y) \le BCF - WHM^\xi\left(\mathfrak{R}'_1, \mathfrak{R}'_2, \ldots, \mathfrak{R}'_y\right)$$

Proof **Omitted.**

Proposition 7.6 (boundedness): Let a gathering of BCFNs $\mathfrak{R}_\sigma = \left(\mathfrak{F}^P_{\mathfrak{R}_\sigma}, \mathfrak{F}^N_{\mathfrak{R}_\sigma}\right) = \left(\mathfrak{F}^{RP}_{\mathfrak{R}_\sigma} + \iota\mathfrak{F}^{IP}_{\mathfrak{R}_\sigma}, \mathfrak{F}^{RN}_{\mathfrak{R}_\sigma} + \iota\mathfrak{F}^{IN}_{\mathfrak{R}_\sigma}\right)\sigma = 1, 2, \ldots, y$. If $\mathfrak{R}^- = \begin{pmatrix} \min\limits_\sigma\{\mathfrak{F}^{RP}_{\mathfrak{R}_\sigma}\} + \iota\min\limits_\sigma\{\mathfrak{F}^{IP}_{\mathfrak{R}_\sigma}\}, \\ \min\limits_\sigma\{\mathfrak{F}^{RN}_{\mathfrak{R}_\sigma}\} + \iota\min\limits_\sigma\{\mathfrak{F}^{IN}_{\mathfrak{R}_\sigma}\} \end{pmatrix}$, and $\mathfrak{R}^+ = \begin{pmatrix} \max\limits_\sigma\{\mathfrak{F}^{RP}_{\mathfrak{R}_\sigma}\} + \iota\max\limits_\sigma\{\mathfrak{F}^{IP}_{\mathfrak{R}_\sigma}\}, \\ \max\limits_\sigma\{\mathfrak{F}^{RN}_{\mathfrak{R}_\sigma}\} + \iota\max\limits_\sigma\{\mathfrak{F}^{IN}_{\mathfrak{R}_\sigma}\} \end{pmatrix}$, then

$$\mathfrak{R}^- \le BCF - WHM^\xi(\mathfrak{R}_1, \mathfrak{R}_2, \ldots, \mathfrak{R}_y) \le \mathfrak{R}^+$$

Proof **Omitted.**

7.2 Bipolar Complex Fuzzy Dual Hamy AOs

We start with the theory of dual HM (DHM) operator deduced by Wu et al. [28].

The DHM operator is demonstrated underneath

$$DHM^\xi(\mathfrak{O}_1, \mathfrak{O}_2, \ldots, \mathfrak{O}_y) = \left(\prod_{1\le\sigma_1<\ldots<\sigma_\xi\le y}\left(\frac{\sum_{\sigma=1}^{\xi}\mathfrak{O}_{\sigma_\varrho}}{\xi}\right)\right)^{\frac{1}{\mathbb{C}_y^\xi}}$$

where, ξ is signified as a parameter and $\xi = 1, 2, \ldots, y$, $\sigma_1, \sigma_2, \ldots, \sigma_\xi$ are ξ integer values taken from the gathering of y integer values i.e., $\{1, 2, \ldots, y\}$ and $\mathbb{C}_y^\xi = \frac{y!}{\xi!(y-\xi)!}$ Identify the binomial coefficient.

The concept of the BCF dual Hamy mean (BCF-DHM) operator is investigated in the underneath definition.

Definition 7.3 The BCF-DHM operator relying on the gathering of BCFNs $\mathfrak{R}_\sigma = \left(\mathfrak{F}^P_{\mathfrak{R}_\sigma}, \mathfrak{F}^N_{\mathfrak{R}_\sigma}\right) = \left(\mathfrak{F}^{RP}_{\mathfrak{R}_\sigma} + \iota\mathfrak{F}^{IP}_{\mathfrak{R}_\sigma}, \mathfrak{F}^{RN}_{\mathfrak{R}_\sigma} + \iota\mathfrak{F}^{IN}_{\mathfrak{R}_\sigma}\right)\sigma = 1, 2, \ldots, y$, is spotted as.

$$BCF-DHM^{\xi}(\mathfrak{R}_1,\mathfrak{R}_2,\ldots,\mathfrak{R}_y)=\left(\underset{1\le\sigma_1<\ldots<\sigma_\xi\le y}{\otimes}\left(\frac{\underset{\varrho=1}{\overset{\xi}{\oplus}}\mathfrak{R}_{\sigma_\varrho}}{\xi}\right)\right)^{\frac{1}{\mathbb{C}_y^{\xi}}} \tag{7.3}$$

In Eq. (7.3), ξ is utilized as a parameter, $\xi = 1, 2, \ldots, y$, $\sigma_1, \sigma_2, \ldots, \sigma_\xi$ are ξ integer values occupied from the gathering of y integer values that is $\{1, 2, \ldots, y\}$ and $\mathbb{C}_y^{\xi} = \frac{y!}{\xi!(y-\xi)!}$ Identify the binomial coefficient.

The underneath Theorem reveals that BCF-DHM again gives a BCFN.

Theorem 7.3 Let a gathering of BCFNs $\mathfrak{R}_\sigma = \left(\mathfrak{F}^{P}_{\mathfrak{R}_\sigma}, \mathfrak{F}^{N}_{\mathfrak{R}_\sigma}\right) = \left(\mathfrak{F}^{RP}_{\mathfrak{R}_\sigma} + \iota\mathfrak{F}^{IP}_{\mathfrak{R}_\sigma}, \mathfrak{F}^{RN}_{\mathfrak{R}_\sigma} + \iota\mathfrak{F}^{IN}_{\mathfrak{R}_\sigma}\right)\sigma = 1, 2, \ldots, y$. The aggregated outcome of Eq. (7.3) is a BCFN and.

$$BCF-DHM^{\xi}(\mathfrak{R}_1,\mathfrak{R}_2,\cdots\mathfrak{R}_y)=\left(\begin{array}{c}\left(\prod_{1\le\sigma_1<\cdots\sigma_\xi\le y}\left(1-\left(\prod_{\varrho=1}^{\xi}\left(1-\mathfrak{F}^{RP}_{\mathfrak{R}_{\sigma_\varrho}}\right)\right)^{\frac{1}{\xi}}\right)\right)^{\frac{1}{\mathbb{C}_y^{\xi}}}\\+\iota\left(\prod_{1\le\sigma_1<\cdots\sigma_\xi\le y}\left(1-\left(\prod_{\varrho=1}^{\xi}\left(1-\mathfrak{F}^{IP}_{\mathfrak{R}_{\sigma_\varrho}}\right)\right)^{\frac{1}{\xi}}\right)\right)^{\frac{1}{\mathbb{C}_y^{\xi}}},\\-1+\left(\prod_{1\le\sigma_1<\cdots\sigma_\xi\le y}\left(1-\left|(-1)^{y+1}\prod_{\varrho=1}^{\xi}\left(\mathfrak{F}^{RN}_{\mathfrak{R}_{\sigma_\varrho}}\right)\right|^{\frac{1}{\xi}}\right)\right)^{\frac{1}{\mathbb{C}_y^{\xi}}}\\+\iota\left(-1+\left(\prod_{1\le\sigma_1<\cdots\sigma_\xi\le y}\left(1-\left|(-1)^{y+1}\prod_{\varrho=1}^{\xi}\left(\mathfrak{F}^{IN}_{\mathfrak{R}_{\sigma_\varrho}}\right)\right|^{\frac{1}{\xi}}\right)\right)^{\frac{1}{\mathbb{C}_y^{\xi}}}\right)\end{array}\right)$$

Proof With the assistance of definition (2.9), we have.

$$\underset{\varrho=1}{\overset{\xi}{\oplus}}\mathfrak{R}_{\sigma_\varrho}=1-\prod_{\varrho=1}^{\xi}\left(1-\mathfrak{F}^{RP}_{\mathfrak{R}_{\sigma_\varrho}}\right)+\iota\left(1-\prod_{\varrho=1}^{\xi}\left(1-\mathfrak{F}^{IP}_{\mathfrak{R}_{\sigma_\varrho}}\right)\right),(-1)^{y+1}\prod_{\varrho=1}^{\xi}\left(\mathfrak{F}^{RN}_{\mathfrak{R}_{\sigma_\varrho}}\right)+\iota\left((-1)^{y+1}\prod_{\varrho=1}^{\xi}\left(\mathfrak{F}^{RN}_{\mathfrak{R}_{\sigma_\varrho}}\right)\right)$$

Consequently,

$$\frac{\underset{\varrho=1}{\overset{\xi}{\oplus}}\mathfrak{R}_{\sigma_\varrho}}{\xi}=\left(\begin{array}{c}1-\left(\prod_{\varrho=1}^{\xi}\left(1-\mathfrak{F}^{RP}_{\mathfrak{R}_{\sigma_\varrho}}\right)\right)^{\frac{1}{\xi}}+\iota\left(1-\left(\prod_{\varrho=1}^{\xi}\left(1-\mathfrak{F}^{IP}_{\mathfrak{R}_{\sigma_\varrho}}\right)\right)^{\frac{1}{\xi}}\right),\\-\left|(-1)^{y+1}\prod_{\varrho=1}^{\xi}\left(\mathfrak{F}^{RN}_{\mathfrak{R}_{\sigma_\varrho}}\right)\right|^{\frac{1}{\xi}}+\iota\left(-\left|(-1)^{y+1}\prod_{\varrho=1}^{\xi}\left(\mathfrak{F}^{IN}_{\mathfrak{R}_{\sigma_\varrho}}\right)\right|^{\frac{1}{\xi}}\right)\end{array}\right)$$

Further,

$$\underset{1\le\sigma_1<\ldots<\sigma_\xi\le y}{\oplus}\left(\frac{\overset{\xi}{\underset{\varrho=1}{\otimes}}\mathfrak{R}_{\sigma_\varrho}}{\xi}\right)=\left(\begin{array}{c}\prod_{1\le\sigma_1<\ldots<\sigma_\xi\le y}\left(1-\left(\prod_{\varrho=1}^{\xi}\left(1-\mathfrak{F}^{RP}_{\mathfrak{R}_{\sigma_\varrho}}\right)\right)^{\frac{1}{\xi}}\right)\\ +\iota\left(\prod_{1\le\sigma_1<\ldots<\sigma_\xi\le y}\left(1-\left(\prod_{\varrho=1}^{\xi}\left(\mathfrak{F}^{IP}_{\mathfrak{R}_{\sigma_\varrho}}\right)\right)^{\frac{1}{\xi}}\right)\right)\\ -1+\prod_{1\le\sigma_1<\ldots<\sigma_\xi\le y}\left(1-\left|(-1)^{y+1}\prod_{\varrho=1}^{\xi}\left(\mathfrak{F}^{RN}_{\mathfrak{R}_{\sigma_\varrho}}\right)\right|^{\frac{1}{\xi}}\right)\\ +\iota\left(-1+\prod_{1\le\sigma_1<\ldots<\sigma_\xi\le y}\left(1-\left|(-1)^{y+1}\prod_{\varrho=1}^{\xi}\left(\mathfrak{F}^{RN}_{\mathfrak{R}_{\sigma_\varrho}}\right)\right|^{\frac{1}{\xi}}\right)\right)\end{array}\right)$$

Next,

$$\left(\underset{1\le\sigma_1<\ldots<\sigma_\xi\le y}{\oplus}\left(\frac{\overset{\xi}{\underset{\varrho=1}{\otimes}}\mathfrak{R}_{\sigma_\varrho}}{\xi}\right)\right)^{\frac{1}{\mathbb{C}_y^\xi}}=\left(\begin{array}{c}\left(\prod_{1\le\sigma_1<\ldots<\sigma_\xi\le y}\left(1-\left(\prod_{\varrho=1}^{\xi}\left(1-\mathfrak{F}^{RP}_{\mathfrak{R}_{\sigma_\varrho}}\right)\right)^{\frac{1}{\xi}}\right)\right)^{\frac{1}{\mathbb{C}_y^\xi}}\\ +\iota\left(\prod_{1\le\sigma_1<\ldots<\sigma_\xi\le y}\left(1-\left(\prod_{\varrho=1}^{\xi}\left(\mathfrak{F}^{IP}_{\mathfrak{R}_{\sigma_\varrho}}\right)\right)^{\frac{1}{\xi}}\right)\right)^{\frac{1}{\mathbb{C}_y^\xi}}\\ -1+\left(\prod_{1\le\sigma_1<\ldots<\sigma_\xi\le y}\left(1-\left|(-1)^{y+1}\prod_{\varrho=1}^{\xi}\left(\mathfrak{F}^{RN}_{\mathfrak{R}_{\sigma_\varrho}}\right)\right|^{\frac{1}{\xi}}\right)\right)^{\frac{1}{\mathbb{C}_y^\xi}}\\ +\iota\left(-1+\left(\prod_{1\le\sigma_1<\ldots<\sigma_\xi\le y}\left(1-\left|(-1)^{y+1}\prod_{\varrho=1}^{\xi}\left(\mathfrak{F}^{IN}_{\mathfrak{R}_{\sigma_\varrho}}\right)\right|^{\frac{1}{\xi}}\right)\right)^{\frac{1}{\mathbb{C}_y^\xi}}\right)\end{array}\right)$$

$$=BCF-DHM^{\xi}(\mathfrak{R}_1,\mathfrak{R}_2,\ldots,\mathfrak{R}_y)$$

Underneath we discuss the Proposition of BCF-DHM operator.

Proposition 7.7 (Idempotency): Let a gathering of BCFNs $\mathfrak{R}_\sigma=\left(\mathfrak{F}^{P}_{\mathfrak{R}_\sigma},\mathfrak{F}^{N}_{\mathfrak{R}_\sigma}\right)=\left(\mathfrak{F}^{RP}_{\mathfrak{R}_\sigma}+\iota\mathfrak{F}^{IP}_{\mathfrak{R}_\sigma},\mathfrak{F}^{RN}_{\mathfrak{R}_\sigma}+\iota\mathfrak{F}^{IN}_{\mathfrak{R}_\sigma}\right)\sigma=1,2,\ldots,y$. If $\forall\sigma\ \mathfrak{R}_\sigma=\mathfrak{R}$, then

$$BCF-DHM^{\xi}(\mathfrak{R}_1,\mathfrak{R}_2,\ldots,\mathfrak{R}_y)=\mathfrak{R}$$

Proof Likewise, Proposition 7.1

Proposition 7.8 (Monotonicity): Let two gatherings of BCFNs $\mathfrak{R}_\sigma=\left(\mathfrak{F}^{P}_{\mathfrak{R}_\sigma},\mathfrak{F}^{N}_{\mathfrak{R}_\sigma}\right)=\left(\mathfrak{F}^{RP}_{\mathfrak{R}_\sigma}+\iota\mathfrak{F}^{IP}_{\mathfrak{R}_\sigma},\mathfrak{F}^{RN}_{\mathfrak{R}_\sigma}+\iota\mathfrak{F}^{IN}_{\mathfrak{R}_\sigma}\right)$ and $\mathfrak{R}'_\sigma=\left(\mathfrak{F}^{P}_{\mathfrak{R}'_\sigma},\mathfrak{F}^{N}_{\mathfrak{R}'_\sigma}\right)=\left(\mathfrak{F}^{RP}_{\mathfrak{R}'_\sigma}+\iota\mathfrak{F}^{IP}_{\mathfrak{R}'_\sigma},\mathfrak{F}^{RN}_{\mathfrak{R}'_\sigma}+\iota\mathfrak{F}^{IN}_{\mathfrak{R}'_\sigma}\right)\sigma=1,2,\ldots,y$. If $\mathfrak{F}^{RP}_{\mathfrak{R}_\sigma}\le\mathfrak{F}^{RP}_{\mathfrak{R}'_\sigma}$, $\mathfrak{F}^{IP}_{\mathfrak{R}_\sigma}\le\mathfrak{F}^{IP}_{\mathfrak{R}'_\sigma}$, $\mathfrak{F}^{RN}_{\mathfrak{R}_\sigma}\le\mathfrak{F}^{RN}_{\mathfrak{r}'_\sigma}$, and $\mathfrak{F}^{IN}_{\mathfrak{R}_\sigma}\le\mathfrak{F}^{IN}_{\mathfrak{R}'_\sigma}$, then

$$BCF-DHM^{\xi}(\mathfrak{R}_1,\mathfrak{R}_2,\ldots,\mathfrak{R}_y)\le BCF-DHM^{\xi}\left(\mathfrak{R}'_1,\mathfrak{R}'_2,\ldots,\mathfrak{R}'_y\right)$$

Proof Likewise, Proposition 7.2

Proposition 7.9 (boundedness): Let a gathering of BCFNs $\mathfrak{R}_\sigma = \left(\mathfrak{F}^{P}_{\mathfrak{R}_\sigma}, \mathfrak{F}^{N}_{\mathfrak{R}_\sigma}\right) = \left(\mathfrak{F}^{RP}_{\mathfrak{R}_\sigma} + \iota\mathfrak{F}^{IP}_{\mathfrak{R}_\sigma}, \mathfrak{F}^{RN}_{\mathfrak{R}_\sigma} + \iota\mathfrak{F}^{IN}_{\mathfrak{R}_\sigma}\right)\sigma = 1, 2, \ldots, y$. If $\mathfrak{R}^{-} = \begin{pmatrix} \min\limits_{\sigma}\left\{\mathfrak{F}^{RP}_{\mathfrak{R}_\sigma}\right\} + \iota \min\limits_{\sigma}\left\{\mathfrak{F}^{IP}_{\mathfrak{R}_\sigma}\right\}, \\ \min\limits_{\sigma}\left\{\mathfrak{F}^{RN}_{\mathfrak{R}_\sigma}\right\} + \iota \min\limits_{\sigma}\left\{\mathfrak{F}^{IN}_{\mathfrak{R}_\sigma}\right\} \end{pmatrix}$, and $\mathfrak{R}^{+} = \begin{pmatrix} \max\limits_{\sigma}\left\{\mathfrak{F}^{RP}_{\mathfrak{R}_\sigma}\right\} + \iota \max\limits_{\sigma}\left\{\mathfrak{F}^{IP}_{\mathfrak{R}_\sigma}\right\}, \\ \max\limits_{\sigma}\left\{\mathfrak{F}^{RN}_{\mathfrak{R}_\sigma}\right\} + \iota \max\limits_{\sigma}\left\{\mathfrak{F}^{IN}_{\mathfrak{R}_\sigma}\right\} \end{pmatrix}$, then

$$\mathfrak{R}^{-} \leq BCF - DHM^{\xi}\left(\mathfrak{R}_1, \mathfrak{R}_2, \ldots, \mathfrak{R}_y\right) \leq \mathfrak{R}^{+}$$

Proof Likewise, Proposition 7.3.

Definition 7.4 The BCF-WDHM operator relying on the gathering of BCFNs $\mathfrak{R}_\sigma = \left(\mathfrak{F}^{P}_{\mathfrak{R}_\sigma}, \mathfrak{F}^{N}_{\mathfrak{R}_\sigma}\right) = \left(\mathfrak{F}^{RP}_{\mathfrak{R}_\sigma} + \iota\mathfrak{F}^{IP}_{\mathfrak{R}_\sigma}, \mathfrak{F}^{RN}_{\mathfrak{R}_\sigma} + \iota\mathfrak{F}^{IN}_{\mathfrak{R}_\sigma}\right)\sigma = 1, 2, \ldots, y$, is spotted as.

$$BCF - WDHM^{\xi}\left(\mathfrak{R}_1, \mathfrak{R}_2, \ldots, \mathfrak{R}_y\right) = \left(\underset{1 \leq \sigma_1 < \ldots < \sigma_\xi \leq y}{\otimes}\left(\frac{\overset{\xi}{\underset{\varrho=1}{\oplus}} \omega_{\sigma_\varrho}\mathfrak{R}_{\sigma_\varrho}}{\xi}\right)\right)^{\frac{1}{\mathbb{C}_y^{\xi}}} \tag{7.4}$$

In Eq. 7.4, $\omega = \left(\omega_{1,}\omega_2, \ldots, \omega_y\right)^T$ is a weight vector and $0 \leq \omega_\sigma \leq 1$ for all σ with $\sum\limits_{\sigma=1}^{y} \omega_\sigma = 1$, ξ is utilized as a parameter, $\xi = 1, 2, \ldots, y$, $\sigma_1, \sigma_2, \ldots, \sigma_\xi$ are ξ integer values occupied from the gathering of y integer values that is $\{1, 2, \ldots, y\}$ and $\mathbb{C}_y^{\xi} = \frac{y!}{\xi!(y-\xi)!}$ Identify the binomial coefficient.

Theorem 7.4 Let a gathering of BCFNs $\mathfrak{R}_\sigma = \left(\mathfrak{F}^{P}_{\mathfrak{R}_\sigma}, \mathfrak{F}^{N}_{\mathfrak{R}_\sigma}\right) = \left(\mathfrak{F}^{RP}_{\mathfrak{R}_\sigma} + \iota\mathfrak{F}^{IP}_{\mathfrak{R}_\sigma}, \mathfrak{F}^{RN}_{\mathfrak{R}_\sigma} + \iota\mathfrak{F}^{IN}_{\mathfrak{R}_\sigma}\right)\sigma = 1, 2, \ldots, y$. The aggregated outcome of Eq. (7.4) is a BCFN and.

$$BCF - WDHM^{\xi}\left(\mathfrak{R}_1, \mathfrak{R}_2, \cdots \mathfrak{R}_y\right) = \begin{pmatrix} \left(\prod\limits_{1 \leq \sigma_1 < \cdots \sigma_\xi \leq y}\left(1 - \left(\prod\limits_{\varrho=1}^{\xi}\left(1 - \mathfrak{F}^{RP}_{\mathfrak{R}_{\sigma_\varrho}}\right)^{\omega_{\sigma_\varrho}}\right)^{\frac{1}{\xi}}\right)\right)^{\frac{1}{\mathbb{C}_y^{\xi}}} \\ +\iota\left(\prod\limits_{1 \leq \sigma_1 < \cdots \sigma_\xi \leq y}\left(1 - \left(\prod\limits_{\varrho=1}^{\xi}\left(1 - \mathfrak{F}^{IP}_{\mathfrak{R}_{\sigma_\varrho}}\right)^{\omega_{\sigma_\varrho}}\right)^{\frac{1}{\xi}}\right)\right)^{\frac{1}{\mathbb{C}_y^{\xi}}}, \\ -1 + \left(\prod\limits_{1 \leq \sigma_1 < \cdots \sigma_\xi \leq y}\left(1 - \left|\prod\limits_{\varrho=1}^{\xi}\left|\mathfrak{F}^{RN}_{\mathfrak{R}_{\sigma_\varrho}}\right|^{\omega_{\sigma_\varrho}}\right|^{\frac{1}{\xi}}\right)\right)^{\frac{1}{\mathbb{C}_y^{\xi}}} \\ +\iota\left(-1 + \left(\prod\limits_{1 \leq \sigma_1 < \cdots \sigma_\xi \leq y}\left(1 - \left|-\prod\limits_{\varrho=1}^{\xi}\left|\mathfrak{F}^{IN}_{\mathfrak{R}_{\sigma_\varrho}}\right|^{\omega_{\sigma_\varrho}}\right|^{\frac{1}{\xi}}\right)\right)^{\frac{1}{\mathbb{C}_y^{\xi}}}\right) \end{pmatrix}$$

Proof With the assistance of definition (2.9), we have.

$$\omega_{\sigma_\varrho}\mathfrak{R}_{\sigma_\varrho} = \left(1 - \left(1 - \mathfrak{F}^{RP}_{\mathfrak{R}_{\sigma_\varrho}}\right)^{\omega_{\sigma_\varrho}} + \iota\left(1 - \left(1 - \mathfrak{F}^{IP}_{\mathfrak{R}_{\sigma_\varrho}}\right)^{\omega_{\sigma_\varrho}}\right), -\left|\mathfrak{F}^{RN}_{\mathfrak{R}_{\sigma_\varrho}}\right|^{\omega_{\sigma_\varrho}} + \iota\left(-\left|\mathfrak{F}^{IN}_{\mathfrak{R}_{\sigma_\varrho}}\right|^{\omega_{\sigma_\varrho}}\right)\right)$$

$$\underset{\varrho=1}{\overset{\xi}{\oplus}} \omega_{\sigma_\varrho}\mathfrak{R}_{\sigma_\varrho} = \left(\begin{array}{c} 1-\prod_{\varrho=1}^{\xi}\left(1-\mathfrak{F}^{RP}_{\mathfrak{R}_{\sigma_\varrho}}\right)^{\omega_{\sigma_\varrho}} + \iota\left(1-\prod_{\varrho=1}^{\xi}\left(1-\mathfrak{F}^{IP}_{\mathfrak{R}_{\sigma_\varrho}}\right)^{\omega_{\sigma_\varrho}}\right), \\ -\prod_{\varrho=1}^{\xi}\left|\mathfrak{F}^{RN}_{\mathfrak{R}_{\sigma_\varrho}}\right|^{\omega_{\sigma_\varrho}} + \iota\left(-\prod_{\varrho=1}^{\xi}\left|\mathfrak{F}^{IN}_{\mathfrak{R}_{\sigma_\varrho}}\right|^{\omega_{\sigma_\varrho}}\right) \end{array} \right)$$

Consequently,

$$\frac{\underset{\varrho=1}{\overset{\xi}{\oplus}} \omega_{\sigma_\varrho}\mathfrak{R}_{\sigma_\varrho}}{\xi} = \left(\begin{array}{c} 1-\left(\prod_{\varrho=1}^{\xi}\left(1-\mathfrak{F}^{RP}_{\mathfrak{R}_{\sigma_\varrho}}\right)^{\omega_{\sigma_\varrho}}\right)^{\frac{1}{\xi}} + \iota\left(1-\left(\prod_{\varrho=1}^{\xi}\left(1-\mathfrak{F}^{IP}_{\mathfrak{R}_{\sigma_\varrho}}\right)^{\omega_{\sigma_\varrho}}\right)^{\frac{1}{\xi}}\right), \\ -\left|-\prod_{\varrho=1}^{\xi}\left|\mathfrak{F}^{RN}_{\mathfrak{R}_{\sigma_\varrho}}\right|^{\omega_{\sigma_\varrho}}\right|^{\frac{1}{\xi}} + \iota\left(-\left|-\prod_{\varrho=1}^{\xi}\left|\mathfrak{F}^{IN}_{\mathfrak{R}_{\sigma_\varrho}}\right|^{\omega_{\sigma_\varrho}}\right|^{\frac{1}{\xi}}\right) \end{array} \right)$$

Further,

$$\underset{1\le\sigma_1<\ldots<\sigma_\xi\le y}{\otimes}\left(\frac{\underset{\varrho=1}{\overset{\xi}{\oplus}} \omega_{\sigma_\varrho}\mathfrak{R}_{\sigma_\varrho}}{\xi}\right)$$

$$= \left(\begin{array}{c} \prod_{1\le\sigma_1<\cdots\sigma_\xi\le y}\left(1-\left(\prod_{\varrho=1}^{\xi}\left(1-\mathfrak{F}^{RP}_{\mathfrak{R}_{\sigma_\varrho}}\right)^{\omega_{\sigma_\varrho}}\right)^{\frac{1}{\xi}}\right) + \iota \prod_{1\le\sigma_1<\cdots\sigma_\xi\le y}\left(1-\left(\prod_{\varrho=1}^{\xi}\left(1-\mathfrak{F}^{IP}_{\mathfrak{R}_{\sigma_\varrho}}\right)^{\omega_{\sigma_\varrho}}\right)^{\frac{1}{\xi}}\right) \\ -1+\prod_{1\le\sigma_1<\cdots\sigma_\xi\le y}\left(1-\left|-\prod_{\varrho=1}^{\xi}\left|\mathfrak{F}^{RN}_{\mathfrak{R}_{\sigma_\varrho}}\right|^{\omega_{\sigma_\varrho}}\right|^{\frac{1}{\xi}}\right) + \iota\left(-1+\prod_{1\le\sigma_1<\cdots\sigma_\xi\le y}\left(1-\left|-\prod_{\varrho=1}^{\xi}\left|\mathfrak{F}^{IN}_{\mathfrak{R}_{\sigma_\varrho}}\right|^{\omega_{\sigma_\varrho}}\right|^{\frac{1}{\xi}}\right)\right) \end{array} \right)$$

Next,

$$\left(\underset{1\le\sigma_1<\ldots<\sigma_\xi\le y}{\oplus}\left(\frac{\underset{\varrho=1}{\overset{\xi}{\otimes}} \mathfrak{R}_{\sigma_\varrho}}{\xi}\right)\right)^{\frac{1}{\mathbb{C}_y^\xi}} = \left(\begin{array}{c} \left(\prod_{1\le\sigma_1<\ldots<\sigma_\xi\le y}\left(1-\left(\prod_{\varrho=1}^{\xi}\left(1-\mathfrak{F}^{RP}_{\mathfrak{R}_{\sigma_\varrho}}\right)^{\omega_{\sigma_\varrho}}\right)^{\frac{1}{\xi}}\right)\right)^{\frac{1}{\mathbb{C}_y^\xi}} \\ +\iota\left(\prod_{1\le\sigma_1<\ldots<\sigma_\xi\le y}\left(1-\left(\prod_{\varrho=1}^{\xi}\left(1-\mathfrak{F}^{IP}_{\mathfrak{R}_{\sigma_\varrho}}\right)^{\omega_{\sigma_\varrho}}\right)^{\frac{1}{\xi}}\right)\right)^{\frac{1}{\mathbb{C}_y^\xi}}, \\ -1+\left(\prod_{1\le\sigma_1<\ldots<\sigma_\xi\le y}\left(1-\left|-\prod_{\varrho=1}^{\xi}\left|\mathfrak{F}^{RN}_{\mathfrak{R}_{\sigma_\varrho}}\right|^{\omega_{\sigma_\varrho}}\right|^{\frac{1}{\xi}}\right)\right)^{\frac{1}{\mathbb{C}_y^\xi}} \\ +\iota\left(-1+\left(\prod_{1\le\sigma_1<\ldots<\sigma_\xi\le y}\left(1-\left|-\prod_{\varrho=1}^{\xi}\left|\mathfrak{F}^{IN}_{\mathfrak{R}_{\sigma_\varrho}}\right|^{\omega_{\sigma_\varrho}}\right|^{\frac{1}{\xi}}\right)\right)^{\frac{1}{\mathbb{C}_y^\xi}}\right) \end{array} \right)$$

$$= BCF-WDHM^{\xi}(\mathfrak{R}_1, \mathfrak{R}_2, \ldots, \mathfrak{R}_y)$$

The BCF-WDHM operator also holds the proposition of idempotency, monotonicity, and boundedness.

7.3 Multi-Attribute Decision-Making Approach

In this Section of the chapter, we interpret an approach of MADM relying on the inaugurated operators are BCF-HM, BCF-WHM, BCF-DHM, and BCF-WDHM operators.

Consider two gatherings, one the gathering of y alternatives i.e., $\mathfrak{R}_{\mathbb{A}} = \{\mathfrak{R}_{\mathbb{A}-1}, \mathfrak{R}_{\mathbb{A}-2}, \ldots, \mathfrak{R}_{\mathbb{A}-y}\}$ and second the gathering of ξ attributes i.e., $\mathfrak{A}_{\mathbb{A}\approx} = \{\mathfrak{A}_{\mathbb{A}\approx-1}, \mathfrak{A}_{\mathbb{A}\approx-2}, \ldots, \mathfrak{A}_{\mathbb{A}\approx-\xi}\}$. Also, consider that $\omega = (\omega_{1,}\omega_2, \ldots, \omega_\xi)^T$ is a weight vector for the attributes and $0 \leq \omega_\varrho \leq 1$ for all ϱ with $\sum_{\varrho=1}^{\xi} \omega_\varrho = 1$. Moreover, assume that $\mathfrak{D} = (\mathfrak{R}_{\sigma\varrho})_{y\times\xi} = \left(\mathfrak{F}^{P}_{\mathfrak{R}_{\sigma\varrho}}, \mathfrak{F}^{N}_{\mathfrak{R}_{\sigma\varrho}}\right)_{y\times\xi} = \left(\mathfrak{F}^{RP}_{\mathfrak{R}_{\sigma\varrho}} + \iota\mathfrak{F}^{IP}_{\mathfrak{R}_{\sigma\varrho}}, \mathfrak{F}^{RN}_{\mathfrak{R}_{\sigma\varrho}} + \iota\mathfrak{F}^{IN}_{\mathfrak{R}_{\sigma\varrho}}\right)_{y\times\xi}$ is a BCF decision matrix. Observe that $\mathfrak{F}^{P}_{\mathfrak{R}_{\sigma\varrho}}$ implies a positive belonging grade of $\mathfrak{R}_\sigma$ with respect to the attribute $\mathfrak{A}_{\mathbb{A}\approx-\varrho}$ established by the decision analyst and $\mathfrak{F}^{N}_{\mathfrak{R}_{\sigma\varrho}}$ implies a negative belonging grade of $\mathfrak{R}_\sigma$ with respect to the attribute $\mathfrak{A}_{\mathbb{A}\approx-\varrho}$ established by the decision analyst. Underneath are the steps, through which the MADM dilemma would be handled.

Step 1: In each MADM dilemma, the attribute can be two types that are benefit type and cost type. In any MADM dilemma the attribute is cost type then the BCF decision-matrix should be normalized by employing the below formula

$$\mathfrak{R}_{\sigma\varrho} = \begin{cases} \left(\mathfrak{F}^{RP}_{\mathfrak{R}_{\sigma\varrho}} + \iota\mathfrak{F}^{RP}_{\mathfrak{R}_{\sigma\varrho}}, \mathfrak{F}^{RN}_{\mathfrak{R}_{\sigma\varrho}} + \iota\mathfrak{F}^{IN}_{\mathfrak{R}_{\sigma\varrho}}\right) & \text{for benefit sort of attributes} \\ \left(1 - \mathfrak{F}^{RP}_{\mathfrak{R}_{\sigma\varrho}} + \iota\left(-1 - \mathfrak{F}^{IP}_{\mathfrak{R}_{\sigma\varrho}}\right), 1 - \mathfrak{F}^{RN}_{\mathfrak{R}_{\sigma\varrho}} + \iota\left(-1 - \mathfrak{F}^{IN}_{\mathfrak{R}_{\sigma\varrho}}\right)\right) & \text{for cost sort of attributes} \end{cases}$$

If there is no cost type of attribute, then the normalization procedure should be skipped.

Step 2: For aggregating this information, any of the inaugurated operators (BCF-HM, BCF-WHM, BCF-DHM, and BCF-WDHM operator) can be employed.

Step 3: Through Eq. (2.5), the score values of the aggregated information would be described.

Step 4: Employing the score values of step 3, the ranking among the alternatives would be demonstrated.

Step 5: End.

7.3.1 Case Study (Classification of Investment Sectors for Multi-national Corporation)

A multinational company with a significant presence in the technology industry, XYZ Corporation, wants to diversify its holdings and grow into other fields. The company's management is convinced that making investments in developing industries will not only spur growth but also establish them as an avant-garde market leader. They have shortlisted four prospective investment areas:

$\mathfrak{R}_{\mathbb{A}-1}$ **(Renewable Energy)**: The XYZ Corporation is aware of the rising demand for sustainable and renewable energy sources worldwide. The company wants to take advantage of the expanding demand for renewable energy solutions while helping to reduce greenhouse gas emissions. Thus, they shortlisted the field of renewable energy to make investments in the research and development of renewable energy technologies, such as solar and wind energy.

$\mathfrak{R}_{\mathbb{A}-2}$ **(Artificial Intelligence (AI) in healthcare)**: XYZ Corporation shortlists the field of AI in healthcare for investment after realizing the game-changing potential of AI in the healthcare industry. In cooperation with healthcare providers and research institutes throughout the world, they intend to create AI algorithms for medical diagnoses, individualized treatment suggestions, and data-driven research.

$\mathfrak{R}_{\mathbb{A}-3}$ **(E-commerce and Logistics)**: XYZ Corporation considers E-commerce and logistics field as an option to invest in after realizing the rapid expansion of e-commerce and the rising demand for effective logistics solutions. To satisfy the changing needs of customers and businesses throughout the world, they seek to improve last-mile delivery services, optimize supply chain management, and streamline the online shopping experience.

$\mathfrak{R}_{\mathbb{A}-4}$ **(Sustainable Agriculture)**: The XYZ Corporation recognizes the value of sustainable agriculture in preserving the environment and maintaining food security. They consider spending money on cutting-edge irrigation systems, precision farming, and vertical farming, among other agricultural innovations as an option. The goal of XYZ Corporation is to increase crop yields and reduce the environmental effect of farming by encouraging sustainable agriculture techniques.

in which one of the areas needs to be finalized. The company develops a set of four evaluation criteria to compare these four fields objectively. These four criteria are $\mathfrak{A}_{\mathbb{A}\approx-1}$ (Market potential), $\mathfrak{A}_{\mathbb{A}\approx-2}$ (Technological feasibility), $\mathfrak{A}_{\mathbb{A}\approx-3}$ (competitive landscape), and $\mathfrak{A}_{\mathbb{A}\approx-4}$ (Financial viability). Based on the predetermined criteria, the evaluation team assesses each field. Each criterion is given a weighted importance value i.e., $(0.42, 0.32, 0.18, 0.08)$, and the assessment values are demonstrated in the structure of BCFNs and interpreted in Table 7.1.

Step 1: In this MADM dilemma, the attributes are benefit types, so we are skipping this step.

Step 2: The aggregating argument of this information is demonstrated in Table 7.2

Step 3: Through Eq. (2.5), the score values of the aggregated information are described in Table 7.3

Step 4: The ranking among the alternatives is demonstrated in Table 7.4.

Table 7.4 reveals that by employing the BCF-HM operator in the MADM technique, the ranking of considered fields is $\mathfrak{R}_{\mathbb{A}-4} > \mathfrak{R}_{\mathbb{A}-3} > \mathfrak{R}_{\mathbb{A}-1} > \mathfrak{R}_{\mathbb{A}-2}$, thus, $\mathfrak{R}_{\mathbb{A}-4}$ (Sustainable Agriculture) is the optimal field for the investment. By employing the BCF-WHM operator in the MADM approach, the ranking is $\mathfrak{R}_{\mathbb{A}-3} > \mathfrak{R}_{\mathbb{A}-4} > \mathfrak{R}_{\mathbb{A}-2} > \mathfrak{R}_{\mathbb{A}-1}$, and hence $\mathfrak{R}_{\mathbb{A}-3}$ (E-commerce and Logistics) is a perfect field for investment. By employing BCF-DHM and BCF-WDHM operators in the technique

Table 7.1 The assessment values, interpreted by the evaluation team

	$\mathfrak{A}_{\mathbb{A}\approx-1}$	$\mathfrak{A}_{\mathbb{A}\approx-2}$	$\mathfrak{A}_{\mathbb{A}\approx-3}$	$\mathfrak{A}_{\mathbb{A}\approx-4}$
$\mathfrak{R}_{\mathbb{A}-1}$	$\begin{pmatrix}0.347+\iota 0.875,\\-0.237-\iota 0.67\end{pmatrix}$	$\begin{pmatrix}0.584+\iota 0.478,\\-0.235-\iota 0.798\end{pmatrix}$	$\begin{pmatrix}0.473+\iota 0.347,\\-0.545-\iota 0.337\end{pmatrix}$	$\begin{pmatrix}0.634+\iota 0.746,\\-0.945-\iota 0.85\end{pmatrix}$
$\mathfrak{R}_{\mathbb{A}-2}$	$\begin{pmatrix}0.984+\iota 0.574,\\-0.242-\iota 0.825\end{pmatrix}$	$\begin{pmatrix}0.745+\iota 0.323,\\-0.235-\iota 0.123\end{pmatrix}$	$\begin{pmatrix}0.577+\iota 0.775,\\-0.247-\iota 0.475\end{pmatrix}$	$\begin{pmatrix}0.245+\iota 0.475,\\-0.394-\iota 0.975\end{pmatrix}$
$\mathfrak{R}_{\mathbb{A}-3}$	$\begin{pmatrix}0.273+\iota 0.845,\\-0.174-\iota 0.28\end{pmatrix}$	$\begin{pmatrix}0.434+\iota 0.343,\\-0.342-\iota 0.283\end{pmatrix}$	$\begin{pmatrix}0.475+\iota 0.674,\\-0.453-\iota 0.346\end{pmatrix}$	$\begin{pmatrix}0.455+\iota 0.675,\\-0.336-\iota 0.66\end{pmatrix}$
$\mathfrak{R}_{\mathbb{A}-4}$	$\begin{pmatrix}0.295+\iota 0.457,\\-0.457-\iota 0.957\end{pmatrix}$	$\begin{pmatrix}0.352+\iota 0.284,\\-0.234-\iota 0.875\end{pmatrix}$	$\begin{pmatrix}0.463+\iota 0.634,\\-0.654-\iota 0.563\end{pmatrix}$	$\begin{pmatrix}0.675+\iota 0.845,\\-0.345-\iota 0.43\end{pmatrix}$

of MADM the ranking of these fields is $\mathfrak{R}_{\mathbb{A}-4} > \mathfrak{R}_{\mathbb{A}-3} > \mathfrak{R}_{\mathbb{A}-2} > \mathfrak{R}_{\mathbb{A}-1}$, and thus, $\mathfrak{R}_{\mathbb{A}-4}$ (Sustainable Agriculture) is the optimal field for the investment.

Step 5: End.

7.4 Comparative Analysis

In this part of the chapter, we are going to compare the inaugurated operators and techniques of MADM with a few prevailing techniques and operators. For this comparison, we considered the underneath operators and techniques based on these operators.

- Dombi AOs and the MADM process were diagnosed by Jana et al. [29] in the structure of BF theory.
- Dombi prioritized AOs and the process of MADM investigated by Jana et al. [30] for the model of BFS.
- Hamacher AOs and MADM process deduced by Wei et al. [4] in the model of BFS.
- The trigonometric AOs and SIR techniques were diagnosed by Riaz et al. [31] for the theory of BFS.
- Arithmetic AOs for CFS deduced by Bi et al. [32]
- Geometric AOs for CFS deduced by Bi et al. [33].

For the comparison, let us consider the information in Table 7.5 and employ the inaugurated and prevailing operators and MADM techniques. The underneath Table 7.5 is displayed the required outcomes.

Table 7.5 demonstrate that the considered prevailing theories of techniques and operators failed to cope with the information described in Table 7.5. The theories investigated by Jana et al. [29], Jana et al. [30], Wei et al. [4] and Riaz et al. [31] are in the model of BFS and we know that the theory of BFS lack 2nd dimension why these

Table 7.2 The aggregating argument of the considered fields ($\xi = 3$)

Operators	$\mathfrak{R}_{\mathbb{A}\approx-1}$	$\mathfrak{R}_{\mathbb{A}\approx-2}$	$\mathfrak{R}_{\mathbb{A}\approx-3}$	$\mathfrak{R}_{\mathbb{A}\approx-4}$
BCF-HM	$\begin{pmatrix} 0.995 + \iota 0.9965, \\ -0.1273 - \iota 0.073 \end{pmatrix}$	$\begin{pmatrix} 0.9787 + \iota 0.9924, \\ -0.2112 - \iota 0.0014 \end{pmatrix}$	$\begin{pmatrix} 0.9848 + \iota 0.9975, \\ -0.0022 - \iota 0.0114 \end{pmatrix}$	$\begin{pmatrix} 0.9942 + \iota 0.9989, \\ -0.0049 - \iota 0.018 \end{pmatrix}$
BCF-WHM	$\begin{pmatrix} 0.9999 + \iota 0.9999, \\ -0.017 - \iota 0.0025 \end{pmatrix}$	$\begin{pmatrix} 0.9993 + \iota 0.9999, \\ -0.0001 - \iota 0.0136 \end{pmatrix}$	$\begin{pmatrix} 0.9997 + \iota 0.99996, \\ -0.0003 - \iota 0.0003 \end{pmatrix}$	$\begin{pmatrix} 0.9999 + \iota 0.9999, \\ -0.0006 - \iota 0.0003 \end{pmatrix}$
BCF-DHM	$\begin{pmatrix} 0.0183 + \iota 0.0325, \\ -0.9988 - \iota 0.9995 \end{pmatrix}$	$\begin{pmatrix} 0.0064 + \iota 0.0164, \\ -0.9592 - \iota 0.9997 \end{pmatrix}$	$\begin{pmatrix} 0.0057 + \iota 0.0347, \\ -0.9676 - \iota 0.9904 \end{pmatrix}$	$\begin{pmatrix} 0.0171 + \iota 0.0646, \\ -0.9771 - \iota 0.994 \end{pmatrix}$
BCF-WDHM	$\begin{pmatrix} 0.0004 + \iota 0.0008, \\ -0.9999 - \iota 0.9999 \end{pmatrix}$	$\begin{pmatrix} 0.00008 + \iota 0.0004, \\ -0.9990 - \iota 0.9999 \end{pmatrix}$	$\begin{pmatrix} 0.0001 + \iota 0.0009, \\ -0.9992 - \iota 0.9998 \end{pmatrix}$	$\begin{pmatrix} 0.0004 + \iota 0.0024, \\ -0.9994 - \iota 0.9998 \end{pmatrix}$

Table 7.3 The score values of the outcome are demonstrated in Table 7.3

Operators	$\mathbb{S}_{\mathbb{SV}}(\mathfrak{R}_{\mathbb{A}-1})$	$\mathbb{S}_{\mathbb{SV}}(\mathfrak{R}_{\mathbb{A}-2})$	$\mathbb{S}_{\mathbb{SV}}(\mathfrak{R}_{\mathbb{A}-3})$	$\mathbb{S}_{\mathbb{SV}}(\mathfrak{R}_{\mathbb{A}-4})$
BCF-HM	0.9478	0.9478	0.9922	0.9925
BCF-WHM	0.9951	0.9964	0.99978	0.99975
BCF-DHM	0.0131	0.016	0.0206	0.0276
BCF-WDHM	0.0003	0.0004	0.0005	0.0009

Table 7.4 The order ranking of the considered fields

Operators	Ranking
BCF-HM	$\mathfrak{R}_{\mathbb{A}-4} > \mathfrak{R}_{\mathbb{A}-3} > \mathfrak{R}_{\mathbb{A}-1} > \mathfrak{R}_{\mathbb{A}-2}$
BCF-WHM	$\mathfrak{R}_{\mathbb{A}-3} > \mathfrak{R}_{\mathbb{A}-4} > \mathfrak{R}_{\mathbb{A}-2} > \mathfrak{R}_{\mathbb{A}-1}$
BCF-DHM	$\mathfrak{R}_{\mathbb{A}-4} > \mathfrak{R}_{\mathbb{A}-3} > \mathfrak{R}_{\mathbb{A}-2} > \mathfrak{R}_{\mathbb{A}-1}$
BCF-WDHM	$\mathfrak{R}_{\mathbb{A}-4} > \mathfrak{R}_{\mathbb{A}-3} > \mathfrak{R}_{\mathbb{A}-2} > \mathfrak{R}_{\mathbb{A}-1}$

operators and techniques did not apply to this information. The theories discussed by Bi et al. [32] and Bi et al. [33] are in the model of CFS and the structure of CFS is lacking negative aspects, thus, these operators did not apply to this information. Only, the inaugurated operators and MADM process are giving us the result. Furthermore, the initiated operators can transform the model of BFS, CFS, and FS as follows,

- By ignoring the unreal parts in both positive and negative belonging grades, the inaugurated operators would be applicable to apply to BF information.
- By ignoring the negative belonging grades, the inaugurated operators would be applicable to apply CF information in the cartesian form.
- By ignoring the unreal parts in positive belonging grades and neglecting the negative belonging grades, the inaugurated operators would be applicable to apply fuzzy information.

As the inaugurated operators can transform to the model of FS, BFS, and CFS, thus the inaugurated MADM process can also transform to these models in a similar pattern.

Table 7.5 The comparison of the score values and order ranking of the inverted theory and a few prevailing theories

Source	$\mathbb{S}_{\mathbb{SV}}(\mathfrak{R}_{\mathbb{A}-1})$	$\mathbb{S}_{\mathbb{SV}}(\mathfrak{R}_{\mathbb{A}-2})$	$\mathbb{S}_{\mathbb{SV}}(\mathfrak{R}_{\mathbb{A}-3})$	$\mathbb{S}_{\mathbb{SV}}(\mathfrak{R}_{\mathbb{A}-4})$	Ranking
Jana et al. [29]	$\upharpoonright\rightarrowtail\upharpoonright\rightarrowtail\upharpoonright$	$\upharpoonright\rightarrowtail\upharpoonright\rightarrowtail\upharpoonright$	$\upharpoonright\rightarrowtail\upharpoonright\rightarrowtail\upharpoonright$	$\upharpoonright\rightarrowtail\upharpoonright\rightarrowtail\upharpoonright$	$\upharpoonright\rightarrowtail\upharpoonright\rightarrowtail\upharpoonright$
Jana et al. [30]	$\upharpoonright\rightarrowtail\upharpoonright\rightarrowtail\upharpoonright$	$\upharpoonright\rightarrowtail\upharpoonright\rightarrowtail\upharpoonright$	$\upharpoonright\rightarrowtail\upharpoonright\rightarrowtail\upharpoonright$	$\upharpoonright\rightarrowtail\upharpoonright\rightarrowtail\upharpoonright$	$\upharpoonright\rightarrowtail\upharpoonright\rightarrowtail\upharpoonright$
Wei et al. [4]	$\upharpoonright\rightarrowtail\upharpoonright\rightarrowtail\upharpoonright$	$\upharpoonright\rightarrowtail\upharpoonright\rightarrowtail\upharpoonright$	$\upharpoonright\rightarrowtail\upharpoonright\rightarrowtail\upharpoonright$	$\upharpoonright\rightarrowtail\upharpoonright\rightarrowtail\upharpoonright$	$\upharpoonright\rightarrowtail\upharpoonright\rightarrowtail\upharpoonright$
Riaz et al. [31]	$\upharpoonright\rightarrowtail\upharpoonright\rightarrowtail\upharpoonright$	$\upharpoonright\rightarrowtail\upharpoonright\rightarrowtail\upharpoonright$	$\upharpoonright\rightarrowtail\upharpoonright\rightarrowtail\upharpoonright$	$\upharpoonright\rightarrowtail\upharpoonright\rightarrowtail\upharpoonright$	$\upharpoonright\rightarrowtail\upharpoonright\rightarrowtail\upharpoonright$
Bi et al. [32]	$\upharpoonright\rightarrowtail\upharpoonright\rightarrowtail\upharpoonright$	$\upharpoonright\rightarrowtail\upharpoonright\rightarrowtail\upharpoonright$	$\upharpoonright\rightarrowtail\upharpoonright\rightarrowtail\upharpoonright$	$\upharpoonright\rightarrowtail\upharpoonright\rightarrowtail\upharpoonright$	$\upharpoonright\rightarrowtail\upharpoonright\rightarrowtail\upharpoonright$
Bi et al. [33]	$\upharpoonright\rightarrowtail\upharpoonright\rightarrowtail\upharpoonright$	$\upharpoonright\rightarrowtail\upharpoonright\rightarrowtail\upharpoonright$	$\upharpoonright\rightarrowtail\upharpoonright\rightarrowtail\upharpoonright$	$\upharpoonright\rightarrowtail\upharpoonright\rightarrowtail\upharpoonright$	$\upharpoonright\rightarrowtail\upharpoonright\rightarrowtail\upharpoonright$
BCF-HM	0.9478	0.9396	0.9922	0.9925	$\mathfrak{R}_{\mathbb{A}-4} > \mathfrak{R}_{\mathbb{A}-3} > \mathfrak{R}_{\mathbb{A}-1} > \mathfrak{R}_{\mathbb{A}-2}$
BCF-WHM	0.9951	0.9964	0.99978	0.99975	$\mathfrak{R}_{\mathbb{A}-3} > \mathfrak{R}_{\mathbb{A}-4} > \mathfrak{R}_{\mathbb{A}-2} > \mathfrak{R}_{\mathbb{A}-1}$
BCF-DHM	0.0131	0.016	0.0206	0.0276	$\mathfrak{R}_{\mathbb{A}-4} > \mathfrak{R}_{\mathbb{A}-3} > \mathfrak{R}_{\mathbb{A}-2} > \mathfrak{R}_{\mathbb{A}-1}$
BCF-WDHM	0.0003	0.0004	0.0005	0.0009	$\mathfrak{R}_{\mathbb{A}-4} > \mathfrak{R}_{\mathbb{A}-3} > \mathfrak{R}_{\mathbb{A}-2} > \mathfrak{R}_{\mathbb{A}-1}$

Chapter 8
WASPAS Technique Based on Power Generalized Heronian Mean Operators for Bipolar Complex Fuzzy Sets

Yager [34] first presented the power average operator in 2001, a new method for aggregating input values that takes the association between the values being aggregated. The aggregated input arguments determine the weight vector in the power operator. It enables collected values to complement and strengthen one another. The Heronian mean operator was created by Beliakov et al. [35] in 2007 to take into account the connections between the input arguments. After that, Skora [36] initiated the generalized Heronian mean operator, which is relying on the HM operator. A helpful AO that captures the relationships between assessment data is the generalized Heronian mean. Two AOs that take into account the interdependent phenomena among the aggregated arguments are the generalized Heronian mean (GHM) and generalized geometric Heronian mean (GGHM) operators. The primary distinction between the generalized Heronian operator and the power operator is that the former represents the subjective features of the input arguments being aggregated while the latter reflects the objective properties. In this chapter, we inaugurate power generalized Heronian mean and power generalized geometric Heronian mean operators in the structure of BCFS.

In Sect. 8.1, of this chapter, we investigate the BCF power generalized Heronian mean (BCFPGHM), BCF weighted power generalized Heronian mean (BCFWPGHM) operators, and the associated results. In Sect. 8.2. we investigate BCF power generalized geometric Heronian mean (BCFPGGHM), BCF weighted power generalized geometric Heronian mean (BCFWPGGHM) operators, and the associated results. Section 8.3, contain a weighted sum model and weighted product model (WASPAS) technique in the setting of BCF information. In Sect. 8.4. the application (supply chain transportation) is investigated. The comparative analysis is in Sect. 8.5.

U. ur Rehman and T. Mahmood, *Bipolar Complex Fuzzy Sets*, Studies in Fuzziness and Soft Computing 439, https://doi.org/10.1007/978-3-032-07388-4_8

8.1 Power Generalized Heronian Mean Operators for BCFNs

We recall the power averaging (PA) operator deduced by Yager [34] in 200.1

Let us assume a gathering of real numbers that is $\Re_{\mathcal{LN}-\sigma}, \sigma = 1, 2, \ldots, y$, then the PA operator is inaugurated underneath

$$PA(\Re_{\mathcal{LN}-1}, \Re_{\mathcal{LN}-2}, \ldots, \Re_{\mathcal{LN}-y}) = \sum_{\sigma=1}^{y}\left(\frac{1 + \underline{T}(\Re_{\mathcal{LN}-\sigma})}{\sum_{\varrho=1}^{y}(1 + \underline{T}(\Re_{\mathcal{LN}-\varrho}))}\Re_{\mathcal{LN}-\sigma}\right)$$

Noted that $\underline{T}(\Re_{\mathcal{LN}-\varrho}) = \sum_{\substack{\varrho=1\\ \varrho\neq\sigma}}^{y} Supp(\Re_{\mathcal{LN}-\sigma}, \Re_{\mathcal{LN}-\varrho})$, $Supp(\Re_{\mathcal{LN}-\sigma}, \Re_{\mathcal{LN}-\varrho})$ has the axioms 1) $Supp(\Re_{\mathcal{LN}-\sigma}, \Re_{\mathcal{LN}-\varrho}) \in [0, 1]$, 2)$Supp(\Re_{\mathcal{LN}-\sigma}, \Re_{\mathcal{LN}-\varrho}) = Supp(\Re_{\mathcal{LN}-\varrho}, \Re_{\mathcal{LN}-\sigma})$, 3) If $d(\Re_{\mathcal{LN}-\sigma}, \Re_{\mathcal{LN}-\varrho}) < \mathfrak{D}_s(\Re_{\mathcal{LN}-\sigma'}, \Re_{\mathcal{LN}-\varrho'})$, then $Supp(\Re_{\mathcal{LN}-\sigma}, \Re_{\mathcal{LN}-\varrho}) > Supp(\Re_{\mathcal{LN}-\sigma'}, \Re_{\mathcal{LN}-\varrho'})$, where $d(\Re_{\mathcal{LN}-\sigma}, \Re_{\mathcal{LN}-\varrho})$ indicates the distance between $\Re_{\mathcal{LN}-\sigma}$ and $\Re_{\mathcal{LN}-\varrho}$, and $Supp(\Re_{\mathcal{LN}-\sigma}, \Re_{\mathcal{LN}-\varrho})$ is identified as the support degree for $\Re_{\mathcal{LN}-\sigma}$ and $\Re_{\mathcal{LN}-\varrho}$.

Sykora [36] investigated the GHM operator by modifying the Heronian mean deduced by Beliakov et al. [35].

Let us consider a gathering of real numbers greater than or equal to zero that is $\Re_{\mathcal{PN}-\sigma}, (\sigma = 1, 2, \ldots, y)$ and $\text{ß}, \mathfrak{L} > 0$. Then the GHM operator is inaugurated as

$$GHM^{\text{ß},\mathfrak{L}}(\Re_{\mathcal{PN}-1}, \Re_{\mathcal{PN}-2}, \ldots, \Re_{\mathcal{PN}-y}) = \left(\frac{2}{y(y+1)}\sum_{\varrho=1}^{y}\sum_{\sigma=1}^{y}\left((\Re_{\mathcal{PN}-\varrho})^{\text{ß}}(\Re_{\mathcal{PN}-\sigma})^{\mathfrak{L}}\right)\right)^{\frac{1}{\text{ß}+\mathfrak{L}}}$$

If $\text{ß} = \mathfrak{L} = \frac{1}{2}$,then the GHM operator transforms to the Heronian mean operator, demonstrated underneath

$$HM^{\frac{1}{2},\frac{1}{2}}(\Re_{\mathcal{PN}-1}, \Re_{\mathcal{PN}-2}, \ldots, \Re_{\mathcal{PN}-y}) = \frac{2}{y(y+1)}\sum_{\varrho=1}^{y}\sum_{\sigma=1}^{y}\left((\Re_{\mathcal{PN}-\varrho}\Re_{\mathcal{PN}-\sigma})^{\frac{1}{2}}\right)$$

In the next definition, we are going to develop the notion of the BCFPGHM operator.

Definition 8.1 The BCFPGHM operator relying on the gathering of BCFNs $\Re_\sigma = (\mathfrak{F}^P_{\Re_\sigma}, \mathfrak{F}^N_{\Re_\sigma}) = (\mathfrak{F}^{RP}_{\Re_\sigma} + \iota\mathfrak{F}^{IP}_{\Re_\sigma}, \mathfrak{F}^{RN}_{\Re_\sigma} + \iota\mathfrak{F}^{IN}_{\Re_\sigma})\sigma = 1, 2, \ldots, y$, is spotted as

$$BCFPGHM^{\text{ß},\mathfrak{L}}(\Re_1, \Re_2, \ldots, \Re_y)$$

$$= \left(\frac{2}{y(y+1)} \overset{y}{\underset{\varrho=1}{\oplus}} \overset{y}{\underset{\sigma=1}{\oplus}} \left(\begin{array}{c} \left(\frac{y(1+\underline{T}(\Re_\varrho))}{\sum_{\rho=1}^{y}(1+\underline{T}(\Re_\rho))} \Re_\varrho \right)^{\text{ß}} \\ \otimes \left(\frac{y(1+\underline{T}(\Re_\sigma))}{\sum_{\rho=1}^{y}(1+\underline{T}(\Re_\rho))} \Re_\sigma \right)^{\mathfrak{L}} \end{array} \right) \right)^{\frac{1}{\text{ß}+\mathfrak{L}}} \tag{8.1}$$

Noticed that $\underline{T}(\Re_\varrho) = \sum_{\substack{\sigma=1 \\ \sigma \neq \varrho}}^{y} Sup(\Re_\varrho, \Re_\sigma)$, $Sup(\Re_\varrho, \Re_\sigma)$ is the support among two BCFNs $\Re_\varrho$, and $\Re_\sigma$ and satisfy the underneath axioms.

1. $Sup(\Re_\varrho, \Re_\sigma) \in [0, 1]$;
2. $Sup(\Re_\varrho, \Re_\sigma) = Sup(\Re_\sigma, \Re_\varrho)$;
3. $Sup(\Re_\varrho, \Re_\sigma) \geq Sup(\Re_{\varrho'}, \Re_{\sigma'})$ if $d_{BCFS}(\Re_\varrho, \Re_\sigma) < d_{BCFS}(\Re'_\varrho, \Re'_\sigma)$, where, d_{BCFS} is any distance among BCFNs.

Further, if we let $\frac{(1+\underline{T}(\Re_\varrho))}{\sum_{\rho=1}^{y}(1+\underline{T}(\Re_\rho))} = \Xi_\varrho$, then Eq. (8.1) will be in the underneath form

$$BCFPGHM^{\text{ß},\mathfrak{L}}(\Re_1, \Re_2, \ldots, \Re_y)$$

$$= \left(\frac{2}{y(y+1)} \overset{y}{\underset{\varrho=1}{\oplus}} \overset{y}{\underset{\sigma=1}{\oplus}} \left(y\Xi_\varrho \Re_\varrho\right)^{\text{ß}} \otimes \left(y\Xi_\sigma \Re_\sigma\right)^{\mathfrak{L}} \right)^{\frac{1}{\text{ß}+\mathfrak{L}}}$$

Noted that $(\Xi_1, \Xi_2, \ldots, \Xi_y)$ is a power-weighted vector, $\Xi_\varrho \in [0, 1]$ and $\sum_{\varrho=1}^{y} \Xi_\varrho = 1$.

The next Theorem will show that the BCFPGHM will give us again a BCFN.

Theorem 8.1 Let a gathering of BCFNs $\Re_\sigma = \left(\mathfrak{F}^{P}_{\Re_\sigma}, \mathfrak{F}^{N}_{\Re_\sigma}\right) = \left(\mathfrak{F}^{RP}_{\Re_\sigma} + \iota\mathfrak{F}^{IP}_{\Re_\sigma}, \mathfrak{F}^{RN}_{\Re_\sigma} + \iota\mathfrak{F}^{IN}_{\Re_\sigma}\right)\sigma = 1, 2, \ldots, y$. The aggregated outcome established by definition 8.2 is still a BCFN that is

$$BCFPGHM^{\text{ß}\mathfrak{L}}(\Re_1, \Re_2, \ldots, \Re_y)$$

$$\left(\begin{array}{l}\left(1-\left(\prod_{\substack{\varrho=1\\ \sigma=1}}^{y}\left(1-\left(1-\left(1-\mathfrak{F}_{\Re_\varrho}^{RP}\right)^{y\Xi_\varrho}\right)^{\text{ß}}\left(1-\left(1-\mathfrak{F}_{\Re_\sigma}^{RP}\right)^{y\Xi_\sigma}\right)^{\mathfrak{L}}\right)\right)^{\frac{2}{y(y+1)}}\right)^{\frac{1}{\text{ß}+\mathfrak{L}}}\\ +\iota\left(1-\left(\prod_{\substack{\varrho=1\\ \sigma=1}}^{y}\left(1-\left(1-\left(1-\mathfrak{F}_{\Re_\varrho}^{IP}\right)^{y\Xi_\varrho}\right)^{\text{ß}}\left(1-\left(1-\mathfrak{F}_{\Re_\sigma}^{IP}\right)^{y\Xi_\sigma}\right)^{\mathfrak{L}}\right)\right)^{\frac{2}{y(y+1)}}\right)^{\frac{1}{\text{ß}+\mathfrak{L}}},\\ -1+\left(1-\left|(-1)^{y+1}\prod_{\substack{\varrho=1\\ \sigma=1}}^{y}\left(-1+\left(1-\left|\mathfrak{F}_{\Re_\varrho}^{RN}\right|^{y\Xi_\varrho}\right)^{\text{ß}}\left(1-\left|\mathfrak{F}_{\Re_\sigma}^{RN}\right|^{y\Xi_\sigma}\right)^{\mathfrak{L}}\right)\right|^{\frac{2}{y(y+1)}}\right)^{\frac{1}{\text{ß}+\mathfrak{L}}}\\ +\iota\left(-1+\left(1-\left|(-1)^{y+1}\prod_{\substack{\varrho=1\\ \sigma=1}}^{y}\left(-1+\left(1-\left|\mathfrak{F}_{\Re_\varrho}^{IN}\right|^{y\Xi_\varrho}\right)^{\text{ß}}\left(1-\left|\mathfrak{F}_{\Re_\sigma}^{IN}\right|^{y\Xi_\sigma}\right)^{\mathfrak{L}}\right)\right|^{\frac{2}{y(y+1)}}\right)^{\frac{1}{\text{ß}+\mathfrak{L}}}\right)\end{array}\right)$$

where, $\Xi_\varrho = \frac{(1+\underline{T}(\Re_\varrho))}{\sum_{\rho=1}^{y}(1+\underline{T}(\Re_\rho))}$.

Proof From the primary operations of BCFNs, we have

$$y\Xi_\varrho\Re_\varrho = \left(\begin{array}{l}1-\left(1-\mathfrak{F}_{\Re_\varrho}^{RP}\right)^{y\Xi_\varrho}+\iota\left(1-\left(1-\mathfrak{F}_{\Re_\varrho}^{IP}\right)^{y\Xi_\varrho}\right),\\ -\left|\mathfrak{F}_{\Re_\varrho}^{RN}\right|^{y\Xi_\varrho}+\iota\left(-\left|\mathfrak{F}_{\Re_\varrho}^{IN}\right|^{y\Xi_\varrho}\right)\end{array}\right)$$

$$y\Xi_\sigma\Re_\sigma = \left(\begin{array}{l}1-\left(1-\mathfrak{F}_{\Re_\sigma}^{RP}\right)^{y\Xi_\sigma}+\iota\left(1-\left(1-\mathfrak{F}_{\Re_\sigma}^{IP}\right)^{y\Xi_\sigma}\right),\\ -\left|\mathfrak{F}_{\Re_\sigma}^{RN}\right|^{y\Xi_\sigma}+\iota\left(-\left|\mathfrak{F}_{\Re_\sigma}^{IN}\right|^{y\Xi_\sigma}\right)\end{array}\right)$$

then,

$$\left(y\Xi_\varrho\Re_\varrho\right)^{\text{ß}} = \left(\begin{array}{l}\left(1-\left(1-\mathfrak{F}_{\Re_\varrho}^{RP}\right)^{y\Xi_\varrho}\right)^{\text{ß}}+\iota\left(\left(1-\left(1-\mathfrak{F}_{\Re_\varrho}^{IP}\right)^{y\Xi_\varrho}\right)\right)^{\text{ß}},\\ -1+\left(1-\left|\mathfrak{F}_{\Re_\varrho}^{RN}\right|^{y\Xi_\varrho}\right)^{\text{ß}}+\iota\left(-1+\left(1-\left|\mathfrak{F}_{\Re_\varrho}^{IN}\right|^{y\Xi_\varrho}\right)^{\text{ß}}\right)\end{array}\right)$$

$$(y\Xi_\sigma \Re_\sigma)^{\mathfrak{L}} = \begin{pmatrix} \left(1-\left(1-\mathfrak{F}^{RP}_{\Re_\sigma}\right)^{y\Xi_\sigma}\right)^{\mathfrak{L}} + \iota\left(1-\left(1-\mathfrak{F}^{IP}_{\Re_\sigma}\right)^{y\Xi_\sigma}\right)^{\mathfrak{L}}, \\ -1+\left(1-\left|\mathfrak{F}^{RN}_{\Re_\sigma}\right|^{y\Xi_\sigma}\right)^{\mathfrak{L}} + \iota\left(-1+\left(1-\left|\mathfrak{F}^{IN}_{\Re_\sigma}\right|^{y\Xi_\sigma}\right)^{\mathfrak{L}}\right) \end{pmatrix}$$

$$\begin{aligned} &\left(y\Xi_\varrho \Re_\varrho\right)^{\text{ß}} \otimes (y\Xi_\sigma \Re_\sigma)^{\mathfrak{L}} \\ &= \begin{pmatrix} \left(1-\left(1-\mathfrak{F}^{RP}_{\Re_\varrho}\right)^{y\Xi_\varrho}\right)^{\text{ß}} + \iota\left(\left(1-\left(1-\mathfrak{F}^{IP}_{\Re_\varrho}\right)^{y\Xi_\varrho}\right)\right)^{\text{ß}}, \\ -1+\left(1-\left|\mathfrak{F}^{RN}_{\Re_\varrho}\right|^{y\Xi_\varrho}\right)^{\text{ß}} + \iota\left(-1+\left(1-\left|\mathfrak{F}^{IN}_{\Re_\varrho}\right|^{y\Xi_\varrho}\right)^{\text{ß}}\right) \end{pmatrix} \\ &\otimes \begin{pmatrix} \left(1-\left(1-\mathfrak{F}^{RP}_{\Re_\sigma}\right)^{y\Xi_\sigma}\right)^{\mathfrak{L}} + \iota\left(1-\left(1-\mathfrak{F}^{IP}_{\Re_\sigma}\right)^{y\Xi_\sigma}\right)^{\mathfrak{L}}, \\ -1+\left(1-\left|\mathfrak{F}^{RN}_{\Re_\sigma}\right|^{y\Xi_\sigma}\right)^{\mathfrak{L}} + \iota\left(-1+\left(1-\left|\mathfrak{F}^{IN}_{\Re_\sigma}\right|^{y\Xi_\sigma}\right)^{\mathfrak{L}}\right) \end{pmatrix} \\ &= \begin{pmatrix} \left(1-\left(1-\mathfrak{F}^{RP}_{\Re_\varrho}\right)^{y\Xi_\varrho}\right)^{\text{ß}} \left(1-\left(1-\mathfrak{F}^{RP}_{\Re_\sigma}\right)^{y\Xi_\sigma}\right)^{\mathfrak{L}} \\ +\iota\left(\left(1-\left(1-\mathfrak{F}^{IP}_{\Re_\varrho}\right)^{y\Xi_\varrho}\right)^{\text{ß}} \left(1-\left(1-\mathfrak{F}^{IP}_{\Re_\sigma}\right)^{y\Xi_\sigma}\right)^{\mathfrak{L}}\right), \\ -1+\left(1-\left|\mathfrak{F}^{RN}_{\Re_\varrho}\right|^{y\Xi_\varrho}\right)^{\text{ß}} \left(1-\left|\mathfrak{F}^{RN}_{\Re_\sigma}\right|^{y\Xi_\sigma}\right)^{\mathfrak{L}} \\ +\iota\left(-1+\left(1-\left|\mathfrak{F}^{IN}_{\Re_\varrho}\right|^{y\Xi_\varrho}\right)^{\text{ß}} \left(1-\left|\mathfrak{F}^{IN}_{\Re_\sigma}\right|^{y\Xi_\sigma}\right)^{\mathfrak{L}}\right) \end{pmatrix} \end{aligned}$$

$$\underset{\varrho=1}{\overset{y}{\oplus}} \underset{\sigma=1}{\overset{y}{\oplus}} \left(y\Xi_\varrho \Re_\varrho\right)^{\text{ß}} \otimes (y\Xi_\sigma \Re_\sigma)^{\mathfrak{L}} = \begin{pmatrix} 1-\prod\limits_{\substack{\varrho=1 \\ \sigma=1}}^{y} \left(1-\left(1-\left(1-\mathfrak{F}^{RP}_{\Re_\varrho}\right)^{y\Xi_\varrho}\right)^{\text{ß}} \left(1-\left(1-\mathfrak{F}^{RP}_{\Re_\sigma}\right)^{y\Xi_\sigma}\right)^{\mathfrak{L}}\right) \\ +\iota 1-\prod\limits_{\substack{\varrho=1 \\ \sigma=1}}^{y} \left(1-\left(1-\left(1-\mathfrak{F}^{IP}_{\Re_\varrho}\right)^{y\Xi_\varrho}\right)^{\text{ß}} \left(1-\left(1-\mathfrak{F}^{IP}_{\Re_\sigma}\right)^{y\Xi_\sigma}\right)^{\mathfrak{L}}\right), \\ (-1)^{y+1} \prod\limits_{\substack{\varrho=1 \\ \sigma=1}}^{y} \left(-1+\left(1-\left|\mathfrak{F}^{RN}_{\Re_\varrho}\right|^{y\Xi_\varrho}\right)^{\text{ß}} \left(1-\left|\mathfrak{F}^{RN}_{\Re_\sigma}\right|^{y\Xi_\sigma}\right)^{\mathfrak{L}}\right) \\ +\iota(-1)^{y+1} \prod\limits_{\substack{\varrho=1 \\ \sigma=1}}^{y} \left(-1+\left(1-\left|\mathfrak{F}^{IN}_{\Re_\varrho}\right|^{y\Xi_\varrho}\right)^{\text{ß}} \left(1-\left|\mathfrak{F}^{IN}_{\Re_\sigma}\right|^{y\Xi_\sigma}\right)^{\mathfrak{L}}\right) \end{pmatrix}$$

And

$$\frac{2}{y(y+1)} \underset{\varrho=1}{\overset{y}{\oplus}} \underset{\sigma=1}{\overset{y}{\oplus}} \left(y\Xi_\varrho \Re_\varrho\right)^{\text{ß}} \otimes (y\Xi_\sigma \Re_\sigma)^{\mathfrak{L}}$$

$$
= \left(\begin{array}{l}
1 - \left(\prod_{\substack{\varrho = 1 \\ \sigma = 1}}^{y} \left(1 - \left(1 - \left(1 - \mathfrak{F}_{\Re_\varrho}^{RP} \right)^{y\Xi_\varrho} \right)^{\text{ß}} \left(1 - \left(1 - \mathfrak{F}_{\Re_\sigma}^{RP} \right)^{y\Xi_\sigma} \right)^{\mathfrak{L}} \right) \right)^{\frac{2}{y(y+1)}} \\
+\iota \left(1 - \left(\prod_{\substack{\varrho = 1 \\ \sigma = 1}}^{y} \left(1 - \left(1 - \left(1 - \mathfrak{F}_{\Re_\varrho}^{IP} \right)^{y\Xi_\varrho} \right)^{\text{ß}} \left(1 - \left(1 - \mathfrak{F}_{\Re_\sigma}^{IP} \right)^{y\Xi_\sigma} \right)^{\mathfrak{L}} \right) \right)^{\frac{2}{y(y+1)}} \right), \\
- \left| (-1)^{y+1} \prod_{\substack{\varrho = 1 \\ \sigma = 1}}^{y} \left(-1 + \left(1 - \left| \mathfrak{F}_{\Re_\varrho}^{RN} \right|^{y\Xi_\varrho} \right)^{\text{ß}} \left(1 - \left| \mathfrak{F}_{\Re_\sigma}^{RN} \right|^{y\Xi_\sigma} \right)^{\mathfrak{L}} \right) \right|^{\frac{2}{y(y+1)}} \\
+\iota \left(- \left| (-1)^{y+1} \prod_{\substack{\varrho = 1 \\ \sigma = 1}}^{y} \left(-1 + \left(1 - \left| \mathfrak{F}_{\Re_\varrho}^{IN} \right|^{y\Xi_\varrho} \right)^{\text{ß}} \left(1 - \left| \mathfrak{F}_{\Re_\sigma}^{IN} \right|^{y\Xi_\sigma} \right)^{\mathfrak{L}} \right) \right|^{\frac{2}{y(y+1)}} \right)
\end{array} \right)
$$

afterwards,

$$
\left(\frac{2}{y(y+1)} \underset{\varrho = 1}{\overset{y}{\oplus}} \underset{\sigma = 1}{\overset{y}{\oplus}} \left(y\Xi_\varrho \Re_\varrho \right)^{\text{ß}} \otimes \left(y\Xi_\sigma \Re_\sigma \right)^{\mathfrak{L}} \right)^{\frac{1}{\text{ß}+\mathfrak{L}}}
$$

$$= \left(\begin{array}{l} \left(\left(1 - \left(\prod_{\substack{\varrho = 1 \\ \sigma = 1}}^{y} \left(1 - \left(1 - \left(1 - \mathfrak{F}_{\mathfrak{R}_\varrho}^{RP} \right)^{y\Xi_\varrho} \right)^{\text{ß}} \left(1 - \left(1 - \mathfrak{F}_{\mathfrak{R}_\sigma}^{RP} \right)^{y\Xi_\sigma} \right)^{\mathfrak{L}} \right) \right)^{\frac{2}{y(y+1)}} \right)^{\frac{1}{\text{ß}+\mathfrak{L}}} \right. \\ \left. + \iota \left(1 - \left(\prod_{\substack{\varrho = 1 \\ \sigma = 1}}^{y} \left(1 - \left(1 - \left(1 - \mathfrak{F}_{\mathfrak{R}_\varrho}^{IP} \right)^{y\Xi_\varrho} \right)^{\text{ß}} \left(1 - \left(1 - \mathfrak{F}_{\mathfrak{R}_\sigma}^{IP} \right)^{y\Xi_\sigma} \right)^{\mathfrak{L}} \right) \right)^{\frac{2}{y(y+1)}} \right)^{\frac{1}{\text{ß}+\mathfrak{L}}} \right), \\ -1 + \left(1 - \left| (-1)^{y+1} \prod_{\substack{\varrho = 1 \\ \sigma = 1}}^{y} \left(-1 + \left(1 - \left| \mathfrak{F}_{\mathfrak{R}_\varrho}^{RN} \right|^{y\Xi_\varrho} \right)^{\text{ß}} \left(1 - \left| \mathfrak{F}_{\mathfrak{R}_\sigma}^{RN} \right|^{y\Xi_\sigma} \right)^{\mathfrak{L}} \right) \right|^{\frac{2}{y(y+1)}} \right)^{\frac{1}{\text{ß}+\mathfrak{L}}} \\ + \iota \left(-1 + \left(1 - \left| (-1)^{y+1} \prod_{\substack{\varrho = 1 \\ \sigma = 1}}^{y} \left(-1 + \left(1 - \left| \mathfrak{F}_{\mathfrak{R}_\varrho}^{IN} \right|^{y\Xi_\varrho} \right)^{\text{ß}} \left(1 - \left| \mathfrak{F}_{\mathfrak{R}_\sigma}^{IN} \right|^{y\Xi_\sigma} \right)^{\mathfrak{L}} \right) \right|^{\frac{2}{y(y+1)}} \right)^{\frac{1}{\text{ß}+\mathfrak{L}}} \right) \end{array} \right)$$

Thus,

$$BCFPGHM^{\text{ß},\mathfrak{L}}(\mathfrak{R}_1, \mathfrak{R}_2, \ldots, \mathfrak{R}_y)$$

$$= \left(\begin{array}{l} \left(\left(1 - \left(\prod_{\substack{\varrho = 1 \\ \sigma = 1}}^{y} \left(1 - \left(1 - \left(1 - \mathfrak{F}_{\mathfrak{R}_\varrho}^{RP} \right)^{y\Xi_\varrho} \right)^{\text{ß}} \left(1 - \left(1 - \mathfrak{F}_{\mathfrak{R}_\sigma}^{RP} \right)^{y\Xi_\sigma} \right)^{\mathfrak{L}} \right) \right)^{\frac{2}{y(y+1)}} \right)^{\frac{1}{\text{ß}+\mathfrak{L}}} \right. \\ \left. + \iota \left(1 - \left(\prod_{\substack{\varrho = 1 \\ \sigma = 1}}^{y} \left(1 - \left(1 - \left(1 - \mathfrak{F}_{\mathfrak{R}_\varrho}^{IP} \right)^{y\Xi_\varrho} \right)^{\text{ß}} \left(1 - \left(1 - \mathfrak{F}_{\mathfrak{R}_\sigma}^{IP} \right)^{y\Xi_\sigma} \right)^{\mathfrak{L}} \right) \right)^{\frac{2}{y(y+1)}} \right)^{\frac{1}{\text{ß}+\mathfrak{L}}} \right), \\ -1 + \left(1 - \left| (-1)^{y+1} \prod_{\substack{\varrho = 1 \\ \sigma = 1}}^{y} \left(-1 + \left(1 - \left| \mathfrak{F}_{\mathfrak{R}_\varrho}^{RN} \right|^{y\Xi_\varrho} \right)^{\text{ß}} \left(1 - \left| \mathfrak{F}_{\mathfrak{R}_\sigma}^{RN} \right|^{y\Xi_\sigma} \right)^{\mathfrak{L}} \right) \right|^{\frac{2}{y(y+1)}} \right)^{\frac{1}{\text{ß}+\mathfrak{L}}} \\ + \iota \left(-1 + \left(1 - \left| (-1)^{y+1} \prod_{\substack{\varrho = 1 \\ \sigma = 1}}^{y} \left(-1 + \left(1 - \left| \mathfrak{F}_{\mathfrak{R}_\varrho}^{IN} \right|^{y\Xi_\varrho} \right)^{\text{ß}} \left(1 - \left| \mathfrak{F}_{\mathfrak{R}_\sigma}^{IN} \right|^{y\Xi_\sigma} \right)^{\mathfrak{L}} \right) \right|^{\frac{2}{y(y+1)}} \right)^{\frac{1}{\text{ß}+\mathfrak{L}}} \right) \end{array} \right)$$

Proposition 8.1 (Idempotency): Let a gathering of BCFNs $\Re_\sigma = \left(\mathfrak{F}^P_{\Re_\sigma}, \mathfrak{F}^N_{\Re_\sigma}\right) = \left(\mathfrak{F}^{RP}_{\Re_\sigma} + \iota\mathfrak{F}^{IP}_{\Re_\sigma}, \mathfrak{F}^{RN}_{\Re_\sigma} + \iota\mathfrak{F}^{IN}_{\Re_\sigma}\right)\sigma = 1, 2, \ldots, y$. If $\forall\sigma\ \Re_\sigma = \Re$, then

$$BCFPGHM^{\text{ß},\mathfrak{L}}\left(\Re_1, \Re_2, \ldots, \Re_y\right) = \Re$$

Proof As $\forall\sigma\ \Re_\sigma = \Re$, then,

$$BCFPGHM^{\text{ß},\mathfrak{L}}(\Re_1, \Re_2, \ldots, \Re_y) = \left(\frac{2}{y(y+1)} \underset{\varrho=1}{\overset{y}{\oplus}} \underset{\sigma=1}{\overset{y}{\oplus}} \left(y\Xi_\varrho\Re_\varrho\right)^{\text{ß}} \otimes \left(y\Xi_\sigma\Re_\sigma\right)^{\mathfrak{L}}\right)^{\frac{1}{\text{ß}+\mathfrak{L}}}$$
$$= \left(\frac{2}{y(y+1)} \underset{\varrho=1}{\overset{y}{\oplus}} \underset{\sigma=1}{\overset{y}{\oplus}} \left(y\Xi_\varrho\Re_\varrho\right)^{\text{ß}} \otimes \left(y\Xi_\sigma\Re_\sigma\right)^{\mathfrak{L}}\right)^{\frac{1}{\text{ß}+\mathfrak{L}}}$$

Proposition 8.2 (Monotonicity): Let two gatherings of BCFNs $\Re_\sigma = \left(\mathfrak{F}^P_{\Re_\sigma}, \mathfrak{F}^N_{\Re_\sigma}\right) = \left(\mathfrak{F}^{RP}_{\Re_\sigma} + \iota\mathfrak{F}^{IP}_{\Re_\sigma}, \mathfrak{F}^{RN}_{\Re_\sigma} + \iota\mathfrak{F}^{IN}_{\Re_\sigma}\right)$ and $\Re_{\sigma'} = \left(\mathfrak{F}^P_{\Re_{\sigma'}}, \mathfrak{F}^N_{\Re_{\sigma'}}\right) = \left(\mathfrak{F}^{RP}_{\Re_{\sigma'}} + \iota\mathfrak{F}^{IP}_{\Re_{\sigma'}}, \mathfrak{F}^{RN}_{\Re_{\sigma'}} + \iota\mathfrak{F}^{IN}_{\Re_{\sigma'}}\right)\sigma = 1, 2, \ldots, y$. If $\mathfrak{F}^{RP}_{\Re_\sigma} \le \mathfrak{F}^{RP}_{\Re_{\sigma'}}$, $\mathfrak{F}^{IP}_{\Re_\sigma} \le \mathfrak{F}^{IP}_{\Re_{\sigma'}}$, $\mathfrak{F}^{RN}_{\Re_\sigma} \le \mathfrak{F}^{RN}_{\Re_{\sigma'}}$, and $\mathfrak{F}^{IN}_{\Re_\sigma} \le \mathfrak{F}^{IN}_{\Re_{\sigma'}}$, then

$$BCFPGHM^{\text{ß},\mathfrak{L}}\left(\Re_1, \Re_2, \ldots, \Re_y\right) \le BCFPGHM^{\text{ß},\mathfrak{L}}\left(\Re_{1'}, \Re_{2'}, \ldots, \Re_{y'}\right)$$

Proof As for any two BCFNs $\Re_{\mathfrak{L}(\sigma)}$ and $\Re_{\mathfrak{L}(\sigma)'}$, we have that $\forall\sigma\ \Re_{\mathfrak{L}(\sigma)} \le \Re_{\mathfrak{L}(\sigma)'}$ which means that $\mathfrak{F}^{RP}_{\Re_\sigma} \le \mathfrak{F}^{RP}_{\Re_{\sigma'}}$, $\mathfrak{F}^{IP}_{\Re_\sigma} \le \mathfrak{F}^{IP}_{\Re_{\sigma'}}$, $\mathfrak{F}^{RN}_{\Re_\sigma} \le \mathfrak{F}^{RN}_{\Re_{\sigma'}}$, and $\mathfrak{F}^{IN}_{\Re_\sigma} \le \mathfrak{F}^{IN}_{\Re_{\sigma'}}$. As $\mathfrak{F}^{RP}_{\Re_\sigma} \le \mathfrak{F}^{RP}_{\Re_{\sigma'}}$ then,

As $\mathfrak{F}^{RP}_{\Re_\sigma} \le \mathfrak{F}^{RP}_{\Re_{\sigma'}}$, then

$$1 - \mathfrak{F}^{RP}_{\Re_\varrho} \ge 1 - \mathfrak{F}^{RP}_{\Re_{\varrho'}}$$

$$\Rightarrow 1 - \left(1 - \mathfrak{F}^{RP}_{\Re_\varrho}\right)^{y\Xi_\varrho} \le 1 - \left(1 - \mathfrak{F}^{RP}_{\Re_{\varrho'}}\right)^{y\Xi_\varrho}$$

then,

$$\left(1-\left(1-\mathfrak{F}^{RP}_{\Re_\varrho}\right)^{y\Xi_\varrho}\right)^{\text{ß}}\left(1-\left(1-\mathfrak{F}^{RP}_{\Re_\sigma}\right)^{y\Xi_\sigma}\right)^{\mathfrak{L}} \le \left(1-\left(1-\mathfrak{F}^{RP}_{\Re_{\varrho'}}\right)^{y\Xi_\varrho}\right)^{\text{ß}}\left(1-\left(1-\mathfrak{F}^{RP}_{\Re_{\sigma'}}\right)^{y\Xi_\sigma}\right)^{\mathfrak{L}}$$
$$\Rightarrow 1-\left(1-\left(1-\mathfrak{F}^{RP}_{\Re_\varrho}\right)^{y\Xi_\varrho}\right)^{\text{ß}}\left(1-\left(1-\mathfrak{F}^{RP}_{\Re_\sigma}\right)^{y\Xi_\sigma}\right)^{\mathfrak{L}} \ge 1-\left(1-\left(1-\mathfrak{F}^{RP}_{\Re_{\varrho'}}\right)^{y\Xi_\varrho}\right)^{\text{ß}}\left(1-\left(1-\mathfrak{F}^{RP}_{\Re_{\sigma'}}\right)^{y\Xi_\sigma}\right)^{\mathfrak{L}}$$

$$\Rightarrow \left(\prod_{\substack{\varrho=1\\ \sigma=1}}^{y} \left(1-\left(1-\left(1-\mathfrak{F}^{RP}_{\Re_\varrho}\right)^{y\Xi_\varrho}\right)^{\text{ß}}\left(1-\left(1-\mathfrak{F}^{RP}_{\Re_\sigma}\right)^{y\Xi_\sigma}\right)^{\mathfrak{L}}\right)\right)^{\frac{2}{y(y+1)}}$$

$$\geq \left(\prod_{\substack{\varrho=1 \\ \sigma=1}}^{y} \left(1 - \left(1 - \left(1 - \mathfrak{F}_{\mathfrak{R}_{\varrho'}}^{RP}\right)^{y\Xi_\varrho}\right)^{ß} \left(1 - \left(1 - \mathfrak{F}_{\mathfrak{R}_{\sigma'}}^{RP}\right)^{y\Xi_\sigma}\right)^{\mathfrak{L}}\right) \right)^{\frac{2}{y(y+1)}}$$

and

$$1 - \left(\prod_{\substack{\varrho=1 \\ \sigma=1}}^{y} \left(1 - \left(1 - \left(1 - \mathfrak{F}_{\mathfrak{R}_{\varrho}}^{RP}\right)^{y\Xi_\varrho}\right)^{ß} \left(1 - \left(1 - \mathfrak{F}_{\mathfrak{R}_{\sigma}}^{RP}\right)^{y\Xi_\sigma}\right)^{\mathfrak{L}}\right) \right)^{\frac{2}{y(y+1)}}$$

$$\leq 1 - \left(\prod_{\substack{\varrho=1 \\ \sigma=1}}^{y} \left(1 - \left(1 - \left(1 - \mathfrak{F}_{\mathfrak{R}_{\varrho'}}^{RP}\right)^{y\Xi_\varrho}\right)^{ß} \left(1 - \left(1 - \mathfrak{F}_{\mathfrak{R}_{\sigma'}}^{RP}\right)^{y\Xi_\sigma}\right)^{\mathfrak{L}}\right) \right)^{\frac{2}{y(y+1)}}$$

This implies that

$$\left(1 - \left(\prod_{\substack{\varrho=1 \\ \sigma=1}}^{y} \left(1 - \left(1 - \left(1 - \mathfrak{F}_{\mathfrak{R}_{\varrho}}^{RP}\right)^{y\Xi_\varrho}\right)^{ß} \left(1 - \left(1 - \mathfrak{F}_{\mathfrak{R}_{\sigma}}^{RP}\right)^{y\Xi_\sigma}\right)^{\mathfrak{L}}\right) \right)^{\frac{2}{y(y+1)}}\right)^{\frac{1}{ß+\mathfrak{L}}}$$

$$\leq \left(1 - \left(\prod_{\substack{\varrho=1 \\ \sigma=1}}^{y} \left(1 - \left(1 - \left(1 - \mathfrak{F}_{\mathfrak{R}_{\varrho'}}^{RP}\right)^{y\Xi_\varrho}\right)^{ß} \left(1 - \left(1 - \mathfrak{F}_{\mathfrak{R}_{\sigma'}}^{RP}\right)^{y\Xi_\sigma}\right)^{\mathfrak{L}}\right) \right)^{\frac{2}{y(y+1)}}\right)^{\frac{1}{ß+\mathfrak{L}}}$$

Similarly, we can get

$$\left(1-\left(\prod_{\substack{\varrho=1\\ \sigma=1}}^{y}\left(1-\left(1-\left(1-\mathfrak{F}_{\Re_{\varrho}}^{IP}\right)^{y\Xi_{\varrho}}\right)^{ß}\left(1-\left(1-\mathfrak{F}_{\Re_{\sigma}}^{IP}\right)^{y\Xi_{\sigma}}\right)^{\mathfrak{L}}\right)\right)^{\frac{2}{y(y+1)}}\right)^{\frac{1}{ß+\mathfrak{L}}}$$

$$\leq\left(1-\left(\prod_{\substack{\varrho=1\\ \sigma=1}}^{y}\left(1-\left(1-\left(1-\mathfrak{F}_{\Re_{\varrho'}}^{IP}\right)^{y\Xi_{\varrho}}\right)^{ß}\left(1-\left(1-\mathfrak{F}_{\Re_{\sigma'}}^{IP}\right)^{y\Xi_{\sigma}}\right)^{\mathfrak{L}}\right)\right)^{\frac{2}{y(y+1)}}\right)^{\frac{1}{ß+\mathfrak{L}}}$$

Further, as $\mathfrak{F}_{\Re_{\sigma}}^{RN}\leq\mathfrak{F}_{\Re_{\sigma'}}^{RN}$, then

$$\left|\mathfrak{F}_{\Re_{\varrho}}^{RN}\right|^{y\Xi_{\varrho}}\geq\left|\mathfrak{F}_{\Re_{\varrho'}}^{RN}\right|^{y\Xi_{\varrho}}$$

$$\Rightarrow 1-\left|\mathfrak{F}_{\Re_{\varrho}}^{RN}\right|^{y\Xi_{\varrho}}\leq 1-\left|\mathfrak{F}_{\Re_{\varrho'}}^{RN}\right|^{y\Xi_{\varrho}}$$

$$\Rightarrow\left(1-\left|\mathfrak{F}_{\Re_{\varrho}}^{RN}\right|^{y\Xi_{\varrho}}\right)^{ß}\left(1-\left|\mathfrak{F}_{\Re_{\sigma}}^{RN}\right|^{y\Xi_{\sigma}}\right)^{\mathfrak{L}}\leq\left(1-\left|\mathfrak{F}_{\Re_{\varrho'}}^{RN}\right|^{y\Xi_{\varrho}}\right)^{ß}\left(1-\left|\mathfrak{F}_{\Re_{\sigma'}}^{RN}\right|^{y\Xi_{\sigma}}\right)^{\mathfrak{L}}$$

then,

$$-1+\left(1-\left|\mathfrak{F}_{\Re_{\varrho}}^{RN}\right|^{y\Xi_{\varrho}}\right)^{ß}\left(1-\left|\mathfrak{F}_{\Re_{\sigma}}^{RN}\right|^{y\Xi_{\sigma}}\right)^{\mathfrak{L}}\leq -1+\left(1-\left|\mathfrak{F}_{\Re_{\varrho'}}^{RN}\right|^{y\Xi_{\varrho}}\right)^{ß}\left(1-\left|\mathfrak{F}_{\Re_{\sigma'}}^{RN}\right|^{y\Xi_{\sigma}}\right)^{\mathfrak{L}}$$

$$\Rightarrow\left|(-1)^{y+1}\prod_{\substack{\varrho=1\\ \sigma=1}}^{y}\left(-1+\left(1-\left|\mathfrak{F}_{\Re_{\varrho}}^{RN}\right|^{y\Xi_{\varrho}}\right)^{ß}\left(1-\left|\mathfrak{F}_{\Re_{\sigma}}^{RN}\right|^{y\Xi_{\sigma}}\right)^{\mathfrak{L}}\right)\right|^{\frac{2}{y(y+1)}}$$

$$\geq\left|(-1)^{y+1}\prod_{\substack{\varrho=1\\ \sigma=1}}^{y}\left(-1+\left(1-\left|\mathfrak{F}_{\Re_{\varrho'}}^{RN}\right|^{y\Xi_{\varrho}}\right)^{ß}\left(1-\left|\mathfrak{F}_{\Re_{\sigma'}}^{RN}\right|^{y\Xi_{\sigma}}\right)^{\mathfrak{L}}\right)\right|^{\frac{2}{y(y+1)}}$$

$$\Rightarrow 1-\left|(-1)^{y+1}\prod_{\substack{\varrho=1\\ \sigma=1}}^{y}\left(-1+\left(1-\left|\mathfrak{F}_{\mathfrak{R}_{\varrho}}^{RN}\right|^{y\Xi_{\varrho}}\right)^{\text{ß}}\left(1-\left|\mathfrak{F}_{\mathfrak{R}_{\sigma}}^{RN}\right|^{y\Xi_{\sigma}}\right)^{\mathfrak{L}}\right)\right|^{\frac{2}{y(y+1)}}$$

$$\leq 1-\left|(-1)^{y+1}\prod_{\substack{\varrho=1\\ \sigma=1}}^{y}\left(-1+\left(1-\left|\mathfrak{F}_{\mathfrak{R}_{\varrho'}}^{RN}\right|^{y\Xi_{\varrho}}\right)^{\text{ß}}\left(1-\left|\mathfrak{F}_{\mathfrak{R}_{\sigma'}}^{RN}\right|^{y\Xi_{\sigma}}\right)^{\mathfrak{L}}\right)\right|^{\frac{2}{y(y+1)}}$$

$$\Rightarrow -1+\left(1-\left|(-1)^{y+1}\prod_{\substack{\varrho=1\\ \sigma=1}}^{y}\left(-1+\left(1-\left|\mathfrak{F}_{\mathfrak{R}_{\varrho}}^{RN}\right|^{y\Xi_{\varrho}}\right)^{\text{ß}}\left(1-\left|\mathfrak{F}_{\mathfrak{R}_{\sigma}}^{RN}\right|^{y\Xi_{\sigma}}\right)^{\mathfrak{L}}\right)\right|^{\frac{2}{y(y+1)}}\right)^{\frac{1}{\text{ß}+\mathfrak{L}}}$$

$$\leq -1+\left(1-\left|(-1)^{y+1}\prod_{\substack{\varrho=1\\ \sigma=1}}^{y}\left(-1+\left(1-\left|\mathfrak{F}_{\mathfrak{R}_{\varrho'}}^{RN}\right|^{y\Xi_{\varrho}}\right)^{\text{ß}}\left(1-\left|\mathfrak{F}_{\mathfrak{R}_{\sigma'}}^{RN}\right|^{y\Xi_{\sigma}}\right)^{\mathfrak{L}}\right)\right|^{\frac{2}{y(y+1)}}\right)^{\frac{1}{\text{ß}+\mathfrak{L}}}$$

In similar manner, we can get

$$-1+\left(1-\left|(-1)^{y+1}\prod_{\substack{\varrho=1\\ \sigma=1}}^{y}\left(-1+\left(1-\left|\mathfrak{F}_{\mathfrak{R}_{\varrho}}^{IN}\right|^{y\Xi_{\varrho}}\right)^{\text{ß}}\left(1-\left|\mathfrak{F}_{\mathfrak{R}_{\sigma}}^{IN}\right|^{y\Xi_{\sigma}}\right)^{\mathfrak{L}}\right)\right|^{\frac{2}{y(y+1)}}\right)^{\frac{1}{\text{ß}+\mathfrak{L}}}$$

$$\leq -1+\left(1-\left|(-1)^{y+1}\prod_{\substack{\varrho=1\\ \sigma=1}}^{y}\left(-1+\left(1-\left|\mathfrak{F}_{\mathfrak{R}_{\varrho'}}^{IN}\right|^{y\Xi_{\varrho}}\right)^{\text{ß}}\left(1-\left|\mathfrak{F}_{\mathfrak{R}_{\sigma'}}^{IN}\right|^{y\Xi_{\sigma}}\right)^{\mathfrak{L}}\right)\right|^{\frac{2}{y(y+1)}}\right)^{\frac{1}{\text{ß}+\mathfrak{L}}}$$

Thus,

$$BCFPGHM^{\text{ß},\mathfrak{L}}(\mathfrak{R}_1,\mathfrak{R}_2,\ldots,\mathfrak{R}_y)\leq BCFPGHM^{\text{ß},\mathfrak{L}}(\mathfrak{R}_{1'},\mathfrak{R}_{2'},\ldots,\mathfrak{R}_{y'}).$$

Proposition 8.3 (Boundedness): Let two gatherings of BCFNs $\mathfrak{R}_\sigma = \left(\mathfrak{F}_{\mathfrak{R}_\sigma}^{P}, \mathfrak{F}_{\mathfrak{R}_\sigma}^{N}\right) = \left(\mathfrak{F}_{\mathfrak{R}_\sigma}^{RP}+\iota\mathfrak{F}_{\mathfrak{R}_\sigma}^{IP}, \mathfrak{F}_{\mathfrak{R}_\sigma}^{RN}+\iota\mathfrak{F}_{\mathfrak{R}_\sigma}^{IN}\right)$ and $\mathfrak{R}_{\sigma'} = \left(\mathfrak{F}_{\mathfrak{R}_{\sigma'}}^{P}, \mathfrak{F}_{\mathfrak{R}_{\sigma'}}^{N}\right) = \left(\mathfrak{F}_{\mathfrak{R}_{\sigma'}}^{RP}+\iota\mathfrak{F}_{\mathfrak{R}_{\sigma'}}^{IP}, \mathfrak{F}_{\mathfrak{R}_{\sigma'}}^{RN}+\iota\mathfrak{F}_{\mathfrak{R}_{\sigma'}}^{IN}\right)\sigma = 1,2,\ldots,y$ and $\mathfrak{R}^- = \left(\min_\sigma\left\{\mathfrak{F}_{\mathfrak{R}_\sigma}^{RP}\right\}+\iota\min_\sigma\left\{\mathfrak{F}_{\mathfrak{R}_\sigma}^{IP}\right\}, \max_\sigma\left\{\mathfrak{F}_{\mathfrak{R}_\sigma}^{RN}\right\}+\iota\max_\sigma\left\{\mathfrak{F}_{\mathfrak{R}_\sigma}^{IN}\right\}\right)$, and $\mathfrak{R}^+ =$

$\left(\max_{\sigma}\left\{\mathfrak{F}^{RP}_{\mathfrak{R}_\sigma}\right\} + \iota \max_{\sigma}\left\{\mathfrak{F}^{IP}_{\mathfrak{R}_\sigma}\right\}, \min_{\sigma}\left\{\mathfrak{F}^{RN}_{\mathfrak{R}_\sigma}\right\} + \iota \min_{\sigma}\left\{\mathfrak{F}^{IN}_{\mathfrak{R}_\sigma}\right\}\right)$, then

$$\mathfrak{R}^- \leq BCFPGHM^{\text{ß},\mathfrak{L}}\left(\mathfrak{R}_1, \mathfrak{R}_2, \ldots, \mathfrak{R}_y\right) \leq \mathfrak{R}^+$$

Proof As $\mathfrak{R}^- = \left(\min_{\sigma}\left\{\mathfrak{F}^{RP}_{\mathfrak{R}_\sigma}\right\} + \iota \min_{\sigma}\left\{\mathfrak{F}^{IP}_{\mathfrak{R}_\sigma}\right\}, \min_{\sigma}\left\{\mathfrak{F}^{RN}_{\mathfrak{R}_\sigma}\right\} + \iota \min_{\sigma}\left\{\mathfrak{F}^{IN}_{\mathfrak{R}_\sigma}\right\}\right)$ and $\mathfrak{R}^+ = \left(\max_{\sigma}\left\{\mathfrak{F}^{RP}_{\mathfrak{R}_\sigma}\right\} + \iota \max_{\sigma}\left\{\mathfrak{F}^{IP}_{\mathfrak{R}_\sigma}\right\}, \max_{\sigma}\left\{\mathfrak{F}^{RN}_{\mathfrak{R}_\sigma}\right\} + \iota \max_{\sigma}\left\{\mathfrak{F}^{IN}_{\mathfrak{R}_\sigma}\right\}\right)$, then using Axiom 2, we get

$$BCFPGHM^{\text{ß},\mathfrak{L}}\left(\mathfrak{R}^-, \mathfrak{R}^-, \ldots, \mathfrak{R}^-\right) \leq BCFPGHM^{\text{ß},\mathfrak{L}}\left(\mathfrak{R}_1, \mathfrak{R}_2, \ldots, \mathfrak{R}_y\right) \leq BCFPGHM^{\text{ß},\mathfrak{L}}\left(\mathfrak{R}^+, \mathfrak{R}^+, \ldots, \mathfrak{R}^+\right)$$

Next, by Proposition 8.1, we get $BCFPGHM^{\text{ß},\mathfrak{L}}\left(\mathfrak{R}^-, \mathfrak{R}^-, \ldots, \mathfrak{R}^-\right) = \mathfrak{R}^-$ and $BCFPGHM^{\text{ß},\mathfrak{L}}\left(\mathfrak{R}^+, \mathfrak{R}^+, \ldots, \mathfrak{R}^+\right) = \mathfrak{R}^+$. Thus,

$$\mathfrak{R}^- \leq BCFPGHM^{\text{ß},\mathfrak{L}}\left(\mathfrak{R}_1, \mathfrak{R}_2, \ldots, \mathfrak{R}_y\right) \leq \mathfrak{R}^+.$$

Now we investigate various cases of BCFPGHM operator.

1. By assuming $\text{ß} = \mathfrak{L} = \frac{1}{2}$, the BCFPGHM operator decreases into BCF basic HM operator

$$BCFPGHM^{\frac{1}{2},\frac{1}{2}}\left(\mathfrak{R}_1, \mathfrak{R}_2, \ldots, \mathfrak{R}_y\right) = \left(\begin{array}{c} \left(\begin{array}{l} 1 - \left(\prod_{\substack{\varrho = 1 \\ \sigma = 1}}^{y} \left(1 - \sqrt{\left(1 - \left(1 - \mathfrak{F}^{RP}_{\mathfrak{R}_\varrho}\right)^{y\Xi_\varrho}\right)\left(1 - \left(1 - \mathfrak{F}^{RP}_{\mathfrak{R}_\sigma}\right)^{y\Xi_\sigma}\right)}\right)\right)^{\frac{2}{y(y+1)}} \\ +\iota\left(1 - \left(\prod_{\substack{\varrho = 1 \\ \sigma = 1}}^{y} \left(1 - \sqrt{\left(1 - \left(1 - \mathfrak{F}^{IP}_{\mathfrak{R}_\varrho}\right)^{y\Xi_\varrho}\right)\left(1 - \left(1 - \mathfrak{F}^{IP}_{\mathfrak{R}_\sigma}\right)^{y\Xi_\sigma}\right)}\right)\right)^{\frac{2}{y(y+1)}}\right) \end{array}\right), \\ \left(\begin{array}{l} -1 + \left(1 - \left|(-1)^{y+1} \prod_{\substack{\varrho = 1 \\ \sigma = 1}}^{y} \left(-1 + \sqrt{\left(1 - \left|\mathfrak{F}^{RN}_{\mathfrak{R}_\varrho}\right|^{y\Xi_\varrho}\right)\left(1 - \left|\mathfrak{F}^{RN}_{\mathfrak{R}_\sigma}\right|^{y\Xi_\sigma}\right)}\right)\right|^{\frac{2}{y(y+1)}}\right) \\ -1 + \left(1 - \left|(-1)^{y+1} \prod_{\substack{\varrho = 1 \\ \sigma = 1}}^{y} \left(-1 + \sqrt{\left(1 - \left|\mathfrak{F}^{RN}_{\mathfrak{R}_\varrho}\right|^{y\Xi_\varrho}\right)\left(1 - \left|\mathfrak{F}^{RN}_{\mathfrak{R}_\sigma}\right|^{y\Xi_\sigma}\right)}\right)\right|^{\frac{2}{y(y+1)}}\right) \end{array}\right) \end{array}\right)$$

$$
= \left(\begin{array}{l}
= 1 - \left(\prod_{\substack{\varrho = 1 \\ \sigma = 1}}^{y} \left(1 - \sqrt{\left(1 - \left(1 - \mathfrak{F}^{RP}_{\Re_\varrho} \right)^{y\Xi_\varrho} \right) \left(1 - \left(1 - \mathfrak{F}^{RP}_{\Re_\sigma} \right)^{y\Xi_\sigma} \right)} \right) \right)^{\frac{2}{y(y+1)}} \\
+ \iota \left(1 - \left(\prod_{\substack{\varrho = 1 \\ \sigma = 1}}^{y} \left(1 - \sqrt{\left(1 - \left(1 - \mathfrak{F}^{IP}_{\Re_\varrho} \right)^{y\Xi_\varrho} \right) \left(1 - \left(1 - \mathfrak{F}^{IP}_{\Re_\sigma} \right)^{y\Xi_\sigma} \right)} \right) \right)^{\frac{2}{y(y+1)}} \right), \\
- \left| (-1)^{y+1} \prod_{\substack{\varrho = 1 \\ \sigma = 1}}^{y} \left(-1 + \sqrt{\left(1 - \left| \mathfrak{F}^{RN}_{\Re_\varrho} \right|^{y\Xi_\varrho} \right) \left(1 - \left| \mathfrak{F}^{RN}_{\Re_\sigma} \right|^{y\Xi_\sigma} \right)} \right) \right|^{\frac{2}{y(y+1)}} \\
+ \iota \left(- \left| (-1)^{y+1} \prod_{\substack{\varrho = 1 \\ \sigma = 1}}^{y} \left(-1 + \sqrt{\left(1 - \left| \mathfrak{F}^{IN}_{\Re_\varrho} \right|^{y\Xi_\varrho} \right) \left(1 - \left| \mathfrak{F}^{IN}_{\Re_\sigma} \right|^{y\Xi_\sigma} \right)} \right) \right|^{\frac{2}{y(y+1)}} \right)
\end{array} \right)
$$

2. By assuming $ß = \mathfrak{L} = 1$, the BCFPGHM operator decreases into BCF power generalized line HM operator as

$$
BCFPGHM^{1,1}(\Re_1, \Re_2, \ldots, \Re_y)
$$

$$
= \left(\begin{array}{l}
\left(1 - \left(\prod_{\substack{\varrho = 1 \\ \sigma = 1}}^{y} \left(1 - \left(1 - \left(1 - \mathfrak{F}^{RP}_{\Re_\varrho} \right)^{y\Xi_\varrho} \right) \left(1 - \left(1 - \mathfrak{F}^{RP}_{\Re_\sigma} \right)^{y\Xi_\sigma} \right) \right) \right)^{\frac{2}{y(y+1)}} \right)^{\frac{1}{2}} \\
+ \iota \left(1 - \left(\prod_{\substack{\varrho = 1 \\ \sigma = 1}}^{y} \left(1 - \left(1 - \left(1 - \mathfrak{F}^{IP}_{\Re_\varrho} \right)^{y\Xi_\varrho} \right) \left(1 - \left(1 - \mathfrak{F}^{IP}_{\Re_\sigma} \right)^{y\Xi_\sigma} \right) \right) \right)^{\frac{2}{y(y+1)}} \right)^{\frac{1}{2}}, \\
-1 + \left(1 - \left| (-1)^{y+1} \prod_{\substack{\varrho = 1 \\ \sigma = 1}}^{y} \left(-1 + \left(1 - \left| \mathfrak{F}^{RN}_{\Re_\varrho} \right|^{y\Xi_\varrho} \right) \left(1 - \left| \mathfrak{F}^{RN}_{\Re_\sigma} \right|^{y\Xi_\sigma} \right) \right) \right|^{\frac{2}{y(y+1)}} \right)^{\frac{1}{2}} \\
+ \iota \left(-1 + \left(1 - \left| (-1)^{y+1} \prod_{\substack{\varrho = 1 \\ \sigma = 1}}^{y} \left(-1 + \left(1 - \left| \mathfrak{F}^{IN}_{\Re_\varrho} \right|^{y\Xi_\varrho} \right) \left(1 - \left| \mathfrak{F}^{IN}_{\Re_\sigma} \right|^{y\Xi_\sigma} \right) \right) \right|^{\frac{2}{y(y+1)}} \right)^{\frac{1}{2}} \right)
\end{array} \right)
$$

3. By assuming $\mathfrak{L} \rightarrow 0$, the BCFPGHM operator decreases to BCF power generalized linear descending weighted operator

$$
\begin{aligned}
&BCFPGHM^{\text{ß},0}(\Re_1, \Re_2, \ldots, \Re_y) \\
&= \left(\begin{array}{l}
\left(\left(1 - \left(\prod_{\substack{\varrho=1 \\ \sigma=1}}^{y} \left(1 - \left(1 - \left(1 - \mathfrak{F}^{RP}_{\Re_\varrho} \right)^{y\Xi_\varrho} \right)^{\text{ß}} \right)^{y+1-\varrho} \right)^{\frac{2}{(y+1)}} \right)^{\frac{1}{\text{ß}}} \right. \\
\left. +\iota \left(1 - \left(\prod_{\substack{\varrho=1 \\ \sigma=1}}^{y} \left(1 - \left(1 - \left(1 - \mathfrak{F}^{IP}_{\Re_\varrho} \right)^{y\Xi_\varrho} \right)^{\text{ß}} \right)^{y+1-\varrho} \right)^{\frac{2}{(y+1)}} \right)^{\frac{1}{\text{ß}}} \right), \\
\left(-1 + \left(1 - \left| (-1)^{y+1} \prod_{\substack{\varrho=1 \\ \sigma=1}}^{y} \left(-1 + \left(1 - \left| \mathfrak{F}^{RN}_{\Re_\varrho} \right|^{y\Xi_\varrho} \right)^{\text{ß}} \right)^{y+1-\varrho} \right|^{\frac{2}{y(y+1)}} \right)^{\frac{1}{\text{ß}}} \right. \\
\left. +\iota \left(-1 + \left(1 - \left| (-1)^{y+1} \prod_{\substack{\varrho=1 \\ \sigma=1}}^{y} \left(-1 + \left(1 - \left| \mathfrak{F}^{IN}_{\Re_\varrho} \right|^{y\Xi_\varrho} \right)^{\text{ß}} \right)^{y+1-\varrho} \right|^{\frac{2}{y(y+1)}} \right)^{\frac{1}{\text{ß}}} \right) \right)
\end{array} \right)
\end{aligned}
$$

4. By assuming $\text{ß} \rightarrow 0$, the BCFPGHM operator decreases to BCF power generalized linear ascending weighted operator

$$BCFPGHM^{0,\mathfrak{L}}(\mathfrak{R}_1, \mathfrak{R}_2, \ldots, \mathfrak{R}_y)$$

$$= \left(\begin{array}{l} \left(\left(1 - \left(\prod_{\substack{\varrho = 1 \\ \sigma = 1}}^{y} \left(1 - \left(1 - (1 - \mathfrak{F}_{\mathfrak{R}_\sigma}^{RP})^{y\Xi_\sigma}\right)^{\mathfrak{L}}\right)^{\varrho} \right)^{\frac{2}{y(y+1)}} \right)^{\frac{1}{\mathfrak{L}}} + \iota \left(1 - \left(\prod_{\substack{\varrho = 1 \\ \sigma = 1}}^{y} \left(1 - \left(1 - (1 - \mathfrak{F}_{\mathfrak{R}_\sigma}^{IP})^{y\Xi_\sigma}\right)^{\mathfrak{L}}\right)^{\varrho} \right)^{\frac{2}{y(y+1)}} \right)^{\frac{1}{\mathfrak{L}}} \right), \\ -1 + \left(1 - \left| (-1)^{y+1} \prod_{\substack{\varrho = 1 \\ \sigma = 1}}^{y} \left(-1 + \left(1 - |\mathfrak{F}_{\mathfrak{R}_\sigma}^{RN}|^{y\Xi_\sigma}\right)^{\mathfrak{L}}\right)^{\varrho} \right|^{\frac{2}{y(y+1)}} \right)^{\frac{1}{\mathfrak{L}}} \\ + \iota \left(-1 + \left(1 - \left| (-1)^{y+1} \prod_{\substack{\varrho = 1 \\ \sigma = 1}}^{y} \left(-1 + \left(1 - |\mathfrak{F}_{\mathfrak{R}_\sigma}^{IN}|^{y\Xi_\sigma}\right)^{\mathfrak{L}}\right)^{\varrho} \right|^{\frac{2}{y(y+1)}} \right)^{\frac{1}{\mathfrak{L}}} \right) \end{array} \right)$$

In the BCFPGHM operator, we merely suppose the weight vector relies on the interrelations of the input BCFNs and PA operator, and the significance of every input BCFN itself is not taken into account in the BCFPGHM operator. But this weight is also a critical factor in a lot of genuine-life procedures of DM. For instance, in a MADM dilemma, the attribute weight will have a significant impact on how the alternatives are ranked. Therefore, next, we investigate the BCFWPGHM operator.

Definition 8.2 The BCFWPGHM operator relying on the gathering of BCFNs $\mathfrak{R}_\sigma = \left(\mathfrak{F}_{\mathfrak{R}_\sigma}^{P}, \mathfrak{F}_{\mathfrak{R}_\sigma}^{N}\right) = \left(\mathfrak{F}_{\mathfrak{R}_\sigma}^{RP} + \iota\mathfrak{F}_{\mathfrak{R}_\sigma}^{IP}, \mathfrak{F}_{\mathfrak{R}_\sigma}^{RN} + \iota\mathfrak{F}_{\mathfrak{R}_\sigma}^{IN}\right)\sigma = 1, 2, \ldots, y$, is spotted as

$$BCFWPGHM^{\text{ß},\mathfrak{L}}\left(\mathfrak{R}_1,\ \mathfrak{R}_2,\ \ldots,\ \mathfrak{R}_y\right)$$

$$
= \left(\frac{2}{\mathrm{y}(\mathrm{y}+1)} \underset{\varrho=1}{\overset{\mathrm{y}}{\oplus}} \underset{\sigma=1}{\overset{\mathrm{y}}{\oplus}} \left(\frac{\mathrm{y}\mathrm{w}_{\varrho}\left(1+\underline{\mathrm{T}}(\Re_{\varrho})\right)}{\sum_{\rho=1}^{\mathrm{y}} \mathrm{w}_{\rho}\left(1+\underline{\mathrm{T}}(\Re_{\rho})\right)} \Re_{\varrho} \right)^{\text{ß}} \otimes \left(\frac{\mathrm{y}\mathrm{w}_{\sigma}\left(1+\underline{\mathrm{T}}(\Re_{\sigma})\right)}{\sum_{\rho=1}^{\mathrm{y}} \mathrm{w}_{\rho}\left(1+\underline{\mathrm{T}}(\Re_{\rho})\right)} \Re_{\sigma} \right)^{\mathfrak{L}} \right)^{\frac{1}{\text{ß}+\mathfrak{L}}}
$$

Noticed that $\mathring{\mathrm{w}} = \left(\underset{1,}{\mathring{\mathrm{w}}} \underset{2}{\mathring{\mathrm{w}}}, \ldots, \underset{y}{\mathring{\mathrm{w}}} \right)^{T}$ is a weight vector, $0 \leq \underset{\sigma}{\mathring{\mathrm{w}}} \leq 1$ for all σ with $\sum_{\sigma=1}^{y} \underset{\sigma}{\mathring{\mathrm{w}}} = 1$.

Theorem 8.2 Let a gathering of BCFNs $\Re_{\sigma} = \left(\mathfrak{F}^{P}_{\Re_{\sigma}}, \mathfrak{F}^{N}_{\Re_{\sigma}}\right) = \left(\mathfrak{F}^{RP}_{\Re_{\sigma}} + \iota\mathfrak{F}^{IP}_{\Re_{\sigma}}, \mathfrak{F}^{RN}_{\Re_{\sigma}} + \iota\mathfrak{F}^{IN}_{\Re_{\sigma}}\right)\sigma = 1, 2, \ldots, y$. The aggregated outcome established by definition 8.2 is still a BCFN that is

$$
BCFWPGHM^{\text{ß},\mathfrak{L}}(\Re_1, \Re_2, \ldots, \Re_y)
$$

$$
= \left(\begin{array}{l}
\left(\left(1 - \left(\prod_{\substack{\varrho=1 \\ \sigma=1}}^{y} \left(1 - \left(1 - \left(1 - \mathfrak{F}^{RP}_{\Re_{\varrho}}\right)^{y\Xi_{\mathring{w}-\varrho}}\right)^{\text{ß}} \left(1 - \left(1 - \mathfrak{F}^{RP}_{\Re_{\sigma}}\right)^{y\Xi_{\mathring{w}-\sigma}}\right)^{\mathfrak{L}} \right) \right)^{\frac{2}{y(y+1)}} \right)^{\frac{1}{\text{ß}+\mathfrak{L}}} \right. \\
\left. +\iota \left(1 - \left(\prod_{\substack{\varrho=1 \\ \sigma=1}}^{y} \left(1 - \left(1 - \left(1 - \mathfrak{F}^{IP}_{\Re_{\varrho}}\right)^{y\Xi_{\mathring{w}-\varrho}}\right)^{\text{ß}} \left(1 - \left(1 - \mathfrak{F}^{IP}_{\Re_{\sigma}}\right)^{y\Xi_{\mathring{w}-\sigma}}\right)^{\mathfrak{L}} \right) \right)^{\frac{2}{y(y+1)}} \right)^{\frac{1}{\text{ß}+\mathfrak{L}}} \right), \\
\left(-1 + \left(1 - \left| (-1)^{y+1} \prod_{\substack{\varrho=1 \\ \sigma=1}}^{y} \left(-1 + \left(1 - \left|\mathfrak{F}^{RN}_{\Re_{\varrho}}\right|^{y\Xi_{\mathring{w}-\varrho}}\right)^{\text{ß}} \left(1 - \left|\mathfrak{F}^{RN}_{\Re_{\sigma}}\right|^{y\Xi_{\mathring{w}-\sigma}}\right)^{\mathfrak{L}} \right) \right|^{\frac{2}{y(y+1)}} \right)^{\frac{1}{\text{ß}+\mathfrak{L}}} \right. \\
\left. +\iota \left(-1 + \left(1 - \left| (-1)^{y+1} \prod_{\substack{\varrho=1 \\ \sigma=1}}^{y} \left(-1 + \left(1 - \left|\mathfrak{F}^{IN}_{\Re_{\varrho}}\right|^{y\Xi_{\mathring{w}-\varrho}}\right)^{\text{ß}} \left(1 - \left|\mathfrak{F}^{IN}_{\Re_{\sigma}}\right|^{y\Xi_{\mathring{w}-\sigma}}\right)^{\mathfrak{L}} \right) \right|^{\frac{2}{y(y+1)}} \right)^{\frac{1}{\text{ß}+\mathfrak{L}}} \right) \right)
\end{array} \right)
$$

where, $\Xi_{\mathring{\mathrm{w}}-\varrho} = \frac{y \underset{\varrho}{\mathring{\mathrm{w}}}\left(1+\underline{\mathrm{T}}(\Re_{\varrho})\right)}{\sum_{\rho=1}^{y} \underset{\rho}{\mathring{\mathrm{w}}}\left(1+\underline{\mathrm{T}}(\Re_{\rho})\right)}$

Proposition 8.4 (Idempotency): Let a gathering of BCFNs $\mathfrak{R}_\sigma = \left(\mathfrak{F}^P_{\mathfrak{R}_\sigma}, \mathfrak{F}^N_{\mathfrak{R}_\sigma}\right) = \left(\mathfrak{F}^{RP}_{\mathfrak{R}_\sigma} + \iota\mathfrak{F}^{IP}_{\mathfrak{R}_\sigma}, \mathfrak{F}^{RN}_{\mathfrak{R}_\sigma} + \iota\mathfrak{F}^{IN}_{\mathfrak{R}_\sigma}\right)\sigma = 1, 2, \ldots, y$. If $\forall\sigma\ \mathfrak{R}_\sigma = \mathfrak{R}$, then

$$BCFWPGHM^{\text{ß},\mathfrak{L}}\left(\mathfrak{R}_1, \mathfrak{R}_2, \ldots, \mathfrak{R}_y\right) = \mathfrak{R}$$

Proof Omitted

Proposition 8.5 (Monotonicity): Let two gatherings of BCFNs $\mathfrak{R}_\sigma = \left(\mathfrak{F}^P_{\mathfrak{R}_\sigma}, \mathfrak{F}^N_{\mathfrak{R}_\sigma}\right) = \left(\mathfrak{F}^{RP}_{\mathfrak{R}_\sigma} + \iota\mathfrak{F}^{IP}_{\mathfrak{R}_\sigma}, \mathfrak{F}^{RN}_{\mathfrak{R}_\sigma} + \iota\mathfrak{F}^{IN}_{\mathfrak{R}_\sigma}\right)$ and $\mathfrak{R}_{\sigma'} = \left(\mathfrak{F}^P_{\mathfrak{R}_{\sigma'}}, \mathfrak{F}^N_{\mathfrak{R}_{\sigma'}}\right) = \left(\mathfrak{F}^{RP}_{\mathfrak{R}_{\sigma'}} + \iota\mathfrak{F}^{IP}_{\mathfrak{R}_{\sigma'}}, \mathfrak{F}^{RN}_{\mathfrak{R}_{\sigma'}} + \iota\mathfrak{F}^{IN}_{\mathfrak{R}_{\sigma'}}\right)\sigma = 1, 2, \ldots, y$. If $\mathfrak{F}^{RP}_{\mathfrak{R}_\sigma} \le \mathfrak{F}^{RP}_{\mathfrak{R}_{\sigma'}}$, $\mathfrak{F}^{IP}_{\mathfrak{R}_\sigma} \le \mathfrak{F}^{IP}_{\mathfrak{R}_{\sigma'}}$, $\mathfrak{F}^{RN}_{\mathfrak{R}_\sigma} \le \mathfrak{F}^{RN}_{\mathfrak{R}_{\sigma'}}$, and $\mathfrak{F}^{IN}_{\mathfrak{R}_\sigma} \le \mathfrak{F}^{IN}_{\mathfrak{R}_{\sigma'}}$, then

$$BCFWPGHM^{\text{ß},\mathfrak{L}}\left(\mathfrak{R}_1, \mathfrak{R}_2, \ldots, \mathfrak{R}_y\right) \le BCFWPGHM^{\text{ß},\mathfrak{L}}\left(\mathfrak{R}_{1'}, \mathfrak{R}_{2'}, \ldots, \mathfrak{R}_{y'}\right)$$

Proof Omitted

Proposition 8.6 (Boundedness): Let two gatherings of BCFNs $\mathfrak{R}_\sigma = \left(\mathfrak{F}^P_{\mathfrak{R}_\sigma}, \mathfrak{F}^N_{\mathfrak{R}_\sigma}\right) = \left(\mathfrak{F}^{RP}_{\mathfrak{R}_\sigma} + \iota\mathfrak{F}^{IP}_{\mathfrak{R}_\sigma}, \mathfrak{F}^{RN}_{\mathfrak{R}_\sigma} + \iota\mathfrak{F}^{IN}_{\mathfrak{R}_\sigma}\right)$ and $\mathfrak{R}_{\sigma'} = \left(\mathfrak{F}^P_{\mathfrak{R}_{\sigma'}}, \mathfrak{F}^N_{\mathfrak{R}_{\sigma'}}\right) = \left(\mathfrak{F}^{RP}_{\mathfrak{R}_{\sigma'}} + \iota\mathfrak{F}^{IP}_{\mathfrak{R}_{\sigma'}}, \mathfrak{F}^{RN}_{\mathfrak{R}_{\sigma'}} + \iota\mathfrak{F}^{IN}_{\mathfrak{R}_{\sigma'}}\right)\sigma = 1, 2, \ldots, y$ and $\mathfrak{R}^- = \left(\min_\sigma\left\{\mathfrak{F}^{RP}_{\mathfrak{R}_\sigma}\right\} + \iota\min_\sigma\left\{\mathfrak{F}^{IP}_{\mathfrak{R}_\sigma}\right\}, \min_\sigma\left\{\mathfrak{F}^{RN}_{\mathfrak{R}_\sigma}\right\} + \iota\min_\sigma\left\{\mathfrak{F}^{IN}_{\mathfrak{R}_\sigma}\right\}\right)$, and $\mathfrak{R}^+ = \left(\max_\sigma\left\{\mathfrak{F}^{RP}_{\mathfrak{R}_\sigma}\right\} + \iota\max_\sigma\left\{\mathfrak{F}^{IP}_{\mathfrak{R}_\sigma}\right\}, \max_\sigma\left\{\mathfrak{F}^{RN}_{\mathfrak{R}_\sigma}\right\} + \iota\max_\sigma\left\{\mathfrak{F}^{IN}_{\mathfrak{R}_\sigma}\right\}\right)$, then

$$\mathfrak{R}^- \le BCFWPGHM^{\text{ß},\mathfrak{L}}\left(\mathfrak{R}_1, \mathfrak{R}_2, \ldots, \mathfrak{R}_y\right) \le \mathfrak{R}^+$$

Proof Omitted.

8.2 Bipolar Complex Fuzzy Power Generalized Geometric AOs

Yu [37] investigated the GGHM operator.

Let us consider a gathering of real numbers greater than or equal to zero that is $\mathfrak{R}_{\mathcal{PN}-\sigma}$, $(\sigma = 1, 2, \ldots, y)$ and ß, $\mathfrak{L} > 0$. Then the GGHM operator is inaugurated as

$$GGHM^{\text{ß},\mathfrak{L}}\left(\mathfrak{R}_{\mathcal{PN}-1}, \mathfrak{R}_{\mathcal{PN}-2}, \ldots, \mathfrak{R}_{\mathcal{PN}-y}\right)$$
$$= \frac{1}{\text{ß} + \mathfrak{L}}\left(\prod_{\varrho=1}^{y}\prod_{\sigma=1}^{y}\left(\text{ß}\mathfrak{R}_{\mathcal{PN}-\varrho} + \mathfrak{L}\mathfrak{R}_{\mathcal{PN}-\sigma}\right)\right)^{\frac{2}{y(y+1)}}$$

If $ß = \mathfrak{L} = \frac{1}{2}$, then the GGHM operator transforms to the geometric Heronian mean operator

In the next definition, we are going to develop the notion of the BCFPGGHM operator.

Definition 8.3 The BCFPGGHM operator relying on the gathering of BCFNs $\Re_\sigma = \left(\mathfrak{F}^P_{\Re_\sigma}, \mathfrak{F}^N_{\Re_\sigma}\right) = \left(\mathfrak{F}^{RP}_{\Re_\sigma} + \iota\mathfrak{F}^{IP}_{\Re_\sigma}, \mathfrak{F}^{RN}_{\Re_\sigma} + \iota\mathfrak{F}^{IN}_{\Re_\sigma}\right)\sigma = 1, 2, \ldots, y$, is spotted as

$$BCFPGGHM^{ß,\mathfrak{L}}(\Re_1, \Re_2, \ldots, \Re_y) = \frac{1}{ß + \mathfrak{L}}\left(\underset{\varrho=1}{\overset{y}{\otimes}}\underset{\sigma=1}{\overset{y}{\otimes}}\left(ß\left(\Re_\varrho\right)^{\frac{y(1+\underline{T}(\Re_\varrho))}{\sum_{\rho=1}^{y}(1+\underline{T}(\Re_\rho))}} \oplus \mathfrak{L}(\Re_\sigma)^{\frac{y(1+\underline{T}(\Re_\sigma))}{\sum_{\rho=1}^{y}(1+\underline{T}(\Re_\rho))}}\right)\right)^{\frac{2}{y(y+1)}} \tag{8.2}$$

Noticed that $\underline{T}(\Re_\varrho) = \sum_{\substack{\sigma=1 \\ \sigma\neq\varrho}}^{y} Sup(\Re_\varrho, \Re_\sigma)$, $Sup(\Re_\varrho, \Re_\sigma)$ is the support among two BCFNs $\Re_\varrho$, and $\Re_\sigma$ and satisfy the underneath axioms.

1. $Sup(\Re_\varrho, \Re_\sigma) \in [0, 1]$;
2. $Sup(\Re_\varrho, \Re_\sigma) = Sup(\Re_\sigma, \Re_\varrho)$;
3. $Sup(\Re_\varrho, \Re_\sigma) \geq Sup(\Re_{\varrho'}, \Re_{\sigma'})$ if $d_{BCFS}(\Re_\varrho, \Re_\sigma) < d_{BCFS}(\Re_{\varrho'}, \Re_{\sigma'})$, where, d_{BCFS} is any distance among BCFNs.

Further, if we let $\frac{(1+\underline{T}(\Re_\varrho))}{\sum_{\rho=1}^{y}(1+\underline{T}(\Re_\rho))} = \Xi_\varrho$, then Eq. (8.2) will be in the underneath form

$$BCFPGGHM^{ß,\mathfrak{L}}(\Re_1, \Re_2, \ldots, \Re_y) = \frac{1}{ß + \mathfrak{L}}\left(\underset{\varrho=1}{\overset{y}{\otimes}}\underset{\sigma=1}{\overset{y}{\otimes}}\left(ß\left(\Re_\varrho\right)^{y\Xi_\varrho} \oplus \mathfrak{L}(\Re_\sigma)^{y\Xi_\sigma}\right)\right)^{\frac{2}{y(y+1)}}$$

Noted that $(\Xi_1, \Xi_2, \ldots, \Xi_y)$ is a power weighted vector, $\Xi_\varrho \in [0, 1]$ and $\sum_{\varrho=1}^{y}\Xi_\varrho = 1$.

Theorem 8.3 Let a gathering of BCFNs $\Re_\sigma = \left(\mathfrak{F}^P_{\Re_\sigma}, \mathfrak{F}^N_{\Re_\sigma}\right) = \left(\mathfrak{F}^{RP}_{\Re_\sigma} + \iota\mathfrak{F}^{IP}_{\Re_\sigma}, \mathfrak{F}^{RN}_{\Re_\sigma} + \iota\mathfrak{F}^{IN}_{\Re_\sigma}\right)\sigma = 1, 2, \ldots, y$. The aggregated outcome established by Definition 8.3 is still a BCFN that is

$$BCFPGGHM^{\beta,\mathfrak{L}}(\mathfrak{R}_1,\mathfrak{R}_2,\ldots,\mathfrak{R}_y)$$

$$=\begin{pmatrix} 1-\left(1-\left(\prod_{\substack{\varrho=1\\ \sigma=1}}^{y}\left(1-\left(1-\left(\mathfrak{F}^{RP}_{\mathfrak{R}_\varrho}\right)^{y\Xi_\varrho}\right)^{\beta}\left(1-\left(\mathfrak{F}^{RP}_{\mathfrak{R}_\sigma}\right)^{y\Xi_\sigma}\right)^{\mathfrak{L}}\right)\right)^{\frac{2}{y(y+1)}}\right)^{\frac{1}{\beta+\mathfrak{L}}} \\ +\iota\left(1-\left(1-\left(\prod_{\substack{\varrho=1\\ \sigma=1}}^{y}\left(1-\left(1-\left(\mathfrak{F}^{IP}_{\mathfrak{R}_\varrho}\right)^{y\Xi_\varrho}\right)^{\beta}\left(1-\left(\mathfrak{F}^{IP}_{\mathfrak{R}_\sigma}\right)^{y\Xi_\sigma}\right)^{\mathfrak{L}}\right)\right)^{\frac{2}{y(y+1)}}\right)^{\frac{1}{\beta+\mathfrak{L}}}\right), \\ -\left|-1+\left(\prod_{\substack{\varrho=1\\ \sigma=1}}^{y}\left(1-\left|-1+\left(1+\mathfrak{F}^{RN}_{\mathfrak{R}_\varrho}\right)^{y\Xi_\varrho}\right|^{\beta}\left|-1+\left(1+\mathfrak{F}^{RN}_{\mathfrak{R}_\sigma}\right)^{y\Xi_\sigma}\right|^{\mathfrak{L}}\right)\right)^{\frac{2}{y(y+1)}}\right|^{\frac{1}{\beta+\mathfrak{L}}} \\ +\iota\left(-\left|-1+\left(\prod_{\substack{\varrho=1\\ \sigma=1}}^{y}\left(1-\left|-1+\left(1+\mathfrak{F}^{IN}_{\mathfrak{R}_\varrho}\right)^{y\Xi_\varrho}\right|^{\beta}\left|-1+\left(1+\mathfrak{F}^{IN}_{\mathfrak{R}_\sigma}\right)^{y\Xi_\sigma}\right|^{\mathfrak{L}}\right)\right)^{\frac{2}{y(y+1)}}\right|^{\frac{1}{\beta+\mathfrak{L}}}\right) \end{pmatrix}$$

where, $\Xi_\varrho=\frac{(1+\underline{T}(\mathfrak{R}_\varrho))}{\sum_{\rho=1}^{y}(1+\underline{T}(\mathfrak{R}_\rho))}$.

Proof From the primary operations of BCFNs, we have

$$(\mathfrak{R}_\varrho)^{y\Xi_\varrho}=\left(\left(\mathfrak{F}^{RP}_{\mathfrak{R}_\varrho}\right)^{y\Xi_\varrho}+\iota\left(\mathfrak{F}^{IP}_{\mathfrak{R}_\varrho}\right)^{y\Xi_\varrho},-1+\left(1+\mathfrak{F}^{RN}_{\mathfrak{R}_\varrho}\right)^{y\Xi_\varrho}+\iota\left(-1+\left(1+\mathfrak{F}^{IN}_{\mathfrak{R}_\varrho}\right)^{y\Xi_\varrho}\right)\right)$$

$$(\mathfrak{R}_\sigma)^{y\Xi_\sigma}=\left(\left(\mathfrak{F}^{RP}_{\mathfrak{R}_\sigma}\right)^{y\Xi_\sigma}+\iota\left(\mathfrak{F}^{IP}_{\mathfrak{R}_\sigma}\right)^{y\Xi_\sigma},-1+\left(1+\mathfrak{F}^{RN}_{\mathfrak{R}_\sigma}\right)^{y\Xi_\sigma}+\iota\left(-1+\left(1+\mathfrak{F}^{IN}_{\mathfrak{R}_\sigma}\right)^{y\Xi_\sigma}\right)\right)$$

Then,

$$\beta\left(\mathfrak{R}_\varrho\right)^{y\Xi_\varrho}=\begin{pmatrix} 1-\left(1-\left(\mathfrak{F}^{RP}_{\mathfrak{R}_\varrho}\right)^{y\Xi_\varrho}\right)^{\beta}+\iota\left(1-\left(1-\left(\mathfrak{F}^{IP}_{\mathfrak{R}_\varrho}\right)^{y\Xi_\varrho}\right)^{\beta}\right), \\ -\left|-1+\left(1+\mathfrak{F}^{RN}_{\mathfrak{R}_\varrho}\right)^{y\Xi_\varrho}\right|^{\beta}+\iota\left(-\left|-1+\left(1+\mathfrak{F}^{IN}_{\mathfrak{R}_\varrho}\right)^{y\Xi_\varrho}\right|^{\beta}\right) \end{pmatrix}$$

$$\mathfrak{L}(\mathfrak{R}_\sigma)^{y\Xi_\sigma}=\begin{pmatrix} 1-\left(1-\left(\mathfrak{F}^{RP}_{\mathfrak{R}_\sigma}\right)^{y\Xi_\sigma}\right)^{\mathfrak{L}}+\iota\left(1-\left(1-\left(\mathfrak{F}^{IP}_{\mathfrak{R}_\sigma}\right)^{y\Xi_\sigma}\right)^{\mathfrak{L}}\right), \\ -\left|-1+\left(1+\mathfrak{F}^{RN}_{\mathfrak{R}_\sigma}\right)^{y\Xi_\sigma}\right|^{\mathfrak{L}}+\iota\left(-\left|-1+\left(1+\mathfrak{F}^{IN}_{\mathfrak{R}_\sigma}\right)^{y\Xi_\sigma}\right|^{\mathfrak{L}}\right) \end{pmatrix}$$

$$\mathfrak{B}\left(\Re_{\varrho}\right)^{y\Xi_{\varrho}} \oplus \mathfrak{L}(\Re_{\sigma}) =$$

$$= \begin{pmatrix} 1-\left(1-\left(\mathfrak{F}^{RP}_{\Re_{\varrho}}\right)^{y\Xi_{\varrho}}\right)^{\mathfrak{B}} + \iota\left(1-\left(1-\left(\mathfrak{F}^{IP}_{\Re_{\varrho}}\right)^{y\Xi_{\varrho}}\right)^{\mathfrak{B}}\right), \\ -\left|-1+\left(1+\mathfrak{F}^{RN}_{\Re_{\varrho}}\right)^{y\Xi_{\varrho}}\right|^{\mathfrak{B}} + \iota\left(-\left|-1+\left(1+\mathfrak{F}^{IN}_{\Re_{\varrho}}\right)^{y\Xi_{\varrho}}\right|^{\mathfrak{B}}\right) \end{pmatrix}$$

$$\oplus \begin{pmatrix} 1-\left(1-\left(\mathfrak{F}^{RP}_{\Re_{\sigma}}\right)^{y\Xi_{\sigma}}\right)^{\mathfrak{L}} + \iota\left(1-\left(1-\left(\mathfrak{F}^{IP}_{\Re_{\sigma}}\right)^{y\Xi_{\sigma}}\right)^{\mathfrak{L}}\right), \\ -\left|-1+\left(1+\mathfrak{F}^{RN}_{\Re_{\sigma}}\right)^{y\Xi_{\sigma}}\right|^{\mathfrak{L}} + \iota\left(-\left|-1+\left(1+\mathfrak{F}^{IN}_{\Re_{\sigma}}\right)^{y\Xi_{\sigma}}\right|^{\mathfrak{L}}\right) \end{pmatrix}$$

$$= \begin{pmatrix} 1-\left(1-\left(\mathfrak{F}^{RP}_{\Re_{\varrho}}\right)^{y\Xi_{\varrho}}\right)^{\mathfrak{B}}\left(1-\left(\mathfrak{F}^{RP}_{\Re_{\sigma}}\right)^{y\Xi_{\sigma}}\right)^{\mathfrak{L}} \\ +\iota\left(1-\left(1-\left(\mathfrak{F}^{IP}_{\Re_{\varrho}}\right)^{y\Xi_{\varrho}}\right)^{\mathfrak{B}}\left(1-\left(\mathfrak{F}^{IP}_{\Re_{\sigma}}\right)^{y\Xi_{\sigma}}\right)^{\mathfrak{L}}\right), \\ -\left(\left(-\left|-1+\left(1+\mathfrak{F}^{RN}_{\Re_{\varrho}}\right)^{y\Xi_{\varrho}}\right|^{\mathfrak{B}}\right)\left(-\left|-1+\left(1+\mathfrak{F}^{RN}_{\Re_{\sigma}}\right)^{y\Xi_{\sigma}}\right|^{\mathfrak{L}}\right)\right) \\ +\iota\left(-\left(\left(-\left|-1+\left(1+\mathfrak{F}^{IN}_{\Re_{\varrho}}\right)^{y\Xi_{\varrho}}\right|^{\mathfrak{B}}\right)\left(-\left|-1+\left(1+\mathfrak{F}^{IN}_{\Re_{\sigma}}\right)^{y\Xi_{\sigma}}\right|^{\mathfrak{L}}\right)\right)\right) \end{pmatrix}$$

$$= \begin{pmatrix} 1-\left(1-\left(\mathfrak{F}^{RP}_{\Re_{\varrho}}\right)^{y\Xi_{\varrho}}\right)^{\mathfrak{B}}\left(1-\left(\mathfrak{F}^{RP}_{\Re_{\sigma}}\right)^{y\Xi_{\sigma}}\right)^{\mathfrak{L}} \\ +\iota\left(1-\left(1-\left(\mathfrak{F}^{IP}_{\Re_{\varrho}}\right)^{y\Xi_{\varrho}}\right)^{\mathfrak{B}}\left(1-\left(\mathfrak{F}^{IP}_{\Re_{\sigma}}\right)^{y\Xi_{\sigma}}\right)^{\mathfrak{L}}\right), \\ -\left(\left|-1+\left(1+\mathfrak{F}^{RN}_{\Re_{\varrho}}\right)^{y\Xi_{\varrho}}\right|^{\mathfrak{B}}\left|-1+\left(1+\mathfrak{F}^{RN}_{\Re_{\sigma}}\right)^{y\Xi_{\sigma}}\right|^{\mathfrak{L}}\right) \\ +\iota\left(-\left(\left|-1+\left(1+\mathfrak{F}^{IN}_{\Re_{\varrho}}\right)^{y\Xi_{\varrho}}\right|^{\mathfrak{B}}\left|-1+\left(1+\mathfrak{F}^{IN}_{\Re_{\sigma}}\right)^{y\Xi_{\sigma}}\right|^{\mathfrak{L}}\right)\right) \end{pmatrix}$$

$$= \left(\begin{array}{l} \prod_{\substack{\varrho=1\\ \sigma=1}}^{y} \left(1-\left(1-\left(\mathfrak{F}_{\Re_\varrho}^{RP}\right)^{y\Xi_\varrho}\right)^{\text{ß}}\left(1-\left(\mathfrak{F}_{\Re_\sigma}^{RP}\right)^{y\Xi_\sigma}\right)^{\mathfrak{L}}\right) \\ +\iota \prod_{\substack{\varrho=1\\ \sigma=1}}^{y} \left(1-\left(1-\left(\mathfrak{F}_{\Re_\varrho}^{IP}\right)^{y\Xi_\varrho}\right)^{\text{ß}}\left(1-\left(\mathfrak{F}_{\Re_\sigma}^{IP}\right)^{y\Xi_\sigma}\right)^{\mathfrak{L}}\right), \\ -1+\prod_{\substack{\varrho=1\\ \sigma=1}}^{y} \left(1-\left|-1+\left(1+\mathfrak{F}_{\Re_\varrho}^{RN}\right)^{y\Xi_\varrho}\right|^{\text{ß}}\left|-1+\left(1+\mathfrak{F}_{\Re_\sigma}^{RN}\right)^{y\Xi_\sigma}\right|^{\mathfrak{L}}\right) \\ +\iota\left(-1+\prod_{\substack{\varrho=1\\ \sigma=1}}^{y} \left(1-\left|-1+\left(1+\mathfrak{F}_{\Re_\varrho}^{IN}\right)^{y\Xi_\varrho}\right|^{\text{ß}}\left|-1+\left(1+\mathfrak{F}_{\Re_\sigma}^{IN}\right)^{y\Xi_\sigma}\right|^{\mathfrak{L}}\right)\right) \end{array} \right)$$

$$\left(\bigotimes_{\varrho=1}^{y} \bigotimes_{\sigma=1}^{y} \left(\text{ß}(\Re_\varrho)^{y\Xi_\varrho} \oplus \mathfrak{L}(\Re_\sigma)^{y\Xi_\sigma}\right)\right)^{\frac{2}{y(y+1)}}$$

$$= \left(\begin{array}{l} \left(\prod_{\substack{\varrho=1\\ \sigma=1}}^{y} \left(1-\left(1-\left(\mathfrak{F}_{\Re_\varrho}^{RP}\right)^{y\Xi_\varrho}\right)^{\text{ß}}\left(1-\left(\mathfrak{F}_{\Re_\sigma}^{RP}\right)^{y\Xi_\sigma}\right)^{\mathfrak{L}}\right)\right)^{\frac{2}{y(y+1)}} \\ +\iota\left(\prod_{\substack{\varrho=1\\ \sigma=1}}^{y} \left(1-\left(1-\left(\mathfrak{F}_{\Re_\varrho}^{IP}\right)^{y\Xi_\varrho}\right)^{\text{ß}}\left(1-\left(\mathfrak{F}_{\Re_\sigma}^{IP}\right)^{y\Xi_\sigma}\right)^{\mathfrak{L}}\right)\right)^{\frac{2}{y(y+1)}}, \\ -1+\left(\prod_{\substack{\varrho=1\\ \sigma=1}}^{y} \left(1-\left|-1+\left(1+\mathfrak{F}_{\Re_\varrho}^{RN}\right)^{y\Xi_\varrho}\right|^{\text{ß}}\left|-1+\left(1+\mathfrak{F}_{\Re_\sigma}^{RN}\right)^{y\Xi_\sigma}\right|^{\mathfrak{L}}\right)\right)^{\frac{2}{y(y+1)}} \\ +\iota\left(-1+\left(\prod_{\substack{\varrho=1\\ \sigma=1}}^{y} \left(1-\left|-1+\left(1+\mathfrak{F}_{\Re_\varrho}^{IN}\right)^{y\Xi_\varrho}\right|^{\text{ß}}\left|-1+\left(1+\mathfrak{F}_{\Re_\sigma}^{IN}\right)^{y\Xi_\sigma}\right|^{\mathfrak{L}}\right)\right)^{\frac{2}{y(y+1)}}\right) \end{array} \right)$$

$$\frac{1}{\text{ß}+\mathfrak{L}}\left(\underset{\varrho=1}{\overset{y}{\otimes}}\ \underset{\sigma=1}{\overset{y}{\otimes}}\left(\text{ß}(\mathfrak{R}_\varrho)^{y\Xi_\varrho}\oplus\mathfrak{L}(\mathfrak{R}_\sigma)^{y\Xi_\sigma}\right)\right)^{\frac{2}{y(y+1)}}$$

$$=\left(\begin{array}{c}
1-\left(1-\left(\prod_{\substack{\varrho=1\\ \sigma=1}}^{y}\left(1-\left(1-\left(\mathfrak{F}^{RP}_{\mathfrak{R}_\varrho}\right)^{y\Xi_\varrho}\right)^{\text{ß}}\left(1-\left(\mathfrak{F}^{RP}_{\mathfrak{R}_\sigma}\right)^{y\Xi_\sigma}\right)^{\mathfrak{L}}\right)\right)^{\frac{2}{y(y+1)}}\right)^{\frac{1}{\text{ß}+\mathfrak{L}}}\\
+\iota\left(1-\left(1-\left(\prod_{\substack{\varrho=1\\ \sigma=1}}^{y}\left(1-\left(1-\left(\mathfrak{F}^{IP}_{\mathfrak{R}_\varrho}\right)^{y\Xi_\varrho}\right)^{\text{ß}}\left(1-\left(\mathfrak{F}^{IP}_{\mathfrak{R}_\sigma}\right)^{y\Xi_\sigma}\right)^{\mathfrak{L}}\right)\right)^{\frac{2}{y(y+1)}}\right)^{\frac{1}{\text{ß}+\mathfrak{L}}}\right),\\
-\left|-1+\left(\prod_{\substack{\varrho=1\\ \sigma=1}}^{y}\left(1-\left|-1+\left(1+\mathfrak{F}^{RN}_{\mathfrak{R}_\varrho}\right)^{y\Xi_\varrho}\right|^{\text{ß}}\left|-1+\left(1+\mathfrak{F}^{RN}_{\mathfrak{R}_\sigma}\right)^{y\Xi_\sigma}\right|^{\mathfrak{L}}\right)\right)^{\frac{2}{y(y+1)}}\right|^{\frac{1}{\text{ß}+\mathfrak{L}}}\\
+\iota\left(-\left|-1+\left(\prod_{\substack{\varrho=1\\ \sigma=1}}^{y}\left(1-\left|-1+\left(1+\mathfrak{F}^{IN}_{\mathfrak{R}_\varrho}\right)^{y\Xi_\varrho}\right|^{\text{ß}}\left|-1+\left(1+\mathfrak{F}^{IN}_{\mathfrak{R}_\sigma}\right)^{y\Xi_\sigma}\right|^{\mathfrak{L}}\right)\right)^{\frac{2}{y(y+1)}}\right|^{\frac{1}{\text{ß}+\mathfrak{L}}}\right)
\end{array}\right)$$

$$= BCFPGGHM^{\text{ß},\mathfrak{L}}(\mathfrak{R}_1, \mathfrak{R}_2, \ldots, \mathfrak{R}_y)$$

Proposition 8.7. (Idempotency): Let a gathering of BCFNs $\mathfrak{R}_\sigma = \left(\mathfrak{F}^P_{\mathfrak{R}_\sigma}, \mathfrak{F}^N_{\mathfrak{R}_\sigma}\right) = \left(\mathfrak{F}^{RP}_{\mathfrak{R}_\sigma} + \iota\mathfrak{F}^{IP}_{\mathfrak{R}_\sigma}, \mathfrak{F}^{RN}_{\mathfrak{R}_\sigma} + \iota\mathfrak{F}^{IN}_{\mathfrak{R}_\sigma}\right)\sigma = 1, 2, \ldots, y$. If $\forall\sigma\ \mathfrak{R}_\sigma = \mathfrak{R}$, then

$$BCFPGGHM^{\text{ß},\mathfrak{L}}(\mathfrak{R}_1, \mathfrak{R}_2, \ldots, \mathfrak{R}_y) = \mathfrak{R}$$

Proof Similar to Proposition 8.2

Proposition 8.8 (Monotonicity): Let two gatherings of BCFNs $\mathfrak{R}_\sigma = \left(\mathfrak{F}^P_{\mathfrak{R}_\sigma}, \mathfrak{F}^N_{\mathfrak{R}_\sigma}\right) = \left(\mathfrak{F}^{RP}_{\mathfrak{R}_\sigma} + \iota\mathfrak{F}^{IP}_{\mathfrak{R}_\sigma}, \mathfrak{F}^{RN}_{\mathfrak{R}_\sigma} + \iota\mathfrak{F}^{IN}_{\mathfrak{R}_\sigma}\right)$ and $\mathfrak{R}_{\sigma'} = \left(\mathfrak{F}^P_{\mathfrak{R}_{\sigma'}}, \mathfrak{F}^N_{\mathfrak{R}_{\sigma'}}\right) = \left(\mathfrak{F}^{RP}_{\mathfrak{R}_{\sigma'}} + \iota\mathfrak{F}^{IP}_{\mathfrak{R}_{\sigma'}}, \mathfrak{F}^{RN}_{\mathfrak{R}_{\sigma'}} + \iota\mathfrak{F}^{IN}_{\mathfrak{R}_{\sigma'}}\right)\sigma = 1, 2, \ldots, y$. If $\mathfrak{F}^{RP}_{\mathfrak{R}_\sigma} \le \mathfrak{F}^{RP}_{\mathfrak{R}_{\sigma'}}$, $\mathfrak{F}^{IP}_{\mathfrak{R}_\sigma} \le \mathfrak{F}^{IP}_{\mathfrak{R}_{\sigma'}}$, $\mathfrak{F}^{RN}_{\mathfrak{R}_\sigma} \le \mathfrak{F}^{RN}_{\mathfrak{R}_{\sigma'}}$, and $\mathfrak{F}^{IN}_{\mathfrak{R}_\sigma} \le \mathfrak{F}^{IN}_{\mathfrak{R}_{\sigma'}}$, then.

$$BCFPGGHM^{\text{ß},\mathfrak{L}}(\mathfrak{R}_1, \mathfrak{R}_2, \ldots, \mathfrak{R}_y) \le BCFPGGHM^{\text{ß},\mathfrak{L}}(\mathfrak{R}_{1'}, \mathfrak{R}_{2'}, \ldots, \mathfrak{R}_{y'})$$

Proof Similar to Proposition 8.3

Proposition 8.9 (Boundedness): Let two gatherings of BCFNs $\Re_{\sigma} = \left(\mathfrak{F}^{P}_{\Re_{\sigma}}, \mathfrak{F}^{N}_{\Re_{\sigma}}\right) = \left(\mathfrak{F}^{RP}_{\Re_{\sigma}} + \iota\mathfrak{F}^{IP}_{\Re_{\sigma}}, \mathfrak{F}^{RN}_{\Re_{\sigma}} + \iota\mathfrak{F}^{IN}_{\Re_{\sigma}}\right)$ and $\Re_{\sigma'} = \left(\mathfrak{F}^{P}_{\Re_{\sigma'}}, \mathfrak{F}^{N}_{\Re_{\sigma'}}\right) = \left(\mathfrak{F}^{RP}_{\Re_{\sigma'}} + \iota\mathfrak{F}^{IP}_{\Re_{\sigma'}}, \mathfrak{F}^{RN}_{\Re_{\sigma'}} + \iota\mathfrak{F}^{IN}_{\Re_{\sigma'}}\right)\sigma = 1, 2, \ldots, y$ and $\Re^{-} = \left(\min_{\sigma}\left\{\mathfrak{F}^{RP}_{\Re_{\sigma}}\right\} + \iota \min_{\sigma}\left\{\mathfrak{F}^{IP}_{\Re_{\sigma}}\right\}, \min_{\sigma}\left\{\mathfrak{F}^{RN}_{\Re_{\sigma}}\right\} + \iota \min_{\sigma}\left\{\mathfrak{F}^{IN}_{\Re_{\sigma}}\right\}\right)$, and $\Re^{+} = \left(\max_{\sigma}\left\{\mathfrak{F}^{RP}_{\Re_{\sigma}}\right\} + \iota \max_{\sigma}\left\{\mathfrak{F}^{IP}_{\Re_{\sigma}}\right\}, \max_{\sigma}\left\{\mathfrak{F}^{RN}_{\Re_{\sigma}}\right\} + \iota \max_{\sigma}\left\{\mathfrak{F}^{IN}_{\Re_{\sigma}}\right\}\right)$, then

$$\Re^{-} \leq BCFPGGHM^{\text{ß},\mathfrak{L}}\left(\Re_1, \Re_2, \ldots, \Re_y\right) \leq \Re^{+}$$

Proof Similar to Proposition 8.3.

Definition 8.4 The BCFWPGGHM operator relying on the gathering of BCFNs $\Re_{\sigma} = \left(\mathfrak{F}^{P}_{\Re_{\sigma}}, \mathfrak{F}^{N}_{\Re_{\sigma}}\right) = \left(\mathfrak{F}^{RP}_{\Re_{\sigma}} + \iota\mathfrak{F}^{IP}_{\Re_{\sigma}}, \mathfrak{F}^{RN}_{\Re_{\sigma}} + \iota\mathfrak{F}^{IN}_{\Re_{\sigma}}\right)\sigma = 1, 2, \ldots, y$, is spotted as

$$BCFPWGGHM^{\text{ß},\mathfrak{L}}\left(\Re_1, \Re_2, \ldots, \Re_y\right)$$

$$= \frac{1}{\text{ß} + \mathfrak{L}}\left(\underset{\varrho=1}{\overset{y}{\otimes}}\ \underset{\sigma=1}{\overset{y}{\otimes}}\left(\text{ß}\left(\Re_{\varrho}\right)^{\frac{y\mathring{\text{w}}_{\varrho}(1+\mathcal{I}(\Re_{\varrho}))}{\sum_{\rho=1}^{y}\mathring{\text{w}}_{\rho}(1+\mathcal{I}(\Re_{\rho}))}} \oplus \mathfrak{L}\left(\Re_{\sigma}\right)^{\frac{y\mathring{\text{w}}_{\sigma}(1+\mathcal{I}(\Re_{\sigma}))}{\sum_{\rho=1}^{y}\mathring{\text{w}}_{\rho}(1+\mathcal{I}(\Re_{\rho}))}}\right)\right)^{\frac{2}{y(y+1)}}$$

Noticed that $\mathring{\text{w}} = \left(\mathring{\text{w}}_1, \mathring{\text{w}}_2, \ldots, \mathring{\text{w}}_y\right)^{T}$ is a weight vector, $0 \leq \mathring{\text{w}}_{\sigma} \leq 1$ for all σ with $\sum_{\sigma=1}^{y}\mathring{\text{w}}_{\sigma} = 1$.

Theorem 8.4 Let a gathering of BCFNs $\Re_{\sigma} = \left(\mathfrak{F}^{P}_{\Re_{\sigma}}, \mathfrak{F}^{N}_{\Re_{\sigma}}\right) = \left(\mathfrak{F}^{RP}_{\Re_{\sigma}} + \iota\mathfrak{F}^{IP}_{\Re_{\sigma}}, \mathfrak{F}^{RN}_{\Re_{\sigma}} + \iota\mathfrak{F}^{IN}_{\Re_{\sigma}}\right)\sigma = 1, 2, \ldots, y$. The aggregated outcome established by Definition 8.4 is still a BCFN that is

$$BCFWPGGHM^{\text{ß},\mathfrak{L}}(\Re_1, \Re_2, \ldots, \Re_y)$$

$$= \left(\begin{array}{l} 1 - \left(1 - \left(1 - \left(\prod_{\substack{\varrho=1 \\ \sigma=1}}^{y} \left(1 - \left(1 - \left(\mathfrak{F}^{RP}_{\Re_\varrho}\right)^{y\Xi_{\mathring{w}-\varrho}}\right)^{\text{ß}} \left(1 - \left(\mathfrak{F}^{RP}_{\Re_\sigma}\right)^{y\Xi_{\mathring{w}-\sigma}}\right)^{\mathfrak{L}}\right)\right)^{\frac{2}{y(y+1)}}\right)^{\frac{1}{\text{ß}+\mathfrak{L}}}\right) \\ +\iota \left(1 - \left(1 - \left(\prod_{\substack{\varrho=1 \\ \sigma=1}}^{y} \left(1 - \left(1 - \left(\mathfrak{F}^{IP}_{\Re_\varrho}\right)^{y\Xi_{\mathring{w}-\varrho}}\right)^{\text{ß}} \left(1 - \left(\mathfrak{F}^{IP}_{\Re_\sigma}\right)^{y\Xi_{\mathring{w}-\sigma}}\right)^{\mathfrak{L}}\right)\right)^{\frac{2}{y(y+1)}}\right)^{\frac{1}{\text{ß}+\mathfrak{L}}}\right), \\ -\left|-1 + \left(\prod_{\substack{\varrho=1 \\ \sigma=1}}^{y} \left(1 - \left|-1 + \left(1 + \mathfrak{F}^{RN}_{\Re_\varrho}\right)^{y\Xi_{\mathring{w}-\varrho}}\right|^{\text{ß}} \left|-1 + \left(1 + \mathfrak{F}^{RN}_{\Re_\sigma}\right)^{y\Xi_{\mathring{w}-\sigma}}\right|^{\mathfrak{L}}\right)\right)^{\frac{2}{y(y+1)}}\right|^{\frac{1}{\text{ß}+\mathfrak{L}}} \\ +\iota \left(-\left|-1 + \left(\prod_{\substack{\varrho=1 \\ \sigma=1}}^{y} \left(1 - \left|-1 + \left(1 + \mathfrak{F}^{IN}_{\Re_\varrho}\right)^{y\Xi_{\mathring{w}-\varrho}}\right|^{\text{ß}} \left|-1 + \left(1 + \mathfrak{F}^{IN}_{\Re_\sigma}\right)^{y\Xi_{\mathring{w}-\sigma}}\right|^{\mathfrak{L}}\right)\right)^{\frac{2}{y(y+1)}}\right|^{\frac{1}{\text{ß}+\mathfrak{L}}}\right) \end{array} \right)$$

where, $\Xi_{\mathring{w}-\varrho} = \frac{y\,\mathring{w}_\varrho(1+\underline{T}(\Re_\varrho))}{\sum_{\rho=1}^{y}\mathring{w}_\rho(1+\underline{T}(\Re_\rho))}$.

Proposition 8.10 (Idempotency): Let a gathering of BCFNs $\Re_\sigma = \left(\mathfrak{F}^{P}_{\Re_\sigma}, \mathfrak{F}^{N}_{\Re_\sigma}\right) = \left(\mathfrak{F}^{RP}_{\Re_\sigma} + \iota\mathfrak{F}^{IP}_{\Re_\sigma}, \mathfrak{F}^{RN}_{\Re_\sigma} + \iota\mathfrak{F}^{IN}_{\Re_\sigma}\right)\sigma = 1, 2, \ldots, y$. If $\forall\sigma\ \Re_\sigma = \Re$, then

$$BCFPWGGHM^{\text{ß},\mathfrak{L}}\left(\Re_1, \Re_2, \ldots, \Re_y\right) = \Re$$

Proposition 8.11 (Monotonicity) Let two gatherings of BCFNs $\Re_\sigma = \left(\mathfrak{F}^{P}_{\Re_\sigma}, \mathfrak{F}^{N}_{\Re_\sigma}\right) = \left(\mathfrak{F}^{RP}_{\Re_\sigma} + \iota\mathfrak{F}^{IP}_{\Re_\sigma}, \mathfrak{F}^{RN}_{\Re_\sigma} + \iota\mathfrak{F}^{IN}_{\Re_\sigma}\right)$ and $\Re_{\sigma'} = \left(\mathfrak{F}^{P}_{\Re_{\sigma'}}, \mathfrak{F}^{N}_{\Re_{\sigma'}}\right) = \left(\mathfrak{F}^{RP}_{\Re_{\sigma'}} + \iota\mathfrak{F}^{IP}_{\Re_{\sigma'}}, \mathfrak{F}^{RN}_{\Re_{\sigma'}} + \iota\mathfrak{F}^{IN}_{\Re_{\sigma'}}\right)\sigma = 1, 2, \ldots, y$. If $\mathfrak{F}^{RP}_{\Re_\sigma} \leq \mathfrak{F}^{RP}_{\Re_{\sigma'}}, \mathfrak{F}^{IP}_{\Re_\sigma} \leq \mathfrak{F}^{IP}_{\Re_{\sigma'}}, \mathfrak{F}^{RN}_{\Re_\sigma} \leq \mathfrak{F}^{RN}_{\Re_{\sigma'}}$, and $\mathfrak{F}^{IN}_{\Re_\sigma} \leq \mathfrak{F}^{IN}_{\Re_{\sigma'}}$, then

$$BCFPWGGHM^{\text{ß},\mathfrak{L}}\left(\Re_1, \Re_2, \ldots, \Re_y\right) \leq BCFPWGGHM^{\text{ß},\mathfrak{L}}\left(\Re_{1'}, \Re_{2'}, \ldots, \Re_{y'}\right)$$

Proposition 8.12 (Boundedness) Let two gatherings of BCFNs $\Re_\sigma = \left(\mathfrak{F}^{P}_{\Re_\sigma}, \mathfrak{F}^{N}_{\Re_\sigma}\right) = \left(\mathfrak{F}^{RP}_{\Re_\sigma} + \iota\mathfrak{F}^{IP}_{\Re_\sigma}, \mathfrak{F}^{RN}_{\Re_\sigma} + \iota\mathfrak{F}^{IN}_{\Re_\sigma}\right)$ and $\Re_{\sigma'} = \left(\mathfrak{F}^{P}_{\Re_{\sigma'}}, \mathfrak{F}^{N}_{\Re_{\sigma'}}\right) = \left(\mathfrak{F}^{RP}_{\Re_{\sigma'}} + \iota\mathfrak{F}^{IP}_{\Re_{\sigma'}}, \mathfrak{F}^{RN}_{\Re_{\sigma'}} + \iota\mathfrak{F}^{IN}_{\Re_{\sigma'}}\right)\sigma = 1, 2, \ldots, y$ and $\Re^{-} = \left(\min_\sigma\left\{\mathfrak{F}^{RP}_{\Re_\sigma}\right\} + \iota\min_\sigma\left\{\mathfrak{F}^{IP}_{\Re_\sigma}\right\}, \min_\sigma\left\{\mathfrak{F}^{RN}_{\Re_\sigma}\right\} + \iota\min_\sigma\left\{\mathfrak{F}^{IN}_{\Re_\sigma}\right\}\right)$, and $\Re^{+} =$

$\left(\max_{\sigma}\{\mathfrak{F}^{RP}_{\mathfrak{R}_\sigma}\} + \iota \max_{\sigma}\{\mathfrak{F}^{IP}_{\mathfrak{R}_\sigma}\}, \max_{\sigma}\{\mathfrak{F}^{RN}_{\mathfrak{R}_\sigma}\} + \iota \max_{\sigma}\{\mathfrak{F}^{IN}_{\mathfrak{R}_\sigma}\}\right)$, then

$$\mathfrak{R}^{-} \leq BCFPWGGHM^{\text{ß},\pounds}(\mathfrak{R}_1, \mathfrak{R}_2, \ldots, \mathfrak{R}_y) \leq \mathfrak{R}^{+}$$

8.3 The Technique of WASPAS for MCDM Dilemmas Under BCF Information

An MCDM tool called the WASPAS method is used to rank and assess alternatives based on many criteria. "Weighted Aggregated Sum Product Assessment" is what it stands for. Zavadskas et al. [38] created the approach and it has since grown in popularity among experts. The WASPAS method's computing process for a DM dilemma has the underneath steps.

Step 1 (Normalization): Divide the assessment scores for each criterion by the total evaluation scores for that criterion to normalize the scores. By doing this, you can guarantee that all the criteria are on the same scale and avoid the domination of the ones with higher values.

Step 2 (Criteria weighting): By giving them weights, you may determine the relative relevance of each criterion. The weights in the DM process represent the importance or order of the criteria. The weights can be determined using procedures or expert judgment.

Step 3 (Compute weighted sum model): Determine the weighted sum model (WSM) of every alternative by multiplying the criteria with corresponding weight and then adding them.

Step 4 (Compute weighted product model): Determine the weighted product model (WPM) of every alternative by powering the criteria with corresponding weight and then multiple them.

Step 5: Compute the WASPAS approach for each alternative by fusing both WPM and WSM and

$$BCF - WASPAS(\mathfrak{R}_{\mathbb{AL}-\sigma}) =^{\circ} \mathrm{F}BCF - WSM(\mathfrak{R}_{\mathbb{AL}-\sigma}) + (1 -^{\circ} \mathrm{F})BCF - WPM(\mathfrak{R}_{\mathbb{AL}-\sigma})$$

where $^{\circ}F \in [0, 1]$ and play a role in aggregating the coefficient of decision accuracy.

Next, we inaugurate a technique of WASPAS for BCF information to cope with MCDM dilemmas.

8.3.1 Clarification of MCDM Problem Under BCFNs

Assume that we have an MCDM dilemma where the number of alternatives $\{\mathfrak{R}_{\mathbb{AL}-1}, \mathfrak{R}_{\mathbb{AL}-2}, \ldots, \mathfrak{R}_{\mathbb{AL}-y}\}$ are y and the number of criteria $\{\mathfrak{A}_{\mathbb{A}\approx-1}, \mathfrak{A}_{\mathbb{A}\approx-2}, \ldots, \mathfrak{A}_{\mathbb{A}\approx-\xi}\}$ are ξ along with $\overset{\circ}{\mathrm{w}} = \left(\underset{1,}{\overset{\circ}{\mathrm{w}}}\underset{2}{\overset{\circ}{\mathrm{w}}}, \ldots, \underset{\xi}{\overset{\circ}{\mathrm{w}}}\right)^T$ as a weight vector, $0 \leq \underset{\varrho}{\overset{\circ}{\mathrm{w}}} \leq 1$ for all ϱ with $\sum_{\varrho=1}^{\xi} \underset{\varrho}{\overset{\circ}{\mathrm{w}}} = 1$. Also assume that in this MCDM dilemma, we have an experts or decision analyst who will assess the alternatives by keeping in view the demonstrated criteria. The assessment value of the decision analyst or expert will be in the structure of BCFN, which will develop a BCF decision matrix i.e., $\underline{D}_{\mathcal{M}} = (\mathfrak{R}_{\sigma\varrho})_{y\times\xi}$, where $\mathfrak{R}_{\sigma\varrho} = \left(\mathfrak{F}^{P}_{\mathfrak{R}_{\sigma\varrho}}, \mathfrak{F}^{N}_{\mathfrak{R}_{\sigma\varrho}}\right) = \left(\mathfrak{F}^{RP}_{\mathfrak{R}_{\sigma\varrho}} + \iota\mathfrak{F}^{IP}_{\mathfrak{R}_{\sigma\varrho}}, \mathfrak{F}^{RN}_{\mathfrak{R}_{\sigma\varrho}} + \iota\mathfrak{F}^{IN}_{\mathfrak{R}_{\sigma\varrho}}\right)$. The BCF decision matrix would be as

$$(\mathfrak{R}_{\sigma\varrho})_{y\times\xi} = \begin{pmatrix} \begin{pmatrix} \mathfrak{F}^{RP}_{\mathfrak{R}_{11}} + \iota\mathfrak{F}^{IP}_{\mathfrak{R}_{11}}, \\ \mathfrak{F}^{RN}_{\mathfrak{R}_{-11}} + \iota\mathfrak{F}^{IN}_{\mathfrak{R}_{-11}} \end{pmatrix} & \begin{pmatrix} \mathfrak{F}^{RP}_{\mathfrak{R}_{12}} + \iota\mathfrak{F}^{IP}_{\mathfrak{R}_{12}}, \\ \mathfrak{F}^{RN}_{\mathfrak{R}_{12}} + \iota\mathfrak{F}^{IN}_{\mathfrak{R}_{12}} \end{pmatrix} & \cdots & \begin{pmatrix} \mathfrak{F}^{RP}_{\mathfrak{R}_{1\xi}} + \iota\mathfrak{F}^{IP}_{\mathfrak{R}_{1\xi}}, \\ \mathfrak{F}^{RN}_{\mathfrak{R}_{1\xi}} + \iota\mathfrak{F}^{IN}_{\mathfrak{R}_{1\xi}} \end{pmatrix} \\ \begin{pmatrix} \mathfrak{F}^{RP}_{\mathfrak{R}_{21}} + \iota\mathfrak{F}^{IP}_{\mathfrak{R}_{21}}, \\ \mathfrak{F}^{RN}_{\mathfrak{R}_{21}} + \iota\mathfrak{F}^{IN}_{\mathfrak{R}_{21}} \end{pmatrix} & \begin{pmatrix} \mathfrak{F}^{RP}_{\mathfrak{R}_{22}} + \iota\mathfrak{F}^{IP}_{\mathfrak{R}_{22}}, \\ \mathfrak{F}^{RN}_{\mathfrak{R}_{22}} + \iota\mathfrak{F}^{IN}_{\mathfrak{R}_{22}} \end{pmatrix} & \cdots & \begin{pmatrix} \mathfrak{F}^{RP}_{\mathfrak{R}_{2\xi}} + \iota\mathfrak{F}^{IP}_{\mathfrak{R}_{2\xi}}, \\ \mathfrak{F}^{RN}_{\mathfrak{R}_{2\xi}} + \iota\mathfrak{F}^{IN}_{\mathfrak{R}_{2\xi}} \end{pmatrix} \\ & . & \cdots & . \\ & . & \cdots & . \\ & . & \cdots & . \\ \begin{pmatrix} \mathfrak{F}^{RP}_{\mathfrak{R}_{y1}} + \iota\mathfrak{F}^{IP}_{\mathfrak{R}_{y1}}, \\ \mathfrak{F}^{RN}_{\mathfrak{R}_{y1}} + \iota\mathfrak{F}^{IN}_{\mathfrak{R}_{y1}} \end{pmatrix} & \begin{pmatrix} \mathfrak{F}^{RP}_{\mathfrak{R}_{y2}} + \iota\mathfrak{F}^{IP}_{\mathfrak{R}_{y2}}, \\ \mathfrak{F}^{RN}_{\mathfrak{R}_{y2}} + \iota\mathfrak{F}^{IN}_{\mathfrak{R}_{y2}} \end{pmatrix} & \cdots & \begin{pmatrix} \mathfrak{F}^{RP}_{\mathfrak{R}_{y\xi}} + \iota\mathfrak{F}^{IP}_{\mathfrak{R}_{y\xi}}, \\ \mathfrak{F}^{RN}_{\mathfrak{R}_{y\xi}} + \iota\mathfrak{F}^{IN}_{\mathfrak{R}_{y\xi}} \end{pmatrix} \end{pmatrix}$$

Underneath are the steps for solving the MCDM dilemma with the assistance of the WASPAS approach.

Step 1: The BCF decision matrix given by a decision analyst can contain both benefit and cost sort of criteria. Thus, normalization is required which would be done as follows.

The normalization of every benefit criteria ϱ would be achieved as

$$\mathfrak{R}^{\propto}_{\sigma\varrho} = \frac{\mathfrak{R}_{\sigma\varrho}}{\max\limits_{\sigma} \mathfrak{R}_{\sigma\varrho}} = \left(\frac{\mathfrak{F}^{RP}_{\mathfrak{R}_{\sigma\varrho}} + \iota\mathfrak{F}^{IP}_{\mathfrak{R}_{\sigma\varrho}}}{\max\limits_{\sigma}\left(\mathfrak{F}^{RP}_{\mathfrak{R}_{\sigma\varrho}} + \iota\mathfrak{F}^{IP}_{\mathfrak{R}_{\sigma\varrho}}\right)}, \frac{\mathfrak{F}^{RN}_{\mathfrak{R}_{\sigma\varrho}} + \iota\mathfrak{F}^{IN}_{\mathfrak{R}_{\sigma\varrho}}}{\max\limits_{\sigma}\left(\mathfrak{F}^{RN}_{\mathfrak{R}_{\sigma\varrho}} + \iota\mathfrak{F}^{IN}_{\mathfrak{R}_{\sigma\varrho}}\right)} \right) \tag{8.3}$$

where, $\max\limits_{\sigma}\left(\mathfrak{F}^{RP}_{\mathfrak{R}_{\sigma\varrho}} + \iota\mathfrak{F}^{IP}_{\mathfrak{R}_{\sigma\varrho}}\right) = \max\limits_{\sigma} \mathfrak{F}^{RP}_{\mathfrak{R}_{\sigma\varrho}} + \iota \max\limits_{\sigma} \mathfrak{F}^{IP}_{\mathfrak{R}_{\sigma\varrho}}$ and $\max\limits_{\sigma}\left(\mathfrak{F}^{RN}_{\mathfrak{R}_{\sigma\varrho}} + \iota\mathfrak{F}^{IN}_{\mathfrak{R}_{\sigma\varrho}}\right) = \max\limits_{\sigma} \mathfrak{F}^{RN}_{\mathfrak{R}_{\sigma\varrho}} + \iota \max\limits_{\sigma} \mathfrak{F}^{IN}_{\mathfrak{R}_{\sigma\varrho}}$ and for every cost criterion ϱ would be achieved as

$$\mathfrak{R}^{\alpha}_{\sigma\varrho} = \frac{\min\limits_{\sigma}\mathfrak{R}_{\sigma\varrho}}{\mathfrak{R}_{\sigma\varrho}} = \left(\frac{\min\limits_{\sigma}\left(\mathfrak{F}^{RP}_{\mathfrak{R}_{\sigma\varrho}}+\iota\mathfrak{F}^{IP}_{\mathfrak{R}_{\sigma\varrho}}\right)}{\mathfrak{F}^{RP}_{\mathfrak{R}_{\sigma\varrho}}+\iota\mathfrak{F}^{IP}_{\mathfrak{R}_{\sigma\varrho}}}, \frac{\min\limits_{\sigma}\left(\mathfrak{F}^{RN}_{\mathfrak{R}_{\sigma\varrho}}+\iota\mathfrak{F}^{IN}_{\mathfrak{R}_{\sigma\varrho}}\right)}{\mathfrak{F}^{RN}_{\mathfrak{R}_{\sigma\varrho}}+\iota\mathfrak{F}^{IN}_{\mathfrak{R}_{\sigma\varrho}}}\right) \tag{8.4}$$

where, $\min\limits_{\sigma}\left(\mathfrak{F}^{RP}_{\mathfrak{R}_{\sigma\varrho}}+\iota\mathfrak{F}^{IP}_{\mathfrak{R}_{\sigma\varrho}}\right) = \min\limits_{\sigma}\mathfrak{F}^{RP}_{\mathfrak{R}_{\sigma\varrho}} + \iota\min\limits_{\sigma}\mathfrak{F}^{IP}_{\mathfrak{R}_{\sigma\varrho}}$ and $\min\limits_{\sigma}\left(\mathfrak{F}^{RN}_{\mathfrak{R}_{\sigma\varrho}}+\iota\mathfrak{F}^{IN}_{\mathfrak{R}_{\sigma\varrho}}\right) = \min\limits_{\sigma}\mathfrak{F}^{RN}_{\mathfrak{R}_{\sigma\varrho}} + \iota\min\limits_{\sigma}\mathfrak{F}^{IN}_{\mathfrak{R}_{\sigma\varrho}}$.

Note that the division in Equations (8.3) and (8.4) will happen as

Suppose two BCFNs $\mathfrak{R}_1 = \left(\mathfrak{F}^{P}_{\mathfrak{R}_1}, \mathfrak{F}^{N}_{\mathfrak{R}_1}\right) = \left(\mathfrak{F}^{RP}_{\mathfrak{R}_1}+\iota\mathfrak{F}^{IP}_{\mathfrak{R}1}, \mathfrak{F}^{RN}_{\mathfrak{R}_1}+\iota\mathfrak{F}^{IN}_{\mathfrak{R}_1}\right)$ and $\mathfrak{R}_2 = \left(\mathfrak{F}^{P}_{\mathfrak{R}_2}, \mathfrak{F}^{N}_{\mathfrak{R}_2}\right) = \left(\mathfrak{F}^{RP}_{\mathfrak{R}_2}+\iota\mathfrak{F}^{IP}_{\mathfrak{R}_2}, \mathfrak{F}^{RN}_{\mathfrak{R}_2}+\iota\mathfrak{F}^{IN}_{\mathfrak{R}_2}\right)$, then

$$\frac{\mathfrak{R}_1}{\mathfrak{R}_2} = \begin{pmatrix} \dfrac{\mathfrak{F}^{RP}_{\mathfrak{R}_1}\mathfrak{F}^{RP}_{\mathfrak{R}_2}}{\mathfrak{F}^{RP}_{\mathfrak{R}_1}+\mathfrak{F}^{RP}_{\mathfrak{R}_2}-\mathfrak{F}^{RP}_{\mathfrak{R}_1}\mathfrak{F}^{RP}_{\mathfrak{R}_2}} + \iota\dfrac{\mathfrak{F}^{IP}_{\mathfrak{R}_1}\mathfrak{F}^{IP}_{\mathfrak{R}_2}}{\mathfrak{F}^{IP}_{\mathfrak{R}_1}+\mathfrak{F}^{IP}_{\mathfrak{R}_2}-\mathfrak{F}^{IP}_{\mathfrak{R}_1}\mathfrak{F}^{IP}_{\mathfrak{R}_2}}, \\ \dfrac{\mathfrak{F}^{RN}_{\mathfrak{R}_1}+\mathfrak{F}^{RN}_{\mathfrak{R}_2}+2\mathfrak{F}^{RN}_{\mathfrak{R}_1}\mathfrak{F}^{RN}_{\mathfrak{R}_2}}{1-\mathfrak{F}^{RN}_{\mathfrak{R}_1}\mathfrak{F}^{RN}_{\mathfrak{R}_2}} + \iota\dfrac{\mathfrak{F}^{IN}_{\mathfrak{R}_1}+\mathfrak{F}^{IN}_{\mathfrak{R}_2}+2\mathfrak{F}^{IN}_{\mathfrak{R}_1}\mathfrak{F}^{IN}_{\mathfrak{R}_2}}{1-\mathfrak{F}^{IN}_{\mathfrak{R}_1}\mathfrak{F}^{IN}_{\mathfrak{R}_2}} \end{pmatrix}$$

Remark If $\mathfrak{F}^{RP}_{\mathfrak{R}_1} = \mathfrak{F}^{RP}_{\mathfrak{R}_2} = 0$, then $\frac{\mathfrak{F}^{RP}_{\mathfrak{R}_1}\mathfrak{F}^{RP}_{\mathfrak{R}_2}}{\mathfrak{F}^{RP}_{\mathfrak{R}_1}+\mathfrak{F}^{RP}_{\mathfrak{R}_2}-\mathfrak{F}^{RP}_{\mathfrak{R}_1}\mathfrak{F}^{RP}_{\mathfrak{R}_2}} = \frac{\mathfrak{F}^{IP}_{\mathfrak{R}_1}\mathfrak{F}^{IP}_{\mathfrak{R}_2}}{\mathfrak{F}^{IP}_{\mathfrak{R}_1}+\mathfrak{F}^{IP}_{\mathfrak{R}_2}-\mathfrak{F}^{IP}_{\mathfrak{R}_1}\mathfrak{F}^{IP}_{\mathfrak{R}_2}} = 0$ and if $\mathfrak{F}^{RN}_{\mathfrak{R}_1} = \mathfrak{F}^{RN}_{\mathfrak{R}_2} = -1$, then $\frac{\mathfrak{F}^{RN}_{\mathfrak{R}_1}+\mathfrak{F}^{RN}_{\mathfrak{R}_2}+2\mathfrak{F}^{RN}_{\mathfrak{R}_1}\mathfrak{F}^{RN}_{\mathfrak{R}_2}}{1-\mathfrak{F}^{RN}_{\mathfrak{R}_1}\mathfrak{F}^{RN}_{\mathfrak{R}_2}} = \frac{\mathfrak{F}^{IN}_{\mathfrak{R}_1}+\mathfrak{F}^{IN}_{\mathfrak{R}_2}+2\mathfrak{F}^{IN}_{\mathfrak{R}_1}\mathfrak{F}^{IN}_{\mathfrak{R}_2}}{1-\mathfrak{F}^{IN}_{\mathfrak{R}_1}\mathfrak{F}^{IN}_{\mathfrak{R}_2}} = 0$.

The required normalized matrix would be designated as

$$\left(\mathfrak{R}^{\alpha}_{\sigma\varrho}\right)_{y\times\xi} = \begin{pmatrix} \begin{pmatrix}\mathfrak{F}^{RP}_{\mathfrak{R}^{\alpha}_{11}}+\iota\mathfrak{F}^{IP}_{\mathfrak{R}^{\alpha}_{11}}, \\ \mathfrak{F}^{RN}_{\mathfrak{R}^{\alpha}_{11}}+\iota\mathfrak{F}^{IN}_{\mathfrak{R}^{\alpha}_{11}}\end{pmatrix} & \begin{pmatrix}\mathfrak{F}^{RP}_{\mathfrak{R}^{\alpha}_{12}}+\iota\mathfrak{F}^{IP}_{\mathfrak{R}^{\alpha}_{12}}, \\ \mathfrak{F}^{RN}_{\mathfrak{R}^{\alpha}_{12}}+\iota\mathfrak{F}^{IN}_{\mathfrak{R}^{\alpha}_{12}}\end{pmatrix} & \cdots & \begin{pmatrix}\mathfrak{F}^{RP}_{\mathfrak{R}^{\alpha}_{1\xi}}+\iota\mathfrak{F}^{IP}_{\mathfrak{R}^{\alpha}_{1\xi}}, \\ \mathfrak{F}^{RN}_{\mathfrak{R}^{\alpha}_{1\xi}}+\iota\mathfrak{F}^{IN}_{\mathfrak{R}^{\alpha}_{1\xi}}\end{pmatrix} \\ \begin{pmatrix}\mathfrak{F}^{RP}_{\mathfrak{R}^{\alpha}_{21}}+\iota\mathfrak{F}^{IP}_{\mathfrak{R}^{\alpha}_{21}}, \\ \mathfrak{F}^{RN}_{\mathfrak{R}^{\alpha}_{21}}+\iota\mathfrak{F}^{IN}_{\mathfrak{R}^{\alpha}_{21}}\end{pmatrix} & \begin{pmatrix}\mathfrak{F}^{RP}_{\mathfrak{R}^{\alpha}_{22}}+\iota\mathfrak{F}^{IP}_{\mathfrak{R}^{\alpha}_{22}}, \\ \mathfrak{F}^{RN}_{\mathfrak{R}^{\alpha}_{22}}+\iota\mathfrak{F}^{IN}_{\mathfrak{R}^{\alpha}_{22}}\end{pmatrix} & \cdots & \begin{pmatrix}\mathfrak{F}^{RP}_{\mathfrak{R}^{\alpha}_{2\xi}}+\iota\mathfrak{F}^{IP}_{\mathfrak{R}^{\alpha}_{2\xi}}, \\ \mathfrak{F}^{RN}_{\mathfrak{R}^{\alpha}_{2\xi}}+\iota\mathfrak{F}^{IN}_{\mathfrak{R}^{\alpha}_{2\xi}}\end{pmatrix} \\ . & . & \cdots & . \\ . & . & \cdots & . \\ . & . & \cdots & . \\ \begin{pmatrix}\mathfrak{F}^{RP}_{\mathfrak{R}^{\alpha}_{y1}}+\iota\mathfrak{F}^{IP}_{\mathfrak{R}^{\alpha}_{y1}}, \\ \mathfrak{F}^{RN}_{\mathfrak{R}^{\alpha}_{y1}}+\iota\mathfrak{F}^{IN}_{\mathfrak{R}^{\alpha}_{y1}}\end{pmatrix} & \begin{pmatrix}\mathfrak{F}^{RP}_{\mathfrak{R}^{\alpha}_{y2}}+\iota\mathfrak{F}^{IP}_{\mathfrak{R}^{\alpha}_{y2}}, \\ \mathfrak{F}^{RN}_{\mathfrak{R}^{\alpha}_{y2}}+\iota\mathfrak{F}^{IN}_{\mathfrak{R}^{\alpha}_{y2}}\end{pmatrix} & \cdots & \begin{pmatrix}\mathfrak{F}^{RP}_{\mathfrak{R}^{\alpha}_{y\xi}}+\iota\mathfrak{F}^{IP}_{\mathfrak{R}^{\alpha}_{y\xi}}, \\ \mathfrak{F}^{RN}_{\mathfrak{R}^{\alpha}_{y\xi}}+\iota\mathfrak{F}^{IN}_{\mathfrak{R}^{\alpha}_{y\xi}}\end{pmatrix} \end{pmatrix}$$

Step 2: The weight can be utilized by using the PA operator or the decision analyst can give his/her choice.

Step 3: The measure of BCF weighted sum model $(BCF-WSM)BCF-WSM\left(\mathfrak{R}_{\mathbb{AL}-\sigma}\right)$ f every alternative would be determined by employing a BCFPGHM operator.

Step 4: The measure of BCF weighted product model $(BCF-WPM)BCF-WPM\left(\mathfrak{R}_{\mathbb{AL}-\sigma}\right)$ of every alternative would be determined by employing the BCFPGGHM operator.

Step 5: The aggregated measure of BCF-WASPAS approach for each alternative $BCF-WASPAS(\Re_{\mathbb{AL}-\sigma})$ by fusing both BCF-WPM and BCF-WSM and the resulting formula is underneath

$$BCF-WASPAS(\Re_{\mathbb{AL}-\sigma}) = {}^{\circ}FBCF-WSM(\Re_{\mathbb{AL}-\sigma}) + (1-{}^{\circ}F)BCF-WPM(\Re_{\mathbb{AL}-\sigma})$$

where ${}^{\circ}F \in [0, 1]$ and play a role in aggregating the coefficient of decision accuracy. This scalar multiplication and addition will be performed by employing part 3 of definition (2.9) and part 1 of definition (2.9).

Step 6: In the last, the score value of every alternative $\Re_{\mathbb{AL}-\sigma}$ will be interpreted for ranking the alternatives. In any case, where the score values of two various alternatives become equal, then the accuracy values will be interpreted for ranking the alternatives.

The flow chart of the initiated BCF-WASPAS approach is demonstrated in Fig. 8.1

8.4 Application (Supply Chain Transportation)

The effective transfer of goods and products from suppliers to consumers is made possible by supply chain transportation, a vital part of the global economy. It includes all of the many tasks, procedures, and mechanisms related to moving raw materials, parts, and completed commodities physically throughout the whole supply chain network. To ensure the smooth flow of goods, the supply chain transportation process involves several stakeholders, including manufacturers, suppliers, distributors, retailers, and logistics providers. For cost reduction and quick customer service, effective supply chain transportation is essential. An efficient transportation system guarantees that goods arrive at their intended locations quickly and affordably, avoiding delays that could result in stockouts, lost sales, and disgruntled customers. This is especially crucial in today's globalized economy, as businesses obtain materials from all over the world and sell their goods to a wide range of customers on several continents.

Concerns about sustainability and the environment are also greatly impacted by supply chain mobility. The utilization of fossil fuels and greenhouse gas emissions rise along with the volume of commodities being carried. Companies are implementing eco-friendly transportation strategies to lessen these negative effects on the environment, including the use of fuel-efficient vehicles, fleet investments in electric or hybrid vehicles, and route optimization to cut down on emissions and fuel use. The development of e-commerce in recent years has highlighted the significance of effective supply chain transportation even more. Customers want quicker deliveries and flexible shipping options as online shopping increases, which puts additional strain on logistical networks. Companies are looking at cutting-edge solutions to meet these demands, including autonomous vehicles, drone deliveries, and last-mile

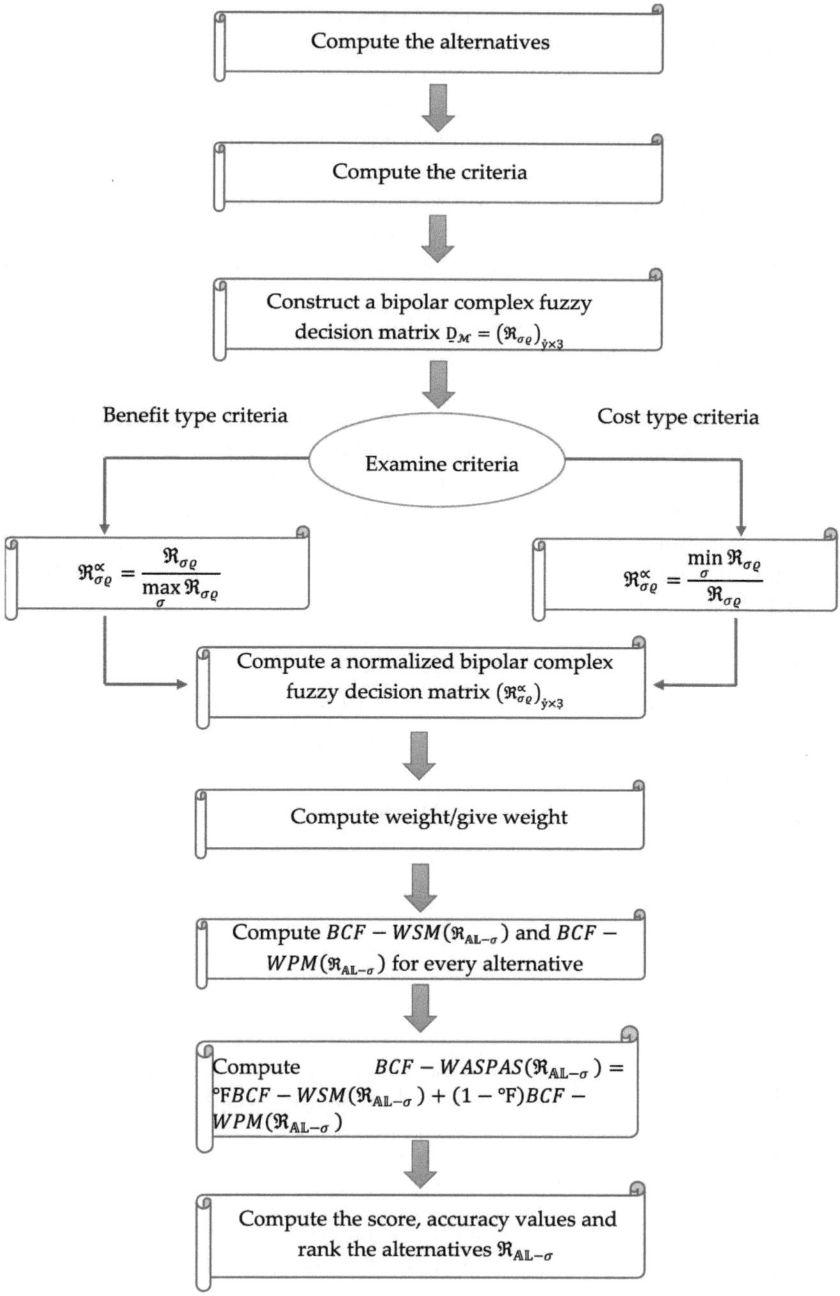

Fig. 8.1 The flow chart of BCF-WASPAS technique

delivery systems. Companies use advanced logistics management systems and technologies to obtain the best supply chain transportation. These systems aid in route planning, real-time shipment tracking, and truck capacity optimization, enabling the most effective use of resources. Additionally, they make it possible for companies to foresee and deal with any transportation-related disruptions like bad weather, strikes, or traffic jams.

In the underneath dilemma, we will discuss the selection of the optimal supply chain transportation for a multi-national corporation.

Scenario: XYZ Electronics - Selecting the Best Mode of Supply Chain Transportation

Leading technological company XYZ Electronics creates and markets a wide range of electronic products globally. They have four options for moving goods in their supply chain: $\mathfrak{R}_{\mathbb{AL}-1}$: Air freight,

$\mathfrak{R}_{\mathbb{AL}-2}$: Ocean freight,$\mathfrak{R}_{\mathbb{AL}-3}$: Rail freight, and the $\mathfrak{R}_{\mathbb{AL}-4}$: Road freight. The business is preparing to introduce the "Smartphone X," a new product, and must choose the optimal mode of transportation for good supply chain management. The company evaluates each method of transportation using a variety of criteria such as $\mathfrak{A}_{\mathbb{A}\approx-1}$: Suitability, $\mathfrak{A}_{\mathbb{A}\approx-2}$: Speed, $\mathfrak{A}_{\mathbb{A}\approx-3}$: Cost and $\mathfrak{A}_{\mathbb{A}\approx-4}$: Capacity and the evaluated values are given in the model of BCFN, which is demonstrated in Table 8.1.

Step 1: In this scenario, we assume that all attributes are benefit types. So, we have.

$\max_{\sigma} \mathfrak{R}_{\sigma 1} = (0.998 + \iota 0.989, -0.183 - \iota 0.112)$, $\max_{\sigma} \mathfrak{R}_{\sigma 2} = (0.771 + \iota 0.309, -0.23 - \iota 0.310), \max_{\sigma} \mathfrak{R}_{\sigma 3} = (0.623 + \iota 0.361, -0.197 - \iota 0.450)$, $\max_{\sigma} \mathfrak{R}_{\sigma 4} = (0.812 + \iota 0.534, -0.737 - \iota 0.248)$ and the normalized BCF decision matrix is devised in Table 8.2

Step 2: The weights by utilizing the PA operator as given below.

$\underset{11}{\overset{\circ}{\mathrm{w}}} = 0.255, \underset{12}{\overset{\circ}{\mathrm{w}}} = 0.279, \underset{13}{\overset{\circ}{\mathrm{w}}} = 0.151, \underset{14}{\overset{\circ}{\mathrm{w}}} = 0.315,$

$\underset{21}{\overset{\circ}{\mathrm{w}}} = 0.223, \underset{22}{\overset{\circ}{\mathrm{w}}} = 0.28, \underset{23}{\overset{\circ}{\mathrm{w}}} = 0.162, \underset{24}{\overset{\circ}{\mathrm{w}}} = 0.335,$

Table 8.1 The evaluated values given by the company

	$\mathfrak{A}_{\mathbb{A}\approx-1}$	$\mathfrak{A}_{\mathbb{A}\approx-2}$	$\mathfrak{A}_{\mathbb{A}\approx-3}$	$\mathfrak{A}_{\mathbb{A}\approx-4}$
$\mathfrak{R}_{\mathbb{AL}-1}$	$\begin{pmatrix} 0.132 + \iota 0.530, \\ -0.987 - \iota 0.778 \end{pmatrix}$	$\begin{pmatrix} 0.112 + \iota 0.309, \\ -0.717 - \iota 0.888 \end{pmatrix}$	$\begin{pmatrix} 0.200 + \iota 0.361, \\ -0.197 - \iota 0.80 \end{pmatrix}$	$\begin{pmatrix} 0.652 + \iota 0.111, \\ -0.987 - \iota 0.265 \end{pmatrix}$
$\mathfrak{R}_{\mathbb{AL}-2}$	$\begin{pmatrix} 0.998 + \iota 0.315, \\ -0.367 - \iota 0.948 \end{pmatrix}$	$\begin{pmatrix} 0.121 + \iota 0.233, \\ -0.537 - \iota 0.768 \end{pmatrix}$	$\begin{pmatrix} 0.427 + \iota 0.132, \\ -0.537 - \iota 0.838 \end{pmatrix}$	$\begin{pmatrix} 0.120 + \iota 0.113, \\ -0.737 - \iota 0.248 \end{pmatrix}$
$\mathfrak{R}_{\mathbb{AL}-3}$	$\begin{pmatrix} 0.112 + \iota 0.741, \\ -0.183 - \iota 0.112 \end{pmatrix}$	$\begin{pmatrix} 0.423 + \iota 0.191, \\ -0.23 - \iota 0.808 \end{pmatrix}$	$\begin{pmatrix} 0.623 + \iota 0.137, \\ -0.38 - \iota 0.450 \end{pmatrix}$	$\begin{pmatrix} 0.812 + \iota 0.534, \\ -0.765 - \iota 0.644 \end{pmatrix}$
$\mathfrak{R}_{\mathbb{AL}-4}$	$\begin{pmatrix} 0.832 + \iota 0.989, \\ -0.777 - \iota 0.626 \end{pmatrix}$	$\begin{pmatrix} 0.771 + \iota 0.111, \\ -0.911 - \iota 0.310 \end{pmatrix}$	$\begin{pmatrix} 0.109 + \iota 0.311, \\ -0.777 - \iota 0.552 \end{pmatrix}$	$\begin{pmatrix} 0.710 + \iota 0.340, \\ -0.811 - \iota 0.478 \end{pmatrix}$

Table 8.2 The normalized BCF decision matrix

	$\mathfrak{A}_{\mathbb{A}\approx-1}$	$\mathfrak{A}_{\mathbb{A}\approx-2}$	$\mathfrak{A}_{\mathbb{A}\approx-3}$	$\mathfrak{A}_{\mathbb{A}\approx-4}$
$\Re_{\mathbb{AL}-1}$	$\begin{pmatrix} 0.132+\iota 0.527, \\ -0.183-\iota 0.784 \end{pmatrix}$	$\begin{pmatrix} 0.112+\iota 0.308, \\ -0.734-\iota 0.89 \end{pmatrix}$	$\begin{pmatrix} 0.2+\iota 0.36, \\ -0.319-\iota 0.805 \end{pmatrix}$	$\begin{pmatrix} 0.651+\iota 0.111, \\ -0.987-\iota 0.327 \end{pmatrix}$
$\Re_{\mathbb{AL}-2}$	$\begin{pmatrix} 0.77+\iota 0.185, \\ -0.468-\iota 0.949 \end{pmatrix}$	$\begin{pmatrix} 0.117+\iota 0.153, \\ -0.593-\iota 0.79 \end{pmatrix}$	$\begin{pmatrix} 0.379+\iota 0.102, \\ -0.593-\iota 0.849 \end{pmatrix}$	$\begin{pmatrix} 0.116+\iota 0.09, \\ -0.756-\iota 0.438 \end{pmatrix}$
$\Re_{\mathbb{AL}-3}$	$\begin{pmatrix} 0.105+\iota 0.321, \\ -0.319-\iota 0.486 \end{pmatrix}$	$\begin{pmatrix} 0.337+\iota 0.143, \\ -0.352-\iota 0.834 \end{pmatrix}$	$\begin{pmatrix} 0.452+\iota 0.11, \\ -0.462-\iota 0.621 \end{pmatrix}$	$\begin{pmatrix} 0.544+\iota 0.275, \\ -0.778-\iota 0.724 \end{pmatrix}$
$\Re_{\mathbb{AL}-4}$	$\begin{pmatrix} 0.698+\iota 0.531, \\ -0.863-\iota 0.667 \end{pmatrix}$	$\begin{pmatrix} 0.654+\iota 0.101, \\ -0.929-\iota 0.438 \end{pmatrix}$	$\begin{pmatrix} 0.106+\iota 0.245, \\ -0.863-\iota 0.61 \end{pmatrix}$	$\begin{pmatrix} 0.61+\iota 0.262, \\ -0.876-\iota 0.555 \end{pmatrix}$

$\underset{31}{\mathring{w}} = 0.222, \underset{32}{\mathring{w}} = 0.274, \underset{33}{\mathring{w}} = 0.161, \underset{34}{\mathring{w}} = 0.344,$

$\underset{41}{\mathring{w}} = 0.226, \underset{42}{\mathring{w}} = 0.257, \underset{43}{\mathring{w}} = 0.145, \underset{44}{\mathring{w}} = 0.372,$

Step 3: Determined the measure of BCF-WSM of every alternative by employing BCFPGHM operator ($\text{ß} = \mathfrak{L} = 3$).

$BCF - WSM(\Re_{\mathbb{AL}-1}) = \begin{pmatrix} 0.834+\iota 0.523, \\ -0.383-\iota 0.472 \end{pmatrix}$, $BCF - WSM(\Re_{\mathbb{AL}-2}) = \begin{pmatrix} 0.835+\iota 0.209, \\ -0.565-\iota 0.534 \end{pmatrix}$.

$BCF - WSM(\Re_{\mathbb{AL}-3}) = \begin{pmatrix} 0.812+\iota 0.389, \\ -0.405-\iota 0.618 \end{pmatrix}$, $BCF - WSM(\Re_{\mathbb{AL}-4}) = \begin{pmatrix} 0.926+\iota 0.515, \\ -0.86-\iota 0.485 \end{pmatrix}$.

Step 4: Determined the measure of BCF-WPM of every alternative by employing BCFPGGHM operator ($\text{ß} = \mathfrak{L} = 3$).

$BCF - WPM(\Re_{\mathbb{AL}-1}) = \begin{pmatrix} 0.171+\iota 0.188, \\ -0.719-\iota 0.8 \end{pmatrix}$, $BCF - WPM(\Re_{\mathbb{AL}-2}) = \begin{pmatrix} 0.158+\iota 0.091, \\ -0.642-\iota 0.843 \end{pmatrix}$.

$BCF - WPM(\Re_{\mathbb{AL}-3}) = \begin{pmatrix} 0.265+\iota 0.16, \\ -0.505-\iota 0.748 \end{pmatrix}$, $BCF - WPM(\Re_{\mathbb{AL}-4}) = \begin{pmatrix} 0.439+\iota 0.189, \\ -0.911-\iota 0.598 \end{pmatrix}$.

Step 5: Computed BCF-WASPAS of each alternative.

$BCF - WASPAS(\Re_{\mathbb{AL}-1}) = \begin{pmatrix} 0.629+\iota 0.378, \\ -0.525-\iota 0.614 \end{pmatrix}$, $BCF - WASPAS(\Re_{\mathbb{AL}-2}) = \begin{pmatrix} 0.627+\iota 0.152, \\ -0.601-\iota 0.671 \end{pmatrix}$.

$BCF - WASPAS(\mathfrak{R}_{\mathbb{AL}-3}) = \begin{pmatrix} 0.628 + \iota 0.,284 \\ -0.452 - \iota 0.68 \end{pmatrix}$, $BCF - WASPAS(\mathfrak{R}_{\mathbb{AL}-4}) = \begin{pmatrix} 0.796 + \iota 0.373, \\ -0.539 - \iota 0.885 \end{pmatrix}$.

Step 6: The score values are.

$\mathcal{S}_{BCFS}(\mathfrak{R}_{\mathbb{AL}-1}) = 0.467$, $\mathcal{S}_{BCFS}(\mathfrak{R}_{\mathbb{AL}-2}) = 0.377$, $\mathcal{S}_{BCFS}(\mathfrak{R}_{\mathbb{AL}-3}) = 0.445$, $\mathcal{S}_{BCFS}(\mathfrak{R}_{\mathbb{AL}-4}) = 0.436$.

and the ranking is

$$\mathfrak{R}_{\mathbb{AL}-1} > \mathfrak{R}_{\mathbb{AL}-4} > \mathfrak{R}_{\mathbb{AL}-3} > \mathfrak{R}_{\mathbb{AL}-2}$$

The optimum method of supply chain transportation would be $\mathfrak{R}_{\mathbb{AL}-1}$ that is air freight. As a high-value product, Smartphone X requires quick and effective shipping to satisfy consumer demand and keep a competitive edge. Air freight is more expensive than other means of shipping, but it is the fastest and most reliable option for getting smartphones to various international markets in time for the launch.

8.5 Comparative Study

The aim of this section in this chapter is to compare the devised operators and BCF-WASPAS technique with certain prevailing AOs and WASPAS approaches.

For this comparison, we consider the prevailing AOs such as BF Dombi devised by Jana et al. [29], BF Dombi prioritized initiated by Jana et al. [30], Hamacher devised by Wei et al. [4], sine trigonometric diagnosed by Riaz et al. [31]. For this let us consider the information in the BCF model which is interpreted in Table 8.3.

To solve the information in Table 8.3, we employ the considered prevailing operators and the initiated operators, and the needed result is interpreted in Table 8.4.

Table 8.3 The considered data for comparison

	$\mathfrak{A}_{\mathbb{A}\approx-1}$	$\mathfrak{A}_{\mathbb{A}\approx-2}$	$\mathfrak{A}_{\mathbb{A}\approx-3}$	$\mathfrak{A}_{\mathbb{A}\approx-4}$
$\mathfrak{R}_{\mathbb{AL}-1}$	$\begin{pmatrix} 0.841 + \iota 0.130, \\ -0.705 - \iota 0.817 \end{pmatrix}$	$\begin{pmatrix} 0.421 + \iota 0.837, \\ -0.745 - \iota 0.598 \end{pmatrix}$	$\begin{pmatrix} 0.120 + \iota 0.173, \\ -0.107 - \iota 0.118 \end{pmatrix}$	$\begin{pmatrix} 0.177 + \iota 0.977, \\ -0.170 - \iota 0.948 \end{pmatrix}$
$\mathfrak{R}_{\mathbb{AL}-2}$	$\begin{pmatrix} 0.752 + \iota 0.123, \\ -0.837 - \iota 0.764 \end{pmatrix}$	$\begin{pmatrix} 0.142 + \iota 0.143, \\ -0.557 - \iota 0.218 \end{pmatrix}$	$\begin{pmatrix} 0.172 + \iota 0.203, \\ -0.177 - \iota 0.298 \end{pmatrix}$	$\begin{pmatrix} 0.932 + \iota 0.143, \\ -0.337 - \iota 0.778 \end{pmatrix}$
$\mathfrak{R}_{\mathbb{AL}-3}$	$\begin{pmatrix} 0.531 + \iota 0.973, \\ -0.956 - \iota 0.858 \end{pmatrix}$	$\begin{pmatrix} 0.124 + \iota 0.243, \\ -0.990 - \iota 0.493 \end{pmatrix}$	$\begin{pmatrix} 0.120 + \iota 0.433, \\ -0.234 - \iota 0.81 \end{pmatrix}$	$\begin{pmatrix} 0.201 + \iota 0.333, \\ -0.888 - \iota 0.381 \end{pmatrix}$
$\mathfrak{R}_{\mathbb{AL}-4}$	$\begin{pmatrix} 0.411 + \iota 0.477, \\ -0.242 - \iota 0.447 \end{pmatrix}$	$\begin{pmatrix} 0.111 + \iota 0.163, \\ -0.209 - \iota 0.756 \end{pmatrix}$	$\begin{pmatrix} 0.754 + \iota 0.289, \\ -0.711 - \iota 0.90 \end{pmatrix}$	$\begin{pmatrix} 0.20 + \iota 0.903, \\ -0.17 - \iota 0.988 \end{pmatrix}$

Table 8.4 The obtained outcome after employing certain prevailing and initiated operators

Source	$\mathbb{S}_{\mathbb{SV}}(\Re_{A-1})$	$\mathbb{S}_{\mathbb{SV}}(\Re_{A-2})$	$\mathbb{S}_{\mathbb{SV}}(\Re_{A-3})$	$\mathbb{S}_{\mathbb{SV}}(\Re_{A-4})$	Ranking
Jana et al. [29]	$\times \rightarrow \times \rightarrow \times$	$\times \rightarrow \times \rightarrow \times$	$\times \rightarrow \times \rightarrow \times$	$\times \rightarrow \times \rightarrow \times$	$\times \rightarrow \times \rightarrow \times$
Jana et al. [30]	$\times \rightarrow \times \rightarrow \times$	$\times \rightarrow \times \rightarrow \times$	$\times \rightarrow \times \rightarrow \times$	$\times \rightarrow \times \rightarrow \times$	$\times \rightarrow \times \rightarrow \times$
Wei et al. [4]	$\times \rightarrow \times \rightarrow \times$	$\times \rightarrow \times \rightarrow \times$	$\times \rightarrow \times \rightarrow \times$	$\times \rightarrow \times \rightarrow \times$	$\times \rightarrow \times \rightarrow \times$
Riaz et al. [31]	$\times \rightarrow \times \rightarrow \times$	$\times \rightarrow \times \rightarrow \times$	$\times \rightarrow \times \rightarrow \times$	$\times \rightarrow \times \rightarrow \times$	$\times \rightarrow \times \rightarrow \times$
BCFPGHM	0.82	0.623	0.669	0.73	$Re_{AL-1} > \Re_{AL-4} > \Re_{AL-3} > \Re_{AL-2}$
BCFPGGHM	0.443	0.278	0.339	0.349	$Re_{AL-1} > \Re_{AL-4} > \Re_{AL-3} > \Re_{AL-2}$
BCFWPGHM	0.271	0.241	0.195	0.289	$Re_{AL-4} > \Re_{AL-1} > \Re_{AL-2} > \Re_{AL-3}$
BCFWPGGHM	0.685	0.658	0.621	0.702	$Re_{AL-4} > \Re_{AL-1} > \Re_{AL-2} > \Re_{AL-3}$

From the above result, it is clear that the considered prevailing operators failed to solve this data because these operators are in the model of BF information which can't cope with 2nd dimension and only consider positive and negative aspects. However, the initiated operators interpreted ranking among the alternative. By employing BCFPGHM and BCFWPGHM operators and the score function, we have the ranking order $\mathfrak{R}_{\mathbb{AL}-1} > \mathfrak{R}_{\mathbb{AL}-4} > \mathfrak{R}_{\mathbb{AL}-3} > \mathfrak{R}_{\mathbb{AL}-2}$ and By employing BCFPGGHM and BCFWPGGHM operators and the score function, we have the ranking order $\mathfrak{R}_{\mathbb{AL}-4} > \mathfrak{R}_{\mathbb{AL}-1} > \mathfrak{R}_{\mathbb{AL}-2} > \mathfrak{R}_{\mathbb{AL}-3}$. In weighted operators, we considered the weight $(0.3, 0.25, 0.2, 0.35)$ and $ß = \mathfrak{L} = 3$.

The BCF-WASPAS strategy can handle BF and complex fuzzy information and is an extension of the fuzzy WASPAS technique. The bipolar, complex, and bipolar complex fuzzy information cannot be handled by the fuzzy WASPAS technique [39], which can only handle fuzzy information. The developed BCF-WASPAS technique can handle data with FS, BFS, and CFS structures. Additionally, there is no WASPAS technique for bipolar fuzzy or complex fuzzy information in the literature, so the developed BCF-WASPAS would be useful in these areas as well.

References

1. Zadeh, L.A.: Fuzzy sets. Inf. Control **8**(3), 338–353 (1965)
2. Zhang, W.R.: Bipolar fuzzy sets and relations: a computational framework for cognitive modeling and multiagent decision analysis. In: NAFIPS/IFIS/NASA'94. Proceedings of the First International Joint Conference of the North American Fuzzy Information Processing Society Biannual Conference. The Industrial Fuzzy Control and Intellige (pp. 305–309). IEEE (1994)
3. Gul, Z.: Some bipolar fuzzy aggregations operators and their applications in multicriteria group decision making. Doctoral dissertation, Hazara University (2015)
4. Wei, G., Alsaadi, F.E., Hayat, T., Alsaedi, A.: Bipolar fuzzy Hamacher aggregation operators in multiple attribute decision making. Int. J. Fuzzy Syst **20**(1), 1–12 (2018)
5. Ramot, D., Milo, R., Friedman, M., Kandel, A.: Complex fuzzy sets. IEEE Trans. Fuzzy Syst. **10**(2), 171–186 (2002)
6. Tamir, D.E., Jin, L., Kandel, A.: A new interpretation of complex membership grade. Int. J. Intell. Syst. **26**(4), 285–312 (2011)
7. Mahmood, T., Ur Rehman, U.: A novel approach towards bipolar complex fuzzy sets and their applications in generalized similarity measures. Int. J. Intell. Syst. **37**(1), 535–567 (2022)
8. Alkouri, A.U.M., Massa'deh, M.O., Ali, M.: On bipolar complex fuzzy sets and its application. J. Intell. Fuzzy Syst. **39**(1), 383–397 (2020)
9. Al-Husban, A., Amourah, A., Jaber, J.J.: Bipolar complex fuzzy sets and their properties. Ital. J. Pure Appl. Math. **43**, 754–761 (2020)
10. Gulistan, M., Yaqoob, N., Elmoasry, A., Alebraheem, J.: Complex bipolar fuzzy sets: an application in a transport's company. J. Intell. Fuzzy Syst. **40**(3), 3981–3997 (2021)
11. Mahmood, T., Rehman, U.U.: A method to multi-attribute decision making technique based on Dombi aggregation operators under bipolar complex fuzzy information. Comput. Appl. Math. **41**(1), 47 (2022)
12. Mahmood, T., Rehman, U.U., Ali, Z., Aslam, M., Chinram, R.: Identification and classification of aggregation operators using bipolar complex fuzzy settings and their application in decision support systems. Mathematics **10**(10), 1726 (2022)
13. Rehman, U.U., Mahmood, T.: The generalized dice similarity measures for bipolar complex fuzzy set and its applications to pattern recognition and medical diagnosis. Comput. Appl. Math. **41**(6), 265 (2022)
14. Rehman, U.U., Mahmood, T., Naeem, M.: Bipolar complex fuzzy semigroups. AIMS Math **8**(2), 3997–4021 (2023)
15. Yang, X., Mahmood, T., ur Rehman, U.: Bipolar complex fuzzy subgroups. Mathematics **10**(16), 2882 (2022)
16. Nadler, S.B., Jr.: Hyperspaces of Sets. Marcel Dekker, New York (1978)

U. ur Rehman and T. Mahmood, *Bipolar Complex Fuzzy Sets*, Studies in Fuzziness and Soft Computing 439, https://doi.org/10.1007/978-3-032-07388-4

17. Zeleny, M.: Multiple Criteria Decision Making. McGraw-Hill, Company (1982)
18. Yu, P.L.: A class of solutions for group decision problems. Manag. Sci. **19**(8), 936–946 (1973)
19. Wang, T.C., Liang, J.L., Ho, C.Y.: Multi-criteria decision analysis by using fuzzy VIKOR. In: 2006 International Conference on Service Systems and Service Management, vol. 2, pp. 901–906. IEEE (2006)
20. Alsolame, B., Alshehri, N.O.: Extension of VIKOR method for MCDM under bipolar fuzzy set. Int. J. Anal. Appl. **18**(6), 989–997 (2020)
21. Rajeshwari, M., Murugesan, R., Venkatesh, K.A.: Distance Between Bipolar Fuzzy Sets
22. Patrascu, V.: Similarity, cardinality and entropy for bipolar fuzzy set in the framework of penta-valued representation. arXiv:1506.02060 (2015)
23. Yoon, K., Hwang, C.L.: TOPSIS (Technique for Order Preference by Similarity to Ideal Solution)–A Multiple Attribute Decision Making, w: Multiple Attribute Decision Making–Methods and Applications, a State-of-the-at Survey. Springer Verlag, Berlin (1981)
24. Keikha, A., Garg, H., Mishmast Nehi, H.: An approach based on combining Choquet integral and TOPSIS methods to uncertain MAGDM problems. Soft Comput. **25**(10), 7181–7195 (2021)
25. Kumar, V., Saxena, P., Garg, H.: Selection of optimal software reliability growth models using an integrated entropy–technique for order preference by similarity to an ideal solution (TOPSIS) approach. Math. Methods Appl. Sci. (2021)
26. Garg, H., Kaur, G.: TOPSIS based on nonlinear-programming methodology for solving decision-making problems under cubic intuitionistic fuzzy set environment. Comput. Appl. Math. **38**, 1–19 (2019)
27. Hara, T., Uchiyama, M., Takahasi, S.E.: A refinement of various mean inequalities. J. Inequalities Appl. **1998**(4), 932025 (1998)
28. Wu, S., Wang, J., Wei, G., Wei, Y.: Research on construction engineering project risk assessment with some 2-tuple linguistic neutrosophic Hamy mean operators. Sustainability **10**(5), 1536 (2018)
29. Jana, C., Pal, M., Wang, J.Q.: Bipolar fuzzy Dombi aggregation operators and its application in multiple-attribute decision-making process. J. Ambient Intell. Humanized Comput. **10**, 3533–3549 (2019)
30. Jana, C., Pal, M., Wang, J.Q.: Bipolar fuzzy Dombi prioritized aggregation operators in multiple attribute decision making. Soft Comput. **24**, 3631–3646 (2020)
31. Riaz, M., Pamucar, D., Habib, A., Jamil, N.: Innovative bipolar fuzzy sine trigonometric aggregation operators and SIR method for medical tourism supply chain. Math. Probl. Eng. (2022)
32. Bi, L., Dai, S., Hu, B., Li, S.: Complex fuzzy arithmetic aggregation operators. J. Intell. Fuzzy Syst. **36**(3), 2765–2771 (2019)
33. Bi, L., Dai, S., Hu, B.: Complex fuzzy geometric aggregation operators. Symmetry **10**(7), 251 (2018)
34. Yager, R.R.: The power average operator. IEEE Trans. Syst. Man Cybern. Part A Syst. Humans **31**(6), 724–731 (2001)
35. Beliakov, G., Pradera, A., Calvo, T.: Aggregation Functions: A Guide for Practitioners, vol. 221. Springer, Heidelberg (2007)
36. Sýkora, S.: Mathematical Means and Averages: Generalized Heronian Means. Stan's Library: Castano Primo, Italy (2009)
37. Yu, D.: Intuitionistic fuzzy geometric Heronian mean aggregation operators. Appl. Soft Comput. **13**(2), 1235–1246 (2013)
38. Zavadskas, E.K., Turskis, Z., Antucheviciene, J., Zakarevicius, A.: Optimization of weighted aggregated sum product assessment. Elektronika ir elektrotechnika **122**(6), 3–6 (2012)
39. Turskis, Z., Goranin, N., Nurusheva, A., Boranbayev, S.: A fuzzy WASPAS-based approach to determine critical information infrastructures of EU sustainable development. Sustainability **11**(2), 424 (2019)
40. Shepard, R.N.: Toward a universal law of generalization for psychological science. Science **237**(4820), 1317–1323 (1987)

MIX
Papier aus verantwortungsvollen Quellen
Paper from responsible sources
FSC® C105338
FSC
www.fsc.org

If you have any concerns about our products,
you can contact us on
ProductSafety@springernature.com

In case Publisher is established outside the EU,
the EU authorized representative is:
Springer Nature Customer Service Center GmbH
Europaplatz 3, 69115 Heidelberg, Germany

Printed by Libri Plureos GmbH
in Hamburg, Germany